An Introduction to
Lifted Probabilistic Inference

An Introduction to
Lifted Probabilistic Inference

Edited by Guy Van den Broeck, Kristian Kersting, Sriraam Natarajan, and David Poole

The MIT Press
Cambridge, Massachusetts
London, England

This book was set in Times-Roman by the editors.

Library of Congress Cataloging-in-Publication Data

Names: Broeck, Guy van den, editor. | Kersting, Kristian, editor. | Natarajan, Sriraam, editor. | Poole, David, editor.
Title: An introduction to lifted probabilistic inference / edited by Guy van den Broeck, Kristian Kersting, Sriraam Natarajan, and David Poole.
Description: Cambridge, Massachusetts : The MIT Press, [2021] | Series: Neural information processing series | Includes bibliographical references and index.
Identifiers: LCCN 2020040684 | ISBN 9780262542593 (paperback)
Subjects: LCSH: Probabilities. | Artificial intelligence. | Heuristic algorithms.
Classification: LCC QA273 .I563 2021 | DDC 519.2–dc23
LC record available at https://lccn.loc.gov/2020040684

Contents

III   APPROXIMATE INFERENCE

List of Figures

Contributors

Babak Ahmadi
C-IAM GmbH, Cologne, Germany

Hendrik Blockeel
Department of Computer Science, KU
Leuven, Belgium
https://people.cs.kuleuven.be/
~hendrik.blockeel/

Hung Bui
VinAI Research, Menlo Park, USA
https://sites.google.com/site/
buihhung/

Yuqiao Chen
Department of Computer Science,
University of Texas at Dallas, USA
https://personal.utdallas.edu/
~yuqiao.chen/

Arthur Choi
Department of Computer Science,
University of Los Angeles, USA
http://web.cs.ucla.edu/~aychoi/

Jaesik Choi
Graduate School of Artificial Intelligence,
KAIST, Republic of Korea
http:
//sailab.kaist.ac.kr/jaesik/

Adnan Darwiche
Department of Computer Science,
University of Los Angeles, USA
http:
//web.cs.ucla.edu/~darwiche/

Jesse Davis
Department of Computer Science, KU
Leuven, Belgium
https:
//people.cs.kuleuven.be/~jesse/

Rodrigo de Salvo Braz
SRI International, Menlo Park, USA
http://www.ai.sri.com/~braz/

Pedro Domingos
Department of Computer Science
University of Washington, Seattle, USA
https://homes.cs.washington.edu/
~pedrod/

Daan Fierens
TenForce, Leuven, Belgium
https://sites.google.com/site/
fierensdaan/home

Martin Grohe
Department of Computer Science, RWTH
Aachen University, Aachen, Germany
https://www.lics.rwth-
aachen.de/go/id/nwej/?lidx=1

Fabian Hadiji
Geodle.io Cologne, Germany

Seyed Mehran Kazemi
Borealis AI, Montreal, Canada
https://mehran-k.github.io/

Roni Khardon
Department of Computer Science, Indiana
University, Bloomington, USA
http://homes.sice.indiana.edu/
rkhardon/

Angelika Kimmig
Department of Computer Science, Cardiff
University, Cardiff, England
http:
//users.cs.cf.ac.uk/KimmigA/

Jacek Kisynski
Visier Inc, Vancouver, Canada

Kristian Kersting
Computer Science Department, TU
Darmstadt, Germany
https://ml-research.github.io/
people/kkersting/index.html

Daniel Lowd
Department of Computer Science,
University of Oregon, Eugene, USA
https://ix.cs.uoregon.edu/~lowd/

Wannes Meert
Department of Computer Science, KU
Leuven, Belgium
https://people.cs.kuleuven.be/
~wannes.meert/

Martin Mladenov
Google Research, Mountainview, USA

Raymond Mooney
Department of Computer Science,
University of Texas at Austin, USA
https:
//www.cs.utexas.edu/~mooney/

Sriraam Natarajan
Department of Computer Science,
University of Texas at Dallas, USA
https://personal.utdallas.edu/
~sriraam.natarajan/

Mathias Niepert
NEC Labs Europe, Heidelberg, Germany
http://www.matlog.net/

David Poole
Department of Computer Science,
University of British Columbia, Vancouver,
Canada
https://www.cs.ubc.ca/~poole/

Scott Sanner
Department of Mechanical and Industrial
Engineering, University of Toronto,
Canada
https://d3m.mie.utoronto.ca/

Pascal Schweitzer
Department of Computer Science, TU
Kaiserslautern, Germany
http://alg.cs.uni-
kl.de/en/team/schweitzer/

Nima Taghipour
Trivago, Amsterdam, Netherlands

Guy Van den Broeck
Department of Computer Science,
University of Los Angeles, USA
http://web.cs.ucla.edu/~guyvdb/

Preface

We are grateful to the entire Statistical Relational AI community for their contribution to lifted learning and inference. This book will not be possible without you. We thank the students of the statistical relational AI labs of the four authors for their help in proof-reading the book. Special thanks to Nandini Ramanan for her help in collating all the references. Thanks to Yuqial Chen, Devendra Dhami, Harsha Kokel, Srijita Das, Navdeep Kaur, Alexander Hayes, Athresh Karanam, Nandini Ramanan, Mike Skinner and Siwen Yan for proof-reading the chapters in the book.

We also thank our families and friends for their support.

Guy, Kristian, Sriraam and David
April 2020

I OVERVIEW

1 Statistical Relational AI: Representation, Inference and Learning

Guy Van den Broeck, Kristian Kersting, Sriraam Natarajan, and David Poole

Abstract. Artificial intelligence (AI) is about creating agents that act in environments (Russell and Norvig, 2010; Poole and Mackworth, 2017). Acting in an environment where there is any partial observability or stochasticity is gambling on the outcomes of actions. Probability and utility are the calculi for gambling; there are numerous results that show that an agent that does not use probability will lose to one that does. The real world is complicated; non-trivial agents need to reason about individuals (things, objects, entities), properties of the individuals and relationships among individuals. **Statistical relational AI** (StaRAI) (De Raedt et al., 2016) is the field that studies the integration of reasoning under uncertainty and reasoning about individuals and relations. The representations used are often called relational probabilistic models.

The integration of uncertainty and relations can be approached from a number of different directions:

- **First-order logic** extends propositional logic with constants and variables that quantify over individuals, and relations among the individuals. Starting with first-order logic, we can add probabilities and utilities to allow for uncertainty about the truth of propositions, as well as the identity and existence of individuals. For example, probabilistic logic programs (Poole, 1993; Sato and Kameya, 1997; De Raedt et al., 2007) can be seen as adding probabilistic inputs to logic programs, which let us define probabilistic models about relations in a Turing-complete language that naturally represents relations.
- Starting from probabilistic models, we can add relations. For example, starting with graphical models (Pearl, 1988), Markov logic (Domingos and Lowd, 2009) allows for (open, where not all variables are quantified) first-order formulae in Markov networks. A weighted formula can contain first-order formulae; replacing the non-quantified (free) variables with constants produces a Markov network.
- Relational embeddings (Nickel et al., 2016) define individuals and relations in terms of embeddings and define the probability of a relation in terms of functions of the embeddings of the individuals and the relations involved.

Many of these models are defined in terms of **grounding** – substituting constants for the variables in all possible ways – thus converting a relational representation into a non-relational one. This allows standard reasoning techniques to be used on the ground representation. These ground representations, however, soon get to be enormous, but have lots of repeated structure, which can be exploited. Lifted inference is about how to exploit the structure inherent in relational probabilistic models, either in the way they are expressed or by extracting structure from observations.

There are numerous examples of reasoning that can be carried out without resorting to grounding:

Example 1.1 Consider determining the guilt of someone fitting the description of a person who committed a crime (Poole, 1993). Suppose the probability of someone at random matching the description was, say, one in a million. The probability this person committed the crime depends on how many people there are. If there were a thousand other people, it is very unlikely there was someone else who committed the crime. If there was a population of 10 million, then we would expect that there would be 10 people who fit the description, and so the probability that this suspect was guilty would be around 10%. We don't need to reason about all of the other individuals separately, but can count over them.

Example 1.2 Suppose someone is giving a presentation, and three people out of 100 people in the audience asked a question (so 97 people were observed to not ask a question). A reasonable model about the eloquence of the speaker might depend on the questions; each of the people who didn't ask a question, their silence might depend on the questions asked, but not on the infinitely many questions not asked. Rather than reasoning separately about each person who was observed to not ask a question and each question, it is reasonable to just count how many of them have various properties.

Example 1.3 The spread of malaria (or other diseases) may depend on the number of people and the number of mosquitoes. Individual mosquitoes are important in such a model, but we do not want to reason about each mosquito separately.

One might be tempted to model these directly in terms of the counts. However, in each of these examples, we want to be able to reason separately about some of the individuals about whom we have observed something special, whether it is a particular person in the audience, or characteristics of the particular mosquitoes that go into a room. Rather than needing the modeller to reason about the counts and the special cases separately, it is better to have the computer do the counting automatically. We want to be able to build the general-purpose models of the world before we know the details of the situation at hand.

1.1   Representations

We begin with a brief overview of the key representations underlying lifted probabilistic inference. These representations will be elaborated on in later chapters.

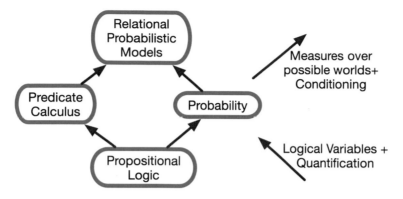

Figure 1.1: First-order logic and probability

Propositional logic is a language of truth values on atoms, with Boolean combinations (using "and/conjunction" ($\wedge$), "or/disjunction" ($\vee$) and "not/negation" ($\neg$), "if" ($\leftarrow$), "implies" ($\rightarrow$), "if-and-only-if" ($\leftrightarrow$)) of atoms.

Relational probabilistic models are based on combinations of **first-order logic** (in particular, **predicate calculus**) and probability. As shown in Figure 1.1, both predicate calculus and probability theory can be derived from propositional calculus, the first by adding logical variables and quantification, and the second by adding measures over possible worlds and probabilistic conditioning. Relational probabilities extend propositional logic by including both.

This particular combination is not arbitrary. Relational representations have been intensely studied in the realms of logic (including relational databases and logic programming as well as the logics studied by philosophers and mathematicians). Decision making also has solid foundations, with Bayesian decision theory – involving both probability and utility – emerging as the predominant framework. The combination of these is more than just their union, as the readers of this book will come to appreciate.

1.1.1 First-order Logic and Logic Programs

First-order logic extends propositional logic to include individuals (sometimes also called objects), relations and variables that quantify over individuals. It is this quantification over individuals that makes the logic first-order.

The first-order predicate calculus extends the propositional calculus by allowing for terms (constants, variables and functions) and predicates. Constants denote particular individuals. A **logical variable**[1] can denote any individual. Functions map a tuple of indi-

---

[1] There is much confusion about the use of the term "variable". Logical variables are very different from random variables. A **random variable** is neither random nor variable. A random variable needs to satisfy the clarity

viduals to another individual. Atoms are of the form $p(t_1, \ldots, t_k)$, where p is a predicate symbol and t_i are terms. The atoms combined with logical connectives and quantification of variables define propositions. For example the formula $r(x, y) \rightarrow s(x)$, where $\rightarrow$ means "implies" and x and y are variables, might have different truth values depending on what x and y denote. To say it is true for all x and y, we can quantify over x and y to give $\forall x \forall y \, r(x, y) \rightarrow s(x)$, which can then be either true or false. Any variable that is not quantified over is a free variable; formulae with free variables are not given a truth value; the free variable needs to be quantified over or substituted by another term for the formula to have a truth value.

One particular subset of first-order logic that has proved to be useful is **logic programs**. This is a language of definitions, where the only formulae allowed are of the form $a \leftrightarrow b_1 \vee \cdots \vee b_k$, where a is an atom and the b_i are conjunctions of **literals** (atoms or the negation of atoms). Such formulae are usually written as separate rules $a \leftarrow b_1, \ldots, a \leftarrow b_k$ with the completion (Clark, 1978) and quantification left implicit. All variables are universally quantified at the top level. Logic programs have an alternative fixed point semantics which lets them represent transitive closure and related notions that are not first-order definable (see Section 2.2.2). If these definitions are acyclic (Apt and Bezem, 1991), so that all recursions eventually reach a base case, the language has many nice properties even though (or because) it cannot represent arbitrary disjunctions. A logic that doesn't incorporate uncertainty is a good candidate for adding probabilities, because the resulting formalism won't need to incorporate multiple forms of uncertainty.

1.1.2   Probability and Graphical Models

Probability extends propositional logic to allow for measures over possible worlds. For the finite (discrete) case, the semantics of probability is simple and straightforward. There is a world for each assignment of a value to each random variable,[2] and a non-negative number (the measure) associated with each world so the sum of the measures is 1 (by convention). The probability of a proposition is the sum of the measures of the worlds in which the proposition is true.

When there are infinitely many worlds, for example when there is a real-valued variable (where atomic propositions are built from statements such as $x < 3$), or there are infinitely many variables, the semantics become more complicated and need a form of measure the-

---

principle (Howard, 1988), which specifies that a random variable needs to have a well-defined meaning. For example, "the age in years of Justin Bieber at the end of 2020" could be a random variable, if we knew the meanings of the words and who Justin Bieber is. If x is a logical variable "the age in years of x at the end of 2020" is not a well-defined random variable unless we know who x denotes.

[2] Equivalently, the possible worlds can be primitive, and a random variable is a function from possible worlds. The range of this function is the set of values the random variable can take.

ory. This measure theory is defined as the limit of the discrete case. The probability need only be defined on sets of worlds that are finitely describable.

Graphical models (Pearl, 1988; Koller and Friedman, 2009) are representations of conditional independence among random variables. The idea is to model direct dependence and make the independence assumption: a variable is independent of the other variables given its neighborhood in the graph. This is useful because the assumptions made are explicit and locality can be exploited for representations, inference and learning.

There are two main classes of graphical models:

· **Undirected graphical models** are defined by an undirected graph where the nodes are random variables; each variable is independent of the other variables given its neighbours in the graph. The probability density can be factored into a product of factors, where a **factor** is a nonnegative function of a subset of the variables. The probability of a world is proportional to the product of the factor values. A **Markov random field** or **Markov network** is a graph where the nodes are random variables, and there is edge between nodes if they are together in a factor. A **factor graph** is a bipartite graph that contains a variable node for each random variable and a factor node for each factor, and there is an edge between a variable node and a factor node if the variable appears in the factor. The product of positive factors is often called a **log-linear model**, as a positive product can be written as an exponential of a sum of logarithms.

· **Directed graphical models** (**Bayesian networks**) provide a directed form of independence: in a total ordering of variables, each variable is independent of its predecessors given some subset of its predecessors (its parents in the directed graph). Thus if $X_1, \ldots, X_n$ are the random variables

$$P(X_1, \ldots, X_n) = \prod_i P(X_i \mid X_1, \ldots, X_{i-1})$$

$$= \prod_i P(X_i \mid parents(X_i))$$

where $parents(X_i)$ is the smallest subset of $\{X_1, \ldots, X_{i-1}\}$ that makes the equality hold.

Directed models are a subset of unordered models (by the Hammersley-Clifford theorem (Hammersley and Clifford, 1971)). They provide a more fine-grained decomposition because each variable is decomposed using only a subset of the other variables. They also have the nice property that (given the total ordering) each conditional probability is unique and can be assessed independently of the other conditional probabilities, which greatly helps in learning. Inference schemes defined for undirected models also work for directed models, but sometimes the directed structure can also be exploited (e.g., in pruning). Learning conditional probabilities where all of the variables of interest are observed – called **supervised learning** – is a mature subject area. There is more recent work on

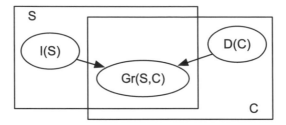

Figure 1.2: Plate notation for graphical models: S and C correspond to logical variables, with shared parameters

learning undirected models because it is much more difficult (in general, each factor cannot be learned independently).

Directed models form a foundation for **causality** (Pearl, 2009). A model in which the variables represent events, and the direct causes of some event are its parents, should follow the independence structure of Bayesian networks. If we knew the direct causes, everything else that happened before is irrelevant, and therefore independent. Many cases of causal cycles, such as working hard causes someone to be rich(er) but being rich causes someone to not work as hard, can be made acyclic by considering time.

1.1.3   Relational Probabilistic Models

The simplest forms of relational probabilistic models extend graphical models (directed or undirected) to have parameter tying. This is equivalent to having a formula such as:

$$\forall S, C, \ P(Gr(S, C)=\text{``B''} \mid I(S)=\textit{false}, D(C)=\textit{true}) = 0.4$$

where "B" is the grade, $Gr(S, C)$ is the grade of student S in course C, and $I(S)$ is a Boolean random variables for each student S, and $D(C)$ is a Boolean random variable for each C. This model implements **parameter sharing** (or **weight tying**) where the conditional probability is the same for all individuals S and C.

These models are often drawn using **plate notation** (Buntine, 1994), where the logical variables correspond to plates as in Figure 1.2. There is a variable $I(S)$ for every student S, a variable $D(C)$ for every course C, and a variable $Gr(S, C)$ for every student-course pair. These all share the same parameters.

Figure 1.3 shows a grounding of the model for 4 students and 4 courses. The conditional probabilities are given in the bottom left. This model can then be conditioned on the observations specified in the database at the bottom right of that figure, to give predictions on how well two students (who both have a B average) would do on a course about which we had no observations.

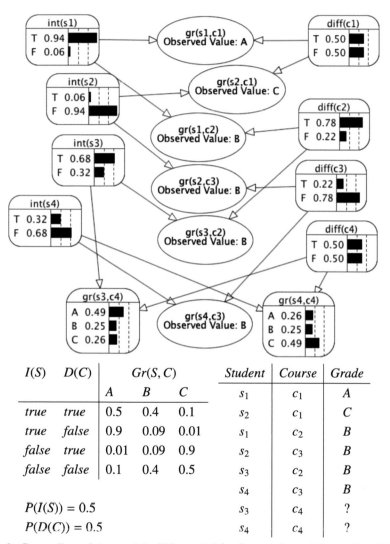

Figure 1.3: Grounding of the model of Figure 1.2 for four students (s1 ... s4) and 4 courses (c1 ... c4), with probabilities on bottom left, and conditioned on the database on bottom right. The nodes with no observed descendents are pruned, as can always be done in a Bayesian network. Figure from the aispace.org Belief and Decision Networks tool.

Plates have been used extensively for **Bayesian learning**, where the parameters of the model can be taken out of the plate. Plates can represent cases where there might be a global parameter or a parameter for each student.

Markov Logic Networks (MLNs) (Domingos and Lowd, 2009) use a similar parameter tying for undirected models. A Markov network can be defined in terms of weighted formulae, where the probability of each complete assignment (the joint distribution) is proportional to the product (or exponential of the sum) of the weights of the formulas that are true for that assignment. Markov logic networks extend this idea to allow for first-order logic rules with open formulae; there is a weighted formula for each substitution of constants for the non-quantified variables. This gives a model defined in terms of weighted logical formulae, where the free variables are quantified outside of the probability. For example, it allows for statements such as

$$\forall S, C, \; weight(Gr(S, C) \vee \neg I(S) \vee D(C)) = 4.3.$$

Directed models are defined in terms of a conditional probability of each variable given its parents. There are many ways to represent conditional probabilities, including tables, decision trees, rules, logistic regression, and neural networks. For relational models, there can be unboundedly many parents in the grounding. For example, in Example 1.2, confidence of the speaker may depend on the members of the audience, which can be arbitrarily large, and the probability of a mosquito bite depends on the number of mosquitoes. The common ways to combine the influence of the parents, which is called **aggregation**, include the following:

- Logic programming models use existential quantification (variables that only appear in the body of a clause can be considered to be existentially quantified in the body). With probabilities, this gives a form of a noisy-or (Pearl, 1988). This model is appropriate for cases where only one cause is needed; for example only one mosquito giving a patient malaria is enough for the patient to get malaria, and each biting mosquito carrying malaria has a probability of passing on malaria.

- **Relational logistic regression** (Kazemi et al., 2014) – which could also be called **conditional Markov logic networks** – is the directed analogue of Markov logic networks; the weighted formulae are used to define conditional probabilities. This is more appropriate than noisy-or when various factors add and subtract to have an effect.

- Many other aggregators, including "sum", "parity" and summary statistics such as "mean" and "mode" have been proposed and used. (Natarajan et al., 2005; Perlich and Provost, 2003).

The properties of relational models are not directly inherited from the grounded versions. For example, directed models cannot be represented in terms of undirected factors (which they can in Bayesian networks by forming a clique tree), because weighted formulae, as used in MLNs, are not adequate to represent conditional probabilities (Buchman and Poole, 2015). The factor that represents a conditional probability, which can be obtained by marrying the parents in undirected models, cannot be represented in many of the representations for undirected models.

1.1.4   Weighted First-order Model Counting

Many of the lifted algorithms are designed for abstractions that other representations are mapped to. MLNs have acted as such an abstraction level (Domingos, 2015). Another common method is weighted model counting, where there are weights on atoms and and constraints formed by logical expressions.

The idea of **model counting** is to count the number of models of a formula (for which the complexity is #P complete, which is harder than determining whether there is a model). This can be extended to having a weighting on the models. The weighting on models can either be done by having weights on formulae (as in MLNs) or by having weights on atoms, where the logic specifies constraints. What has become known as **weighted model counting** (Chavira and Darwiche, 2008) has nonnegative weights on atoms. The weight of a model is the product of the weights of the atoms that are true in the model. The formulae act as constraints to remove worlds in which a formula is false.

In **weighted first-order model counting** (Van den Broeck, 2016), there are (nonnegative) weights on atoms, with arbitrary logical formulae as constraints. The weight of a model is the product of the weights of the atoms that are true in the model. A model must satisfy the constraints. Given some observations, or evidence, e, the probability of some hypothesis h given e can be computed by

$$P(h \mid e) = \frac{weight(h \wedge e)}{weight(e)}$$

where $weight(f)$ is the sum of the weights of the models of the background theory conjoined with formula f.

1.1.5   Observations and Queries

Given a model and observations e, there are a number of quantities that could be computed:

- The **marginal inference** $P(h \mid e)$ for an arbitrary hypotheses h. It is called marginal inference because the other variables are marginalized (summed out). This is typically the most useful, as it lets us compute expectations for decision making.
- The **most probable explanation (MPE)** is an assignment a to the variables that maximizes $P(a \mid e)$. This is much less useful than marginal inference, as the MPE assignment might still be very unlikely and not representative of the population. For example, consider the case with 1000 independent variables, each with probability 0.7; the MPE assignment has 1000 variables true, whereas almost all of the assignments have around 700 variables true.

- The **maximum a-posteriori probability (MAP)** is a mix of the previous two. Given the evidence, compute the most likely assignment to some other variables, marginalizing the remaining variables. This can be extended to the top-n explanations. This is useful if we want to show the most likely diagnoses (marginalizing out the details) or the most likely sentences given an utterance (as in a phone app that suggests words).
- A directed model could be cyclic, and thus induce a **Markov chain**. The aim could be to reason about the equilibrium (stationary) distribution of the Markov chain. This is the idea behind a **dependency network** (Heckerman et al., 2000), which can be extended into **relational dependency networks** (Neville and Jensen, 2007).
- For making decisions, we want to compute the expected utility. Computing expectations is a generalization of marginal inference for discrete variables, as the expected value of a Boolean variable is its probability. However, the expected value does not characterize the distribution for continuous variables. For sequential decisions, discounting the future is often used, which allows for the infinite future to have only a finite effect on the present value.

1.2   Inference

Statistical relational AI is about the mix of probability and logic, both of which have techniques for exploiting structure. Logic has resolution and search variants and unification for handling quantification. Probabilistic inference has techniques to exploit structure, and well-defined notions of approximation.

Although exact inference in graphical models and model counting is intractable in general (#P hard), we can exploit structure. Figure 1.4 shows how a belief network can induce a factorization by distributing out of a sum any factor that does not depend on the variable being summed out. The factorization does not depend on the factors being conditional probabilities, but can use arbitrary factors on subsets of variables.

Variable elimination (Zhang and Poole, 1994; Dechter, 1996) is a dynamic programming method that computes the sum from the inside out, storing the factors as needed. **Recursive conditioning** (Darwiche, 2001) is a seach-based algorithm that computes the sum from the outside-in, caching as necessary. The variable ordering affects the efficiency, but not the result. Variable elimination and recursive conditioning do exactly the same additions and multiplications (given the variable order). Recursive conditioning has the advantage that it requires fewer data structures; it evaluates existing factors rather than constructing new ones, and so naturally works with structured representations. The analogy in theorem proving is the DPLL algorithm (Davis et al., 1962), which does "or" instead of sum and "and" instead of product (Dechter, 2003). These algorithms have complexity exponential in the tree-width (Bodlaender, 1993) of the induced graph, and so work well for sparse graphs. Nevertheless, tree-width does not take structure of the individual factors into account, which can be additionally exploited for faster inference.

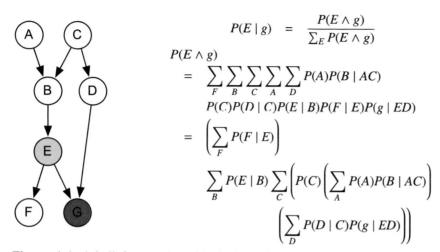

$$P(E \mid g) \;=\; \frac{P(E \wedge g)}{\sum_E P(E \wedge g)}$$

$$P(E \wedge g)$$
$$= \sum_F \sum_B \sum_C \sum_A \sum_D P(A)P(B \mid AC)$$
$$P(C)P(D \mid C)P(E \mid B)P(F \mid E)P(g \mid ED)$$

$$= \left(\sum_F P(F \mid E) \right)$$

$$\sum_B P(E \mid B) \sum_C \left(P(C) \left(\sum_A P(A)P(B \mid AC) \right) \right.$$

$$\left. \left(\sum_D P(D \mid C)P(g \mid ED) \right) \right)$$

Figure 1.4: A belief network and its induced factorization for computing $P(E \mid g)$ for a given elimination ordering (D, A, C, B, F)

Lifted variable elimination is discussed in Chapter 4 and lifted recursive conditioning is discussed in Chapter 5.

Inference can also be carried out by **stochastic simulation**, including particle filtering and Markov chain Monte Carlo (MCMC) (Doucet et al., 2001). The idea is to draw samples from the posterior distribution of interest. Exact inference is still useful; the Rao-Blackwell theorem (Rao et al., 1973) shows that the more exact inference (by summing out variables, rather than sampling them), the less the error. As we can do more exact inference more efficiently, we can potentially do stochastic simulation more effectively. Lifted MCMC is discussed in Chapter 10.

Another common method is **variational inference** (Jordan et al., 1997) where the idea is to find the closest tractable model to the actual model, and carry out inference in the tractable model. At one extreme, one can find the closest disconnected graphical model, in which inference is trivial, and the main problem is in optimizing to find the closest such model. Variational approximations can be improved by having better methods for exact inference. Lifted variational inference is discussed in Chapters 11, 12 and 13

In logic, theorem proving can be lifted into the relational case by using unification, and reasoning at the lifted level (Chang and Lee, 1973; Loveland, 1978). This is the basis for languages such as Prolog (Kowalski, 2014), which, although defined at the ground level, reasons as much as possible at the lifted level. Lifted probabilistic inference takes inspiration from such methods.

1.3   Learning

Learning relational probabilistic models can take various forms:

- The most basic task is learning probabilities and/or weights. This is typically done using counting, or gradient descent on the parameter space. The problem of learning conditional probabilities (where all variables are observed) is called **supervised learning**. Often models include unobserved properties of individuals; the reason that properties are invented is to make models simpler. When there is missing data, it is often assumed to be missing at random, but that is often a misleading assumption, and it is important to model why the data is missing (Mohan and Pearl, 2014).
- Having relational representations allows for learning about individuals and classes of individuals. Many methods let us learn about individuals. For example, consider movie recommendation, which is about learning a single binary relation ($rating(P, M)$ for person P and movie M that encodes whether the person rated the movie as well as the actual rating). A common method is matrix factorization (Koren et al., 2009), which learns properties of the individual people and movies that best predict the ratings.
- The problem of structure learning is to learn the structure of models, which is essentially learning how the world works. Unfortunately we cannot learn how the world works by observations alone (Pearl, 2009); we need causal models that infer the effect of actions by agents (interventions) and models of hypotheticals (often called counterfactuals).

1.4   Book Details

This book is a followup to *Statistical Relational Learning* (Getoor and Taskar, 2007), which primarily focused on the representations and learning inside relational probabilistic models. Given the recent significant advances in the area of lifted inference, there was a need for an unifying book that covers most of the related research in this direction. We have organized the book as follows: after covering the necessary foundations on probabilistic graphical models, relational probabilistic models and learning inside these models, we proceed to discuss lifted inference. The lifted inference chapters are organized based on whether they are used for exact inference or approximate inference. In addition, we consider the theory of liftability and acting in relational domains which will allow connecting learning and reasoning in relational domains.

2 Modeling and Reasoning with Statistical Relational Representations

Angelika Kimmig and David Poole

Abstract. The first part of this chapter provides an introduction to two main streams of representation languages for statistical relational domains, namely lifted (undirected) graphical models, focusing on parameterized factors and Markov Logic Networks, and probabilistic logic programming, focusing on ProbLog. It also summarizes the main inference tasks considered for such languages. The second part of the chapter provides a more abstract overview of statistical relational representations. It introduces a number of desirable but potentially competing high level properties for such representation languages, and discusses possible features of representation languages related to these properties.

Over the years, a wide range of statistical relational representation languages, defining **relational probabilistic models**, have been developed, with the goal of compactly representing probability distributions over relational domains. In the same way as first-order logic lifts propositional logic by making statements about all entities using logical variables, statistical relational languages define random variables and their correlations on the level of all entities (of the same type) rather than for each individual entity. Furthermore, all members of such a group share the same **tied parameters** in the probabilistic model, thus allowing for flexible numbers of entities while keeping the number of parameters fixed. Such languages thus exploit both the relational structure of first-order logic and the independence structure of graphical models.

Taking inspiration from graphical models, we can divide the models into two main streams: directed models that define conditional probabilities, and undirected models that define distributions over worlds in other ways.

Directed models can be represented as deterministic systems with (independent) noise. Pearl attributes to Laplace (1814) that the "conception of natural phenomena, according to which nature's laws are deterministic and randomness surfaces owing merely to our ignorance of the underlying boundary conditions (...)" (Pearl, 2009). This is the basis for **probabilistic programming** and **causal reasoning**. The first explicit probabilistic programming languages were based on logic programming (Poole, 1991, 1993; Sato, 1995). Logic programming is an appropriate choice as it is Turing complete, and extends both relational databases and a simple first-order logic. We use ProbLog (De Raedt et al., 2007) as the representative example language in this class.

Undirected models are defined using factors that induce distributions over worlds. **Relational undirected models** include logical variables in the factors, and instances of a logical variable share the same parameters. We first discuss the general form of these, namely parameterized factors or parfactors. We then describe Markov logic networks, where the parfactors are defined in terms of weighted first-order logic sentences. The logical formulation provides a more natural representation and finer-grained structure that can be exploited for inference and learning.

In the second part of the chapter, we consider statistical relational representations from a more abstract point of view, discussing both high level properties such languages should satisfy (Section 2.3.1) and lower level design choices they can make (Section 2.3.2). In doing so, we build on prior work in this direction, including (Bruynooghe et al., 2009; Kimmig et al., 2015; De Raedt and Kimmig, 2015; De Raedt et al., 2016).

2.1   Preliminaries: First-order Logic

We briefly summarize the key concepts of first-order logic (FOL), the representation language that relational probabilistic models build on. FOL distinguishes four kinds of symbols: constants, variables, predicates, and functors. Elements of the domain of interest we call **individuals**, but they are also called **entities**, things or **objects**[3]. **Constants**, denoted here by upper case letters refer to particular individuals. **Logical variables**, denoted here by lower case letters, refer to arbitrary rather than particular individuals in the domain. **Predicates**, denoted by atoms starting with an upper-case letter, represent attributes or relationships between individuals. **Functors**, or function symbols, also denoted by atoms starting with an upper-case letter, evaluate to an individual in the domain when applied to one or more entities. The number of arguments of a predicate or a functor is called its *arity*. A **term** is a constant, a variable, or of the form $f(t_1, \ldots, t_n)$ where f is a functor and $t_1, \ldots, t_n$ are terms (parentheses are omitted when $n = 0$).

An **atom** of the form $f(t_1, \ldots, t_n)$ where f is a predicate and $t_1, \ldots, t_n$ are terms (parentheses are omitted when $n = 0$). Terms and atoms are **ground** if they do not contain logical variables. Atoms are also called **positive literals**, and atoms preceded by the negation operator $\neg$ are called **negative literals**. A **formula** is a (positive or negative) literal or constructed from smaller formulas using conjunction ($\wedge$), disjunction ($\vee$) negation ($\neg$),

---

[3] Individuals in logic are different from objects in object-oriented programming (OOP). In object-oriented programming an object is something stored in a computer, whereas an individual is typically something external. The individual Justin Bieber could refer to the real person, rather that the internal representation of the singer. In Wikidata (https://www.wikidata.org/), for example, Justin Bieber is referred to by the constant Q3099714. One of the facts is that the name (in English) of Q3099714 is the string "Justin Bieber".

quantification (for-all ∀, and exists ∃) and other operations that can be defined in terms of these. A first-order **theory** is a finite set of formulas.

An **interpretation** specifies a mapping from the constants to the individuals (specifying which individual each constant denotes) and assigns *true* or *false* to each ground atom in a theory. If we say an atom is *true* or *false*, we mean with respect to an interpretation (which may be implicit from the context). A conjunctive formula $f_1 \wedge f_2$ evaluates to *true* in an interpretation iff both f_1 and f_2 evaluate to *true* in that interpretation. A disjunctive formula $f_1 \vee f_2$ evaluates to *true* iff at least one of the formulas f_1 and f_2 evaluates to *true*. A negated formula $\neg f_1$ evaluates to *true* iff f_1 evaluates to *false*. Implication $f_1 \Rightarrow f_2$, or equivalently $f_2 \leftarrow f_1$, is defined as $\neg f_1 \vee f_2$. Variables in formulas are quantified, either by an existential quantifier (∃) or by a universal quantifier (∀). A formula $\forall x\, f_1$ is *true* iff f_1 is *true* for every assignment of an individual to variable x. A formula $\exists x\, f_1$ is *true* iff f_1 is *true* for at least one assignment of an individual to variable x. A common assumption is that when no quantifier is specified for a variable, ∀ is understood by default.

An interpretation that makes all formulas in a first-order theory *true* is called a *model* of the theory. A formula expressed as a disjunction with at most one positive literal is called a *Horn clause*; if a Horn clause contains exactly one positive literal, then it is a *definite clause*. Using the laws of first-order logic, a definite clause $\neg b_1 \vee \ldots \vee \neg b_n \vee h$ can also be written as an implication $b_1 \wedge \ldots \wedge b_n \Rightarrow h$ or sometimes as $h \leftarrow b_1 \wedge \ldots \wedge b_n$. The conjunction $b_1 \wedge \ldots \wedge b_n$ is called the *body*, the single atom h the *head* of the clause.

Grounding or *instantiating* a formula is done by replacing each occurrence of a variable with the same ground term, for all variables in the formula. This is done in all possible ways to find the grounded theory.

2.2 Lifted Graphical Models and Probabilistic Logic Programs

We start with a discussion of the two main classes of statistical relational learning, *lifted graphical models* and *probabilistic logic programs*. For lifted undirected graphical models, following Kimmig et al. (2015), we first introduce the general notion of a parameterized factor (Poole, 2003), which allows us to generalize factor graphs to relational representations, and then discuss how Markov Logic Networks (Richardson and Domingos, 2006) implement this idea. For probabilistic logic programs, following De Raedt and Kimmig (2015), we focus on the general idea of adding probabilistic choices to logic programs, which we detail for the probabilistic logic programming language ProbLog (De Raedt et al., 2007). We conclude this part with an overview of common inference tasks.

2.2.1 Parameterized Factors: Markov Logic Networks

A **Markov random field** is defined by a set of random variables and a set of **factors**, where a factor is a nonnegative potential function of a subset of the variables. The probability of a complete assignment to the random variables is proportional to the product of the value

of the factors applied to that assignment. Positive factors (with no zeros) can be written in
log-linear form as exponentials of sums of weighted formulae of variable assignments.

A **Markov network** is a network where the nodes are random variables, and there is an
edge between random variables that are in the same factor; the factors are implicit. Often
we assume there is a factor for each **clique** (maximal fully connected set of variables).
Note that having the factors implicit loses information. For example, a Markov random
field with a factor between each pair of variables has the same Markov network as a one
with a single factor on all variables, but has $O(n^2)$ parameters instead of 2^n free parameters
in a clique.

A key concept in the first stream of statistical relational representations is that of a **parameterized factor (parfactor)** (Poole, 2003). Parfactors provide a relational language
to compactly specify sets of factors for which sets of random variables share the same
structure and potential function.

A **parameterized random variable** or **par-RV** has the same syntax as a functor or predicate symbol. Instead of having a particular value, the representations define a distribution
on the values of the par-RVs.

Formally, a parfactor is a triple $\Phi = (\mathbf{A}, \phi, C)$, where $\mathbf{A}$ is a vector of par-RVs, ϕ is a
function from a tuple of values of RVs instantiating these par-RVs to the non-negative real
numbers, and C is a set of constraints on how the par-RVs may be instantiated. Inequality
constraints arise naturally in many models (e.g., a model of how x likes y will be different
when x is the same individual as y from when x and y are different individuals). For typed
relational languages, type constraints can be included in C.

To emphasize the connection to factor graphs, we refer to a set of parfactors $\mathcal{F} =$
$\{(\mathbf{A}_i, \phi_i, C_i)\}$ as a **parfactor graph**. Parfactor graphs lift factor graphs analogously to how
first-order logic lifts propositional logic. In particular, a parfactor represents **parameter
sharing**; there is a shared value for all groundings that satisfy the constraint. A **grounding**
(page 17) is obtained by replacing each variable by a constant denoting a particular entity.

Example 2.1 The left part of Figure 2.1 shows a parfactor graph with a single parfactor
over par-RVs Age(x), the age of person x, Genre(y), the genre of movie y, and Likes(x, y),
which represents whether x likes y. This can represent the dependence between whether a
person likes a movie, the person's age and the movie's genre. The right part of Figure 2.1
shows the corresponding factor graph instantiating the par-RVs for x ∈ {Ann, Bob, Carl}
and y ∈ {GodFather, RainMaker}. There are six instances of Likes; one for each combination of person and movie.

The parfactor graph $\mathcal{F}$ defines a probability distribution as follows, where X is the vector
of all RVs that instantiate par-RVs in $\mathcal{F}$:

$$P(X = x) = \frac{1}{Z} \prod_{(\mathbf{A}_i, \phi_i, C_i) \in \mathcal{F}} \prod_{\theta \in \mathcal{I}(\mathbf{A}_i, C_i)} \phi_i \theta \qquad (2.1)$$

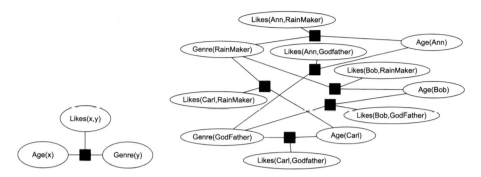

Figure 2.1: Example of a parfactor graph and the corresponding factor graph obtained by instantiating parfactor on Age(x), Genre(y) and Likes(x, y) for x ∈ {Ann, Bob, Carl} and y ∈ {GodFather, RainMaker}.

where $\mathcal{I}(\mathbf{A}_i, \mathbf{C}_i)$ denote the ground instantiations of $\mathbf{A}_i$ under constraints C_i. $\phi_i\theta$ is the value of ϕ_i for the assignment given by θ.

That is, the probability of a complete assignment to the groundings of the par-RVs is proportional to the product of the groundings of all of the parfactors. All the factors that are instantiations of the same parfactor share common structure and parameters. Especially in the context of parameter learning, those **shared parameters** are also called **tied parameters**.

In the grounding, each random variable is either observed or not; there is no data in the traditional sense where there might be multiple instances of a random variable from which to learn a pattern. Parameter tying enables generalization when learning. It combines a flexible number of random variables with a fixed number of parameters. Parfactor graphs thus exploit both probabilistic and relational structure to compactly represent probability distributions.

A traditional approach to inference in these models is to construct the underlying factor graph, on which standard inference approaches can be used. Knowledge-based model construction (KBMC) is one of the earliest techniques used to efficiently obtain the factor graph by dynamically instantiating the model only to the extent necessary to answer a particular query of interest (Wellman et al., 1992). These approaches suffer from the fact that the relational nature of the model causes potentially large amounts of repeated computations on the ground level, which can be avoided by using the lifted representation in lifted inference.

Markov Logic Networks   Markov Logic Networks (MLNs) (Richardson and Domingos, 2006) are the focus of much of the research in lifted inference. MLNs consist of a set of weighted first-order logic formulae, where all ground instances of a weighted formula

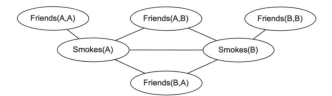

Figure 2.2: Markov network of MLN example.

share the same weight. A weighted formula, represented as $w : f$, where weight w is real number, and f is a logical formula, means that there is a factor for each assignment of an individual to a free variable in f, and all of these factors share the parameter w. It is traditional for the weight to be exponentiated, so that the probability of a complete assignment is proportional to the product of the exponentiated weights of the formulae that are true in the complete assignment, which is the same as the exponential of the sum of the weights.

Example 2.2 As an illustration, we present an example from (Richardson and Domingos, 2006), in which the patterns of human interactions and smoking habits are considered.

The following weighted formula encodes that friends have correlated smoking habits, e.g., if two people are friends, then they tend to either both be smokers or both be non-smokers.

$$w : \mathsf{Friends}(x, y) \Rightarrow (\mathsf{Smokes}(x) \Leftrightarrow \mathsf{Smokes}(y))$$

The parfactor defined by this rule is

$$\mathbf{A} = (\langle \mathsf{Friends}(x, y), \mathsf{Smokes}(x), \mathsf{Smokes}(y) \rangle, \phi, \{\})$$

where ϕ is the function that returns e^w when the first argument is false or the other arguments are equal, and 1 otherwise. Every possible instantiation of these par-RVs establishes a clique in the instantiated Markov network. For example, if there are only two entities, A and B, the instantiated Markov network is shown in Figure 2.2.

So far, we have not discussed how MLNs specify the constraints C of a parfactor. MLNs do not have a special mechanism for describing constraints, but constraints can be implicit in the formula structure. Two ways of doing this are as follows.

First, we can constrain groundings by providing constants as arguments of par-RVs. For example, continuing Example 2.2, writing $\mathsf{Friends}(A, y) \Rightarrow (\mathsf{Smokes}(A) \Leftrightarrow \mathsf{Smokes}(y))$ for constant A results in the subset of groundings of the formula above where $x = A$.

Second, when computing conditional probabilities, we can treat some predicates as background knowledge that is given at inference time rather than as definitions of random variables. For example, suppose we know that at inference time we will observe as evidence the truth values of all groundings of Friends atoms, and the goal will be to infer people's smoking habits. Then, the formula Friends(x, y) ⇒ (Smokes(x) ⇔ Smokes(y)) can be seen as setting up a clique between the Smokes values only of entities that are friends. If Friends(X, Y) is false for a particular pair of entities X and Y, then the corresponding instantiation of the formula is trivially satisfied, regardless of assignments to groundings of Smokes.

2.2.2 Probabilistic Logic Programs: ProbLog

The key idea behind the second stream of statistical relational representations is that of adding independent probabilistic choices to a relational formalism such as a relational database or a programming language. We focus here on logic programming as the host formalism, a choice that has been made many times, cf. (Poole, 1991; Dantsin, 1991; Poole, 1993; Sato, 1995; Fuhr, 2000; Poole, 2000; Sato and Kameya, 2001; Dalvi and Suciu, 2004; De Raedt et al., 2007). The semantics of these languages generalizes the least model semantics of logic programming – where each atom is false unless there is a clause to imply it is true – to represent a distribution over the least models[4].

Prominent examples of Prolog-based languages based on this semantics include ICL (Poole, 1993, 1997), PRISM (Sato and Kameya, 2001) and ProbLog (De Raedt et al., 2007; Kimmig et al., 2011); there exist subtle differences between these languages (see De Raedt and Kimmig (2015)). While the semantics can be defined for a countably infinite set of random variables and a general class of distributions, we focus on the finite case here, discussing the two most popular instances of the semantics, based on a set of independent random variables and independent probabilistic choices, respectively, and refer to Sato (1995) for details on the general case.

Boolean Random Variables   The most basic instance of this semantics, as realized in ProbLog, uses a finite set of Boolean random variables that are all assumed to be independent.

---

[4] For simplicity, we focus on definite clauses here, even though many probabilistic logic programming languages, including ProbLog, allow negation as failure as long as every truth value assignment to probabilistic facts has a unique model, cf., e.g., (Poole, 2000; Sato et al., 2005; Fierens et al., 2015; De Raedt and Kimmig, 2015), but differ in the set of facts

Example 2.3 We use the following running example inspired by Pearl's Bayesian network (Pearl, 1988):[5]

$$0.1 :: \text{burglary.} \quad 0.7 :: \text{hears\_alarm(mary).}$$
$$0.2 :: \text{earthquake.} \quad 0.4 :: \text{hears\_alarm(john).}$$
$$\text{alarm :– earthquake.} \tag{2.2}$$
$$\text{alarm :– burglary.}$$
$$\text{calls(X) :– alarm, hears\_alarm(X).}$$
$$\text{call :– calls(X).}$$

The program consists of a set R of definite clauses or *rules*, written as in Prolog as head :– body with comma denoting conjunction, and a set F of ground facts f, each of them labeled with a probability p, written as $p :: f$. We call such labeled facts *probabilistic facts*. Rules encode logical consequences, i.e., whenever the body of a rule is true, its head has to be true as well. Each probabilistic fact corresponds to a Boolean random variable that is *true* with probability p and *false* with probability $1 - p$. In the running example, we use b, e, hm and hj to denote the random variables corresponding to burglary, earthquake, hears_alarm(mary) and hears_alarm(john), respectively. We assume that all such random variables are independent given no evidence. Thus we obtain the following probability distribution P_F over truth value assignments to these random variables and their corresponding sets of ground facts $F' \subseteq F$:

$$P_F(F') = \prod_{f_i \in F'} p_i \cdot \prod_{f_i \in F \backslash F'} (1 - p_i) \tag{2.3}$$

Take for example the truth value assignment burglary $=$ *true*, earthquake $=$ *false*, hears_alarm(mary) $=$ *true*, hears_alarm(john) $=$ *false*, which we will abbreviate as $b \wedge \neg e \wedge hm \wedge \neg hj$, corresponds to the chosen facts {burglary, hears_alarm(mary)}, and has probability $0.1 \cdot (1 - 0.2) \cdot 0.7 \cdot (1 - 0.6) = 0.0336$. The corresponding logic program obtained by adding the set of rules R to the set of chosen facts, also called a *possible world*,

---

[5] For probabilistic logic programs, we follow the Prolog convention of names of predicates and constants starting with lower case letters, and variables starting with upper case letters.

world	calls(john)	probability
$b \wedge \neg e \wedge hm \wedge \neg hj$	*false*	$0.1 \cdot (1 - 0.2) \cdot 0.7 \cdot (1 - 0.4) = 0.0336$
$b \wedge \neg e \wedge hm \wedge hj$	*true*	$0.1 \cdot (1 - 0.2) \cdot 0.7 \cdot 0.4 = 0.0224$
$b \wedge e \wedge hm \wedge \neg hj$	*false*	$0.1 \cdot 0.2 \cdot 0.7 \cdot (1 - 0.4) = 0.0084$
$b \wedge e \wedge hm \wedge hj$	*true*	$0.1 \cdot 0.2 \cdot 0.7 \cdot 0.4 = 0.0056$
$\neg b \wedge e \wedge hm \wedge \neg hj$	*false*	$(1 - 0.1) \cdot 0.2 \cdot 0.7 \cdot (1 - 0.4) = 0.0756$
$\neg b \wedge e \wedge hm \wedge hj$	*true*	$(1 - 0.1) \cdot 0.2 \cdot 0.7 \cdot 0.4 = 0.0504$

Figure 2.3: The possible worlds of Example 2.3 where `calls(mary)` is true.

is

$$\text{burglary.}$$
$$\text{hears\_alarm(mary).}$$
$$\text{alarm :- earthquake.} \tag{2.4}$$
$$\text{alarm :- burglary.}$$
$$\text{calls(X) :- alarm, hears\_alarm(X).}$$
$$\text{call :- calls(X).}$$

As each logic program obtained by choosing a truth value for every probabilistic fact has a unique least Herbrand model (a least model using only symbols from the program), P_F can be used to define the *success probability* $P(q)$ of a query q, as the sum over all programs that entail q:

$$P(q) = \sum_{\substack{F' \subseteq F \\ \exists \theta F' \cup R \models q\theta}} P_F(F') \tag{2.5}$$

$$= \sum_{\substack{F' \subseteq F \\ \exists \theta F' \cup R \models q\theta}} \prod_{f_i \in F'} p_i \cdot \prod_{f_i \in F \backslash F'} (1 - p_i) . \tag{2.6}$$

Naively, the success probability can thus be computed by enumerating all sets $F' \subset F$, for each of them checking whether the corresponding possible world entails the query, and summing the probabilities of those that do. As fixing the set of facts yields an ordinary logic program, the entailment check can use any reasoning technique for such programs.

Going over all sixteen possible worlds of our example program (2.2) in this way, we obtain the success probability of calls(mary), $P(\text{calls(mary)}) = 0.196$, as the sum of the probabilities of six possible worlds (listed in Table 2.3).

Clearly, enumerating all possible worlds is infeasible for larger programs. Traditional techniques for exact inference in probabilistic logic programs can be viewed as operating (implicitly or explicitly) on a propositional logic representation of all possible worlds that entail the query q of interest. Note that this set of possible worlds is given explicitly by the following formula in disjunctive normal form (DNF)

$$DNF(q) = \bigvee_{\substack{F' \subseteq F \\ \exists \theta F' \cup R \models q\theta}} \left(\bigwedge_{f_i \in F'} f_i \wedge \bigwedge_{f_i \in F \setminus F'} \neg f_i \right) \tag{2.7}$$

and that the structure of this formula exactly mirrors that of Equation (2.6) defining the success probability, where we replace summation by disjunction, multiplication by conjunction, and probabilities by truth values of random variables (or facts). Actual inference approaches do not build this explicit formula, but an equivalent, more compact representation, typically based on grounding the program using logical inference, and then using techniques from weighted model counting (WMC) to compute the probability of the formula being true (De Raedt et al., 2007; Fierens et al., 2015). A detailed discussion of these approaches can be found in (De Raedt and Kimmig, 2015).

For ease of modeling (and to allow for countably infinite sets of probabilistic facts), probabilistic languages such as ICL and ProbLog use *non-ground probabilistic facts* to define sets of random variables. All ground instances of such a fact are mutually independent and share the same probability value. As an example, consider a simple coin game which can be won either by throwing two times heads or by cheating. This game can be modeled by the program below. The probability to win the game is then defined by the success probability $P(\text{win})$.

0.5 :: heads(X).   0.2 :: cheat.

win :– cheat.

win :– heads(1), heads(2).

Legal groundings of such facts can also be restricted by providing a domain, as in the following variant of our alarm example where all persons have the same probability of hearing the alarm if there is an alarm, and they call independently given an alarm.

0.1 :: burglary.   0.2 :: earthquake

0.7 :: hears_alarm(X) :– person(X).

person(mary). person(john). person(bob). person(ann).

alarm :– earthquake.

alarm :– burglary.

calls(X) :– alarm, hears_alarm(X).

call :– calls(X).

It is often assumed that probabilistic facts do not unify with other probabilistic facts or heads of rules. This ensures that the label of a probabilistic fact equals the fact's success probability, and achieves a clean separation between the facts F used to define the distribution P_F and their logical consequences given by the set of rules R.

Probabilistic Choices As noted by Poole (1993) and Sato (1995), probabilistic facts (or binary switches) are expressive enough to represent a wide range of models, including Bayesian networks, Markov chains and hidden Markov models. However, for ease of modeling, it is often more convenient to use multi-valued random variables instead of binary ones. The concept commonly used to realize such variables in probabilistic logic programs is a probabilistic choice, that is, a finite set of ground atoms exactly one of which is true in any possible world. Examples of primitives implementing the concept of a probabilistic choice are the **probabilistic alternatives** of probabilistic Horn abduction (PHA) (Poole, 1993), the independent choice logic (ICL) (Poole, 1997), the *multi-ary random switches* of PRISM (Sato and Kameya, 2001), the *probabilistic clauses* of stochastic logic programs (SLPs) (Muggleton, 1996), and the *annotated disjunctions* of logic programs with annotated disjunctions (LPADs) (Vennekens et al., 2004), or the *CP-events* of CP-logic (Vennekens et al., 2009). We restrict the following discussion to annotated disjunctions (Vennekens et al., 2004), using the notation introduced below.

An **annotated disjunction** (AD) is an expression of the form

$$p_1 :: h_1; \ldots ; p_N :: h_N :- b_1, \ldots, b_M.$$

where $b_1, \ldots, b_M$ is a possibly empty conjunction of literals, the p_i are probabilities and $\sum_{i=1}^{N} p_i \leq 1$. Considered in isolation, an annotated disjunction states that if the body $b_1, \ldots, b_M$ is true at most one of the h_i is true as well, where the choice is governed by the probabilities (see below for interactions between multiple ADs with unifying atoms in the head). If the p_i in an annotated disjunction do not sum to 1, there is also the case that nothing is chosen. The probability of this event is $1 - \sum_{i=1}^{n} p_i$. A probabilistic fact is thus a special case of an AD with a single head atom and empty body.

For instance, consider the following program:

$$0.4 :: \text{draw}.$$

$$\frac{1}{3} :: \text{color(green)}; \frac{1}{3} :: \text{color(red)}; \frac{1}{3} :: \text{color(blue)} :- \text{draw}.$$

The probabilistic fact states that we draw a ball from an urn with probability 0.4, and the annotated disjunction states that if we draw a ball, the color is picked uniformly among green, red and blue. The program thus has four possible worlds, the empty one with no draw (with probability 0.6), and three that each contain draw and one of the color atoms (each with probability 0.4/3). Similarly to probabilistic facts, a non-ground AD denotes the set of all its groundings. For each such grounding, choosing one of its head atoms to be true can be seen as an event that is independent of other events. For example, the annotated disjunction

$$\frac{1}{3} :: \text{color}(\text{B}, \text{green}); \frac{1}{3} :: \text{color}(\text{B}, \text{red}); \frac{1}{3} :: \text{color}(\text{B}, \text{blue}) :- \text{ball}(\text{B}).$$

defines an independent probabilistic choice of color for each ball B.

As noted by Vennekens et al. (2004), the probabilistic choice over head atoms in an annotated disjunction can equivalently be expressed using a set of logical clauses, one for each head, and a probabilistic choice over facts added to the bodies of these clauses, e.g.

$$\text{color}(\text{B}, \text{green}) :- \text{ball}(\text{B}), \text{choice}(\text{B}, \text{green}).$$
$$\text{color}(\text{B}, \text{red}) :- \text{ball}(\text{B}), \text{choice}(\text{B}, \text{red}).$$
$$\text{color}(\text{B}, \text{blue}) :- \text{ball}(\text{B}), \text{choice}(\text{B}, \text{blue}).$$
$$\frac{1}{3} :: \text{choice}(\text{B}, \text{green}); \frac{1}{3} :: \text{choice}(\text{B}, \text{red}); \frac{1}{3} :: \text{choice}(\text{B}, \text{blue}).$$

This example illustrates that annotated disjunctions define a distribution P_F over basic facts as required in the distribution semantics, but can simplify modeling by directly expressing probabilistic consequences.

Independent Causes   Some languages assume that head atoms in the same or different annotated disjunctions cannot unify with one another, while others, e.g., LPADs (Vennekens et al., 2004), do not make this restriction, but instead view each annotated disjunction as an independent cause for the conclusions to hold. In that case, the structure of the program defines the combined effect of these causes, similarly to how the two clauses for alarm in our earlier example (2.2) combine the two causes burglary and earthquake. We illustrate this on the Russian roulette example of (Vennekens et al., 2009), which involves two guns.

$$\frac{1}{6} :: \text{death} :- \text{pull\_trigger}(\text{left\_gun}).$$
$$\frac{1}{6} :: \text{death} :- \text{pull\_trigger}(\text{right\_gun}).$$

Each gun is an independent cause for death. Pulling both triggers will result in death being true with a probability of $1 - (1 - \frac{1}{6})^2$, which corresponds to the probability of death being proven via the first or via the second annotated disjunction (or both). Assuming independent causes closely corresponds to the noisy-or combining rule that is often employed in the Bayesian network literature (Pearl, 1988); see Section 2.3.2.3.

2.2.3   Inference Tasks

While the two types of models we discussed above differ in how they specify a probability distribution $P(X = x)$ over ground logical atoms, the most common reasoning tasks in both cases include the following:

- In **marginal inference**, the *MARG*$(Q \mid e)$ task, a par-RV Q, the query, and a ground query e, the evidence, are given. The task is to compute the marginal probability distribution of instances q of Q given the evidence,

$$MARG(Q \mid e) = P(q \mid e) = \frac{P(q \wedge e)}{P(e)}.$$

It is called marginal inference as it involves marginalizing (summing out) all of the non-observed, non-query variables. If q is not ground, all instances have the same probability. The special case with ground Q and $e = true$ (and thus $P(e) = 1$) is also known as computing the **success probability** of query q in probabilistic logic programming.

· The **maximum a-posteriori probability (MAP)** task, written $MAP(Q \mid e)$, is to find the most likely truth-assignment v to the atoms in Q given the evidence e, marginalizing other variables, that is, to compute

$$MAP(Q \mid e) = \arg\max_{v} P(Q = v \mid e)$$

· The **most probable explanation (MPE)** task, $MPE(e)$, is to find the most likely world (assignment to all of the variables) where the given evidence query e holds. Let U be the set of all ground atoms that do not occur in e. The task is to compute the most likely truth-assignment u to the atoms in U,

$$MPE(e) = MAP(U \mid e).$$

· Define an **explanation** (Poole, 1993) of formula q to be a minimal set of assignment to groundings of probabilistic atoms from which q can be inferred. In the **Viterbi task**, written $VIT(q)$ task, considered mostly in probabilistic logic programming, a query q (but no evidence) is given, and the task is to find a most likely explanation of q. Let $E(q)$ be the set of all explanations q. The Viterbi task is to compute

$$VIT(q) = \arg\max_{X \in E(q)} P(\bigwedge_{f \in X} f).$$

Example 2.4   To illustrate, consider our initial alarm example (2.2) with $e = $ calls(mary) and $Q = \{$burglary, calls(john)$\}$. The worlds where the evidence holds are listed in Table 2.3, together with their probabilities. The answer to the MARG task is

$$P(\text{burglary} \mid \text{calls(mary)}) = 0.07/0.196 = 0.357$$
$$P(\text{calls(john)} \mid \text{calls(mary)}) = 0.0784/0.196 = 0.4$$

The answer to the MAP task is burglary=*false*, calls(john)=*false*, as its probability 0.0756/0.196 is higher than 0.028/0.196 (for *true, true*), 0.042/0.196 (for *true, false*) and 0.0504/0.196 (for *false, true*). The world returned by MPE is the one corresponding to the set of facts {earthquake, hears_alarm(mary)}. Finally, the Viterbi proof of query calls(john), which does not take into account evidence, is $e \wedge hj$, as $0.2 \cdot 0.4 > 0.1 \cdot 0.4$ (for $b \wedge hj$), whereas the Viterbi proof for query burglary is its only proof b.

2.3   Statistical Relational Representations

Our discussion of MLNs and ProbLog has provided some insight into commonalities and differences that can exist between representation languages. Given the variety of languages developed so far, a full overview is beyond the scope of this chapter. In the following, we therefore take a step back from existing languages. Instead, we first provide a brief overview of desirable properties of statistical relational representations, and then discuss possible features of representation languages related to these properties.

2.3.1   Desirable Properties of Representation Languages

When designing a new representation language or choosing an existing one for modeling a specific problem or solving a specific task, a number of desirable but potentially competing properties are of interest, including the following (De Raedt et al., 2016):

Expressivity and ease of modeling: Can all relevant knowledge be (easily) expressed in the language? If we can't represent some knowledge, we can't learn it. What can be inferred is constrained by what can be represented.

Understandability and explainability: Are probabilistic models written in the language easy to understand and explain? Often the inputs and the outputs can be interpreted, but the intermediate variables are much more challenging to understand. For example, **sigmoid belief networks** (Saul et al., 1996) are like neural networks with sigmoid activation functions, but where the intermediate values can be interpreted as Boolean random variables. This makes inference and learning much less efficient, but aids in interpretability. Similarly, **probabilistic soft logic (PSL)** (Kimmig et al., 2012; Bach et al., 2017) is to MLNs what neural networks are to sigmoid belief networks. Inference and learning can be done much more efficiently, but the intermediate variables cannot be interpreted as probabilities. How much one is willing to trade off explainability for efficiency depends on the application.

Suitability for different tasks: What tasks (e.g., computing marginals, finding explanations, predicting, or planning) can be performed for models expressed in the language? For some languages, different types of inference become more natural or efficient.

Efficient inference: Do efficient inference methods exist for probabilistic models and tasks expressed in the language? Efficient algorithms depend on exploiting structure in the problem, which may or may not be exposed by the representation. Note that this is different from expressivity; restricting a language may change what can be represented without affecting how efficiently queries can be answered.

Learnability: Can the parameters and the structure of models be learned from data, in such a way that the learned model generalizes beyond the initial observations to new sets of individuals? Can models with latent variables – variables not observed in the data – be learned? For example, probabilistic soft logic has a different semantics to MLNs which enables learning to be done more efficiently, at the expense of explainability.

Modularity: Does the language support easy combination or decomposition of models, such that parts of models can be developed and used individually? For example, Buchman and Poole (2015) show that MLNs cannot modularly represent conditional probabilities: if M is some model represented as an MLN which does not contain atom a, we cannot modularly define $P(a \mid M)$ using an MLN, except for trivial cases, without changing the distribution on M.

Compactness: Can knowledge be represented succinctly?

Prior knowledge: Can informed prior knowledge be easily incorporated in models, in such a way that the prior is used when there no evidence, but it is consistent (in the statistical sense) in that the model will eventually adapt to any data?

Heterogenous data: Can models combine information from multiple, heterogenous data sources that provide different levels of abstraction (using of more general or more specific terms) or detail (in terms of parts and subparts), different amounts of information for different objects (including missing data), and different structural knowledge (e.g., ontologies)?

Support for additional concepts: Does the language support concepts beyond basic probabilistic models, such as time, utilities, or similarity functions?

2.3.2 Design Choices

Given the properties just outlined, we now discuss a number of design choices for statistical relational representations. In doing so, we focus on representations based on first-order logic, which are the most prominent in lifted inference. We first consider the key question of the semantics used on the logical side (first-order logic or logic programming) and on the probabilistic side (directed or undirected models). Then, we discuss further issues related to how parameters are combined and specified, and to handling open universe models.

2.3.2.1 First-Order Logic and Logic Programming While classical first-order logic and logic programming are closely related, their fundamental difference is on the semantical level: a first-order theory can have many models, whereas a logic program has a single model. This difference has direct consequences for the expressivity of statistical relational representations building on these languages. As an illustrative example (from De Raedt and Kimmig, 2015), consider the following logic program

```
edge(1, 2).
path(A, C) :- edge(A, C).
path(A, C) :- edge(A, B), path(B, C).
```

which due to its least models semantics inductively defines path as the transitive closure of the relation edge, that is, its unique model is {edge(1, 2), path(1, 2)}. The transitive closure of a binary relation cannot be represented in first-order logic, but requires a least models semantics or second-order constructs; see (Huth and Ryan, 2004) for a detailed formal discussion.

We could use Clark's 1978 completion, giving the definitions:

$$\forall X \forall Y \; \text{edge}(X, Y) \Leftrightarrow X = 1 \wedge Y = 2.$$

$$\forall A \forall C \; \text{path}(A, C) \Leftrightarrow \text{edge}(A, C) \vee \exists B \; \text{edge}(A, B) \wedge \text{path}(B, C).$$

which has the same logical consequences as the logic program (for the atoms using edge or path).

Interpreting the program naively as the following set of first-order clauses

$$\text{Edge}(1, 2)$$

$$\text{Edge}(a, c) \Rightarrow \text{Path}(a, c)$$

$$\text{Edge}(a, b) \wedge \text{Path}(b, c) \Rightarrow \text{Path}(a, c)$$

results in a total of six models where(assuming Edge(1, 2) is the only edge atom):

{Edge(1, 2),	Path(1, 2)}			
{Edge(1, 2),	Path(1, 2),			Path(1, 1)}
{Edge(1, 2),	Path(1, 2),		Path(2, 1),	Path(1, 1)}
{Edge(1, 2),	Path(1, 2),	Path(2, 2)}		
{Edge(1, 2),	Path(1, 2),	Path(2, 2),		Path(1, 1)}
{Edge(1, 2),	Path(1, 2),	Path(2, 2),	Path(2, 1),	Path(1, 1)}

Note that an MLN using this first-order theory as hard constraints would (in the absence of more information) assign equal probabilities to these six models, whereas ProbLog, defined in terms of logic programming, would assign a probability of one to the single model of the logic program. The details of the semantics are important to understand, particularly if they are used by non-sophisticated users.

A second choice relating directly to the first-order logic level is whether the representation language allows **functors** (function symbols) or not. Functors allow for easy and compact representation of structured objects, such as lists or trees, but result in infinite sets of ground atoms, and thus potentially infinite models even for finite sets of constants. Having functors makes logic programs Turing-complete, so subject to the halting problem and semi-decidability. In addition to functors, logic programming based representation languages also provide additional modeling flexibility through their programming language features, e.g., built-in arithmetic functions, support for including libraries, or second-order predicates for creating sets. In both cases, the increased modeling flexibility requires more elaborate inference approaches.

2.3.2.2   Directed and Undirected Models   As in the propositional case, we can distinguish between directed and undirected models. The former often have a generative flavour, with models describing the process of generating objects and their relations probabilistically, whereas the latter often can be seen as introducing soft constraints on possible worlds.

Both directed and undirected models can use weighted formulae. **Relational logistic regression** (RLR) (Kazemi et al., 2014) is the directed analogue of **Markov logic networks**. It uses the same weighted formulae to define conditional probabilities. They are identical models when there is a single target variable and all other variables are observed. However they have very different properties when not all variables are observed (Poole et al., 2014).

A **canonical representation** is one where there is a unique set of parameters for a model. Non-canonical representations tend to be more difficult to interpret because the parameters cannot be understood locally, independently of the other parameters. They also tend to be more difficult to to learn because there is typically no global minimum parameter assignment; many different parameter assignments may give the same value. Undirected models, such as MLNs are not canonical, their influence on the probability of certain ground atoms often cannot be understood locally, but only in the context of all other weights. In many directed models, parameters have a local interpretation as (conditional) probabilities, and are canonical so they can be interpreted in isolation. In undirected models one needs to find efficient ways of computing, or estimating, the normalization constant Z (cf. Equation (2.1)). Tha analogy in directed models is to compute the prob ability of the evidence, which can often be done without considering the whole model.

Relational domains frequently contain cyclic dependencies, for instance, if a property of a certain object is recursively defined in terms of the same property on related objects (such as people being more likely to smoke the more of their friends are smoking). Undirected relational models naturally handle such cyclic dependencies, as they simply use the corresponding ground formulas as features of a Markov network. In directed models, a decision needs to be made on how to handle cycles in the model, which influences the expressivity of the language. One option is to explicitly prohibit cyclic dependencies between random variables, as is done for instance in Bayesian logic programs, where cycles are allowed on the predicate level, but not on the level of ground atoms (Kersting et al., 2006). A second option is to allow such dependencies, even if they do not necessarily define a coherent probability distribution, and instead interpret models in terms of the equilibrium of a Markov chain, as done in relational dependency networks (Neville and Jensen, 2007). Third, in probabilistic logic programming languages such as ProbLog, cyclic dependencies between ground atoms are allowed as long as they have a well-defined interpretation in logic programming terms, i.e., the logic program has a unique model for each combination of probabilistic choices that can be made. This, however, means that not all probability distributions can be represented (Buchman and Poole, 2016).

2.3.2.3   Combining Rules and Aggregation   In directed relational models, when the parents contain a logical variable that is not in the child, there are an unbounded number of parents of the child node in the grounding (polynomial in the population size). In this case, the conditional probability cannot be represented as a table, but needs some form of **aggregation** to combine the parents. Sometimes the number of parents of a node in the grounding depends on the particular instantiation of the node.

There are a number of aggregation methods that have been proposed:

Noisy-or   When there are Boolean variables, a noisy-or is the case where a child is true if there exists a true value for one the parents in the grounding. For instance, as illustrated in Section 2.2.2, ProbLog's logic programming based semantics implicitly corresponds to a noisy-or combination of several rules defining the same random variables, and the existential quantification of variables in the body provides an aggregator. In particular, $h \leftarrow b(X)$ means $h \leftarrow \exists X\, b(X)$. If there is a probabilistic atom $n(X)$ such that $h \leftarrow b(X) \wedge n(X)$, then h is noisy-or of the b's; the existential providing an "or" and the instances of $n(X)$ providing the noise. If $P(n(X)) = p$ and the domain is $i_1 \ldots i_k$, and *obs* is an assignment of *true* or *false* to each of $b(i_1) \ldots b(i_k)$

$$P(h \mid obs) = 1 - (1 - p)^t$$

where t is the number of $b(i_1) \ldots b(i_k)$ that are true in *obs*. Often there is an "leak probability" for when none of the b are true, provided by an extra rule.

One problem with the use of the noisy-or is that $P(h) \rightarrow 1$ as the population grows (Poole et al., 2014). The generalization of noisy-or to ordinal variables is called noisy-max (Díez and Galán, 2003).

Log-linear models   Weighted formula can also be used as an aggregator, forming a log-linear model. In particular, **relational logistic regression** (RLR) (Kazemi et al., 2014) is the directed version of Markov logic networks.

For example, consider the weighted formulae:

$$\omega_0 : q$$
$$\omega_1 : q \wedge \neg r(x)$$
$$\omega_2 : q \wedge r(x)$$
$$\omega_3 : r(x)$$

If *obs* consists of an observation of true or false for $r(x)$ for every individual x:

$$P(q \mid obs) = sigmoid(\omega_0 + n_F \omega_1 + n_T \omega_2)$$

where n_T is number of individuals for which $r(x)$ is true and n_F is number of individuals for which $r(x)$ is false. This computation, with *obs* observed, holds for both MLNs and RLR, however RLR uses this as the definition of the conditional probability, but MLNs treat this as a constraint on worlds. In weighted logistic regression, a set of weighted rules for a proposition results in the conditional probability as the sigmoid of a polynomial of the parents (Kazemi et al., 2014).

Aggregates Perlich and Provost (2003) propose to first aggregate the values of all parent variables of the same type into a single value, and provide the conditional probability of the child variable given these aggregated values. For example, one could use two aggregate functions agg_Y and agg_Z, together with a conditional probability distribution P', to define $P(X \mid Y_1, Z_1, \ldots, Y_n, Z_n) = P'(X \mid agg_Y(Y_1, \ldots, Y_n), agg_Z(Z_1, \ldots, Z_n))$.

Combining rules It is also possible to explicit rules that are used to combine the parents (Jaeger, 1997; Natarajan et al., 2005). An approach based on combining rules, would specify the conditional probability distribution for the child variable for $n = 1$ as well as a function that computes a single conditional distribution from n conditional distributions. In the example, one would thus use a distribution $P''(X \mid Y, Z)$ and a combining function f, and define $P(X \mid Y_1, Z_1, \ldots, Y_n, Z_n) = f(P''(X \mid Y_1, Z_1), \ldots, P''(X \mid Y_n, Z_n))$. In Bayesian logic programs, combining rules are an explicit part of the modeling language Kersting et al. (2006).

Thus, modeling languages can either provide explicit constructs for specifying such operations, or implicitly provide them via other language constructs.

Undirected models, such as **MLNs,** do not require explicit aggregation, as varying numbers of groundings simply contribute corresponding numbers of factors to the unnormalized probability distribution. However, there is an implicit aggregation of the neighbours in the grounding.

Combining rules and aggregation are concerned with a variable number of parents for different groundings of a logical atom typically assuming a fixed model and set of constants. However, we often want to apply the same relational model for different sets of constants of varying size. However, it is not always evident how a relational model behaves in such different settings. For instance, as discussed by Poole et al. (2014), models that encode a dependency on the *number* of related objects often implicitly encode dependencies on the entire population size as well, and thus may behave differently for different numbers of objects. As an example, suppose we want to model that someone's happiness depends on the number of his friends coming to a party. If a person has ten friends, predicting happiness if at least five friends come to the party, if fewer than five friends do not come to the party, or if at least half of the friends come to the party will all have the same effect, but this no longer holds for a different number of friends. Moreover, for MLNs, and RLR, the probability happiness must approach 0 or 1 as the number of friends approaches infinity (Poole et al., 2014). For Problog the probability approches 1.

2.3.2.4 Parameterizing Relational Models Besides the high level choices of which logic semantics to use and whether to base the representation on directed or undirected probabilistic models, a number of more low level choices on how to parameterize a model are of interest as well.

Domains of Random Variables   So far, we have focused on random variables with Boolean or categorical domains, as those are most common in logic-based representations. However, certain problem domains may require random variables that take values from infinite or even continuous domains, under for instance Poisson distributions or Gaussians. The choice of whether to include such distributions in a language has direct consequences on both the expressivity of the language and the ways inference can be performed. While exact inference approaches are often restricted to finite domains, sampling-based approximate inference naturally handles continuous distributions (Chapter 10), and we can exploit analytical solutions as much as possible for hybrid models (Chapter 14).

Grounding-Specific Parameters   To achieve compact models, it can be interesting to allow for parameters whose values are determined at runtime rather than provided as constants. One way to achieve this are the flexible probabilities used in probabilistic logic programming, which make it possible to label a probabilistic fact with a logical variable whose value is computed by the program itself. For instance,

$$W :: \text{heads}(C) :- \text{weight}(C, W).$$

compactly specifies that the probability of any coin C coming up heads is given by the weight W associated with C in the weight predicate, which may itself be defined by arbitrary ProbLog code. As another example,

$$\text{Prob} :: \text{cheap}(H) :- \text{numLuxuryFeatures}(H, N), \text{baseProb}(P), \text{Prob is } P**N$$

with suitable definitions of the predicates in the body, specifies that the probability that a house with n of luxury features is cheap is p^n.

Independence and Determinism   Another strength of logic based modeling languages is their ability to expose independence and determinism in the model structure to the inference engine for increased efficiency, while at the same time achieving compact representations. For instance, consider the following ProbLog program fragment, where not denotes negation as failure:

$$0.4 :: \text{hears\_alarm}(X) :- \text{at\_home}(X), \text{music\_on}(X).$$

$$0.7 :: \text{hears\_alarm}(X) :- \text{at\_home}(X), \text{not music\_on}(X), \text{in\_basement}(X).$$

$$0.9 :: \text{hears\_alarm}(X) :- \text{at\_home}(X), \text{not music\_on}(X), \text{not in\_basement}(X).$$

This models the conditional probability of someone hearing the alarm depending on whether that person is at home, listening to music, and in the basement or not. It exposes conditional independence by only including relevant literals in the rule bodies, and omits rules with zero probabilities (specifically, for people not being at home). A full representation with one rule per truth value assignment for the parents would require eight rules. For parfactors, similar effects can be obtained by including suitable constraints on the instantiations.

Furthermore, as noted by (Poole, 1993), independent choices of probabilistic atoms together with a logical theory describing their consequences, are sufficient to represent any discrete conditional probability distribution. Thus, it is not necessary to consider parameters associated with arbitrary formulas, although this can simplify modeling (cf. the discussion of annotated disjunctions in Section 2.2.2).

Persistent Random Values   One may want to distinguish between random variables that refer to persistent properties, which do not change within a possible world, such as the height of a person, and random variables that may appear as multiple copies with independent values, such as repeated flips of a coin or which way to go when repeatedly reaching the same node on a random walk. Relational models based on a possible worlds semantics often use persistent (or memoized) values by default, and can include additional arguments (or trial identifiers) when distinct copies are needed. An exception is the probabilistic logic programming language PRISM (Sato and Kameya, 2001), which uses a fresh, independent copy of a random variable on every use, much like in stochastic grammars or random walk models.

The difference between the two approaches can be explained using the following example of drawing balls from an urn. For the probabilistic choice

$$\frac{1}{3} :: \text{color(green)}; \frac{1}{3} :: \text{color(red)}; \frac{1}{3} :: \text{color(blue)},$$

there are three answers to the goal (color(X),color(Y)), one answer X = Y = c for each color c with probability $\frac{1}{3}$, as exactly one of the facts color(c) is true in each possible world when color is persistent (as in ProbLog and ICL). Asking the same question when color is not persistent (as in PRISM) results in 9 possible answers with probability $\frac{1}{9}$ each. The query then – implicitly – corresponds to an ICL or ProbLog query (color(X,id1), color(Y,id2)), where the choice above is replaced by a nonground variant

$$\frac{1}{3} :: \text{color(green, ID)}; \frac{1}{3} :: \text{color(red, ID)}; \frac{1}{3} :: \text{color(blue, ID)}$$

and id1 and id2 are identifiers that are unique to the call. Essentially we have reified each draw for the urn.

Avoiding the memoization of probabilistic facts is useful in order to model stochastic automata, probabilistic grammars, or stochastic logic programs (Muggleton, 1996) under the distribution semantics without needing to create extra probabilistic atoms. There, a new rule is chosen randomly for each occurrence of the same nonterminal state/symbol/predicate within a derivation, and each such choice contributes to the probability of the derivation. The rules for a nonterminal thus form a family of independent identically distributed random variables, and each choice is automatically associated with one variable from this family.

The choice between persistent and non-persistent random variables (or a combination of both, as in the probabilistic functional programming language Church (Goodman et al., 2008)) influences both what can be modeled easily in the language, and whether efficient inference for certain models is possible (cf. also the discussion of time below). For instance, probabilistic grammars and random walks may be more easily expressed in languages with non-persistent values, such as PRISM, whereas languages with persistent values, such as ProbLog, may be better suited when modeling distributions over subgraphs of a graph where each edge independently exists with an associated probability, or over instances of a probabilistic database, where each tuple has an associated probability of being part of an instance.

Models where the values are persistent can represent non-persistence by reifying coin tosses and other non-persistent events, and adding an extra argument to the predicates that refer to a particular toss.

Time and Dynamics   Among the most popular probabilistic models are those that deal with dynamics and time such as Hidden Markov Models (HMMs) and Dynamic Bayesian Networks. Most relational formalisms can naturally represent models that involve a temporal aspect or whose population or structure evolve over time simply by means of an additional argument. Naively using such a time argument with exact inference results in exponential running times (in the number of time steps), though this can often be avoided using dynamic programming approaches and principles, as shown by the PRISM system, which achieves the same time complexity for HMMs as corresponding special-purpose algorithms (Sato and Kameya, 2001). Relational approaches that provide explicit support for modeling and inference in temporal domains include Logical HMMs (Kersting et al., 2006; Natarajan et al., 2009), a language for modeling HMMs with structured states, CPT-L (Thon et al., 2011), a dynamic version of CP-logic, and the work on a particle filter for dynamic distributional clauses (Nitti et al., 2013).

2.3.2.5   Open Universe Models   While all relational models are necessarily expressed through a finite number of statements, this does not mean that they have to express distributions over finite spaces. Even simple probabilistic models such as Hidden Markov models or probabilistic grammars encode distributions over structures of arbitrary size, using a fixed number of parameters. Similarly, parameter tying in relational models with functors allows for finite specifications of distributions over infinite models, though inference typically requires queries to have finite support, i.e., to depend on a finite subset of all ground atoms only. For instance, in the context of an HMM, it suffices to consider sequences up to the length of the longest sequence in the query or evidence. On the other hand, non-parametric Bayesian approaches (such as (Carbonetto et al., 2005)) allow one to specify both the structure of a model and its parameters as stochastic processes, thus supporting unbounded numbers of individuals as well as unbounded numbers of parameters. Again, this is a choice that influences both expressivity and inference, as unbounded numbers of parameters call for approximate inference.

In logic-based relational models, constants or ground terms are used to refer to individuals in the domain of interest. This is straightforward if we both know which individuals exist and can uniquely identify them, as we may then use the same term for every mention of an individual. However, in general, we may have **existence uncertainty** where we not know which individuals exist (e.g., a house may have no, one, or several balconies), or **identity uncertainty** where we do not know whether two descriptions refer to the same or different individuals (e.g., whether Dr. Smith, Mary's neighbour and Sam refer to one, two or three different individuals).

In generative models, **existence uncertainty** has been addressed by explicitly modeling distributions over the number of objects that fit a description as well as the correspondence between terms in the model and these objects (typically by using a fixed naming scheme, such as $Ball(1), Ball(2), \ldots$). This is a key feature of Bayesian logic (BLOG) (Milch et al., 2005), but can also be achieved in languages such as ICL and ProbLog, which combine flexible probabilities and Prolog code (Poole, 2008). Alternatively, such a model may introduce objects sequentially, randomly deciding for each description whether it refers to a new object or to an already existing one (and if so, to which one), as done for instance in nonparametric Bayesian logic (NP-BLOG) (Carbonetto et al., 2005).

Combining identity uncertainty and existence uncertainty where there can be complex descriptions can be semantically tricky (Poole, 2007). For example, the existence of a balcony might depend on the room it is attached to and the height of the floor, and the existence of other balconies. The probability that Sam likes an apartment might depend on the existence of a bedroom for her and the existence of a bedroom for her child, with associated properties and relationships, but apartments do not come labelled with the roles that Sam may like to use them for. We have to consider the case where Sam and her child have to share a bedroom (so Sam's bedroom is the same as her child's bedroom).

While relational approaches can often represent models for identity uncertainty, as witnessed by such models for citation matching or entity resolution, e.g., (Pasula et al., 2003; Singla and Domingos, 2006a), inference in this context is inherently expensive for large domains. The reason is that this requires to either explicitly model pairwise equality between terms as a binary relation with reflexivity, symmetry and transitivity constraints, or else to reason over partitions of the names such that each partition contains the descriptions of the same object. Searching the partition space is easier than searching over equality (because the equality implies partitions) but is still prohibitive as there are more than an exponential number of partitions (the Bell number). A standard way to implement this is to run MCMC on the partitioning (Pasula et al., 2003).

2.4   Conclusions

The first part of this chapter has provided an introduction to the two main streams of representation languages for statistical relational domains, namely lifted graphical models, focusing on parameterized factors and Markov Logic Networks, and probabilistic logic programming, focusing on ProbLog. In the second part, we have considered statistical relational representations from a more abstract point of view, discussing both high level properties such languages should satisfy and lower level design choices we may face when selecting or developing such a language. While we have pointed to some existing representations making certain choices, a full inventory of existing languages based on the criteria we have seen is beyond the scope of this chapter, and a thorough understanding of the full design landscape and the interaction of its different dimensions is a largely open question.

3 Statistical Relational Learning

Sriraam Natarajan, Jesse Davis, Kristian Kersting, Daniel Lowd, Pedro Domingos, and Raymond Mooney

Abstract. This chapter presents a brief introduction to Statistical Relational Learning which combines the power of relational and logical representations with the ability of probability theory to model uncertainty. We present the different models and focus mainly on the most popular formalism and outline the different learning methods for learning this model. We also address some interesting extensions of these algorithms for transfer learning.

Keywords: statistical relational learning, probabilistic logic learning, relational probabilistic models

3.1 Introduction

Over the last couple of decades, Artificial Intelligence (AI) has matured into a sophisticated scientific field. Coupled with advancements in computational hardware and powerful statistical approaches, AI is being successfully applied in robotics, computer vision, medical diagnosis, games, search, and many other problems arising in industry, military, and medical applications. Yet, much remains to be done to reach the potential AI has to offer as a transformative technology with high societal and economic impact.

One of the key issues with using the current statistical learning techniques is the inability to handle complex relational data. These algorithms overuse the simplifying independent and identically distributed (iid) assumption which states that instances (examples) are drawn from the same distribution independently, which curtails the ability to handle complex interacting objects in the domain (Friedman et al., 1999; Getoor et al., 2001; Pfeffer et al., 1999). Real-world domains contain inherent structure, and learning algorithms must be capable of representing and reasoning with structure inside the domain.

In the last decade, a great deal of progress has been made in combining statistical methods with relational or logical models, i.e., statistical relational AI (De Raedt et al., 2016). At the core of statistical relational AI is Statistical Relational Learning (SRL) (Getoor and Taskar, 2007), which addresses the challenge of applying statistical inference and learning approaches to problems that involve rich collections of objects linked together in complex, stochastic and relational worlds. These approaches can be broadly classified into directed models (Poole, 1993; Ngo and Haddawy, 1995; Jaeger, 1997; Friedman et al., 1999; Pfeffer et al., 1999; Sato and Kameya, 2001; Getoor et al., 2001; Kersting and De Raedt, 2001; Heckerman et al., 2004; Milch et al., 2005; Neville and Jensen, 2007; De Raedt et al., 2007; Ramanan et al., 2018) and undirected models (Taskar et al., 2002; Richardson and Domingos, 2006; Gutmann and Kersting, 2006). The advantage of these models is that they can succinctly represent probabilistic dependencies among the attributes of different related objects, leading to a compact representation of learned models.

In this chapter, we present a high-level picture of these models and then present one specific model that is widely employed. We also present the recent advancement of the parameter and structure learning algorithms. We finally conclude the chapter by presenting some recent transfer learning algorithms that are inspired by SRL algorithms.

3.2   Statistical Relational Learning Models

While **statistical relational models** are indeed highly attractive due to their compactness and comprehensibility, learning them is typically much more demanding than learning propositional ones. Most of these methods essentially use first-order logic to capture domain knowledge and soften the rules using probabilities or weights. At least this was the goal of the initial models developed in this area. This is due to the fact that the field of Inductive Logic Programming (Muggleton and de Raedt, 1994), which is mainly concerned with learning first-order rules from data, was evolving at that time. Consequently, early systems simply used a logic learner under the hood to learn the rules and employed probabilistic learning techniques such as maximum likelihood estimation (for complete data) and EM (for incomplete data) to learn the parameters (i.e., weights or probabilities) for these rules. This is due to the fact learning structure of an SRL model requires learning the parameters repeatedly in the inner loop which in turn can sometimes require probabilistic inference in its inner loop. This problem is exacerbated for SRL models since the predicates can allow for arbitrary combinations of variables or constants as arguments.[6]

To summarize, many early SRL methods rely on the structure as well as the parameters of the models being specified by an expert. With the availability of training data, it became possible to learn the parameters for an expert-defined model. *Parameter learning* reduced the effort needed from the expert and potentially improved the accuracy of the model (by relying on data to correct mistakes made by experts), but also increased the computational time. For some domains, the structure of the model may be non-trivial, not known or insufficient. As a result, both the structure and parameters of the model need to be learned from the data. Although *structure learning* reduces the expert's effort, it can be computationally intensive due to the large space of possible structures while including parameter learning

---

[6] For instance, when learning to predict if someone (say x) is popular, it is possible to use the predicate *Friends* in several ways. Some possible ways are *Friends(x,y), Friends(y,x), Friends(x,"Erdos")* and *Friends("Erdos",x)*. Of course, the constant "Erdos" can be replaced with all possible constants in the data base

as a sub-task. We next discuss parameter learning of SRL models at a fairly high-level and will present a structure learning algorithm later in this chapter.

3.3   Parameter Learning of SRL models

Since the parameters of a model are defined with respect to a model structure, parameter-learning approaches assume that the structure is already provided.

Directed Models: Directed SRL models are essentially the logical/relational counterparts of propositional graphical models. For example, **Bayesian Logic Programs** (BLP) (Kersting and De Raedt, 2001) extend Bayesian networks to the logical setting where the random variables are replaced by first-order logic predicates.

In case of directed models, it is assumed that the parents of every logical predicate is known. Similar to Bayesian networks, the problem of parameter learning in these models can be viewed as learning the conditional distributions for each predicate. The standard approach in this research is to formulate the optimization problem as either maximizing the log-likelihood or minimizing the mean squared error when given some training data ((Natarajan et al., 2005; Kersting and De Raedt, 2001; Getoor et al., 2001)). Then either gradient-descent or the expectation maximization (EM) algorithm is adapted to fill in the parameter values. Since relational models can have multiple instantiations of a logical variable, either combining rules or aggregations are used to handle this issue. The algorithms are capable of learning the parameters of these combining rules as well.

We outline the most popular undirected formalism of Markov Logic Networks next and present the parameter and structure learning algorithms for this model.

3.4   Markov Logic Networks

A popular SRL representation is **Markov Logic Networks** (MLNs) (Richardson and Domingos, 2006). An MLN consists of a set of formulas in first-order logic and their real-valued weights, $\{(w_i, f_i)\}$. The higher the weight of the rule, the more likely it is to be true in the world. An example MLN for predicting smokers in a group is shown below. The first rule states that a friend of a smoker is likely to be a smoker whereas the second rule is used to specify a prior distribution over a person smoking.

$$1.0 \quad Friends(\mathsf{X}, \mathsf{Y}), Smoker(\mathsf{X}) \rightarrow Smoker(\mathsf{Y})$$

$$-1.0 \quad Smoker(\mathsf{X})$$

Together with a set of constants, we can instantiate an MLN as a Markov network with a variable node for each ground predicate and a factor for each ground formula. All factors corresponding to groundings of the same formula are assigned the same potential function

$(\exp(w_i))$, leading to the following joint probability distribution over all atoms:

$$P(\mathbf{X} = \mathbf{x}) = \frac{1}{Z} \exp\left(\sum_i w_i n_i(\mathbf{x})\right) \tag{3.1}$$

where $n_i(\mathbf{x})$ is the number of times the ith formula is satisfied by the instantiation of the variable nodes, $\mathbf{x}$ and Z is a normalization constant (as in Markov networks). Intuitively, an instantiation where formula f_i is true one more time than a different possible instantiation is e^{w_i} times as probable, all other things being equal. For a detailed description of MLNs, we refer to the book (Richardson and Domingos, 2006).

3.5   Parameter and Structure Learning of Markov Logic Networks

There are two primary learning tasks for MLNs: **parameter learning** and structure learning. Next, we provide a brief overview of the key algorithms used to address each of these tasks.

3.5.1   Parameter Learning

The goal of parameter learning for MLNs is to learn from data the weights associated with each formula. It can be defined as follows:

Given: Set of formulas $f_1, \ldots, f_m$ and a set of training databases D
Learn: The weight w_i associated with each formula f_i.

This problem boils down to selecting weights that optimize some objective function. Broadly speaking, the types of functions considered can be divided into two standard settings: generative learning and discriminative learning.

3.5.1.1   Generative Learning   One natural setting for weight learning is **generative learning**, where the goal is to find the model that is most likely to generate the observed data. Thus the goal is to find weights that maximize the likelihood of the model given the data. Fortunately, for MLNs the log-likelihood is a convex function of the weights which can be solved using standard convex optimization techniques such as L-BFGS. The derivative of the log-likelihood with respect to the jth feature is (Richardson and Domingos, 2006):

$$\frac{\partial}{\partial w_j} \log \Pr_w(db) = n_j(db) - \mathbb{E}_w[n_j] \tag{3.2}$$

where $n_j(db)$ is the number of true groundings of F_j in the training data and $\mathbb{E}_w[n_j]$ is computed using the current weight vector. The jth component of the gradient is simply the difference between the empirical counts of the jth feature in the data and its expectation according to the current model. Thus, each iteration of weight learning must perform inference on the current model to compute the expectations. This is often computationally infeasible.

Given the computational challenges associated with learning maximum likelihood weights, a common alternative objective function is the pseudo-likelihood (Besag, 1974) (PLL), which is much more efficient to compute. The pseudo-likelihood is defined as

$$\overset{\bullet}{\Pr_w}(\mathbf{x}) = \prod_{j=1}^{|\mathbf{X}|} \Pr_w(X_j = \mathbf{x}_j \mid MB_j = \mathbf{x}_{MB_j}),$$

where $|\mathbf{X}|$ is the number of random variables, $\mathbf{x}_j$ is the state of the jth variable in $\mathbf{x}$, MB_j is the Markov blanket of the jth variable, and $\mathbf{x}_{MB_j}$ is the state of that Markov blanket in $\mathbf{x}$. The pseudo-likelihood is also a convex function of the weights for MLNs, so they are also learned via convex optimization.

3.5.1.2   Discriminative Learning   Generative learning finds parameters for a probability distribution over all atoms. However, in many settings, the goal is to predict an output given an input. For example, in natural language processing, we might wish to determine the meaning of a sentence given the words; in computer vision, we might wish to determine the objects in a picture and their locations given the pixels of an image; and in computational biology, we might wish to infer the structure of a protein from its amino acid sequence. In such settings, we do not need a probability distribution over all atoms; we only need a *conditional* probability distribution of the output atoms given the input atoms. This is often much easier, since we do not need the probability distribution over inputs explicitly.

In **discriminative learning**, the goal is to find formula weights that maximize the conditional log-likelihood of the output (or query) variables Q given the input (or evidence) variables E. Derivatives of the conditional log-likelihood are identical to those of the log-likelihood, except that expectations are done over the conditional probability distribution given the evidence:

$$\frac{\partial}{\partial w_j} \log \Pr_w(Q = q \mid E = e) = n_j(q, e) - \mathbb{E}_{q\ \Pr_w(Q|e)}[n_j] \tag{3.3}$$

As in the generative setting, this is a convex optimization problem that requires performing inference in the model. Inference is typically computationally intractable, but can be approximated by various approximate inference algorithms. This approach works better for discriminative learning because the conditional distribution is highly constrained by the evidence, making the inference task substantially easier. Nonetheless, there are several key challenges to optimizing conditional likelihood. First, inference is noisy, so L-BFGS and other methods that rely on line search are likely to get stuck and fail. Second, some formulas have many more groundings than others, leading to a highly ill-conditioned gradient. For example, in a social network domain with 1000 people, a transitivity formula (friends of friends are friends) has 1 million groundings, while a formula relating smoking to cancer has only 1000 groundings. The derivative of the conditional log-likelihood with respect to each weight is the difference in the number of observed and expected satisfied groundings of a formula, so formulas with different numbers of groundings will have gradients of very different magnitudes. This renders standard gradient and stochastic gradient approaches ineffective, since no learning rate works well for all weights.

To handle these issues, there are several specialized algorithms for discriminative parameter learning in MLNs, the most popular of which is preconditioned scaled conjugate gradient descent (PSCG) (Lowd and Domingos, 2007). PSCG is based on conjugate gradient, a well-known optimization method that accelerates gradient descent by using conjugate search directions. To handle noisy gradients, PSCG chooses step sizes using a trust region approach instead of line searches. In a trust region approach, the CLL is locally approximated as a quadratic function; when this approximation is good, the step size is increased, and when it is poor, the step size is decreased. To handle ill-conditioned gradients, PSCG uses a preconditioner which divides each dimension of the gradient by its variance. When the gradient is estimated using sampling methods, the variance can also be estimated and used to help normalize the scale in each direction.

Other approaches to discriminative parameter learning include the voted perceptron algorithm, which takes gradient steps based on approximate maximum a posteriori (MAP) inference and averages the weights to reduce the effect of inference noise (Richardson and Domingos, 2006). Max-margin objectives have also been used, which can lead to more accurate predictions at the cost of poor probability estimates (Huynh and Mooney, 2008).

3.5.2   Structure Learning

Structure learning aims to learn both the formulas and their corresponding weights from data:

Given: A set of training databases D, and a set of predicates

Learn: A set of formulas $f_1, \ldots, f_m$ and the weights $w_1, \ldots, w_m$ associated with each formula.

The first popular MLN structure learning (MSL) (Kok and Domingos, 2005) is a canonical example of a top-down approach and it builds on standard ideas for clausal discovery in relational domains. MSL begins by adding all length-one clauses to the MLN, which capture the marginal probability of a given grounding of each predicate being true. Then, MSL constructs all clauses of length two. It then runs a beam search to find the current best clause and adds it to the network. In each iteration, MSL constructs new candidate clauses by adding literals to the best clauses in the beam. The search iterates until none of the extended clauses improves the score of the MLN. To score each clause, MSL uses weighted pseudo-log-likelihood (WPLL), which simply divides the PLL of a predicate by its number of possible groundings. This normalization prevents predicates with large numbers of groundings from dominating the score function (Kok and Domingos, 2005). To avoid overfitting, each clause receives a penalty term proportional to the number of literals in the clause. To control the search space, a user typically provides a maximum number of variables and literals that can appear in a clause.

The second category of structure learners adopts a bottom-up approach, using the data to restrict the search space. BUSL (Mihalkova et al., 2007) is a two-step algorithm that follows this paradigm. In the first step, it constructs a template Markov network from a ground Markov network by discovering recurring paths of true atoms in the data. In the second step, it transforms the template Markov network into candidate clauses by replacing specific constants with variables. It greedily iterates through the set of candidate clauses, and scores each clause by how much it improves the model's WPLL. In each iteration, it adds the clause to the MLN that most improves the score of the model. The search terminates when no clause improves the model's score.

Neither of these approaches can efficiently learn long, complex clauses, since the number of possible paths and clauses grows exponentially with clause length. LSM solves this by using random walks in place of exhaustive search (Kok and Domingos, 2010). Specifically, LSM constructs a hypergraph in which each node is a constant from the training database and each hyperedge is a true ground atom. Then LSM uses random walks to estimate the hitting time from each node to all other nearby nodes. Nodes with similar hitting times are clustered together and generalized to generate a structural motif. Each motif summarizes the structure among a group of densely connected objects. Motifs that are common in the data are used to generate candidate clauses that are scored with WPLL and greedily added to the network.

3.6   Boosting-based Learning of SRL Models

Another recent advancement is triggered by the insight that finding many rough rules of thumb for how to change our probabilistic relational models locally can be a lot easier than finding a single, highly accurate local model.

Consider for example **relational dependency networks** (Neville and Jensen, 2007). Instead of learning the conditional probability distribution associated with each predicate using relational tree learning in single shot manner, one can represent it as a weighted sum of regression models grown in a stage-wise optimization using **gradient boosting** (Natarajan et al., 2012). This functional gradient approach has successfully been used to train conditional random fields for labeling relational sequences (Gutmann and Kersting, 2006) and to learn relational policies (Kersting and De Raedt, 2008). For implementation as well as further material on how to use it we refer to http://starling.utdallas.edu/ software/boostsrl/wiki/

The key idea in this line of work is to represent each conditional distribution as a set of relational regression trees (RRT) (Blockeel and De Raedt, 1998), see Figure 3.1 for

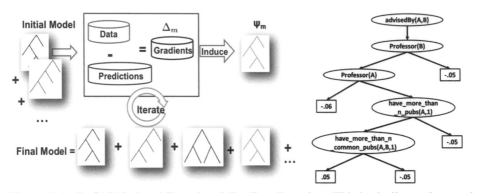

Figure 3.1: (Left) Relational Functional Gradient Boosting. This is similar to the standard gradient-boosting where trees are induced in stage-wise manner. At every iteration, the gradients are computed as the difference between observed and predicted probabilities of each example. A new regression tree is fitted to the gradients (residual difference of these examples). (Right) Example of a Relational Regression Tree. The goal is to predict if a person*A* is advisedBy *B* (where *A* and *B* are logical variables). At each node a (set of) predicate(s) is evaluated. The leaves denote regression values that are exponentiated and normalized to obtain the probabilities.

an example of RRT. The goal is to predict if person A is *advised by* person B given the properties and relations of people at a university. The second branch from the left states that if *B* is a professor, *A* is not a professor, *A* has more than 1 publication, and has more than 1 common publication with *B*, then the regression value is 0.05. These regression values are then exponentiated and normalized to compute the probability.

To learn this set of RRTs for each conditional distribution, the learning approach was adapted based on the gradient boosting technique and was called relational functional-gradient boosting (RFGB) (Natarajan et al., 2012, 2015). Assume that the training examples are of the form $(\mathbf{x}_i, y_i)$ for $i = 1, ..., N$ and $y_i \in \{1, ..., K\}$. Let us use $\mathbf{x}$ to denote the set of non-target predicates (features) and y_i to denote the current target predicate. Then the goal is to fit a model $P(y \mid \mathbf{x}) \propto e^{\psi(y, \mathbf{x})}$. The key idea is to compute the gradient (weight) for each example separately and fit a regression tree over all the weighted examples. The sum of these local gradients approximates the global gradient. The functional gradient of each example $\langle \mathbf{x}_i, y_i \rangle$ w.r.t likelihood $(\psi(y_i = 1; \mathbf{x}_i))$ is

$$\frac{\partial \log P(y_i; \mathbf{x_i})}{\partial \psi(y_i = 1; \mathbf{x_i})} = I(y_i = 1; \mathbf{x_i}) - P(y_i = 1; \mathbf{x_i}) \tag{3.4}$$

where I is the indicator function that is 1 if $y_i = 1$ and 0 otherwise. The expression is simply the adjustment required to match the predicted probability with the true label of the example. If the example is positive and the predicted probability is less than 1, this gradient is positive indicating that the predicted probability should move towards 1. Conversely, if the example is negative and the predicted probability is greater than 0, the gradient is negative, driving the value the other way. We use RRTs to fit the gradient function for every training example.

Each RRT can be viewed as defining several new feature combinations, each corresponding to one of the paths from the root to a leaf. The resulting potential functions from all these different RRTs still have the form of a linear combination of features but the features can be quite complex. This idea is illustrated in Figure 3.1.1. While originally developed for relational dependency networks (Natarajan et al., 2012), this work was later extended to learning MLNs by simply optimizing the pseudo-likelihood (Khot et al., 2011) and to learning both RDNs and MLNs in the presence of hidden data (Khot et al., 2015). Based on (Ravkic et al., 2015), who recently showed how to learn hybrid relational dependency networks, this may also be extended to hybrid domains.

Using the connection between RDNs and MLNs, one could also learn MLNs by first learning a (parametrized) Bayesian network (BN) and then transform the Bayesian network to an MLN (Khosravi et al., 2012).

The benefits of a boosted learning approach are manifold. First, being a nonparametric approach the number of parameters grows with the number of training episodes. In turn, interactions among random variables are introduced only as needed, so that the potentially large search space is not explicitly considered. Second, such an algorithm is fast and straightforward to implement. Existing off-the-shelf regression learners can be used to deal with propositional, continuous, and relational domains in a unified way. Third, it learns the structure and parameters simultaneously, which is an attractive feature as learning probabilistic relational models is computationally quite expensive.

3.7 Deep Transfer Learning Using SRL Models

Statistical relational learning (SRL) also provides a natural formalism for addressing the problem of **deep transfer.** Typically, we are interested in using a learning algorithm to address tasks in multiple different domains. However, this requires learning a model from scratch for each separate task as most inductive learning algorithms assume that training instances and test instances are drawn from the same distribution. For example, a model learned on physics data would not be applicable to finance data, because the variables in the two domains are different. In contrast, people are able to reapply knowledge learned in one domain to an entirely different one. For example, financial firms actively recruit physicists. Skills, such as solving differential equations and performing Monte Carlo simulations, acquired during training as a physicist, can be used to address financial problems.

Transfer learning addresses this problem by explicitly assuming that the source and target problems are different. Most transfer learning approaches focus on *shallow* transfer, which typically entails modeling either a change of distributions over the same variables or minor variations of the same domain (e.g., different numbers of objects). One example would be adapting a model learned on one hospital's ICU data to another hospital that serves a different patient population. In *deep transfer*, the goal is to generalize across different domains, that is, between domains that are described by different objects, classes, properties and relations. SRL is particularly suited to addressing the problem of deep transfer for two reasons. First, it is relational so that it can capture properties among different predicates. Second, it is probabilistic and can handle in a principled way the uncertainty inherent in transfer.

Several different approaches (Mihalkova et al., 2007; Davis and Domingos, 2009; Van Haaren et al., 2015; Kumaraswamy et al., 2015) have been proposed that make use of SRL techniques for transfer learning. While these approaches are typically based on Markov logic, they differ along key lines such as what is transferred between domains and how the transferred knowledge is used in the target domain. Next, we briefly describe each approach.

3.7.1   TAMAR

TAMAR (Mihalkova et al., 2007) is an approach to transfer learning that works by transferring an entire learned model from one domain to another. At a high level, TAMAR works in two phases: model translation and model refinement. In the first step, TAMAR receives a learned model, which it translates to the target domain. The translation is performed by mapping each clause in the given learned model to the target domain. The key challenge is establishing a mapping between the predicate names in the source and target domains. As opposed to working with one fixed global mapping, TAMAR establishes a separate local mapping between predicates for each clause. To determine the local mapping, TAMAR performs a search that replaces the source predicate names in a clause with the predicate names from the target domain in all possible ways. It scores each mapping by computing the WPLL of an MLN consisting of only the mapped clause and selects the highest scoring mapping.

In the second step, TAMAR refines the translated model using data from the target domain. It considers two types of refinements. First, it performs an analysis of each transferred clause in an attempt to detect if it can be improved by either lengthening or shortening it. Then it employs a relational path finding approach to learn new clauses to add to the MLN.

3.7.2   DTM

DTM (Davis and Domingos, 2009, 2011) is an approach to deep transfer that is based on second-order Markov logic, where formulas contain predicate variables (Kok and Domingos, 2007). The use of predicate variables allows DTM to represent high-level structural regularities in a domain-independent fashion by capturing common structures among first-order formulas. To illustrate the intuition behind DTM consider the following two formulas from a molecular biology domain:

$$\text{Complex}(z, y) \land \text{SameFunction}(x, z) \Rightarrow \text{Complex}(x, y)$$
$$\text{Location}(z, y) \land \text{Interacts}(x, z) \Rightarrow \text{Location}(x, y).$$

Both are instantiations of the formula:

$$r(z, y) \land s(x, z) \Rightarrow r(x, y),$$

where r and s are predicate variables.

DTM transfers knowledge from one domain to another using second-order cliques, which are sets of literals with predicate variables representing a set of formulas. A clique groups formulas with related effects into one structure. For example, the clique { $R(X, Y)$, $R(Y, X)$ }, where R is a predicate variable and X and Y are object variables, gives rise to the second-order formulas:

$$R(X, Y) \land \quad R(Y, X),$$
$$R(X, Y) \land \neg R(Y, X),$$
$$\neg R(X, Y) \land \quad R(Y, X),$$
$$\neg R(X, Y) \land \neg R(Y, X).$$

By replacing the predicate variables with predicate names, a second-order clique can give rise to multiple first-order formulas.

Second-order cliques can be constructed using any standard search procedure. DTM uses data from the source domain to evaluate each second-order clique by checking which cliques represent regularities whose probability deviates significantly from independence among their subcliques. It selects the top k highest-scoring second-order cliques to transfer to the target domain. Finally, the highest-scoring cliques are transferred to the target domain. For a given clique, each of its formulas is instantiated with the target predicate names in all ways that respect the arity and type constraints of the target domain. Learning a target model proceeds in two steps. First, DTM learns a model that only involve the formulas from the transferred cliques. Second, it refines the model by either modifying the selected clauses or learning new clauses.

3.7.3 TODTLER

At a high level, TODTLER (Van Haaren et al., 2015) approaches the problem of transfer learning as the process of learning a declarative bias in the source domain and transferring it to the target domain. It views MLN learning through a generative learning lens, where each first-order MLN can be seen as an instantiation of a set of second-order templates or clauses expressed in a language called SOLT. A clause in SOLT is simply a clause that contains predicate variables in place of predicate names. Hence, this is similar to the second-order cliques used by DTM except that each template corresponds to a single clause whereas a clique represents multiple related clauses. TODTLER's main insight is that transferring knowledge amounts to acquiring a posterior over the sets of second-order templates by learning in the source domain and using this posterior when learning in the target setting.

TODTLER's generative model assumes that each domain is characterized by a second-order model of the data, denoted as $M^{(2)}$, that consists of a set of SOLT clauses. $M^{(2)}$ is sampled from a prior $P(M^{(2)})$ induced by independently including each template T expressible in SOLT into $M^{(2)}$ with some probability p_T. Given a second-order model $M^{(2)}$, a first-order (MLN) model $M^{(1)}$ is generated by instantiating all templates in $M^{(2)}$ with the set of predicates relevant to the data at hand in all possible ways. Thus, given data D, the model yields a joint probability density $p(D, M^{(1)}, M^{(2)})$ that factorizes as

$$p(D, M^{(1)}, M^{(2)}) = p(D \mid M^{(1)})p(M^{(1)} \mid M^{(2)})P(M^{(2)}) \tag{3.5}$$

where $p(D \mid M^{(1)})$ is probability of observed data D given the first-order model $M^{(1)}$ $P(M^{(2)})$ is given by the probabilities p_T of including each template T into the second-order model, and $p(M^{(1)} \mid M^{(2)})$ is positive if the set of clauses in $M^{(1)}$ is the *complete instantiation* of $M^{(2)}$ and 0 otherwise.

The source data D_s is used to infer the posterior $P_s(M^{(2)} \mid D_s)$. This posterior can then serve as prior in the target domain, where the search is biased towards clauses that originate from a SOLT clause that was found to be useful in the source data. As computing this posterior exactly would be computationally infeasible, TODTLER approximates these quantities. Namely, in the source data, it uses the WPLL to estimate the usefulness of each first-order instantiation of a SOLT clause, which is then scaled to be between 0 and 1. A template T's score $\hat{p}_{T,s}$ is obtained by averaging the score of each of its instantiations. In the target domain, the goal is to rank clauses, not templates. Each first-order formula in the target domain $F_{T,t}$ has an associated SOLT template T. The clause is scored using the same procedure as in the source domain, except that its final score is the product of its target domain score and the score $\hat{p}_{T,s}$, which serves to bias the model towards containing clauses from templates previously found to be useful. Then TODTLER creates a rank-ordered list of clauses. It walks down the list and attempts to add each clause to the target MLN. A clause is only retained if it improves the target MLN's WPLL.

There are several important differences between TODTLER and DTM. First, TODTLER transfers fine-grained knowledge because it performs transfer on a per-template basis instead of a per-clique basis. DTM's transfer mechanism is based on second-order cliques, where each second-order clique gives rise to many first-order formulas. Within a clique, only a small subset of these formulas will be helpful for modeling the target domain. Second, TODTLER transfers both individual second-order templates (i.e., structural regularities) as well as information about their usefulness (i.e., the posterior of the formulas, $P_s(M^{(2)} \mid D_s)$). In contrast, DTM just transfers the second-order cliques and the target data is used to assess whether the regularities are important. Finally, TODTLER transfers a more diversified set of regularities whereas DTMis restricted to a smaller set of user-selected cliques. This increases the chance that TODTLER transfers something of use for modeling the target domain.

3.7.4 LTL

Language-Bias Transfer Learning (LTL) (Kumaraswamy et al., 2015), utilizes language bias from the knowledge of the source domain to guide construction of knowledge in the target domain. Since the knowledge is represented in First Order Logic (FOL), the flow of types in the source knowledge provides us with the knowledge structure of the source, which is in turn exploited during knowledge construction/learning in the target domain.

Inspired by the research in Inductive Logic Programming, LTL approach performs "mode-matching" that compares the modes in both predicates.[7] Mode-matching compares the types of the arguments in the predicates of the source and target domain to identify potentially similar groups across the domains. Once the match is obtained, we perform a mode-based tree construction that allows us to construct the clauses in the target domain. This matching of the modes and the construction of the initial knowledge in the target domain can be seen as introduction of a *language-bias* for the target domain. Hence, this algorithm is called language-bias transfer learning (LTL).

Once the potential clauses are constructed, LTL employs two different types of refinements. The first is by exploiting the power of PLMs in allowing the *softening* of the rules. LTL learns new parameters for these rules from the training data in the target domain. LTL considers both weights (as with MLNs (Richardson and Domingos, 2006)) and conditional probabilities (Kersting and De Raedt, 2001) with combining rules (Natarajan et al., 2005). Some of the rules can be spurious or inapplicable for the target, because they are simply constructed from predicate matching. Hence, softening them using training data decreases the weights of the rules that are not supported by data. It was shown that empirically, with fewer data, softening is an easier task than learning both the rules and the parameters. The second improvement is employing *theory refinement* (Richards and Mooney, 1995) by which LTL modifies some of the rules based on the training data. Specifically, it considers dropping a predicate or adding an attribute of one of the objects already present in the rule. For instance, if the rule mentions an organization and an employee, then LTL will search the space to potentially add an attribute of these two objects (say department) based on the improvement in the likelihood of the data.

---

[7] Note the difference between modes in ILP and modes of probability distributions. Modes inside ILP define the argument types of a predicate and help in the inductive search of the rules

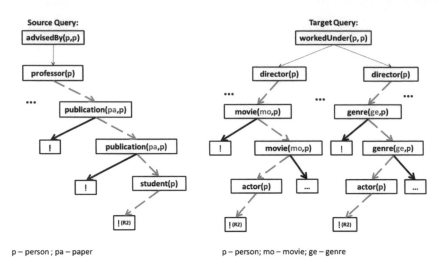

Figure 3.2: Examples of a simplified mode-based matching between the UW-CSE and IMDb datasets. It represents the flow of types which is shown with colors. Note that in the right figure, we present only a part of the search tree for brevity. Paths in the tree representing clauses are shown as shown as !.

Illustrative Example: LTL method relies on creating *mode-matched trees* (M^2Ts) in both the source (say UW-CSE (Richardson and Domingos, 2006)) and target (say IMDb) domains. The simplified M^2Ts for the UW-CSE domain that models the relationship *advisedBy* is shown in Figure 3.2-*left*, and is generated from the rules in the source domain model. The rules correspond to predicting the advisor of each student, both of which are represented by the type *person*. The colors represent the parameter tying of the types in the clause, and the leaf nodes of the tree denote the rule - represented by a path from root to that leaf, if any, that has been learned or provided by an expert. Note that every path from the root to the leaf is a rule or clause in FOL. Based on the search process over these parameter types in the source, clauses with a similar structure are constructed in the target domain where the model predicts the relationship *workedUnder*. This is shown in Figure 3.2-*right*. Here, the rules predict if an actor has worked for a director, both of which are represented by the type *person*. As can be seen, for one source clause we get multiple target clauses because each source predicate is mapped to multiple target predicates, for example, *publication* is automatically mapped to *movie* and *genre*, as shown in the figure. We refer to the original paper for theory and empirical results behind this approach (Kumaraswamy et al., 2015).

3.8   Conclusion

We have presented the different learning formalisms inside statistical relational AI. Specifically, we considered the formalism of Markov Logic Networks and discussed the parameter and structure learning methods inside this formalism. We then outlined other methods such as Boosting for SRL and discussed the transfer learning methods which make the application of SRL very effective in many domains. Of course, this is by no means a comprehensive list of methods. This book focuses on lifted inference and hence we restricted the chapter to focus on a couple of methods for the most popular models. Learning other models such as Problog (De Raedt et al., 2007) or PSL (Bach et al., 2017; Bröcheler et al., 2010) are interesting and have been used significantly.

The other direction of learning is that of Bayesian learning. From a Bayesian point of view, learning is just a case of inference: we condition on all of the observations (all of the data), and determine the posterior distribution over some hypotheses or any query of interest. Starting from the work of (Buntine, 1994), there has been considerable work in using relational models for **Bayesian learning** (Jordan, 2010). This work uses parameterized random variables (or the equivalent plates) and the probabilistic parameters are real-valued random variables (perhaps parameterized). Dealing with real-valued variables requires sophisticated reasoning techniques, often in terms of MCMC and stochastic processes. Although these methods use relational probabilistic models for learning, the representations learned are typically not relational probabilistic models although the approach is also popular for learning functional probabilistic programs (Pfeffer, 2007; Goodman et al., 2008). It is still an open challenge to bring these two threads together, mainly because of the difficulty of inference in these complex models.

II   EXACT INFERENCE

4 Lifted Variable Elimination

Nima Taghipour, Daan Fierens, Jesse Davis, Hendrik Blockeel, and Rodrigo de Salvo Braz

Abstract. This chapter describes the GC-FOVE algorithm, a generalization of the well-known Variable Elimination algorithm for probabilistic inference. GC-FOVE exploits the symmetric structure of relational probabilistic models to eliminate entire groups of random variables in a single operation. Such groups are specified via a constraint language. GC-FOVE is described in a way that is neutral with respect to the constraint language used, and one possible language is suggested.

Keywords: lifted first-order variable elimination, constraint language, unification, GC-FOVE, extensionally complete constraint language

4.1 Introduction

A major challenge when dealing with **statistical relational learning** (SRL) models (Getoor and Taskar, 2007; De Raedt et al., 2008) is how to perform inference efficiently. Initially, the typical approach to inference in SRL was to first convert the model into a propositional graphical model and then employ a standard propositional inference algorithm. This contrasts with inference in first-order logic, which can reason on the level of logical variables: if a model states that for all x, $P(x)$ implies $Q(x)$, then whenever $P(x)$ is known to be true, one can infer $Q(x)$, without knowing what x stands for. This capacity was lost in the initial SRL inference algorithms which led to repeating the same inference steps for each different value x of x, instead of once for all x.

To address this problem, Poole (2003) proposed a **lifted variable elimination** algorithm in order to try to exploit the symmetries present in SRL models. Its key idea was to group together interchangeable objects, and perform the inference operations once for each group instead of once for each object. Since this initial idea, there has been substantial work on how to improve this first approach to lifted variable elimination (de Salvo Braz et al., 2005; Milch et al., 2008; Sen et al., 2009a; Choi et al., 2010; Apsel and Brafman, 2011; Taghipour et al., 2013b,a).

This chapter describes the **GC-FOVE** algorithm for lifted variable elimination (Taghipour and Davis, 2012; Taghipour et al., 2013c). GC-FOVE focuses on the central role that constraints play in defining a group of interchangeable objects. The type of constraints that are allowed, and the way in which they are handled, directly influence the granularity of the grouping, and hence, the efficiency of the subsequent lifted inference (Kisynski and Poole, 2009). Initially, the most commonly used constraint class was conjunctions of pairwise (in)equalities, which is the bare minimum requirement for performing lifted inference. In contrast, we will present an algorithm for lifted variable elimination that uses a constraint language that is ***extensionally complete***. That is, for any group of variables a constraint exists that defines exactly that group, which allows the inference algorithm to capture a broader set of symmetries in the model. To this aim, the algorithm's constraint manipulation is defined in terms of relational algebra operators. This decouples the lifted inference algorithm from the constraint representation mechanism. Consequently, any constraint language that is closed under these operators can be plugged into the algorithm to obtain a working system.

This chapter will begin by providing a gentle introduction to the principles of lifted variable elimination through an example. It will then provide the necessary background information for understanding lifted variable elimination. Next, it will give a high-level outline of the lifted variable elimination algorithm and describe all the operators that the algorithm uses. Finally, it will briefly discuss an efficient representation for the constraints themselves.

4.2   Lifted Variable Elimination by Example

Although lifted variable elimination builds on simple intuitions, it is relatively complicated, and an accurate description of it requires a level of technical detail that is not conducive to a clear understanding. For this reason, we first illustrate the basic principles of lifted inference on a simple example, and without referring to the technical terminology that is introduced later. We start with describing the example; next, we illustrate variable elimination on this example, and show how it can be lifted.

4.2.1   The Workshop Example

This example is from Milch et al. (2008). Suppose a new workshop is organized. If the workshop is popular (that is, many people attend), it may be the start of a series. Whether a person is likely to attend depends on the topic.

We introduce a random variable T, indicating the topic of the workshop, and a random variable S, indicating whether the workshop becomes a series. We consider N people, and for each person i, we include a random variable A_i that indicates whether i attends. Each random variable has a finite domain from which it takes on values, i.e., $\{SRL, DB, \dots\}$ for T, $\{Yes, No\}$ for S, and $\{true, false\}$ for each A_i.

The joint probability distribution of these variables can be specified by an undirected graphical model. A set of *factors* captures dependencies between the random variables in such a model. In our model, there are two kinds of factors. For each person i, there is a factor $\phi_1(A_i, S)$ that states how having a series depends on whether person i attends, and a

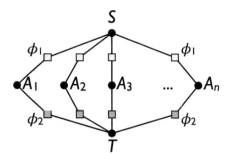

Figure 4.1: A factor graph for the workshop example. Square nodes represent factors, round nodes variables. Variables are labeled with their name, factors with their potential function.

factor $\phi_2(T, A_i)$ that states how i's attendance depends on the topic. Note that all N factors of the first type have the same potential function ϕ_1, and all factors of the second type have potential function ϕ_2. This imposes a certain symmetry on the model: it implies that S depends on each person's attendance in exactly the same way, and every person have the same topic preferences.

The model defines a joint probability distribution over the variables that is the normalized product of the factors (normalized such that all joint probabilities sum to one):

$$Pr(T, S, A_1, \ldots, A_N) = \frac{1}{Z} \prod_{i=1}^{n} \phi_1(A_i, S) \prod_{i=1}^{n} \phi_2(T, A_i)$$

where Z is the normalization constant.

Undirected graphical models can be visualized as *factor graphs* (Kschischang et al., 2001), which have a node for each random variable and each factor, and an edge between a factor and a random variable if that variable occurs in the factor. Figure 4.1 shows a factor graph for our example.

4.2.2   Variable Elimination

From now on, we refer to the values taken by a variable by the corresponding lowercase symbols (e.g., a_i as shorthand for $A_i = a_i$).

Suppose we want to compute the marginal probability distribution $Pr(S)$.

$$Pr(S) = \sum_{T} \sum_{A_1} \cdots \sum_{A_N} Pr(T, S, A_1, \ldots, A_N) \qquad (4.1)$$

$$= \frac{1}{Z} \sum_{T} \sum_{A_1} \cdots \sum_{A_N} \prod_{i=1}^{N} \phi_1(A_i, S) \prod_{i=1}^{N} \phi_2(T, A_i) \qquad (4.2)$$

Usually, the normalization constant Z is ignored during the computation, and normalization happens only at the very end. So, we can focus on how to compute

$$\tilde{Pr}(S) = \sum_T \sum_{A_1} \cdots \sum_{A_N} \prod_{i=1}^{N} \phi_1(A_i, S) \prod_{i=1}^{N} \phi_2(T, A_i). \qquad (4.3)$$

A straightforward way of computing $\tilde{Pr}(S)$ is to compute $\tilde{Pr}(s)$ for each possible value s of S, and tabulate the results. We can compute $\tilde{Pr}(true)$ by iterating over all possible value combinations $(t, a_1, \ldots, a_n)$ of $(T, A_1, \ldots, A_n)$ and computing $\prod_{i=1}^{N} \phi_1(a_i, true) \prod_{i=1}^{N} \phi_2(t, a_i)$ for each combination, and similarly for $\tilde{Pr}(false)$. If all variables are binary, there are 2^{N+1} such combinations, and for each combination $2N - 1$ multiplications are performed. This clearly does not scale.

However, we can improve efficiency by rearranging the computation. In the above computation, the same multiplications are performed repeatedly. Since $\phi_1(A_1, S)$ and $\phi_2(T, A_1)$ are constant in all A_i except A_1, they can be moved out of the summations over A_i, $i > 1$, so the right hand side of Equation 4.3 becomes:

$$\sum_T \sum_{A_1} \phi_1(A_1, S)\phi_2(T, A_1) \sum_{A_2} \cdots \sum_{A_N} \prod_{i=2}^{N} \phi_1(A_i, S) \prod_{i=2}^{N} \phi_2(T, A_i) \qquad (4.4)$$

Conversely, the factor starting with $\sum_{A_2}$ is independent of A_1, so it can be moved outside of the summation over A_1, giving

$$\sum_T \left(\sum_{A_2} \cdots \sum_{A_N} \prod_{i=2}^{N} \phi_1(A_i, S) \prod_{i=2}^{N} \phi_2(T, A_i) \right)\left(\sum_{A_1} \phi_1(A_1, S)\phi_2(T, A_1) \right) \qquad (4.5)$$

Repeating this for each A_i eventually yields

$$\sum_T \left(\sum_{A_1} \phi_1(A_1, S)\phi_2(T, A_1) \right) \ldots \left(\sum_{A_N} \phi_1(A_N, S)\phi_2(T, A_N) \right) \qquad (4.6)$$

which shows that for each A_i, the product $\phi_1(A_i, S)\phi_2(T, A_i)$ needs to be computed only once for each combination of values for (T, S, A_i). When T is binary, there are eight such combinations, reducing the total number of multiplications to $8N$.

Note that the result of Formula 4.6 is a function of S; it can yield a different value for each value s of S. In other words, it is a factor over S. Similarly, the result of $\phi_1(A_1, S) \cdot \phi_2(T, A_1)$ depends on the values of S, T and A_1 (is a factor over these variables), and after summation over A_1 a factor over S and T is obtained. Thus, the multiplications and summations in Formula 4.6 are best seen as operating on factors, not individual numbers. Figure 4.2 illustrates the process of multiplying and summing factors.

The result of Formula 4.6 can be computed as follows. First, multiply the factors $\phi_1(A_1, S)$ and $\phi_2(T, A_1)$ for each value of A_1, and sum out A_1 from the product. This is exactly the computation illustrated in Figure 4.2. After summing over all values of A_1, the result de-

$$\phi_1(A_1, S) \qquad\qquad \phi_2(T, A_1) \qquad\qquad \phi_{12}(T, A_1, S)$$

A_1	S	ϕ_1
true	true	1
false	true	2
true	false	2
false	false	1

$\otimes$

T	A_1	ϕ_2
SRL	true	3
SRL	false	1
DB	true	2
DB	false	2

$=$

T	A_1	S	ϕ_{12}
SRL	true	true	3
SRL	true	false	6
SRL	false	true	2
SRL	false	false	1
DB	true	true	2
DB	true	false	4
DB	false	true	4
DB	false	false	2

$$\phi'_{12}(T, S)$$

$$\sum_{A_1} \phi_{12}(T, A_1, S) \quad = $$

T	S	ϕ'_{12}
SRL	true	5
SRL	false	7
DB	true	6
DB	false	6

Figure 4.2: Two example factors, their product, and the result of summing out A_1 from the product. The values of ϕ_1 and ϕ_2 are chosen arbitrarily for this illustration.

pends on T and S only; A_1 no longer occurs in this factor, nor in any other factors. We say that A_1 has been eliminated. Note that the elimination consisted of first gathering all factors containing A_1, multiplying them, then summing over all possible values of A_1.

After eliminating A_1, we can repeat the process for all other A_i, each time obtaining a factor over T and S. All those factors can then be multiplied and the result summed over T, at which point a single factor over S is obtained. This factor equals $\tilde{Pr}(S)$.

The above computation is exactly what Variable Elimination (VE) does. Generally, VE works as follows. It considers one variable at a time, in an order called the *elimination order*. For each considered variable V, VE first retrieves all factors that involve V, then multiplies these factors together into a single joint factor, and finally sums out V, thereby eliminating it from the factor. Hence, in each step, the number of remaining variables strictly decreases (by 1) and also the number of factors decreases (because the set of factors involving x is replaced by a single factor).

The elimination order can heavily influence runtime. Unfortunately, finding the optimal order is NP-hard. In the above example, the elimination order was $A_1, A_2, \ldots, A_N, T$, and this resulted in a computation with $8N$ multiplications, which is $O(N)$.

4.2.3   Lifted Inference: Exploiting Symmetries among Factors

In the above example, by avoiding many redundant computations, VE obtained an exponential speedup compared to the naive computation discussed before, reducing computation time from $O(2^N)$ to $O(N)$. N can still be large. Even more efficiency can be gained when we know that certain factors have the same potential function.

In our example, VE computes the same product N times: in Expression 4.6, factors $\phi_1(A_i, S)$ and $\phi_2(T, A_i)$ are the same for all i, and so is their product $\phi_{12}(A_i, S, T) = \phi_1(A_i, S)\phi_2(T, A_i)$. It also computes the sum $\sum_{A_i} \phi_{12}(A_i, S, T)$ N times. This redundancy arises because in our probabilistic model all N people behave in the same way, i.e., all A_i are interchangeable. The idea behind lifted inference is to exploit such symmetries, and compute the product and sum only once. From the algorithmic perspective, lifted variable elimination eliminates only one A_i variable, then *exponentiates* the resulting factor (see formula below), and then sums out T. Mathematically, Expression 4.6 is computed as follows:

$$\sum_T \left(\sum_{A_1} (\phi_1(A_1, S)\phi_2(T, A_1)) \right)^N \tag{4.7}$$

The way in which lifted variable elimination manipulates the set of variables $\{A_1, \ldots, A_N\}$ is called **lifted multiplication** and **lifted summing-out** (a.k.a. lifted elimination). Note that the number of operations required is now constant in N. Assuming N is already known, the main operation here is computing the N-th power, which is $O(\log N)$ (logarithmic in N if exact arithmetic is used, constant for floating point arithmetic). Thus, lifted variable elimination runs in $O(\log N)$ time in this case.

4.2.4   Lifted Inference: Exploiting Symmetries within Factors

Now consider a second elimination order, where we first eliminate T and then the A_i:

$$\tilde{Pr}(S) = \sum_{A_1,\ldots,A_N} \sum_T \prod_{i=1}^N \phi_1(A_i, S) \prod_{i=1}^N \phi_2(T, A_i) = \sum_{A_1,\ldots,A_N} \prod_{i=1}^N \phi_1(A_i, S) \left(\sum_T \prod_{i=1}^N \phi_2(T, A_i) \right)$$
$$\tag{4.8}$$

With this order, regular variable elimination works as follows. The inner summation (elimination of T) first multiplies all factors $\phi_2(T, A_i)$ into a factor $\phi_3(T, A_1, \ldots, A_N)$, and then sums out T:

$$\sum_T \prod_{i=1}^N \phi_2(T, A_i) = \sum_T \phi_3(T, A_1, \ldots, A_N) = \phi_3'(A_1, \ldots, A_N)$$

Note that ϕ_3 is a function of $N + 1$ binary variables, so its tabular representation has 2^{N+1} entries, which makes the cost of this elimination $O(2^{N+1})$. Substituting the computed ϕ_3' into Equation (4.8) yields:

$$\tilde{P}r(S) = \sum_{A_1,\dots,A_N} \left(\prod_{i=1}^{N} \phi_1(A_i, S) \right) \phi_3'(A_1, \dots, A_N)$$

Now we can multiply $\phi_3'(A_1, \dots, A_N)$ by $\phi_1(A_1, S)$ and sum out A_1, then multiply the result by $\phi_1(A_2, S)$ and sum out A_2, and so on, until we obtain a factor $\phi_4'(S)$:

$$\tilde{P}r(S) = \phi_4'(S)$$

This again involves N multiplications and summations with exponential complexity. In summary, variable elimination computes the result in $O(2^{N+1})$.

This elimination order also has symmetries that lifted inference can exploit. Let us examine $\phi_3(T, A_1, \dots, A_N)$, the product of factors $\phi_2(T, A_i)$. For each assignment $T = t$ and $(A_1, \dots, A_N) = (a_1, \dots a_N) \in \{true, false\}^N$:

$$\phi_3(t, a_1, \dots, a_N) = \phi_2(t, a_1) \dots \phi_2(t, a_N)$$

Note that, since each $a_i \in \{true, false\}$, the multiplicands on the right hand side can have only one of two values, $\phi_2(t, true)$ or $\phi_2(t, false)$. That is, for each $a_i = true$ there is a $\phi_2(t, true)$, and similarly for each $a_i = false$, a $\phi_2(t, false)$. This means that, with $\mathcal{A}_t = \{A_i \mid a_i = true\}$ and $\mathcal{A}_f = \{A_i \mid a_i = false\}$, we can rewrite the above expression as:

$$\phi_3(t, a_1, \dots, a_N) = \prod_{a_i \in \mathcal{A}_t} \phi_2(t, true) \prod_{a_i \in \mathcal{A}_f} \phi_2(t, false) = \phi_2(t, true)^{|\mathcal{A}_t|} \phi_2(t, false)^{|\mathcal{A}_f|}.$$

This shows that to evaluate $\phi_3(T, A_1, \dots, A_N)$ it suffices to know how many A_i are true (call this number n_t) and false (n_f); we do not need to know the value of each individual A_i. We can therefore restate ϕ_3 in terms of a new variable #[$\mathcal{A}$], called a ***counting variable***, the value of which is the two-dimensional vector (n_t, n_f). Generally, #[$\mathcal{A}$] can take any value (x, y) with $x, y \in \mathbb{N}$ and $x + y = N$. We call such a value a *histogram*. It captures the distribution of values among $\mathcal{A} = \{A_1, \dots, A_N\}$. The reformulation of a factor in terms of a counting variable is called ***counting conversion***. Rewriting $\phi_3(T, A_1, \dots, A_N)$ as $\phi_3^*(T, \#[\mathcal{A}])$, we have

$$\phi_3^*(t, (n_t, n_f)) = \phi_2(t, true)^{n_t} \phi_2(t, false)^{n_f}.$$

ϕ_3^* has $2(N + 1)$ possible input combinations (two values for t and $N + 1$ values for (n_t, n_f), since $n_t + n_f = N$). It can be tabulated in time $O(N)$, using the recursive formula $\phi_3^*(t, (n_t + 1, n_f - 1)) = \phi_3^*(t, (n_t, n_f)) \cdot \phi_1(t, true)/\phi_2(t, false)$. Note that VE's computation of ϕ_3 was $O(2^N)$.

Because ϕ_3^* has only $2(N+1)$ possible input states, instead of 2^{N+1}, we can now eliminate T in $O(N)$:

$$\sum_T \prod_{i=1}^N \phi_2(T, A_i) = \sum_T \phi_3^*(T, \#[\mathcal{A}]) = \phi_3'(\#[\mathcal{A}])$$

Using this result, we continue with the elimination:

$$\tilde{P}r(S) = \sum_{A_1,\ldots,A_N} \prod_{i=1}^N \phi_1(A_i, S) \, \phi_3'(\#[\mathcal{A}])$$

Using counting conversion a second time, we can reformulate the result of $\prod_{i=1}^N \phi_1(A_i, S)$ as $\phi_4(\#[\mathcal{A}], S)$, which gives:

$$\tilde{P}r(S) = \sum_{A_1,\ldots,A_N} \phi_4(\#[\mathcal{A}], S) \, \phi_3'(\#[\mathcal{A}]) = \sum_{A_1,\ldots,A_N} \phi_{43}(\#[\mathcal{A}], S) \tag{4.9}$$

In itself, the final summation still enumerates all 2^N joint states of variables $\mathcal{A}$, computes the histogram (n_t, n_f) and $\phi_{43}((n_t, n_f), S)$ for each state, and adds up all the ϕ_{43}. But we can do better: all states that result in the same histogram (n_t, n_f) have the same value for $\phi_{43}((n_t, n_f), S)$, and we know exactly how many such joint states there are, namely $\binom{N}{n_t} = \frac{N!}{n_t! n_f!}$. We will call this the *multiplicity* of the histogram (n_t, n_f), denoted $\mathrm{MUL}((n_t, n_f))$. Thus, we can compute $\phi_{43}((n_t, n_f), S)$ just once for each histogram (n_t, n_f) and multiply it by its multiplicity:

$$\sum_{A_1,\ldots,A_N} \phi_{43}(\#[\mathcal{A}], S) = \sum_{\#[\mathcal{A}]} \mathrm{MUL}(\#[\mathcal{A}]) \cdot \phi_{43}(\#[\mathcal{A}], S)$$

This way we enumerate over $N+1$ possible values of $\#[\mathcal{A}]$ instead of 2^N possible states of $\mathcal{A}$. To summarize, we can reformulate Equation (4.9) as

$$\tilde{P}r(S) = \sum_{A_1,\ldots,A_N} \phi_{43}(\#[\mathcal{A}], S) = \sum_{\#[\mathcal{A}]} \mathrm{MUL}(\#[\mathcal{A}]) \cdot \phi_{43}(\#[\mathcal{A}], S) = \phi_5(S)$$

which shows that $\#[\mathcal{A}]$ can be eliminated with $O(N)$ operations.

The whole computation of $\tilde{P}r(S)$ thus has complexity $O(N)$, instead of $O(2^N)$ for VE with this elimination order. This reduction in complexity is possible due to symmetries in the model that allow us to treat all variables $\mathcal{A}$ as one unit $\#[\mathcal{A}]$.

4.2.5   Splitting Overlapping Groups and Its Relation to Unification

Sometimes the groups of random variables expressed in a problem are not symmetric and need to be split into smaller, symmetric ones. Consider a problem involving some of the same variables as the workshop example from the previous section, but now with a potential ϕ_1 applied to (A_i, S), $i = 1, \ldots, N$, and a potential ϕ_2 applying to A_4 alone. Potential ϕ_1 can be interpreted as a general association between attendance of each person and whether the workshop is a series, whereas ϕ_2 encodes knowledge about the attendance about person 4 specifically. We then have

$$\tilde{Pr}(S) = \sum_{A_1} \cdots \sum_{A_N} \left(\prod_{i=1}^{N} \phi_1(A_i, S) \right) \phi_2(A_4).$$

Note that ϕ_2 breaks the symmetry because A_4 has a factor on it that other A variables do not. Therefore we cannot directly exploit symmetry as done in equations (4.3)-(4.6). First, we need to break the group of factors on (A_i, S) so that each group is symmetric.

$$\tilde{Pr}(S) = \sum_{A_1} \cdots \sum_{A_3} \sum_{A_5} \cdots \sum_{A_N} \sum_{A_4} \left(\prod_{i\in\{1,\ldots,N\}\backslash\{4\}} \phi_1(A_i, S) \right) \phi_1(A_4, S)\phi_2(A_4) \qquad (4.10)$$

$$= \sum_{A_1} \cdots \sum_{A_3} \sum_{A_5} \cdots \sum_{A_N} \left(\prod_{i\in\{1,\ldots,N\}\backslash\{4\}} \phi_1(A_i, S) \right) \sum_{A_4} \phi_1(A_4, S)\phi_2(A_4)$$

at which point the random variables groups are symmetric and the same operations as in last sub-section can be applied.

It is interesting to think about how this type of splitting relates to **unification** in logic reasoning. Assume

$$\phi_1(A_i, S) = \begin{cases} 1, \text{if } A_i \Rightarrow S \\ 0, \text{otherwise} \end{cases} \qquad \phi_2(A_4) = \begin{cases} 1, \text{if } A_4 \\ 0, \text{otherwise.} \end{cases}$$

Then this model is deterministic, and equivalent to the logic theory

$$\forall i \in \{1, \ldots, N\} . A(i) \Rightarrow S$$

$$A(4)$$

(where we use a more logic-compatible relational notation rather than index notation).

If a logic inference algorithm using unification is applied to this theory, it will unify atoms $A(i)$ and $A(4)$, obtain the unifier $i = 4$, and use this unifier to instantiate the first sentence into $A(4) \Rightarrow S$. It is easy to see that a similar instantiation was performed in (4.10). In fact, the GC-FOVE operations described in Section 4.5.4 end up reducing to the mechanism of logic unification, when the given model is a deterministic one equivalent to to a logic theory. Moreover, it *generalizes* logic inference when given non-deterministic factors.

There is one caveat to this correspondence, however. A logic inference algorithm would *not* instantiate $\forall i \in \{1, \ldots, N\} \backslash \{4\} . A(i) \Rightarrow S$, because it is redundant with the theory's first sentence, which is kept. However, we can see that (4.10) does instantiate the corresponding group of factors (called the **residual factors**), while no longer representing the original group of factors directly. The reason for this discrepancy is the need to avoid over-counting factors. In logic, there is no harm in keeping the original universally quantified formula and its instantiation, even though their groundings will share some of the same grounded formulas, because they are semantically combined by conjunction, an idempotent operation insensitive to their number of occurrences. In probabilistic inference, on the other hand, factors are combined by non-idempotent multiplication, so it is important to keep the exact same set of grounded factors at each step. For this reason, we must exclude the original set of factors (so that it does not overlap with the set instantiated with $i = 4$), which in turn forces us to keep the residual set of factors that would disappear otherwise.

This ends our informal introduction to lifted variable elimination. In the following sections, we first introduce formal notation and terminology, then present the algorithm in more detail.

4.3   Representation

This section introduces representation concepts needed for lifted inference. These concepts have been introduced in earlier work (de Salvo Braz et al., 2007; Milch et al., 2008), but differences arise in terminology and notation as we emphasize the constraint part.

4.3.1   A Constraint-based Representation Formalism

An undirected model is a factorization of a joint distribution over a set of random variables (Kschischang et al., 2001). Given a set of random variables $§ = \{x_1, x_2, \ldots, x_n\}$, a factor consists of a potential function ϕ and an assignment of a random variable to each of ϕ's inputs. For instance, the factorization $f(x_1, x_2, x_3) = \phi(x_1, x_2)\phi(x_2, x_3)$ contains two different factors (even if their potential functions are the same).

Likewise, in our probabilistic-logical representation framework, a model is a set of factors. The random variables they operate on are properties of, and relationships between, objects in the universe. We now introduce some terminology to make this more concrete. We assume familiarity with set and relational algebra (union $\cup$, intersection $\cap$, difference $\setminus$, set partitioning, selection σ_C, projection π_X, attribute renaming ρ, join $\bowtie$); see, for instance, the work of Ramakrishnan and Gehrke (2003).

The term "variable" can be used in both the logical and probabilistic context. To avoid confusion, we use the term **logvar** to refer to logical variables, and **randvar** to refer to random variables. We write variable names in lowercase, and their values in uppercase. Sets or sequences of logvars are written in boldface, sets or sequences of randvars in calligraphic; their values are written in boldface lowercase.

The vocabulary of our representation includes a finite set of predicates and a finite set of constants. A *constant* represents an object in our universe. A *term* is either a constant or a logvar. A *predicate P* has an arity n and a finite range (*range(P)*); it is interpreted as a mapping from n-tuples of objects (constants) to the range. An *atom* is of the form $P(t_1, t_2, \ldots, t_n)$, where the t_i are terms. A *ground atom* is an atom where all t_i are constants.

A ground atom represents a random variable; this implies that its interpretation, an element of *range(P)*, corresponds to the assignment of a value to the random variable. Hence, the range of a predicate corresponds to the range of the random variables it can represent, and is not limited to {*true*, *false*} as in logic.

Logvars have a finite domain, which is a set of constants. The domain of a logvar x is denoted $D(x)$. A ***constraint*** is a relation defined on a set of logvars, i.e., it is a pair $(\mathbf{x}, C_{\mathbf{x}})$, where $\mathbf{x} - (x_1, x_2, \ldots, x_n)$ is a tuple of logvars, and $C_{\mathbf{x}}$ is a subset of $D(\mathbf{x}) = \times_i D(x_i)$ (Dechter, 2003). Hence, $C_{\mathbf{x}}$ is a set, whose elements (tuples) indicate the allowed combinations of value assignments for the variables in $\mathbf{x}$. For ease of exposition, we identify a constraint with its relation $C_{\mathbf{x}}$, and write C instead of $C_{\mathbf{x}}$ when $\mathbf{x}$ is apparent from the context. We assume an implicit ordering of values in $C_{\mathbf{x}}$'s tuples according to the order of logvars in $\mathbf{x}$. For instance with $\mathbf{x} = (x_1, x_2)$, the constraint $C_{\mathbf{x}} = \{(A, B), (C, D)\}$ indicates that there are two possibilities: either $x_1 = A$ and $x_2 = B$, or $x_1 = C$ and $x_2 = D$. A constraint that contains only one tuple is called *singleton*.

A constraint may be defined extensionally, by listing the tuples that satisfy it, or intensionally, by means of some logical condition, expressed in a ***constraint language***. We call a constraint language $\mathcal{L}$ ***extensionally complete*** if it can express any relation over logvars $\mathbf{x}$, i.e., for any subset of $D(\mathbf{x})$, there is a constraint $C_{\mathbf{x}} \in \mathcal{L}$ whose extension is exactly that subset.

A *constrained atom* $P(\mathbf{x}) \mid C$, where $P(\mathbf{x})$ is an atom and C is a constraint on $\mathbf{x}$, represents a set of ground atoms $\{P(\mathbf{x}) \mid \mathbf{x} \in C\}$, and hence a set of randvars. For consistency with the literature, we call such a constrained atom a ***parametrized randvar*** (PRV), and use calligraphic notation to denote it. Given a PRV $\mathcal{V}$, we use $RV(\mathcal{V})$ to denote the set of randvars it represents; we also say these randvars are *covered* by $\mathcal{V}$.

A *valuation* of a randvar (set of randvars) is an assignment of a value to the randvar (an assignment of values to all randvars in the set).

Example 4.1 The PRV $\mathcal{V} = Smokes(x) \mid C$, with $C = \{x_1, \ldots, x_n\}$, represents n randvars $\{Smokes(x_1), \ldots, Smokes(x_n)\}$.

A *factor* $f = \phi_f(\mathcal{A}_f)$ consists of a sequence of randvars $\mathcal{A}_f = (A_1, \ldots, A_n)$ and a *potential function* $\phi_f : \times_{i=1}^{n} range(A_i) \to \mathbb{R}^+$. The product of two factors, $f_1 \otimes f_2$, is defined as follows. Factor $f = \phi(\mathcal{A})$ is the product of $f_1 = \phi_1(\mathcal{A}_1)$ and $f_2 = \phi_2(\mathcal{A}_2)$ if and only if $\mathcal{A} = \mathcal{A}_1 \cup \mathcal{A}_2$ and for all $\mathbf{a} \in D(\mathcal{A})$: $\phi(\mathbf{a}) = \phi_1(\mathbf{a}_1)\phi_2(\mathbf{a}_2)$ with $\pi_{\mathcal{A}_i}(\mathbf{a}) = \mathbf{a}_i$ for $i = 1, 2$. That is, $\mathbf{a}$ assigns to each randvar in $\mathcal{A}_i$ the same value as $\mathbf{a}_i$. We use $\prod$ to denote multiplication of multiple factors. Multiplying a factor by a scalar c means replacing its potential ϕ by $\phi' : x \mapsto c \cdot \phi(x)$.

An *undirected model* is a set of factors F. It represents a probability distribution $\mathcal{P}_F$ on randvars $\mathcal{A} = \bigcup_{f \in F} \mathcal{A}_f$ as follows: $\mathcal{P}_F(\mathcal{A}) = \frac{1}{z} \prod_{f \in F} \phi_f(\mathcal{A}_f)$, with Z a normalization constant such that $\sum_{\mathbf{a} \in range(\mathcal{A})} \mathcal{P}_F(\mathbf{a}) = 1$.

A *parametric factor* or *parfactor* has the form $\phi(\mathcal{A}) \mid C$, with $\mathcal{A} = \{A_i\}_{i=1}^n$ a sequence of atoms, ϕ a potential function on $\mathcal{A}$, and C a constraint on the logvars appearing in $\mathcal{A}$.[8] The set of logvars occurring in $\mathcal{A}$ is denoted $logvar(\mathcal{A})$; the set of logvars in C is denoted $logvar(C)$. A factor $\phi(\mathcal{A}')$ is a *grounding* of a parfactor $\phi(\mathcal{A})$ if $\mathcal{A}'$ can be obtained by instantiating $\mathbf{x} = logvar(\mathcal{A})$ with some $\mathbf{x} \in C$. The set of groundings of a parfactor g is denoted $gr(g)$.

Example 4.2   Parfactor $g_1 = \phi_1(Smokes(x)) \mid x \in \{x_1, \ldots, x_n\}$ represents the set of factors $gr(g_1) = \{\phi_1(Smokes(x_1)), \ldots, \phi_1(Smokes(x_n))\}$.

A set of parfactors G is a compact way of defining a set of factors $F = \{f \mid f \in gr(g) \wedge g \in G\}$ and the corresponding probability distribution $\mathcal{P}_G(\mathcal{A}) = \frac{1}{z} \prod_{f \in F} \phi_f(\mathcal{A}_f)$.

4.3.2   Counting Formulas

Milch et al. (2008) introduced the idea of counting formulas and (parametrized) counting randvars.

A *counting formula* is a syntactic construct of the form $\#_{x_i \in C}[P(\mathbf{x})]$, where $x_i \in \mathbf{x}$ is called the *counted logvar*.

A *grounded counting formula* is a counting formula in which all arguments of the atom $P(\mathbf{x})$, except for the counted logvar, are constants. It defines a *counting randvar* (CRV), the meaning of which is as follows. First, we define the set of randvars it *covers* as $RV(\#_{x \in C}[P(\mathbf{x})]) = RV(P(\mathbf{x}) \mid x \in C)$. The value of the CRV is determined by the values of the randvars it covers. More specifically, it is a *histogram* that indicates, given a valuation of $RV(P(\mathbf{x}) \mid x \in C)$, how many different values of x occur for each $r \in range(P)$. Thus, its value is of the form $\{(r_1, n_1), (r_2, n_2), \ldots, (r_k, n_k)\}$, with $r_i \in range(P)$ and n_i the corresponding count. Given a histogram h, we will also write $h(v)$ for the count of v in h. Note that the range of a CRV, i.e., the set of all possible histograms it can take as a value, is determined by $k = |range(P)|$ and $|C|$.

Example 4.3 $\#_{x \in \{x_1, x_2, x_3\}}[P(x, y, z)]$ is a grounded counting formula. It covers the randvars $P(x_1, y, z)$, $P(x_2, y, z)$ and $P(x_3, y, z)$. It defines a CRV, the value of which is determined by the values of these three randvars; if $P(x_1, y, z) = true$, $P(x_2, y, z) = false$ and $P(x_3, y, z) = true$, the CRV takes the value $\{(true, 2), (false, 1)\}$.

The concept of a CRV is somewhat complicated. A CRV behaves like a regular randvar in some ways, but not all. It is a construct that can occur as an argument of a factor, like regular randvars, but in that role it actually stands for a set of randvars, all of which are arguments of the factor. A factor of the form $\phi^*(\cdots, \#_{x \in C}[P(\mathbf{x})], \cdots)$ is equivalent to a factor of the form $\phi(\cdots, P(\mathbf{x}_1), P(\mathbf{x}_2), \ldots, P(\mathbf{x}_k), \cdots)$, with $P(\mathbf{x}_i)$ all the instantiations of $\mathbf{x}$

---

[8] We use the definition of Kisynski and Poole (2009) for parfactors, as it allows us to simplify the notation.

obtainable by instantiating x with a value from C, and with ϕ returning for any valuation of the $P(\mathbf{x}_i)$ the value that ϕ^* returns for the corresponding histogram.

Example 4.4 The factor $\phi^*(\#_{x \in \{x_1, x_2, x_3\}}[P(x, y, z)])$ is equivalent to a factor $\phi(P(x_1, y, z), P(x_2, y, z), P(x_3, y, z))$. If $\phi^*(\{(true, 2), (false, 1)\}) = 0.3$, this implies that $\phi(false, true, true) = \phi(true, false, true) = \phi(true, true, false) = 0.3$.

As illustrated in Section 4.2.4, counting formulas are useful for capturing symmetries within a potential function. Recall the workshop example. Whether a person attends a workshop depends on its topic, and this dependence is the same for each person. We can represent this with a single parfactor $\phi(T, A(x)) \mid x \in \{x_1, \ldots, x_n\}$ that represents n ground factors. Eliminating T requires multiplying these n factors into a single factor $\phi'(T, A(x_1), A(x_2), \ldots, A(x_n))$ before summing out T. The potential function ϕ' is high-dimensional, so a tabular representation for it would be very costly. However, it contains a certain symmetry: ϕ' depends only on how many times each possible value for $A(x_i)$ occurs, not on where exactly these occur. By representing the factor using a potential function ϕ^* that has only two arguments, T and the CRV $\#_{x \in \{x_1, \ldots, x_n\}}[A(x)]$, it can be represented more concisely, and computed more efficiently. For instance, to sum out $A(x)$, we do not need to enumerate all possible (2^n) value combinations of the $A(x_i)$ and sum the corresponding $\phi'(T, A(x_1), \ldots, A(x_n))$, we just need to enumerate all possible $(n + 1)$ values for the histogram of $\#_{x \in \{x_1, \ldots, x_n\}}[A(x)]$ and sum the corresponding $\phi^*(T, \#_{x \in \{x_1, \ldots, x_n\}}[A(x)])$, each multiplied by its multiplicity.

Note the complementarity between PRVs and CRVs. While the randvars covered by a PRV occur in different factors, the randvars covered by a CRV occur in one and the same factor. Thus, PRVs impose a symmetry among different factors, whereas CRVs impose a symmetry within a single factor.

A *parametrized counting randvar* (PCRV) is of the form $\#_x[P(\mathbf{x})] \mid C_{\mathbf{x}}$. In this notation we write the constraint on the counted logvar x as part of the constraint $C_{\mathbf{x}}$ on all variables in $\mathbf{x}$. Similar to the way in which a PRV defines a set of randvars through its groundings, a PCRV defines a set of CRVs through its groundings of all variables in $\mathbf{x} \setminus \{x\}$.

Example 4.5 $\#_y[Friend(x, y)] \mid C$ represents a set of CRVs, one for each $x \in \pi_X(C)$, indicating the number of friends x has. If $C = D(x) \times D(y)$ with $D(x) = D(y) = \{Ann, Bob, Carl\}$, we might for instance have $\#_y[Friend(Ann, y)] \mid C = \{(true, 1), (false, 2)\}$ (Ann has one friend, and two people are not friends with her).

Some definitions from the previous section need to be extended slightly in order to accommodate PCRVs. First, because CRVs are not regular randvars, they are not included in the set of randvars covered by the PCRV; that is, $RV(\#_{x_i}[P(\mathbf{x})] \mid C) = RV(P(\mathbf{x}) \mid C)$. Second, since a counting formula "binds" the counted logvar (it is no longer a parameter of the resulting PCRV), we define $logvar(\#_{x_i}[P(\mathbf{x})]) = \mathbf{x} \setminus \{x_i\}$. Thus, generally, $logvar(\mathcal{A})$

refers to all the logvars occurring in $\mathcal{A}$, *excluding* the counted logvars. Note that *logvar(C)* remains unchanged: it refers to all logvars in C, whether they appear as counted or not.

We end this section with two definitions that will be useful later on.

Definition 4.1 (count function) *Given a constraint $C_\mathbf{x}$, for any $\mathbf{y} \subseteq \mathbf{x}$ and $\mathbf{z} \subseteq \mathbf{x} - \mathbf{y}$, the function* $\mathrm{COUNT}_{\mathbf{y}|\mathbf{z}} : C_\mathbf{x} \to \mathbb{N}$ *is defined as follows:*

$$\mathrm{COUNT}_{\mathbf{y}|\mathbf{z}}(t) = |\pi_\mathbf{y}(\sigma_{\mathbf{z}=\pi_\mathbf{z}(t)}(C_\mathbf{x}))|$$

That is, for any tuple t, this function tells us how many values for $\mathbf{y}$ co-occur with t's value for $\mathbf{z}$ in the constraint. We define $\mathrm{COUNT}_{\mathbf{y}|\mathbf{z}}(t) = 1$ *when* $\mathbf{y} = \emptyset$.

Definition 4.2 (count-normalized constraint) *For any constraint $C_\mathbf{x}$, $\mathbf{y} \subseteq \mathbf{x}$ and $\mathbf{z} \subseteq \mathbf{x} - \mathbf{y}$, $\mathbf{y}$ is* count-normalized *w.r.t. $\mathbf{z}$ in $C_\mathbf{x}$ if and only if*

$$\exists n \in \mathbb{N} : \forall t \in C_\mathbf{x} : \mathrm{COUNT}_{\mathbf{y}|\mathbf{z}}(t) = n.$$

When such an n exists, we call it the conditional count *of $\mathbf{y}$ given $\mathbf{z}$ in $C_\mathbf{x}$, and denote it* $\mathrm{COUNT}_{\mathbf{y}|\mathbf{z}}(C_\mathbf{x})$.

Example 4.6 Let $\mathbf{x}$ be $\{p, c\}$ and let the constraint $C_\mathbf{x}$ be $(p, c) \in \{(Ann, Eric), (Bob, Eric),$ $(Carl, Finn), (Debbie, Finn), (Carl, Gemma), (Debbie, Gemma)\}$, representing the parent relationship: Ann is a parent of Eric, etc. Then $\{p\}$ is count-normalized w.r.t. $\{c\}$ because all children (i.e., all values of c in $C_\mathbf{x}$: *Eric*, *Finn* and *Gemma*) have two parents according to $C_\mathbf{x}$, or formally, for all tuples $t \in C_\mathbf{x}$ it holds that $\mathrm{COUNT}_{\{p\}|\{c\}}(t) = 2$. Conversely, $\{c\}$ is not count-normalized w.r.t. $\{p\}$ because not all parents have equally many children. For instance, $\mathrm{COUNT}_{\{c\}|\{p\}}((Ann, Eric)) = 1$ (Ann has 1 child), but $\mathrm{COUNT}_{\{c\}|\{p\}}((Carl, Finn)) = 2$ (Carl has 2 children).

4.4 The GC-FOVE Algorithm: Outline

We now turn to the problem of performing lifted inference on models specified using the above representation. The algorithm we introduce for this is called **GC-FOVE** (for Generalized **C-FOVE**).

GC-FOVE visits PRVs in a particular order and elimates them one at at time. Ideally, it eliminates each PRV by multiplying the parfactors in which it occurs into one parfactor, then summing out the PRV, using the *lifted multiplication* and *lifted summing-out* operators. However, these operators are not always immediately applicable: it may be necessary to refine the involved parfactors and PRVs to make them so. This is done using other operators, which we call *enabling operators*.[9]

---

[9] Technically speaking, multiplication is also an enabling operator as summing-out can only be applied after multiplication.

GC-FOVE

Inputs:

G: a model

Q: the query randvar

Algorithm:

while G contains other randvars than Q:

    if there is a PRV $\mathcal{V}$ that can be eliminated by lifted summing-out

        $G \leftarrow$ apply SUM-OUT to eliminate $\mathcal{V}$ in G

    else apply an enabling operator (one of MULTIPLY, COUNT-CONVERT, EXPAND,

           COUNT-NORMALIZE, SPLIT or GROUND-LOGVAR) on some parfactors in G

end while

return G

Figure 4.3: Outline of the GC-FOVE algorithm.

A high-level description of GC-FOVE is shown in Algorithm 4.3. It makes use of a number of operators, and repeatedly selects and performs one of the possible operators on one or more parfactors. It uses a greedy heuristic, choosing the operation with the minimum cost, where the cost of each operation is defined as the total size (number of rows in tabular form) of all the potentials it creates.

4.5   GC-FOVE's Operators

This section provides detailed information on GC-FOVE's operators. These can conceptually be split into two categories: operators that manipulate potential functions, and operators that refine the model so that the first type of operators can be applied. We will start with three operators that belong to the first category: *lifted multiplication, lifted summing-out* and *counting conversion*. Next, we discuss *splitting, shattering, expansion*, and *count normalization*. Finally, we discuss *grounding*.

In the following, G refers to a model (i.e., a set of parfactors), and $G_1 \sim G_2$ means that models G_1 and G_2 define the same probability distribution.

4.5.1   Lifted Multiplication

The lifted multiplication operator multiplies whole parfactors at once, instead of separately multiplying the ground factors they cover (Poole, 2003; de Salvo Braz et al., 2007; Milch et al., 2008). Figure 4.4 illustrates this for two parfactors $g_1 = \phi_1(S(x)) \mid C$ and $g_2 = \phi_2(S(x), A(x)) \mid C$, where $C = (x \in \{x_1, \ldots, x_n\})$. Lifted multiplication is equivalent to n multiplications on the ground level.

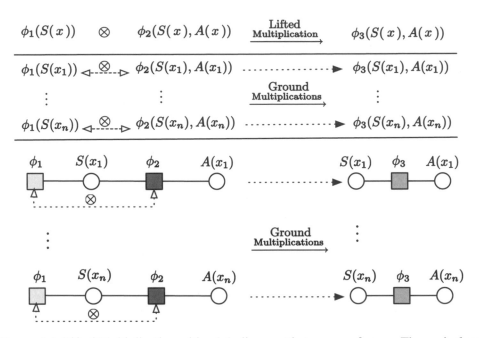

Figure 4.4: Lifted Multiplication with a 1:1 alignment between parfactors. The equivalent of the lifted operation (top), is shown at the level of ground factors (middle), and also in terms of factor graphs (bottom). ⊗ denotes (par)factor multiplication.

The above illustration is deceptively simple, for several reasons. First, the naming of the logvars suggests that logvar x in g_1 corresponds to x in g_2. In fact, g_2 could have multiple logvars, with different names. An *alignment* between the parfactors is necessary, showing how logvars in different parfactors correspond to each other (de Salvo Braz et al., 2007). The alignment must constrain the aligned logvars to exactly the same values in g_1 and g_2 (otherwise, they cannot give identical PRVs in both parfactors). We formalize this as follows.

Definition 4.3 (substitution) *A substitution $\theta = \{x_1 \rightarrow t_1, \ldots, x_n \rightarrow t_n\} = \{\mathbf{x} \rightarrow \mathbf{t}\}$ maps each logvar x_i to a term t_i, which can be a constant or a logvar. When all t_i are constants, θ is called a grounding substitution, and when all are different logvars, a renaming substitution. Applying a substitution θ to an expression α means replacing each occurrence of x_i in α with t_i; the result is denoted $\alpha\theta$.*

Definition 4.4 (alignment) *An alignment θ between two parfactors $g = \phi(\mathcal{A}) \mid C$ and $g' = \phi'(\mathcal{A}') \mid C'$ is a one-to-one substitution $\{\mathbf{x} \rightarrow \mathbf{x}'\}$, with $\mathbf{x} \subseteq logvar(\mathcal{A})$ and $\mathbf{x}' \subseteq logvar(\mathcal{A}')$, such that $\rho_\theta(\pi_{\mathbf{x}}(C)) = \pi_{\mathbf{x}'}(C')$ (with ρ the attribute renaming operator).*

An alignment tells the multiplication operator that two atoms in two different parfactors represent the same PRV, so it suffices to include it in the resulting parfactor only once. Including it twice is not wrong, but less efficient: some structure in the parfactor is then lost. For this reason, it is useful to look for "maximal" alignments which map as many PRVs to each other as possible.

Example 4.7 Consider $g_1 = \phi_1(S(x), F(x, y)) \mid C_{x,y}$ and $g_2 = \phi_2(S(x'), F(x', y')) \mid C_{x',y'}$ with $C_{x,y} = C_{x',y'} = \{x_i\}_1^n \times \{y_j\}_1^m$. Using the maximal alignment $\{x \to x', y \to y'\}$, we get the product parfactor $\phi_3(S(x), F(x, y)) \mid C_{x,y}$. This alignment establishes a 1:1 association between each ground factor $\phi_1(S(x_i), F(x_i, y_j))$ and the corresponding $\phi_2(S(x_i), F(x_i, y_j))$. If, however, we multiply g_1 and g_2 with the alignment $\{x \to x'\}$, the result is a parfactor $\phi_3'(S(x), F(x, y), F(x, y')) \mid (x, y, y') \in \{x_i\}_1^n \times \{y_j\}_1^m \times \{y_k\}_1^m$, which for each x_i unnecessarily multiplies each factor $\phi_1(S(x_i), F(x_i, y_j))$ with all factors $\phi_2(S(x_i), F(x_i, y_k)), k = 1, \ldots, m$. In other words, it unnecessarily creates a direct dependency between all pairs of randvars $F(x_i, y_j), F(x_i, y_k)$.

A second complication is that a single randvar may participate in multiple factors within a certain parfactor, and the number of such factors it appears in may differ across parfactors. Consider parfactors $g_1 = \phi_1(S(x)) \mid x \in \{x_i\}_1^n$ and $g_2 = \phi_2(S(x), F(x, y)) \mid (x, y) \in \{x_i\}_1^n \times \{y_i\}_1^m$. For each x_i, $\phi_1(S(x_i))$ shares randvar $S(x_i)$ with m factors $\phi_2(S(x_i), F(x_i, y_j)), j = 1, \ldots, m$. Multiplication should result in a single parfactor $\phi_3(S(x_i), F(x_i, y)) \mid y \in \{y_i\}_1^m$ that covers m factors $\phi_3(S(x_i), F(x_i, y_j))$, and is equivalent to the product of one factor $\phi_1(S(x_i))$ and m factors $\phi_2(S(x_i), F(x_i, y_j))$. This means we must find a ϕ_3 such that $\forall v, w : \phi_3(v, w)^m = \phi_1(v) \prod_{i=1}^m \phi_2(v, w)$. This gives $\phi_3(v, w) = \phi_1(v)^{1/m} \phi_2(v, w)$. The exponentiation of ϕ_1 to the power $1/m$ is called *scaling*. The result of this multiplication for a single x_i is the same regardless of x_i, so finally, the product of the parfactors g_1 and g_2 will be the parfactor

$$\phi_3(S(x), F(x, y)) = \phi_1(S(x))^{1/m} \cdot \phi_2(S(x), F(x, y)) \mid (x, y) \in \{x_i\}_1^n \times \{y_j\}_1^m.$$

Figure 4.5 illustrates this multiplication graphically.

An alignment between parfactors is called 1:1 if all non-counted logvars in the parfactors are mapped to each other, and is called *m:n* otherwise. Multiplication based on an *m:n* alignment involves scaling, and requires that the non-aligned logvars be count-normalized (Definition 4.2, p. 70) with respect to the aligned logvars in the constraints (otherwise there is no single scaling exponent that is valid for the whole parfactor).

Figure 4.6 formally defines the lifted multiplication. Note that this definition does not assume any specific format for the constraints.

4.5.2 Lifted Summing-out

Once a PRV occurs in only one parfactor, it can be summed out from that parfactor (Milch et al., 2008). We begin with an example of lifted summing-out, which will help motivate the formal definition of the operator.

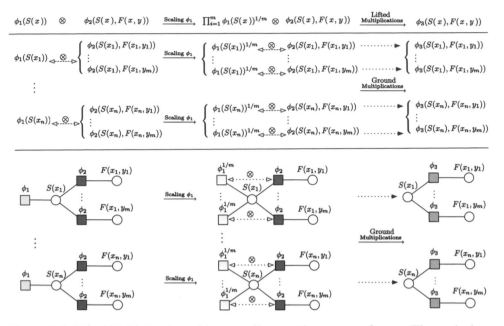

Figure 4.5: Lifted Multiplication with a *m:n* alignment between parfactors. The equivalent of the lifted operation (top), is shown at the level of ground factors (middle), and also in terms of factor graphs (bottom).

Example 4.8 Consider parfactor $g = \phi(S(x), F(x, y)) \mid C$, in which $C = \{(x_i, y_{i,j}) : i \in \{1, \dots, n\}, j \in \{1, \dots, m\}\}$ (Figure 4.7). Note that y is count-normalized w.r.t x in C. Assume we want to sum out randvars $F(x_i, y_{i,j}) \in RV(F(x, y) \mid C)$ on the ground level. Each randvar $F(x_i, y_{i,j})$ appears in exactly one ground factor $\phi(S(x_i), F(x_i, y_{i,j}))$ (see Figure 4.7 (middle)). We can therefore sum out each $F(x_i, y_{i,j})$ from its factor independently from the others, obtaining a factor $\phi'(S(x_i)) = \sum_{F(x_i, y_{i,j})} \phi(S(x_i), F(x_i, y_{i,j}))$. Since the m ground factors $\phi(S(x_i), F(x_i, y_{i,j}))$ have the same potential ϕ, summing out their second argument always results in the same potential ϕ', so we can compute ϕ' just once and, instead of storing m copies of the resulting factor $\phi'(S(x_i))$, store a single factor $\phi''(S(x_i)) = \phi'(S(x_i))^m$. In the end, we obtain n such factors, one for each $S(x_i)$, $i = 1, \dots, n$. We can represent this result using a single parfactor $g' = \phi''(S(x)) \mid C'$, with $C' = \{x_1, \dots, x_n\} = \pi_X(C)$. Lifted summing-out directly computes g' from g in one operation. Note that to have a single exponent for all ϕ'', y must be count-normalized w.r.t. x in C.

Lifted summing-out operator requires a one-to-one mapping between summed-out randvars and factors; that is, each summed-out randvar appears in exactly one factor, and all these factors are different. This is guaranteed when the eliminated atom contains all the logvars of the parfactor, since there is a different ground factor for each instantiation of the

---

Operator MULTIPLY

Inputs:

(1) $g_1 = \phi_1(\mathcal{A}_1) \mid C_1$: a parfactor in G

(2) $g_2 = \phi_2(\mathcal{A}_2) \mid C_2$: a parfactor in G

(3) $\theta = \{\mathbf{x}_1 \rightarrow \mathbf{x}_2\}$: an alignment between g_1 and g_2

Preconditions:

(1) for $i = 1, 2$: $\mathbf{y}_i = logvar(\mathcal{A}_i) \setminus \mathbf{x}_i$ is count-normalized w.r.t. $\mathbf{x}_i$ in C_i

Output: $\phi(\mathcal{A}) \mid C$, with

(1) $C = \rho_\theta(C_1) \bowtie C_2$.

(2) $\mathcal{A} = \mathcal{A}_1\theta \cup \mathcal{A}_2$, and

(3) for each valuation $\mathbf{a}$ of $\mathcal{A}$, with $\mathbf{a}_1 = \pi_{\mathcal{A}_1\theta}(\mathbf{a})$ and $\mathbf{a}_2 = \pi_{\mathcal{A}_2}(\mathbf{a})$:

$$\phi(\mathbf{a}) = \phi_1^{1/r_2}(\mathbf{a}_1) \cdot \phi_2^{1/r_1}(\mathbf{a}_2), \text{ with } r_i = \text{COUNT}_{\mathbf{y}_i|\mathbf{x}_i}(C_i)$$

Postcondition: $G \sim G \setminus \{g_1, g_2\} \cup \{\text{MULTIPLY}(g_1, g_2, \theta)\}$

Figure 4.6: Lifted multiplication algorithm. The algorithm assumes, without loss of generality, that the logvars in the parfactors are standardized apart, i.e., the two parfactors do not share variable names (this can always be achieved by renaming logvars).

logvars. Further, lifted summing-out may result in identical factors on the ground level, which is exploited by computing one factor and exponentiating. This is the case when there is a logvar that occurs only in the eliminated atom, but not in the other atoms (such as y in $F(x, y)$ in the above example).

As already illustrated in Section 4.2.4, counting randvars require special attention in lifted summing-out. A formula like $\phi(\#_x[P(x)]) \mid x \in \{x_1, \ldots, x_k\}$ is really a shorthand for a factor $\phi(P(x_1), P(x_2), \ldots, P(x_k))$ whose value depends only on how many arguments take particular values. In principle, we need to sum out over all combinations of values of $P(x_i)$. We can replace this by summing out over all values of $\#_x[P(x)]$, on the condition that we take the multiplicities of the latter into account. The multiplicity of a histogram $h = \{(r_1, n_1), (r_2, n_2), \ldots, (r_k, n_k)\}$ is a multinomial coefficient, defined as

$$\text{MUL}(h) = \frac{n!}{\prod_{i=1}^{k} n_i!}.$$

As multiplicities should only be taken into account for (P)CRVs, never for regular PRVs, we define for each PRV A and for each value $v \in range(A)$: $\text{MUL}(A, v) = 1$ if A is a regular PRV, and $\text{MUL}(A, v) = \text{MUL}(v)$ if A is a PCRV. This MUL function is identical to Milch et al. (2008)'s NUM-ASSIGN.

With all this in mind, the formal definition of the lifted summing-out in the operator in Figure 4.8 is mostly self-explanatory. Precondition (1) ensures that all randvars in the summed-out P(C)RV occur exclusively in this parfactor. Precondition (2) ensures that

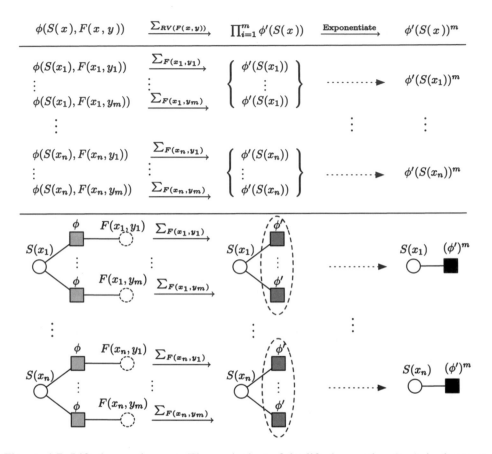

Figure 4.7: Lifted summing-out. The equivalent of the lifted operation (top), is shown at the level of ground factors (middle), and also in terms of factor graphs (bottom).

each summed out randvar occurs in exactly one, separate, ground factor. Precondition (3) ensures that logvars occurring exclusively in the eliminated PRV are count-normalized with respect to the other logvars in that PRV, so that there is one unique exponent for exponentiation.

4.5.3   Counting Conversion

Counting randvars may be present in the original model, but they can also be introduced into parfactors by an operation called *counting conversion* (Milch et al., 2008) (see also Section 4.2.4). To see why this is useful, consider a parfactor $g = \phi(S(x), F(x, y)) \mid C$, with $C = \{x_i\}_{i=1}^{n} \times \{y_j\}_{j=1}^{m}$, and assume we want to eliminate $S(x) \mid C$. To do that, we first need to make sure each $S(x_i)$ occurs in only one factor. On the ground level, this can be

---

Operator SUM-OUT

Inputs:

(1) $g = \phi(\mathcal{A}) \mid C$: a parfactor in G

(2) A_i: an atom in $\mathcal{A}$, to be summed out from g_1

Preconditions

(1) For all PRVs $\mathcal{V}$, other than $\Lambda_i \mid C$, in model G: $RV(\mathcal{V}) \cap RV(A_i \mid C) = \emptyset$

(2) A_i contains all the logvars $x \in logvar(\mathcal{A})$ for which $\pi_X(C)$ is not singleton.

(3) $\mathbf{x}^{excl} = logvar(A_i) \setminus logvar(\mathcal{A} \setminus A_i)$ is count-normalized w.r.t.

$\quad \mathbf{x}^{com} = logvar(A_i) \setminus \mathbf{x}^{excl}$ in C

Output: $\phi'(\mathcal{A}') \mid C'$, such that

(1) $\mathcal{A}' = \mathcal{A} \setminus A_i$

(2) $C' = \pi_{\mathbf{x}^{com}}(C)$

(3) for each assignment $\mathbf{a}' = (\ldots, a_{i-1}, a_{i+1}, \ldots)$ to $\mathcal{A}'$,

$\quad \phi'(\ldots, a_{i-1}, a_{i+1}, \ldots) = \sum_{a_i \in range(A_i)} \text{MUL}(A_i, a_i)\, \phi(\ldots, a_{i-1}, a_i, a_{i+1}, \ldots)^r$

$\quad$ with $r = \text{COUNT}_{\mathbf{x}^{excl} \mid \mathbf{x}^{com}}(C)$

Postcondition: $\mathcal{P}_{G \setminus \{g\} \cup \{\text{SUM-OUT}(g, A_i)\}} = \sum_{RV(A_i \mid C)} \mathcal{P}_G$.

Figure 4.8: The lifted summing-out algorithm.

achieved for a given $S(x_i)$ by multiplying all factors $\phi(S(x_i), F(x_i, y_j))$ in which it occurs. This results in a single factor $\phi'(S(x_i), F(x_i, y_1), \ldots, F(x_i, y_m)) = \prod_j \phi(S(x_i), F(x_i, y_j))$ (see Figure 4.9). This is a high-dimensional factor, but because it equals a product of identical potentials ϕ, its $F(x_i, y_j)$ arguments are mutually interchangeable: all that matters is how often values $v_1, v_2, \ldots$ occur among them, not where they occur. This is exactly the kind of symmetry that CRVs aim to exploit. The factor $\phi'(S(x_i), F(x_i, y_1), \ldots, F(x_i, y_m))$ can therefore be replaced by a two-dimensional $\phi''(S(x_i), h)$ with h a histogram that indicates how often each possible value in the range of $F(x_i, y_j)$ occurs. Thus, by introducing a CRV, we can define a two-dimensional ϕ'' with that CRV as an argument, as opposed to the high-dimensional ϕ'. As argued in Section 4.2.4, this reduces the size of the potential function, and hence computational complexity, exponentially.

In many situations where lifted elimination cannot immediately be applied, counting conversion makes it applicable. The conditions of the SUM-OUT operator (Section 4.5.2) state that an atom A_i can only be eliminated from a parfactor g if A_i has all the logvars in g. When an atom has fewer logvars than the parfactor, counting conversion modifies the parfactor by replacing another atom A_j by a counting formula, which removes this counted logvar from $logvar(\mathcal{A})$. For instance, in the above example, $S(x)$ does not have the logvar y in $g = \phi(S(x), F(x, y)) \mid C$ and cannot be eliminated from the original parfactor g, but

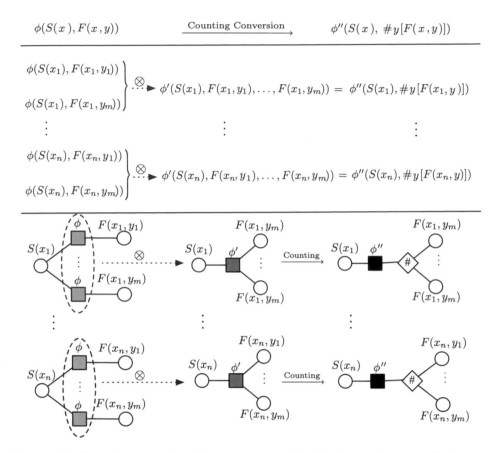

Figure 4.9: Counting conversion. The equivalent of the lifted operation (top), is shown at the level of ground factors (middle), and also in terms of factor graphs (bottom).

a counting conversion on y replaces $F(x,y)$ with $\#_y[F(x,y)]$, allowing us to sum out $S(x)$ from the new parfactor $g' = \phi(S(x), \#_y[F(x,y)]) \mid C$.

The operator in Figure 4.10 formally defines counting conversion. It is mostly self-explanatory, apart from the preconditions. Precondition 1 makes sure that counting conversion, on the ground level, corresponds to multiplying factors that only differ in one randvar (i.e., are the same up to their instantiation of the counted logvar). Precondition 2 guarantees that the resulting histograms have the same range. Precondition 3 is more difficult to explain. It imposes a kind of independence between the logvar to be counted and already occurring counted logvars.

To see why precondition 3 is necessary, consider the parfactor $g = \phi(S(x), \#_y[A(y)]) \mid (x,y) \in \{(x_1, y_2), (x_1, y_3), (x_2, y_1), (x_2, y_3), (x_3, y_1), (x_3, y_2)\}$, which does not satisfy it. This parfactor represents three factors of the form $\phi(S(x_i), \#_y[A(y)]) \mid y \in \{y_1, y_2, y_3\} \setminus \{y_i\}$, which contribute to the joint distribution with the product

$$\phi(S(x_1), \#_{y \in \{y_2, y_3\}}[A(y)]) \cdot \phi(S(x_2), \#_{y \in \{y_1, y_3\}}[A(y)]) \cdot \phi(S(x_3), \#_{y \in \{y_1, y_2\}}[A(y)]).$$

Counting conversion on logvar x turns g into a factor of the form

$$\phi'(\#_x[S(x)], \#_y[A(y)])$$

that should be equivalent. Note that ϕ' depends only on $\#_x[S(x)]$ and $\#_y[A(y)]$.

Now consider valuations V_1: $[S(x_1), S(x_2), S(x_3), A(y_1), A(y_2), A(y_3)] = [true, true, false, true, true, false]$ and V_2: $[S(x_1), S(x_2), S(x_3), A(y_1), A(y_2), A(y_3)] = [true, true, false, true, false, true]$. For both valuations, $\#_x[S(x)] = (2, 1)$ and $\#_y[A(y)] = (2, 1)$, so $\phi'(\#_x[S(x)], \#_y[A(y)])$ must return the same value under V_1 and V_2. The original parfactor, however, returns $\phi(S(true), (1, 1)) \cdot \phi(S(true), (1, 1)) \cdot \phi(S(false), (2, 0))$ under V_1, and $\phi(S(true), (1, 1)) \cdot \phi(S(true), (2, 0)) \cdot \phi(S(false), (1, 1))$ under V_2, which may be different. Since the original parfactor can distinguish valuations that no factor of the form $\phi'(\#_x[S(x)], \#_y[A(y)])$ can, counting conversion cannot be applied in this case.

In contrast, consider $g' = \phi(S(x), \#_y[A(y)]) \mid (x, y) \in \{x_1, x_2, x_3\} \times \{y_1, y_2, y_3\}$, which is similar to g, except that its constraint satisfies precondition 3. All three factors represented by g' differ only in their first argument, randvar $S(x_i)$; they have the same counting randvar $\#_y[A(y)] \mid y \in \{y_1, y_2, y_3\}$ as their second argument (this was not the case for g). Their product, thus, can be represented by a parfactor $\phi'(\#_x[S(x)], \#_y[A(y)]) \mid (x, y) \in \{x_1, x_2, x_3\} \times \{y_1, y_2, y_3\}$, which is derived from g' by a counting conversion.

4.5.4   Splitting and Shattering

When the preconditions for lifted multiplication, lifted summing-out and counting conversion are not fulfilled, it is necessary to reformulate the model in terms of parfactors that do fulfill them. For instance, if $g_1 = \phi_1(S(x)) \mid x \in \{x_1, x_2, x_3\}$ and $g_2 = \phi_2(S(x)) \mid x \in \{x_1, x_2, x_3, x_4, x_5\}$, we cannot multiply g_1 and g_2 directly without creating unwanted dependencies. However, we can replace g_2 with $g_{2a} = \phi_2(S(x)) \mid x \in \{x_1, x_2, x_3\}$ and $g_{2b} = \phi_2(S(x)) \mid x \in \{x_4, x_5\}$. The resulting model is equivalent, but in this new model, we can multiply g_1 with g_{2a}, resulting in $g_3 = \phi_3(S(x)) \mid x \in \{x_1, x_2, x_3\}$.

The above is a simple case of *splitting* parfactors (Poole, 2003; de Salvo Braz et al., 2007; Milch et al., 2008). Basically, splitting two parfactors partitions each parfactor into a part that is shared with the other parfactor, and a part that is disjoint. The goal is to rewrite the P(C)RVs and parfactors into a *proper* form. Two P(C)RVs $(\mathcal{V}_1, \mathcal{V}_2)$ are proper if $RV(\mathcal{V}_1)$ and $RV(\mathcal{V}_2)$ are either identical or disjoint; two parfactors are proper if all their pairs of P(C)RVs are proper. A pair of parfactors can be written into proper form by applying the following procedure, until all their P(C)RVs are proper. Choose a P(C)RV $\mathcal{V}_1$ from one parfactor, compare it to a P(C)RV $\mathcal{V}_2$ from the other, and rewrite the first parfactor such

---

Operator COUNT-CONVERT

Inputs:

(1) $g = \phi(\mathcal{A}) \mid C$: a parfactor in G

(2) x: a logvar in $logvar(\mathcal{A})$

Preconditions

(1) there is exactly one atom $A_i \in \mathcal{A}$ with $x \in logvar(A_i)$

(2) x is count-normalized w.r.t $logvar(\mathcal{A}) \setminus \{x\}$ in C

(3) for all counted logvars $x^{\#}$ in g: $\pi_{x,x^{\#}}(C) = \pi_x(C) \times \pi_{x^{\#}}(C)$

Output: $\phi'(\mathcal{A}') \mid C$, such that

(1) $\mathcal{A}' = \mathcal{A} \setminus \{A_i\} \cup \{A_i'\}$ with $A_i' = \#_x[A_i]$

(2) for each assignment $\mathbf{a}'$ to $\mathcal{A}'$ with $a_i' = h$:

$$\phi'(\ldots, a_{i-1}, h, a_{i+1}, \ldots) = \prod_{a_i \in range(A_i)} \phi(\ldots, a_{i-1}, a_i, a_{i+1}, \ldots)^{h(a_i)}$$

with $h(a_i)$ denoting the count of a_i in histogram h

Postcondition: $G \sim G \setminus \{g\} \cup \{\text{COUNT-CONVERT}(g, x)\}$.

Figure 4.10: The counting conversion algorithm.

that $\mathcal{V}_1$ is split into two parts: one that is disjoint from $\mathcal{V}_2$ and one that is shared with $\mathcal{V}_2$. All the parfactors in the model can be made proper w.r.t. each other by repeatedly applying this rewrite until convergence. This is called *shattering* the model.

It is simpler to rewrite a PRV into the proper form than a PCRV. We describe the operator that handles PRVs, namely SPLIT, in this section and discuss the operator that handles PCRVs, namely EXPAND, in the following section. Before defining the SPLIT operator, we provide the following auxiliary definitions, which will also be used later on.

Definition 4.5 (Splitting on overlap) *Splitting a constraint C_1 on its **y**-overlap with C_2, denoted $C_1/_{\mathbf{y}}C_2$, partitions C_1 into two subsets, containing all tuples for which the **y** part occurs or does not occur, respectively, in C_2. $C_1/_{\mathbf{y}}C_2 = \{\{t \in C_1 \mid \pi_{\mathbf{y}}(t) \in \pi_{\mathbf{y}}(C_2)\}, \{t \in C_1 \mid \pi_{\mathbf{y}}(t) \notin \pi_{\mathbf{y}}(C_2)\}\}$.*

Definition 4.6 (Parfactor partitioning) *Given a parfactor $g = \phi(\mathcal{A}) \mid C$ and a partition $\mathbb{C} = \{C_i\}_{i=1}^{n}$ of C, $\text{PARTITION}(g, \mathbb{C}) = \{\phi(\mathcal{A}) \mid C_i\}_{i=1}^{n}$.*

The operator in Figure 4.11 defines splitting of parfactors. Note that, in the operator definition, for simplicity, we assume that $A = A' = P(\mathbf{y})$, which means that the logvars used in A and A' must be the same, in the same order. We can always rewrite the model such that any two PRVs with the same predicate are in this form. For this, we rewrite the parfactors as follows: (i) if the parfactors share logvars, we first standardize apart the logvars between two parfactors, (ii) *linearize* each atom in which some logvar occurs more than once, i.e., rewrite it such that it has a distinct logvar in each argument, and

---

Operator SPLIT
Inputs:
(1) $g = \phi(\mathcal{A}) \mid C$: a parfactor in G
(2) $A = P(\mathbf{y})$: an atom in $\mathcal{A}$
(3) $A' = P(\mathbf{y}) \mid C'$ or $\#_y[P(\mathbf{y})] \mid C'$
Output: PARTITION$(g, \mathbb{C})$, with $\mathbb{C} = C/_y C' \setminus \{\emptyset\}$
Postcondition $G \sim G \setminus \{g\} \cup$ SPLIT(g, A, A')

Figure 4.11: The split operator.

(iii) apply a renaming substitution on the logvars such that the concerned atoms have the same logvars. For instance, consider the two parfactors $g_1 = \phi_1(P(x, x)) \mid x \in C_1$ and $g_2 = \phi_2(P(y, z)) \mid (y, z) \in C_2$. The logvars of the two parfactors are already different, so there is no need for standardizing them apart. However, the atom $P(x, x)$ in g_1 is not linearized yet. To linearize it, we rewrite g_1 into the form $\phi_1(P(x, x')) \mid (x, x') \in C_1'$, where $C_1' = \{(x, x) \mid x \in C_1\}$. Finally, we rename the logvars x and x' to y and z, respectively, to derive $\phi_1(P(y, z)) \mid (y, z) \in C_1'$. This brings the atom $P(x, x)$ into the desired form $P(y, z)$.

For ease of exposition, we will not explicitly mention this linearization and renaming; whenever two PRVs from different parfactors are compared, any notation suggesting that they have the same logvars is to be interpreted as "have the same logvars after linearization and renaming".

When GC-FOVE wants to multiply two parfactors, it first checks for all pairs $A_1 \mid C_1$, $A_2 \mid C_2$ (one from each parfactor) whether they are proper. If a pair is found that is not proper, this means A_1 and A_2 are both of the form $P(\mathbf{y})$, with different (but overlapping) instantiations for $\mathbf{y}$ in C_1 and C_2. The pair is then split on $\mathbf{y}$.

Example 4.9 Consider $g_1 = \phi_1(N(x, y), R(x, y, z)) \mid C_1$ with $C_1 = (x, y, z) \in \{x_i\}_{i=1}^{50} \times \{y_i\}_{i=1}^{50} \times \{z_i\}_{i=1}^{5}$, and $g_2 = \phi_2(N(x, y)) \mid C_2$ with $C_2 = (x, y) \in \{x_{2i}\}_{i=1}^{25} \times \{y_i\}_{i=1}^{50}$. First, we compare the PRVs $N(x, y) \mid C_1$ and $N(x, y) \mid C_2$. These PRVs partially overlap, so splitting is necessary. To split the parfactors, we split C_1 and C_2 on their (x,y)-overlap. This partitions C_1 into two sets: $C_1^{com} = \{x_{2i}\}_{i=1}^{25} \times \{y_i\}_{i=1}^{50} \times \{z_i\}_{i=1}^{5}$, and $C_1^{excl} = C_1 \setminus C_1^{com} = \{x_{2i-1}\}_{i=1}^{25} \times \{y_i\}_{i=1}^{50} \times \{z_i\}_{i=1}^{5}$. C_2 does not need to be split, as it has no tuples for which the (x,y)-values do not occur in C_1. After splitting the constraints, we split the parfactors accordingly: g_1 is split into two parfactors $g_1^{com} = \phi(N(x, y), R(x, y, z)) \mid C_1^{com}$ and $g_1^{excl} = \phi(N(x, y), R(x, y, z)) \mid C_1^{excl}$, and parfactor g_2 remains unmodified.

Our splitting procedure splits any two PRVs into at most two partitions each. Similarly, the involved parfactors are split into at most two partitions each. This strongly contrasts with C-FOVE's approach to splitting. C-FOVE operates per logvar, and splits off each value in a separate partition (*splitting based on substitution*) (Poole, 2003; Milch et al.,

2008). Thus, it may require many splits where GC-FOVE requires just one. In the above example, instead of $g_1^{excl} = \phi(N(x,y), R(x,y,z)) \mid C_1^{excl}$, C-FOVE ends up with 1250 parfactors $\phi(N(x_1,y_1), R(x_1,y_1,z)) \mid \{z_i\}_{i=1}^5$, $\phi(N(x_1,y_2), R(x_1,y_2,z)) \mid \{z_i\}_{i=1}^5$, ..., $\phi(N(x_3,y_1)$, $R(x_3,y_1,z)) \mid \{z_i\}_{i=1}^5$, ..., $\phi(N(x_{49},y_{50}), R(x_{49},y_{50},z)) \mid \{z_i\}_{i=1}^5$.

4.5.5   Expansion of Counting Formulas

When handling parfactors with counting formulas, to rewrite a P(C)RV into the proper form, we employ the operation of *expansion* (Milch et al., 2008). When we split one group of randvars $RV(\mathcal{V})$ into a partition $\{RV(\mathcal{V}_i)\}_{i=1}^m$, any counting randvar γ that counts the values of $RV(\mathcal{V})$ needs to be *expanded*, i.e., replaced by a group of counting randvars $\{\gamma_i\}_{i=1}^m$, where each γ_i counts the values of randvars in $RV(\mathcal{V}_i)$. In parallel with this, the potential that originally had $\mathcal{V}$ as an argument must be replaced by a potential that has all the $\mathcal{V}_i$ as arguments; we call this *potential expansion*.

Example 4.10 Suppose we need to split $g_1 = \phi_1(\#_x[S(x)]) \mid C_1$ and $g_2 = \phi_2(S(x)) \mid C_2$, with $C_1 = \{x_1, \ldots, x_{100}\}$ and $C_2 = \{x_1, \ldots, x_{40}\}$. C_1 is split into $C_1^{com} = C_1 \cap C_2 = \{x_1, \ldots x_{40}\}$ and $C_1^{excl} = C_1 \setminus C_2 = \{x_{41}, \ldots x_{100}\}$. Consequently, the original group of randvars in parfactor g_1, namely $\{S(x_1), \ldots S(x_{100})\}$, is partitioned into $\mathcal{V}_1^{com} = \{S(x_1), \ldots S(x_{40})\}$ and $\mathcal{V}_1^{excl} = \{S(x_{41}), \ldots S(x_{100})\}$. To preserve the semantics of the original counting formula, we now need two separate counting formulas, one for $\mathcal{V}_1^{com}$ and one for $\mathcal{V}_1^{excl}$, and we need to replace the original potential $\phi_1(\#_x[S(x)])$ by $\phi_1'(\#_{x_{com}}[S(x_{com})], \#_{x_{excl}}[S(x_{excl})])$, where $\phi_1'()$ depends only on the sum of the two new counting randvars $\#_{x_{com}}[S(x_{com})]$ and $\#_{x_{excl}}[S(x_{excl})]$. The end effect is that the parfactor g_1 is replaced by the new parfactor $\phi_1'(\#_{x_{com}}[S(x_{com})], \#_{x_{excl}}[S(x_{excl})]) \mid C_1'$, where $C_1' = C_1^{com} \times C_1^{excl}$.

To explain expansion, we begin with the case of (non-parametrized) CRVs and then move to the general case of expansion for PCRVs.

4.5.5.1   Expansion of CRVs   First consider the simplest possible type of CRV: $\#_x[P(x)] \mid C$. It counts for how many values of x in C, $P(x)$ has a certain value. When C is partitioned, x must be counted within each subset of the partition.

In the following, we assume C is partitioned into two non-empty subsets C_1 and C_2. If one of them is empty, the other equals C, which means the CRV can be kept as is and no expansion is needed.

In itself, splitting $\#_x[P(x)] \mid C$ into $\#_x[P(x)] \mid C_1$ and $\#_x[P(x)] \mid C_2$ is trivial, but a problem is that both of the resulting counting formulas will occur in one single parfactor, and a constraint is always associated with a parfactor, not with a particular argument of a parfactor. Thus, we need to transform $\phi(\#_x[P(x)]) \mid C$ into a parfactor of the form $\phi'(\#_{x_1}[P(x_1)], \#_{x_2}[P(x_2)]) \mid C'$, where the single constraint C' expresses that x_1 can take only values in C_1, and x_2 only values in C_2. It is easily seen that $C' = \rho_{x \to x_1} C_1 \times \rho_{x \to x_2} C_2$ satisfies this condition. Further, to preserve the semantics, ϕ' should, for any count of x_1 and x_2, give the same result as ϕ with the corresponding count of x. The function $\phi'(h_1, h_2) = \phi(h_1 \oplus h_2)$, with $\oplus$ denoting summation of histograms, has this property. Indeed, the histogram for x_1 (resp. x_2) in C' is equal to that for x in C_1 (resp. C_2), and since $\{C_1, C_2\}$ is a partition of C, the sum of these histograms equals the histogram for x in C.

More generally, consider a non-parametrized CRV $\#_x[P(\mathbf{x})] \mid C$, with $x \in \mathbf{x}$ meaning that $\pi_{\mathbf{x} \setminus \{x\}}(C)$ is singleton. The constraint $C' = \pi_{\mathbf{x} \setminus \{x\}}(C) \times (\pi_{x_1}(\rho_{x \to x_1} C_1) \times \pi_{x_2}(\rho_{x \to x_2} C_2))$ joins this singleton with the Cartesian product of $\pi_x(C_1)$ and $\pi_x(C_2)$, and is equivalent to the constraint $\rho_{x \to x_1}(C_1) \bowtie \rho_{x \to x_2}(C_2)$. The result is again such that counting x_1 (x_2) in C' is equivalent to counting x in C_1 (C_2), while the constraint on all other variables remains unchanged. This shows that a parfactor $\phi(\mathcal{A}, \#_x[P(\mathbf{x})]) \mid C$, for any partition $\{C_1, C_2\}$ of C with C_1 and C_2 non-empty, can be rewritten in the form $\phi'(\mathcal{A}, \#_{x_1}[P(\mathbf{x})], \#_{x_2}[P(\mathbf{x})]) \mid C'$, where $C' = \rho_{x \to x_1}(C_1) \bowtie \rho_{x \to x_2}(C_2)$.

Note that the ranges of the counting formulas in ϕ' (the h_i arguments) depend on the cardinality of C_1 and C_2, which we will further denote as n_1 and n_2 respectively.

4.5.5.2 Expansion of PCRVs
Consider the case where $\pi_{\mathbf{x} \setminus \{x\}}(C)$ is not a singleton, i.e., we have a *parametrized* CRV $\mathcal{V}$ that represents a *group* of CRVs, each counting the values of a *subset* of $RV(\mathcal{V})$. Given a partitioning of the constraint C, we need to expand each underlying CRV and the corresponding potential. The constraint $C' = \rho_{x \to x_1}(C_1) \bowtie \rho_{x \to x_2}(C_2)$ remains correct (for non-empty C_1, C_2), even when $\pi_{\mathbf{x} \setminus \{x\}}(C)$ is no longer singleton: it associates the correct values of x_1 and x_2 with each tuple in $\pi_{\mathbf{x} \setminus \{x\}}(C)$. However, because the result of potential expansion depends on the size of the partitions, n_1 and n_2, which are a function of $\mathbf{x} \setminus \{x\}$, only those CRVs that have the same (n_1, n_2) result in identical potentials after expansion, and can be grouped in one parfactor. To account for this, PCRV expansion first splits the PCRV into groups of CRVs that have the same "joint count" (n_1, n_2), then applies for each group the corresponding potential expansion.

To formalize this, we first provide the following auxiliary definitions.

Definition 4.7 (Group-by) *Given a constraint C and a function $f : C \to R$, GROUP-BY$(C, f) = C/ \sim_f$, with $x \sim_f y \Leftrightarrow f(x) = f(y)$ and $/$ denoting set quotient. That is, GROUP-BY(C, f) partitions C into subsets of elements that have the same result for f.*

Definition 4.8 (Joint-count) *Given a constraint C over variables $\mathbf{x}$, partitioned into $\{C_1, C_2\}$, and a counted logvar $x \in \mathbf{x}$; then for any $t \in C$, with $L = \mathbf{x} \setminus \{x\}$ and $l = \pi_L(t)$,*

$$\text{JOINT-COUNT}_{x, \{C_1, C_2\}}(t) = (|\pi_x(\sigma_{L=l}(C_1))|, |\pi_x(\sigma_{L=l}(C_2))|).$$

When a PCRV $\mathcal{V} = \#_{x_i}[P(\mathbf{x})] \mid C$ in a parfactor g partially overlaps with another PRV $A' \mid C'$ in the model, expansion performs the following on g: (1) partition C on its $\mathbf{x}$-overlap with C', resulting in $C/_{\mathbf{x}}C'$; (2) partition C into $\mathbb{C} = $ GROUP-BY$(C, \text{JOINT-COUNT}_{x, C/_{\mathbf{x}}C'})$ (this corresponds to a partition of $\mathcal{V}$ into CRVs that have the same number of randvars in each of the common and exclusive partitions in $C/_{\mathbf{x}}C'$); (3) split g, based on

---

Operator EXPAND

Inputs:

(1) $g = \phi(\mathcal{A}) \mid C$: a parfactor in G

(2) $A = \#_x[P(\mathbf{x})]$: a counting formula in $\mathcal{A}$

(3) $A' = P(\mathbf{x}) \mid C'$ or $\#_y[P(\mathbf{x})] \mid C'$

Output: $\{g_i = \phi'_i(\mathcal{A}'_i) \mid C'_i\}_{i=1}^n$ where

(1) $C/_{\mathbf{x}}C' = \{C^{com}, C^{excl}\}$

(2) $\{C_1, \ldots, C_n\} = \text{GROUP-BY}(C, \text{JOINT-COUNT}_{x,C/_{\mathbf{x}}C'})$

(3) for all i where $C_i \bowtie C^{com} = \emptyset$ or $C_i \bowtie C^{excl} = \emptyset$: $\phi'_i = \phi$, $\mathcal{A}'_i = \mathcal{A}$, $C'_i = C_i$

(4) for all other i:

    (5) $C'_i = \pi_{logvar(\mathcal{A})}(C_i) \bowtie (\rho_{x \to x_{com}}(C^{com}) \bowtie \rho_{x \to x_{excl}}(C^{excl}))$

    (6) $\mathcal{A}'_i = \mathcal{A} \setminus \{A\} \cup \{A\theta_{com}, A\theta_{excl}\}$ with $\theta_{com} = \{x \to x_{com}\}$, $\theta_{excl} = \{x \to x_{excl}\}$

    (7) for each valuation $(\mathbf{l}, h_{com}, h_{excl})$ of $\mathcal{A}'_i$, $\phi'_i(\mathbf{l}, h_{com}, h_{excl}) = \phi(\mathbf{l}, h_{com} \oplus h_{excl})$

Postcondition $G \sim G \setminus \{g\} \cup \text{EXPAND}(g, A, A')$

Figure 4.12: The expansion operator.

$\mathbb{C} = \{C_1, \ldots, C_n\}$, resulting in parfactors $g_1, \ldots, g_n$ that each require a distinct expanded potential; (4) in each g_i, replace potential ϕ with its expanded version. The formal definition of expansion is given in the operator in Figure 4.12.

Example 4.11 Suppose we need to split parfactors $g = \phi(\#_y[F(x, y)]) \mid C$ and $g' = \phi'(F(x, y)) \mid C'$, with $C = \{Ann, Bob, Carl\} \times \{Dave, Ed, Fred, Gina\}$ and $C' = \{Ann, Bob\} \times \{Dave, Ed\}$. Assume F stands for friendship; $\#_y[F(x, y)] \mid C$ counts the number of friends and non-friends each x has in C. The random variables covered by PCRV $\#_y[F(x, y)] \mid C$ partially overlap with those of $F(x, y) \mid C'$. If we need to split C on overlap with C', yielding C^{com} and C^{excl}, we need to replace the original PCRV with separate PCRVs for C^{com} and C^{excl}. But PCRVs require count-normalization, and the fact that y is count-normalized w.r.t. x in C does not necessarily imply that the same holds in C^{com} and C^{excl}. That is why, in addition to the split on overlap, we need an orthogonal partitioning of C according to the joint counts. Within a subset C_i of this partitioning, y will be count-normalized w.r.t. x in C_i^{com} and in C_i^{excl}.

We follow the four steps outlined above. Figure 4.13 illustrates these steps. First, we find the partition $C/_{x,y}C' = \{C^{com}, C^{excl}\}$ with $C^{com} = \{Ann, Bob\} \times \{Dave, Ed\}$ and $C^{excl} = \{Ann, Bob\} \times \{Fred, Gina\} \cup \{Carl\} \times \{Dave, Ed, Fred, Gina\}$. Inspecting the joint counts, we see that C^{com} contains 2 possible friends for Ann or Bob (namely Dave and Ed), but 0 for Carl, whereas C^{excl} contains 2 possible friends for Ann or Bob and 4 for Carl. Formally, JOINT-COUNT$_{y,C/_{x,y}C'}(t)$ equals (2,2) for $\pi_X(t) = Ann$ or $\pi_X(t) = Bob$, and equals (0,4) for $\pi_X(t) = Carl$. So, within C^{com} and C^{excl}, y is no longer count-normalized with respect to x. This is why Operator 4.12 partitions C into subsets $\{C_1, C_2\} =$ GROUP-BY$(C, \text{JOINT-COUNT}_{y, C/_{x,y}C'})$, which gives $C_1 = \{Ann, Bob\} \times \{Dave, Ed, Fred, Gina\}$ and $C_2 = \{Carl\} \times \{Dave, Ed, Fred, Gina\}$. For each C_i, we can now construct a C_i' that allows for counting the friends in C_i^{com} and in C_i^{excl} separately, using the series of joins discussed earlier. Where both C_i^{com} and C_i^{excl} are non-empty, the original PCRV $\#_y[F(x, y)] \mid C$ is replaced by two PCRVs per C_i, $\#_{Y_{com}}[F(x, Y_{com})] \mid C_i$ and $\#_{Y_{excl}}[F(x, Y_{excl})] \mid C_i$, and the new potential ϕ' is defined such that $\phi'(h_{com}, h_{excl}) = \phi(h_{com} \oplus h_{excl})$.

4.5.6 Count Normalization

Lifted multiplication, summing-out and counting conversion all require certain variables to be count-normalized (recall Definition 4.2, p. 70). When this property does not hold, it can be achieved by *normalizing* the involved parfactor, which amounts to splitting the parfactor into parfactors for which the property does hold (Milch et al., 2008). Concretely, when **y** is not count-normalized given **z** in a constraint C, then C is simply partitioned into $\mathbb{C} = \text{GROUP-BY}(C, \text{COUNT}_{\mathbf{y}|\mathbf{z}})$, with COUNT$_{\mathbf{y}|\mathbf{z}}$ as defined in Definition 4.1; next, the parfactor is split according to $\mathbb{C}$. The formal definition of count normalization is shown in Figure 4.14.

Example 4.12 Consider the parfactor g with $\mathcal{A} = (Prof(P), Supervises(P, S))$ and constraint $C = \{(p_1, s_1), (p_1, s_2), (p_2, s_2), (p_2, s_3), (p_3, s_5), (p_4, s_3), (p_4, s_4), (p_5, s_6)\}$. Lifted elimination of $Supervises(P, S)$ requires logvar S (student) to be count-normalized with respect to logvar P (professor). Intuitively, we need to partition the professors into groups such that all professors in the same group supervise the same number of students. In our example, C needs to be partitioned into two, namely $C_1 = \sigma_{P \in \{p_3, p_5\}}(C) = \{(p_3, s_5), (p_5, s_6)\}$ (tuples involving professors with 1 student) and $C_2 = \sigma_{P \in \{p_1, p_2, p_4\}}(C) = \{(p_1, s_1), (p_1, s_2), (p_2, s_2), (p_2, s_3), (p_4, s_3), (p_4, s_4)\}$ (professors with 2 students). Next, the parfactor g is split accordingly into two parfactors g_1 and g_2 with constraints C_1 and C_2. These parfactors are now ready for lifted elimination of $Supervises(P, S)$.

4.5.7 Grounding a Logvar

There is no guarantee that the enabling operators eventually result in PRVs and parfactors that allow for any of the lifted operators. To illustrate this, consider a model consisting of a single parfactor $\phi(R(x, y), R(y, z), R(x, z)) \mid C$, which expresses a probabilistic variant of transitivity. Since there is only one factor, no multiplications are needed before starting to eliminate variables. Yet, because of the structure of the parfactor, no single PRV can be eliminated (the preconditions for lifted summing out and counting conversion are not fulfilled, and none of the other operators can change that). In cases like this, when no other operators can be applied, lifted VE can always resort to a last operator: grounding a logvar x in a parfactor g (de Salvo Braz et al., 2007; Milch et al., 2008). Given a parfactor $g = \phi(\mathcal{A}) \mid C$ and a logvar $x \in logvar(\mathcal{A})$ with $\pi_x(C) = \{x_1, \ldots, x_n\}$, grounding x replaces

C

Ann	Dave
Ann	Ed
Ann	Fred
Ann	Gina
Bob	Dave
Bob	Ed
Bob	Fred
Bob	Gina
Carl	Dave
Carl	Ed
Carl	Fred
Carl	Gina

C'

Ann	Dave
Ann	Ed
Bob	Dave
Bob	Ed

$C/_{x,y} C'$

Ann	Dave
Ann	Ed
Bob	Dave
Bob	Ed
Ann	Fred
Ann	Gina
Bob	Fred
Bob	Gina
Carl	Dave
Carl	Ed
Carl	Fred
Carl	Gina

GROUP-BY(C, JOINT-COUNT$_{y,C/_{x,y}C'}$)

C_1	Ann	Dave
	Ann	Ed
	Ann	Fred
	Ann	Gina
	Bob	Dave
	Bob	Ed
	Bob	Fred
	Bob	Gina
C_2	Carl	Dave
	Carl	Ed
	Carl	Fred
	Carl	Gina

C'_1

x	Y_{com}	Y_{excl}
Ann	Dave	Fred
Ann	Dave	Gina
Ann	Ed	Fred
Ann	Ed	Gina
Bob	Dave	Fred
Bob	Dave	Gina
Bob	Ed	Fred
Bob	Ed	Gina

C'_2

x	y
Carl	Dave
Carl	Ed
Carl	Fred
Carl	Gina

Figure 4.13: Illustration of the PCRV expansion operator. (1) y is count-normalized w.r.t. x in C (with each x, four y values are associated). Splitting C on overlap with C' results in subsets in which y is no longer count-normalized w.r.t. x: the joint counts of y for both subsets are (2,2) for *Ann* and *Bob*, and (0,4) for *Carl*. To obtain count-normalized subsets, we need to partition C into a subset C_1 for *Ann* and *Bob*, and C_2 for *Carl*; this is what the GROUP-BY construct does. For each of the subsets, a split on overlap with C' will yield subsets in which y is count-normalized w.r.t. x. C'_1 is the result of joining the common and exclusive parts according to the join construct motivated earlier. C'_2 equals C_2 because C_2 has no overlap with C' and hence need not be split.

Operator COUNT-NORMALIZE

Inputs:

(1) $g = \phi(\mathcal{A}) \mid C$: a parfactor in G

(2) $\mathbf{y} \mid \mathbf{z}$: sets of logvars indicating the desired normalization property in C

Preconditions

(1) $\mathbf{y} \subset logvar(\mathcal{A})$ and $\mathbf{z} \subseteq logvar(\mathcal{A}) \setminus \mathbf{y}$

Output: PARTITION(g, GROUP-BY(C, COUNT$_{\mathbf{y}|\mathbf{z}}$))

Postconditions $G \sim G \setminus \{g\} \cup$ COUNT-NORMALIZE($g, \mathbf{y} \mid \mathbf{z}$)

Figure 4.14: The count-normalization operator.

Operator GROUND-LOGVAR

Inputs:

(1) $g = \phi(\mathcal{A}) \mid C$: a parfactor in G

(2) x: a logvar in $logvar(\mathcal{A})$

Output: PARTITION(g, GROUP-BY(C, π_X))

Postcondition

$G \sim G \setminus \{g\} \cup$ GROUND-LOGVAR(g, x)

Figure 4.15: Grounding.

g with the set of parfactors $\{g_1, \ldots, g_n\}$ with $g_i = \phi(\mathcal{A}) \mid \sigma_{x=x_i}(C)$. This is equivalent to splitting g based on the partition GROUP-BY(C, π_X), which yields the definition shown in Figure 4.15. Note that in each resulting parfactor g_i, logvar x can only take on a single value x_i, so in practice x can be replaced by the constant x_i and removed from the set of logvars.

Grounding can significantly increase the granularity of the model and decrease the opportunities for performing lifted inference: in the extreme case where all logvars are grounded, inference is performed at the propositional level. It is therefore best used only as a last resort. In practice, GC-FOVE's heuristic for selecting operators, which relies on the size of the resulting factors, automatically has this effect.

Calling the GROUND-LOGVAR operator should not be confused with the event of obtaining a ground model. GROUND-LOGVAR grounds only one logvar, and does not necessarily result in a ground model. Conversely, one may arrive at a ground model without ever calling GROUND-LOGVAR, simply because the splitting continues up to the singleton level.

5 Search-Based Exact Lifted Inference

Seyed Mehran Kazemi, Guy Van den Broeck, and David Poole

The problem of lifted inference was first explicitly proposed and studied by Poole (2003) who developed a lifted version of the variable elimination algorithm that could exploit exchangeability to some extent. Since then, several other works extended Poole's proposal to enable exploiting more exchangeability (de Salvo Braz et al., 2005; Milch et al., 2008; Taghipour et al., 2013c). Other works looked into using lifted inference with factors representing aggregation (Kisynski and Poole, 2009; Choi et al., 2011a); see Chapter 6. A major issue with these early works was that they were based on variable elimination, and much of the work is in designing intermediate representations that are closed under observing and marginalization. These intermediate representations become quite complicated.

Tackling the issues with variable elimination-based lifted inference algorithms, another thread of research extends recursive conditioning (Darwiche, 2001) to develop search-based lifted inference algorithms (Jha et al., 2010; Gogate and Domingos, 2010; Van den Broeck et al., 2011; Poole et al., 2011). Besides having the advantage that conditioning simplified the representations for these algorithms, they also enabled exploiting determinism and context-specific independence (Boutilier et al., 1996). The basis for these algorithms is to recursively select a lifted inference operation/rule that is applicable to the input model, apply the selected rule and get one or more simplified sub-models, and continue the process for each sub-model until evaluating the sub-models becomes trivial.

Since the advent of search-based lifted inference algorithms, several such rules have been proposed (or extended from variable elimination-based lifted inference algorithms) for exact marginal lifted inference (de Salvo Braz et al., 2005; Milch et al., 2008; Poole et al., 2011; Choi et al., 2011a; Jha et al., 2010; Gogate and Domingos, 2010; Van den Broeck et al., 2011, 2014; Kazemi et al., 2016), often providing exponential speedups for specific models. These rules provided the foundation for some variational and **over-symmetric approximations** (Van den Broeck and Darwiche, 2013; Venugopal and Gogate, 2014a) and a rich literature on approximate lifted inference and learning (Singla and Domingos, 2008; Kersting et al., 2009; Niepert, 2012b; Bui et al., 2013b; Venugopal and Gogate, 2014b; Kopp et al., 2015; Ahmadi et al., 2012; Jernite et al., 2015). Kazemi and Poole (2014) showed that the order of applying the rules can substantially affect the running time of inference and proposed heuristics for selecting an ordering for the rules.

In this chapter, we provide a detailed description of search-based lifted inference algorithms. The study will be based upon the **weighted first-order model counting** (WFOMC) formulation of StarAI models (Van den Broeck et al., 2011), an intermediate language that simplifies the development of lifted inference algorithms.

5.1   Background and Notation

In this section, we define our notation and provide necessary background for readers to follow the rest of the chapter. We only consider finite cases, with finitely many random variables, each with finite ranges.

5.1.1   Random Variables, Independence, and Symmetry

Probability can be defined either with random variables as primitive, where each random variable has a range and a possible world is an assignment of a value to each random variable, or where possible worlds are primitive and a **random variable** is a function from worlds to some set of values; the **range** of the random variable (Halpern, 2003). A Boolean formula over assignments to random variables is a proposition that is true or false in each possible world. In either case, a measure over possible worlds (a non-negative function that sums to 1) induces a probability over propositions where the probability of any proposition is the measure of the set of worlds in which the proposition is true. A probability distribution over a random variable is a function from range of the random variable into non-negative reals. For example, suppose the worlds consist of a sequence of coin tosses. An example of a random variable is the outcome of the 7th coin toss, $Cointoss_7$ whose range is {*Heads, Tails*}. A probability distribution over this random variable $P(Cointoss_7=Heads) = 0.5$ and $P(Cointoss_7=Tails) = 0.5$, representing a fair coin. In this chapter, we use capital letters to show random variables and, unless otherwise specified, we assume all random variables are *Boolean*, i.e. their range is {*True, False*}.

A **joint probability distribution** $P(X_1, \ldots, X_m)$ is a probability distribution over the possible outcomes of the set {$X_1, \ldots, X_m$} of all of the primitive random variables. Thus, a joint probability distribution gives the probability of each possible world. A **conditional probability distribution** $P(X_1, \ldots, X_m \mid Y_1, \ldots, Y_l)$ is a probability distribution over a set of random variables {$X_1, \ldots, X_m$} when the value of the random variables $Y_1, \ldots, Y_l$ is given.

Two random variables X and Y are **marginally independent** of each other if knowing about the outcome of one does not change our belief about the outcome of the other. More formally, X and Y are independent of each other if $P(X \mid Y) = P(X)$ (and $P(Y \mid X) = P(Y)$) whenever the conditional is defined. A $Cointoss_7$ and a $Diceroll$ are an example of two marginally independent random variables. That is because knowing the value of the Cointoss$_7$ is *Heads* (or *Tails*) does not change our belief about what the value of the Diceroll is, and vice versa. Cloudy and Rainy are an example of two dependent random variables. That is because observing Cloudy = *true* (or = *false*) changes our belief about Rainy being *true* and vice versa.

Two random variables X and Y are **conditionally independent** of each other given a set $\{Z_1, \ldots, Z_m\}$ of random variables if $P(X \mid Z_1, \ldots, Z_m, Y) = P(X \mid Z_1, \ldots, Z_m)$ (and $P(Y \mid Z_1, \ldots, Z_m, X) = P(Y \mid Z_1, \ldots, Z_m)$). BloodType of two siblings (who are not identical twins) is an example of two random variables that are conditionally independent given the BloodType of their parents.

Let $X_1, \ldots, X_m$ be m random variables and $F(X_1, \ldots, X_m)$ be a function over these random variables. Let π be a permutation function over m random variables. F is **symmetric** if $F(X_1, \ldots, X_m) = F(X_{\pi(1)}, \ldots, X_{\pi(m)})$ for any permutation function π. That is, F is symmetric if its value does not depend on the order of random variables. It is clear that if F is symmetric, its value only depends on the number of random variables that are *true* (or that are *false*).

5.1.2   Finite-domain Function-free First-order Logic

The first-order logic theories that we use in this chapter correspond to the finite-domain function-free subset of first-order predicate calculus. Below we describe this subset.

For every object, we assume there exists a unique *constant* denoting that object. Constants are represented using upper-case letters. A **population** is a set of constants. A **logical variable (logvar)**, represented with a lower-case letter, is typed with a population. The population associated with a logvar x is represented as Δ_x and the cardinality of Δ_x is represented as $|\Delta_x|$. As an example, Δ_x can be $\{X_1, \ldots, X_n\}$ in which case $|\Delta_x| = n$. We use x $\in \Delta_x$ as a shorthand for instantiating x with one of the X_i. An **atom** is of the form $F(t_1, \ldots, t_k)$ where F is a predicate symbol and each t_i is a logvar or a constant. If $k = 0$, then the atom is a random variable and we drop the parenthesis. A **grounding** of an atom is obtained by replacing each of its logvars x by one of the individuals in Δ_x. A ground atom corresponds to a random variable; in the rest of the chapter, we use the terms atom and random variable interchangeably.

A **literal** is an atom or its negation. A **formula** φ is a literal, a disjunction $\varphi_1 \vee \varphi_2$ of formulae, a conjunction $\varphi_1 \wedge \varphi_2$ of formulae, or a quantified formula $\forall x \in \Delta_x : \varphi$ or $\exists x \in \Delta_x : \varphi$ where x appears in φ. For brevity, for a tuple of logvars $\mathbf{x} = \{x_1, \ldots, x_k\}$, we denote $\forall x_1 \in \Delta_{x_1}, \ldots, x_k \in \Delta_{x_k}$ with $\forall \mathbf{x} \in \Delta_{\mathbf{x}}$. A formula is **propositional** if none of its atoms contain logvars. A **weighted formula** is a pair $\langle \varphi, w \rangle$ where φ is a formula and w is a real number. An **instance** of a formula φ is obtained by replacing any free logvar x in φ with a constant X from its domain. A **sentence** is a formula with all logvars quantified. A **clause** is a disjunction of literals. A **theory** is a set of sentences. A theory is **clausal** if all its sentences are clauses. A theory is **propositional** if all its formulae are propositional. An **interpretation** (or **world**) is an assignment of values to all ground atoms in a theory. We

	Cloudy	Rainy	$\eta(\varphi_1, I)$	$\eta(\varphi_2, I)$	$\eta(\varphi_3, I)$	$\prod_{\langle \varphi_i, w_i \rangle \in \psi} \exp(w_i * \eta(\varphi_i, I))$	Probability
I_1	T	T	1	1	1	$\exp(1.1 + 1.4 + 2.0) \approx 90.0$	0.615
I_2	T	F	1	0	1	$\exp(1.1 + 2) \approx 22.2$	0.152
I_3	F	T	0	1	0	$\exp(1.4) \approx 4.1$	0.028
I_4	F	F	0	1	1	$\exp(1.4 + 2.0) \approx 30.0$	0.205
	-	-	-	-	-	$Z = 146.2$	-

Figure 5.1: Normalization constant (Z) and joint probabilities for the log-linear model in Example 5.1. φ_1 corresponds to Cloudy, φ_2 corresponds to Cloudy $\implies$ Rainy, and φ_3 corresponds to ¬Cloudy $\implies$ ¬Rainy. T represents *true* and F represents *false*.

use the terms interpretation and world interchangeably. For a formula φ and interpretation I, $\eta(\varphi, I)$ is the number of instances of φ that are *true* given the value assignments in I. If φ is propositional, $\eta(\varphi, I)$ is either 0 or 1. A **model** of a theory T is an interpretation I such that any sentence in T evaluates to *true* given the value assignments in I. We use $I \models T$ to denote that I is a model of T.

5.1.3   Log-linear Models

We describe log-linear models as an example of a propositional probabilistic model.

A **log-linear model** (Bishop et al., 2007) defining the joint probability distribution of a set of random variables $\{V_1, V_2, \ldots, V_m\}$ consists of a set of weighted propositional formulae $\psi = \{\langle \varphi_1, w_1 \rangle, \ldots, \langle \varphi_k, w_k \rangle\}$, where the variables of the formulae come from $\{V_1, V_2, \ldots, V_m\}$, and induces the following probability distribution:

$$Prob_\psi(I) = \frac{1}{Z} \prod_{\langle \varphi_i, w_i \rangle \in \psi} \exp(w_i * \eta(\varphi_i, I)) \qquad (5.1)$$

$$= \frac{1}{Z} \exp(\sum_{\langle \varphi_i, w_i \rangle \in \psi} w_i * \eta(\varphi_i, I))$$

where I is an interpretation/world assigning truth values to all random variables in $\{V_1, V_2, \ldots, V_m\}$ and $Z = \sum_{I'} \prod_{\langle \varphi_i, w_i \rangle \in \psi} \exp(w_i * \eta(\varphi_i, I')))$ is a normalization constant. Note that since φ_is are propositional, $\eta(\varphi_i, I)$ is either 0 or 1.

Example 5.1 Consider the log-linear model over variables {Cloudy, Rain} with the weighted formulae:

$$\psi = \{\langle \text{Cloudy}, 1.1 \rangle, \langle \text{Cloudy} \implies \text{Rain}, 1.4 \rangle, \langle \neg\text{Cloudy} \implies \neg\text{Rain}, 2.0 \rangle\}$$

For this log-linear model, there are four possible joint assignment of truth values to the two random variables. Fig 5.1 represents these four joint assignments. For each assignment I, $\eta(\text{Cloudy}, I)$ and $\eta(\text{Cloudy} \implies \text{Rainy}, I)$, etc. have been calculated in the middle columns. Then $\prod_{\langle \varphi_i, w_i \rangle \in \psi} \exp(w_i * \eta(\varphi_i, I))$ has been calculated in the second-to-right col-

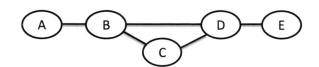

Figure 5.2: Dependency graph for the log-linear model in Example 5.2

umn. The sum of these values gives the normalization constant Z. The joint probability of each assignment can be calculated by dividing the value in the second-to-right column by Z, as shown in the rightmost column.

Log-linear models are closed under evidence. That is, if the value of a random variable in a log-linear model is observed, replacing the random variable with its observed value results in a new log-linear model.

Note that log-linear models are a restricted case of undirected models, where the joint is a product of factors. In particular, log-linear models cannot represent any models where any assignments have zero probability. Log-linear models can represent any positive distribution (no zero probabilities).

5.1.4 Graphical Notation of Dependencies in Log-linear Models

The dependencies (and independencies) for a log-linear model can be represented as a graph. The nodes in this graph correspond to the random variables of the log-linear model, and there is an edge between two nodes if the corresponding random variables appear in at least one weighted formula together. Once the graph is constructed, the **Markov blanket** for a random variable V is defined as the set of random variables that are directly connected to V in the graph. V is conditionally independent of all random variables in the log-linear model given its Markov blanket.

Example 5.2 Consider a log-linear model with the following weighted formulae: $\langle A \vee B, w_1 \rangle$, $\langle B \vee C \vee D, w_2 \rangle$ and $\langle D \vee E, w_3 \rangle$. The graphical representation of dependencies for this log-linear model is in Fig 5.2. From the figure, it can be inferred that the Markov blanket for C is {B, D}. Therefore, given B and D, C is conditionally independent of A and E.

5.1.5 Markov Logic Networks

Markov logic networks (MLNs) (Domingos et al., 2008) are relational probabilistic models that extend log-linear models to allow more general weighted formulae. In particular, they allow for weighted finite-domain, function-free, first-order logic formulae. Given a set of weighted formulae ψ and an interpretation/world I, the probability distribution induced by an MLN is defined exactly as in Equation 5.1. While in log-linear models $\eta(\varphi, I)$ always evaluates to either 0 or 1 (because the formulae are propositional), in MLNs, $\eta(\varphi, I)$ can be more than 1 as it corresponds to the number of instances of φ that hold according to I.

Example 5.3 Let $\Delta_x = \{Alice, Bob\}$ and consider an MLN having two weighted formulae as follows:

$$\psi = \{\langle \mathsf{Smokes(x)} \wedge \mathsf{Friend(x, y)} \implies \mathsf{Smokes(y)}, 1.1 \rangle,$$
$$\langle \mathsf{Smokes(x)} \implies \mathsf{Cancer(x)}, 1.5 \rangle\}$$

There are 8 ground atoms in this MLN: two corresponding to smoking, two corresponding to cancer, and four corresponding to friendships. Since all ground atoms are binary, there are 2^8 possible assignments of truth values for these ground atoms. In order to calculate the normalization constant Z of this MLN, for any assignment I' of these 2^8 possible assignments, we have to evaluate $\prod_{\langle \varphi_i, w_i \rangle \in \psi} \exp(w_i * \eta(\varphi_i, I'))$ and sum the results. Consider, for example, the interpretation where all ground atoms are assigned *true*. For this interpretation, all four instances of the first formula and both instances of the second formula hold, so the weight for this interpretation is $\exp(1.1 * 4) * \exp(1.5 * 2) \approx 1636$.

Similar to log-linear models, MLNs are closed under evidence. That is, if we observe the value of a ground atom, we can replace that ground atom with its observed value (and do some shattering operations as will be described later) and get a new MLN.

5.1.6   From Normalization Constant to Probabilistic Inference

Consider a log-linear model M (e.g., the model in Example 5.1) and a marginal inference query $Prob(\mathbf{Q} \mid \mathbf{E})$ (e.g., $Prob(\mathsf{Cloudy} = true \mid \mathsf{Rain} = false)$ for the model in Example 5.1). In order to compute this query, we construct two log-linear models M_1 and M_2. M_1 is identical to M except that any random variable in $\mathbf{E}$ is replaced by its observed value. M_2 is identical to M except that any random variable in $\mathbf{E}$ or $\mathbf{Q}$ is replaced by its observed or queried value respectively. Then $Prob(\mathbf{Q} \mid \mathbf{E}) = \frac{Z(M_2)}{Z(M_1)}$, where $Z(M_1)$ and $Z(M_2)$ represent the normalization constant of M_1 and M_2 respectively.

The same procedure can be employed to answer marginal inference queries for many other (relational) probabilistic models such as Markov networks, Bayesian networks, factor graphs, parfactor graphs (Poole, 2003), MLNs, etc. According to this procedure, in order to compute marginal inference queries on these models, all we need to know is how to compute the normalization constant of a given model. In the next section, we will describe weighted model counting, a procedure for efficiently calculating the partition function of several relational models.

5.2   Weighted Model Count

Instead of developing specific inference algorithms for each probabilistic model, one of the goals of AI is to develop general purpose inference algorithms that can be used by many (or ideally all) probabilistic models, thus separating the model construction and inference algorithm development. Separating these two components has several key advantages: 1- it allows for faster development of probabilistic models (since the separation obviates the need for developing new inference algorithms for each new probabilistic model), 2-any algorithmic advancement in probabilistic inference can be potentially leveraged by many probabilistic models instead of just one, 3- it provides the grounds for probabilistic inference researchers to all work on one problem and make progress much faster that each group working on a specific inference algorithm for a specific probabilistic model.

Weighted model counting (Chavira and Darwiche, 2008) is one general approach to probabilistic inference. Inference for several probabilistic models including Bayesian networks (Chavira and Darwiche, 2008), relational Bayesian networks (Chavira et al., 2006), dynamic Bayesian networks (Vlasselaer et al., 2016b), Markov networks, factor graphs (Kisa et al., 2014), log-linear models, probabilistic databases (Van den Broeck and Suciu, 2017), and probabilistic programs (Fierens et al., 2011; Holtzen et al., 2018, 2019b) can be converted into calculating weighted model counts, even for the purpose of approximate inference (Friedman and Van den Broeck, 2018). In this chapter, we use weighted model counting for probabilistic inference as it abstracts away the probabilistic model being used.

5.2.1 Formal Definition

Let T be a theory in propositional logic. The **model count (MC)** of T corresponds to the number of interpretations of T that are models of T. That is:

$$MC(T) = \sum_{I \models T} 1$$

Let $atoms(T)$ denote the atoms in theory T and $\Phi : atoms(T) \to \mathbb{R}$ and $\overline{\Phi} : atoms(T) \to \mathbb{R}$ be two functions that map each atom $V \in atoms(T)$ to a weight. These functions associate a weight with assigning *true* or *false* to V. For an interpretation I of T, let $I^{true} = \{V \in atoms(T) \mid I \models V\}$ be the set of atoms assigned *true*, and $I^{false} = \{V \in atoms(T) \mid I \models \neg V\}$ the atoms assigned *false*. The **weight** of I is:

$$\omega(I) = \prod_{V \in I^{true}} \Phi(V) \cdot \prod_{V \in I^{false}} \overline{\Phi}(V)$$

The **weighted model count (WMC)** of T for weight functions Φ and $\overline{\Phi}$ is the sum of the weights of the models of T. That is:

$$\text{WMC}(T, \Phi, \overline{\Phi}) = \sum_{I \models T} \omega(I)$$

Example 5.4 Consider a propositional theory T with two sentences: $A \vee B$ and $B \vee \neg C$. Fig 5.3 represents different assignments of truth values to A, B and C, whether each assignment is a model of the theory or not, and the model count of the theory.

Let $\Phi(A) = 0.1$, $\Phi(B) = 0.9$ and $\Phi(C) = 2.0$, and $\overline{\Phi}(A) = 1.0$, $\overline{\Phi}(B) = 1.1$ and $\overline{\Phi}(C) = 5.0$. Fig 5.3 also shows the weight of each model and the WMC of the theory with weight functions Φ and $\overline{\Phi}$.

	A	B	C	A ∨ B	B ∨ ¬C	Model?	Weight
I_1	*T*	*T*	*T*	*T*	*T*	Yes	0.1 * 0.9 * 2.0 = 0.18
I_2	*T*	*T*	*F*	*T*	*T*	Yes	0.1 * 0.9 * 5.0 = 0.45
I_3	*T*	*F*	*T*	*T*	*F*	No	-
I_4	*T*	*F*	*F*	*T*	*T*	Yes	0.1 * 1.1 * 5.0 = 0.55
I_5	*F*	*T*	*T*	*T*	*T*	Yes	1.0 * 0.9 * 2.0 = 1.8
I_6	*F*	*T*	*F*	*T*	*T*	Yes	1.0 * 0.9 * 5.0 = 4.5
I_7	*F*	*F*	*T*	*F*	*F*	No	-
I_8	*F*	*F*	*F*	*F*	*T*	No	-
-	-	-	-	-	-	$MC(T) = 5$	$WMC = 7.48$

Figure 5.3: Model count and weighted model count for the theory in Example 5.4. *T* represents *true* and *F* represents *false*.

5.2.2   Converting Inference into Calculating WMCs

Inference for many probabilistic models can be converted into calculating WMCs. Here we show the conversion for a log-linear model (Bishop et al., 2007). For many other probabilistic models (e.g., Bayesian networks, Markov networks, and factor graphs), the conversion can be done similarly.

As explained earlier, in order to compute marginal inference queries on a log-linear model M, all we need to calculate is the normalization constant Z of M. Suppose M has k weighted formulae $\{\langle \varphi_1, w_k \rangle, \langle \varphi_2, w_k \rangle, \ldots, \langle \varphi_k, w_k \rangle\}$. Then $Z(M)$ is equal to $\mathsf{WMC}(T, \Phi, \overline{\Phi})$ for theory T and weight functions Φ and $\overline{\Phi}$ constructed as follows. For every weighted formula $\langle \varphi_i, w_i \rangle$ in M, T contains a sentence $\mathsf{AUX}_i \iff \varphi_i$ where AUX_i is an auxiliary atom not in M. $\Phi(AUX_i) = \exp(w_i)$, $\overline{\Phi}(AUX_i) = 1$, and Φ and $\overline{\Phi}$ of all other atoms are 1.

Note that this construction increases the number of interpretations, but not the number of models, because, for each assignment to the original atoms, there is only one consistent assignment of truth values of the auxiliary atoms.

Example 5.5 Consider a log-linear model $\{\langle \mathsf{A} \vee \mathsf{B}, 2.0 \rangle, \langle \mathsf{A} \vee \neg \mathsf{C} \vee \mathsf{D}, 1.1 \rangle\}$. The normalization constant of this model is equal to the WMC of a theory with sentences $\mathsf{X}_1 \iff \mathsf{A} \vee \mathsf{B}, \mathsf{X}_2 \iff \mathsf{A} \vee \neg \mathsf{C} \vee \mathsf{D}$ where $\Phi(\mathsf{X}_1) = \exp(2.0)$, $\Phi(\mathsf{X}_2) = \exp(1.1)$, Φ of the other atoms is 1, and $\overline{\Phi}$ of all atoms is 1.

5.2.3   Efficiently Finding the WMC of a Theory

To describe how WMC of a theory can be computed, we assume in the rest of this chapter that the input theory is clausal. If it is not, it can be converted into a clausal theory by distributing or by using the Tseitin transformation (Tseitin, 1968). We can then treat a clausal theory as a set of sets of literals, and use set notation for operations.

Algorithm 1 WMC($T, \Phi, \overline{\Phi}$)

Input: A clausal theory T and two weight functions Φ and $\overline{\Phi}$
Output: Weighted model count of T with weights Φ and $\overline{\Phi}$
 1: **if** $T = \{\}$ **then**
 2:     **return** 1
 3: **if** $\{false\} \in T$ **then**
 4:     **return** 0
 5: **if** $T \in Cache$ **then**
 6:     **return** $Cache[T]$
 7: **if** $\exists \delta \in T$ such that $\delta = \cdots \vee true \vee \ldots$ **then**
 8:     $FAs = atoms(T) \setminus atoms(T \setminus \{\delta\})$
 9:     **return** WMC($T \setminus \{\delta\}, \Phi, \overline{\Phi}$) $* \prod_{V \in FAs}(\Phi(V) + \overline{\Phi}(V))$
10: **if** $\exists \delta \in T$ such that $\delta = \cdots \vee false \vee \ldots$ **then**
11:     **return** WMC($T \setminus \{\delta\} \cup \{\delta \setminus \{false\}\}, \Phi, \overline{\Phi}$)
12: **if** $\exists \delta \in T$ such that $\delta = \neg A$ **then**
13:     **return** $\overline{\Phi}(A) *$ WMC($T_{A/false}, \Phi, \overline{\Phi}$)
14: **if** $\exists \delta \in T$ such that $\delta = A$ **then**
15:     **return** $\Phi(A) *$ WMC($T_{A/true}, \Phi, \overline{\Phi}$)
16: **if** $|CC = connected\_components(T)| > 1$ **then**
17:     **return** $\prod_{cc \in CC}$ WMC($cc, \Phi, \overline{\Phi}$)
18: Select $A \in atoms(T)$
19: $Cache[T] = \Phi(A) *$ WMC($T_{A/true}, \Phi, \overline{\Phi}$) $+ \overline{\Phi}(A) *$ WMC($T_{A/false}, \Phi, \overline{\Phi}$)
20: **return** $Cache[T]$

Suppose we would like to find WMC($T, \Phi, \overline{\Phi}$) for some theory T and weight functions Φ and $\overline{\Phi}$. The naive approach is to enumerate all possible interpretations for T, determine if they are models of T, find the weight of the models according to Φ and $\overline{\Phi}$, and sum the weights. This approach, however, does not take advantage of the independencies and other structures inherent in the theory. Instead, we find the WMC of the theory by applying some rules to the theory that do take advantage of the inherent structure of the theory. Algorithm 1 demonstrates how WMC of a clausal theory can be computed. We describe the rules used in this algorithm on an example theory.

As a running example, consider a theory T with sentences:

$$A \vee B \vee C$$
$$C \vee D \vee E \vee F$$
$$C \vee \neg D$$

And consider two weight functions Φ and $\overline{\Phi}$.

Case analysis: (Lines 18 and 19 of Algorithm 1) This rule selects an atom A and constructs two theories from T: the theory T_1 in which A is replaced with *true* (denoted as $T_{A/true}$) and the theory T_2 in which A is replaced with *false* (denoted as $T_{A/false}$). Then $\text{WMC}(T, \Phi, \overline{\Phi}) = \Phi(A) * \text{WMC}(T_1, \Phi, \overline{\Phi}) + \overline{\Phi}(A) * \text{WMC}(T_2, \Phi, \overline{\Phi})$.

In the running example, let us select C for case analysis. Then T_1 will be:

$$A \vee B \vee true$$
$$true \vee D \vee E \vee F$$
$$true \vee \neg D$$

and T_2 will be:

$$A \vee B \vee false$$
$$false \vee D \vee E \vee F$$
$$false \vee \neg D$$

Note that neither contain C.

Simplifying and Forgetting: (Lines 7-11) When atoms *true* and *false* are introduced into a theory (as in T_1 and T_2 above), one can simplify the sentences. For instance for T_1, all sentences evaluate to *true* and in T_2, *false*s can be removed from the sentences as they have no effect on the truth of the sentences. When simplifying sentences, one has to be careful about the forgetting effect. In T_1, if we remove all sentences from the theory (as they all evaluate to *true*), then we forget about all 5 atoms. That is, the resulting theory T_3 (having no sentences) has 5 atoms fewer then the unsimplified theory.

To see how we can account for these forgotten atoms, let us apply case-analysis on T_1 for one of the atoms to be forgotten, e.g., A. Let T_{11} represent the theory T_1 in which A has been assigned *true* and T_{12} represent the theory T_1 in which A has been assigned *false*. According to the case-analysis rule, $\text{WMC}(T_1, \Phi, \overline{\Phi}) = \Phi(A) * \text{WMC}(T_{11}, \Phi, \overline{\Phi}) + \overline{\Phi}(A) * \text{WMC}(T_{12}, \Phi, \overline{\Phi})$. Both T_{11} and T_{12} have the same WMC, so we can rewrite the case-analysis rule as $\text{WMC}(T_1, \Phi, \overline{\Phi}) = (\Phi(A) + \overline{\Phi}(A)) * \text{WMC}(T_{11}, \Phi, \overline{\Phi})$ (or we could alternatively use T_{12} instead of T_{11}). $\text{WMC}(T_{11}, \Phi, \overline{\Phi})$ shows the WMC of the theory T_1 where A has been forgotten, and $(\Phi(A) + \overline{\Phi}(A))$ shows how to account for forgetting A.

Suppose δ is the clause to be removed. *FAs*, the set of forgotten atoms, is *FAs* $= atoms(T) \setminus atoms(T \setminus \{\delta\})$. With the above explanation, we can write $\text{WMC}(T_1, \Phi, \overline{\Phi}) = \text{WMC}(T_3, \Phi, \overline{\Phi}) * \prod_{V \in FAs}(\Phi(V) + \overline{\Phi}(V))$.

Empty theory: (Lines 1-2) If a theory has no sentences (like T_3), its WMC is 1.

Contradiction: (Lines 3-4) If one of the clauses is *false*, the theory has no model and so its WMC is 0.

Unit propagation: (Lines 12-15) Consider the theory T_2. As explained, we can remove all *false*s from the formulae resulting in theory T_4 as follows:

$$\mathsf{A} \vee \mathsf{B}$$
$$\mathsf{D} \vee \mathsf{E} \vee \mathsf{F}$$
$$\neg \mathsf{D}$$

According to the last sentence, in any model of theory T_4, D must be assigned *false*. The last sentence is called a *unit clause*. *Unit propagation* is a rule that propagates the effect of unit clauses throughout the theory. For T_4, for instance, applying *unit propagation* on the last sentence replaces D with *false* throughout the theory resulting in theory T_5:

$$\mathsf{A} \vee \mathsf{B}$$
$$\textit{false} \vee \mathsf{E} \vee \mathsf{F}$$
$$\textit{true}$$

where $\mathsf{WMC}(T_4, \Phi, \overline{\Phi}) = \overline{\Phi}(D) * \mathsf{WMC}(T_5, \Phi, \overline{\Phi})$. T_5 will be simplified to T_6:

$$\mathsf{A} \vee \mathsf{B}$$
$$\mathsf{E} \vee \mathsf{F}$$

Decomposition: (Lines 16-17) When the theory has disconnected components, they can be solved separately, and the model counts multiplied. In the running example, the sentences in T_6 can be grouped into two sub-theories T_7 and T_8 (T_7 containing only the first sentence and T_8 containing only the second sentence) such that the atoms in the two sub-theories are mutually exclusive. In such cases, $\mathsf{WMC}(T_6, \Phi, \overline{\Phi}) = \mathsf{WMC}(T_7, \Phi, \overline{\Phi}) * \mathsf{WMC}(T_8, \Phi, \overline{\Phi})$.

Caching: (Lines 5-6) Suppose we follow applying our rules until we find the WMC of a sub-theory T_i that we generated in our recursive procedure. We keep this value in a cache so that if we happened to need the WMC of the same theory T_i somewhere else, we avoid computing it again by reusing the value stored in the cache.

5.2.4   Order of Applying Rules

One of the important steps in Algorithm 1 is the choice of the next atom in line 18 for applying case analysis on it. It is possible that one order may result in linear (in the number of atoms) time complexity for a model, whereas another order may result in exponential time complexity.

Finding the optimal order for atoms is NP-complete (Arnborg et al., 1987; Koller and Friedman, 2009), so in practice, heuristics are used to choose the order. Among these heuristics, min-fill, min-size (aka min-neighbours) and min-weight are known to produce good orderings (Sato and Tinney, 1963; Kjaerulff, 1990; Dechter, 2003; Darwiche, 2009). Other methods in the literature to choose an order include maximum cardinality search (Tarjan and Mihalis, 1984), LEX M (Rose et al., 1976) and MCS-M (Berry et al., 2004). There are also heuristics with local search methods such as simulated annealing (Kjaerulff, 1990), genetic algorithm (Larranaga et al., 1997) and tabu search (Clautiaux et al., 2004).

Figure 5.4: A simple example taken from Van den Broeck (2015) indicating the effects of not exploiting exchangeability.

5.3   Lifting Search-Based Inference

Consider the following example by Van den Broeck (2015) indicating why WMC should be extended with the ability to exploit symmetry.

Example 5.6   Consider a deck of cards where all 52 cards are faced down on the ground as in Fig 5.4(a), and assume we need to answer queries such as the probability of the first card being the *queen of hearts (QofH)*, or the probability of the second card being a *hearts* given that the first card is the *QofH*. The answer to these two queries are obviously $\frac{1}{52}$ and $\frac{12}{51}$.

Suppose we decide to build a log-linear model for answering such queries automatically, in which we consider one variable for each card whose domain is the set of all suit-number pairs, and impose a probability distribution on these variables. The dependency graph for this model is a fully connected graph as in Fig 5.4(b). That is, no pair of variables in this model are marginally or conditionally independent of each other. Using traditional WMC engines, reasoning for this example requires on the order of 52^{52} computations.

The reason why us humans can answer such queries more easily than the traditional reasoning algorithms lies in the notion of symmetry. In Fig 5.4(a), the probability of the first random variable (card) being the *QofH* is the same as the probability of the second, third, or any other variable (card) being the *QofH*. Swapping/renaming two random variables in Fig 5.4(b) will result in the exact same probability distribution. That is, the random variables are pair-wise symmetric. When a group of random variables are pair-wise symmetric, one can exploit this symmetry to speed up reasoning by treating the whole group identically rather than treating each member of the group separately. Symmetry appears in many relational models; objects about which we have the same information can be assumed symmetric. Reasoning algorithms must be extended with the capability to exploit symmetry to answer such queries in polynomial time.

The aim of lifted inference is to develop probabilistic inference algorithms that not only exploit independencies among random variables, but also symmetry. We describe lifted inference through weighted model counting over first-order theories (Van den Broeck, 2013), as inference for several (relational) models with symmetry can be converted to weighted model counting, and the use of weighted model counting abstracts away the (relational) model being used. Furthermore, a recent comparative study (Riguzzi et al., 2017) shows the superiority of weighted model counting formulation compared to some other approaches (Milch et al., 2008; Kisynski and Poole, 2009; Gomes and Costa, 2012; Bellodi et al., 2014).

Examples of models for which inference can be converted into calculating WMCs over first-order theories include Markov logic networks (Richardson and Domingos, 2006; Gogate and Domingos, 2011; Van den Broeck et al., 2011), parfactor graphs (Poole, 2003), probabilistic logic programs (De Raedt et al., 2007; Fierens et al., 2015; Meert et al., 2014; Vlasselaer et al., 2016a), and probabilistic databases (Suciu et al., 2011; Gribkoff et al., 2014a,b; Van den Broeck and Suciu, 2017). First-order model counting is also the inference framework that underlies open-world probabilistic databases (Ceylan et al., 2016; Friedman and Van den Broeck, 2019; Grohe and Lindner, 2019; Borgwardt et al., 2017), scalable probabilistic rule learners (Jain et al., 2019), hashing-based samplers (van Bremen and Kuzelka, 2020) and techniques for querying relational embedding models (Friedman and Van den Broeck, 2020). Moreover, recent hybrid approaches to probabilistic-logical inference perform Weighted Model Integration (WMI) (Belle et al., 2015a,b) – a form of weighted model counting on logical sentences in SAT Modulo Theories (SMT), which is a specific form of first-order logic. Solvers for the WMI task either enumerate models (Dos Martires et al., 2019; Morettin et al., 2019), perform search-based inference (Zeng and Van den Broeck, 2019), or a form of lifted variable elimination (de Salvo Braz et al., 2016). Here, however, we will focus on the WMC task for classical first-order logic.

5.3.1 Weighted Model Counting for First-order Theories

Let $\mathcal{F}(T)$ be the set of predicate symbols in theory T, and $\Phi : \mathcal{F}(T) \to \mathbb{R}$ and $\overline{\Phi} : \mathcal{F}(T) \to \mathbb{R}$ be two functions that map each predicate F to weights. Let I^{True} and I^{False} represent the set of ground atoms assigned *true* and *false* respectively for an interpretation I of T. The weight of I is given by:

$$\omega(I) = \prod_{F(C_1,...,C_k) \in I^{True}} \Phi(F) \cdot \prod_{F(C_1,...,C_{k'}) \in I^{False}} \overline{\Phi}(F)$$

Given a first-order theory T and two functions Φ and $\overline{\Phi}$, the **weighted model count** **(WMC)** of the theory is: $\mathsf{WMC}(T, \Phi, \overline{\Phi}) = \sum_{I \models T} \omega(I)$. It is also sometimes called the weighted first-order model count (WFOMC) or lifted WMC (LWMC).

In this chapter, we assume that all first-order theories are clausal. If they are not, they can be converted into clausal theories by distributing or by using the Tseitin transformation (Tseitin, 1968; Meert et al., 2016). Please see Section 6.4 for a detailed description of Skolemization. When a clause mentions two logvars x_1 and x_2 with the same population

	$H(A)$	$H(B)$	$D(A)$	$D(B)$	$\neg H(A) \vee D(A)$	$\neg H(B) \vee D(B)$	Model?	Weight
I_1	T	T	T	T	T	T	Yes	0.2 * 0.2 * 0.8 * 0.8
I_2	T	T	T	F	T	F	No	-
I_3	T	T	F	T	T	T	Yes	0.2 * 0.2 * 1.2 * 0.8
I_4	T	T	F	F	T	T	Yes	0.2 * 0.2 * 1.2 * 1.2
I_5	T	F	T	T	F	T	No	-
I_6	T	F	T	F	F	F	No	-
I_7	T	F	F	T	F	T	No	-
I_8	T	F	F	F	F	T	No	-
I_9	F	T	T	T	T	T	Yes	0.5 * 0.2 * 0.8 * 0.8
I_{10}	F	T	T	F	T	F	No	-
I_{11}	F	T	F	T	T	T	Yes	0.5 * 0.2 * 1.2 * 0.8
I_{12}	F	T	F	F	T	T	Yes	0.5 * 0.2 * 1.2 * 1.2
I_{13}	F	F	T	T	T	T	Yes	0.5 * 0.5 * 0.8 * 0.8
I_{14}	F	F	T	F	T	F	No	-
I_{15}	F	F	F	T	T	T	Yes	0.5 * 0.5 * 1.2 * 0.8
I_{16}	F	F	F	F	T	T	Yes	0.5 * 0.5 * 1.2 * 1.2
-	-	-	-	-	-	-	$MC = 9$	WMC=1.1856

Figure 5.5: Model count and weighted model count for the theory in Example 5.7. *T* represents *true* and *F* represents *false*.

Δ_x, or a logvar x with population Δ_x and a constant $C \in \Delta_x$, we assume they refer to different objects.[10]

Example 5.7 Consider the theory $\forall x \in \Delta_x : \neg \mathsf{Happy(x)} \vee \mathsf{Dances(x)}$ having only one clause and assume $\Delta_x = \{A, B\}$. Fig 5.5 gives the truth assignments to the groundings of Happy and Dances, whether the truth assignment is a model or not, the weight of the models, and the (weighted) model count of the first order theory.

---

[10] This does not restrict what can be represented. To convert a theory that does not make this assumption, split a clauses with logvars x_1 and x_2 into the cases of the variables being equal (in which case x_2 can be replaced by x_1) or unequal, and similarly for a clause with a variable and a constant. This step is super-exponential (the Bell number) in the number of logical variables and constants in a clause, but this is usually small. It is also often the appropriate model, for example, in a theory about whether x likes y, we usually want to have different theories for when someone likes themselves and when someone likes someone else.

5.3.2 Converting Inference for Relational Models into Calculating WMCs of First-order Theories

For many relational models, (lifted) inference can be converted into calculating WMCs of first-order theories. As an example, consider an MLN with the weighted formulae $\langle \varphi_1, w_1 \rangle, \ldots, \langle \varphi_k, w_k \rangle$. As explained in the background section, all we need to know to do inference for an MLN is to be able to compute the normalization constant of any given MLN. We construct a theory T and two weight functions Φ and $\overline{\Phi}$ for a given MLN such that $\text{WMC}(T, \Phi, \overline{\Phi})$ corresponds to the normalization constant of the MLN. To construct such a theory, for every weighted formula $\langle \varphi_i, w_i \rangle$ of this MLN, we let theory T have a sentence $\text{AUX}_i(\mathbf{x}) \Leftrightarrow \varphi_i$ such that AUX_i is a predicate and $\mathbf{x}$ is the logvars appearing in φ_i. For the weight functions, we let $\Phi(\text{AUX}_i) = \exp(w_i)$, $\overline{\Phi}(\text{AUX}_i) = 1$, and Φ and $\overline{\Phi}$ be 1 for the other predicates. The *partition function* of the MLN is provably equal to $\text{WMC}(T, \Phi, \overline{\Phi})$ (Van den Broeck, 2013).

5.3.3 Efficiently Calculating the WMC of a First-order Theory

We develop a lifted version of WMC algorithm, called *lifted weighted model count (LWMC)* to efficiently find the WMC of a first-order logic theory. Suppose we would like to find $\text{WMC}(T, \Phi, \overline{\Phi})$ for some first-order theory T and weight functions Φ and $\overline{\Phi}$. A naive approach is to ground all atoms, enumerate all possible interpretations (assignments of truth values to ground atoms), determine if they are models of T, find the weight of the models according to Φ and $\overline{\Phi}$, and sum the weights. Not taking advantage of the independencies, symmetries, and the other structures inherent in the theory, this naive approach may only be practical for very small models over very small populations. Instead, LWMC finds the WMC of first-order theories by utilizing all the rules introduced earlier for propositional theories that take advantage of the independencies, as well as some new rules that take advantage of the symmetries. Algorithm 2 demonstrates LWMC. We describe the rules used in this algorithm on an example theory.

Consider two weight functions Φ and $\overline{\Phi}$ and the following theory T:

$$\forall x \in \Delta_x, \exists y \in \Delta_y : \neg R(x, y)$$

$$\forall x \in \Delta_x, y_1, y_2 \subset \Delta_y : R(x, y_1) \vee T(x, y_2)$$

$$\forall x_1, x_2 \in \Delta_x, y \in \Delta_y : \neg Q(x_1, y) \vee \neg Q(x_2, y)$$

$$\forall x \in \Delta_x, y_1, y_2 \in \Delta_y : \neg Q(x, y_1) \vee \neg Q(x, y_2)$$

Skolemization: (Lines 9-10 of Algorithm 2, and Algorithm 3) When a theory contains existential quantifiers, almost all current state-of-the-art algorithms (with one exception that will be explained later) remove these quantifiers in a preprocessing step using a technique called Skolemization (Van den Broeck et al., 2014). The idea behind this technique is to convert the input theory and the two weight functions into new theory and weight functions with equal LWMC, such that the new theory has no existential quantifiers. The following theorem describes this conversion. We refer to Van den Broeck et al. (2014); Beame et al. (2015) for the proof and details, or to Section 6.4 for a detailed discussion.

Theorem 5.1 *Let* Φ *and* $\overline{\Phi}$ *be two weight functions, T be a theory, and* δ *be a sentence in T where* δ *has a subexpression of the form* $\exists x, \varphi(\mathbf{y_1}, x, \mathbf{y_2})$. *Let* $Z(\mathbf{y_1}, \mathbf{y_2})$ *and* $S(\mathbf{y_1}, \mathbf{y_2})$ *be two auxiliary atoms (i.e.* Z *and* S *are not in T). Let T' be a theory having all sentences in T except* δ, *plus four more sentences: one sentence* δ' *obtained by replacing the subexpression* $\exists x, \varphi(\mathbf{y_1}, x, \mathbf{y_2})$ *in* δ *with* $Z(\mathbf{y_1}, \mathbf{y_2})$, *and the following three sentences:*

$$\forall \mathbf{y_1} \in \Delta_{\mathbf{y_1}}, x \in \Delta_x, \mathbf{y_2} \in \Delta_{\mathbf{y_2}} : Z(\mathbf{y_1}, \mathbf{y_2}) \vee \neg\varphi(\mathbf{y_1}, x, \mathbf{y_2})$$

$$\forall \mathbf{y_1} \in \Delta_{\mathbf{y_1}}, \mathbf{y_2} \in \Delta_{\mathbf{y_2}} : Z(\mathbf{y_1}, \mathbf{y_2}) \vee S(\mathbf{y_1}, \mathbf{y_2})$$

$$\forall \mathbf{y_1} \in \Delta_{\mathbf{y_1}}, x \in \Delta_x, \mathbf{y_2} \in \Delta_{\mathbf{y_2}} : S(\mathbf{y_1}, \mathbf{y_2}) \vee \neg\varphi(\mathbf{y_1}, x, \mathbf{y_2})$$

Let Φ' *and* $\overline{\Phi'}$ *be two weight functions identical to* Φ *and* $\overline{\Phi}$ *except that* $\Phi'(Z) = \Phi'(S) = \overline{\Phi'}(Z) = 1$ *and* $\overline{\Phi'}(S) = -1$. *Then* $\mathrm{WMC}(T, \Phi, \overline{\Phi}) = \mathrm{WMC}(T', \Phi', \overline{\Phi'})$ *and T' has one less existential quantifier than T.*

All existential quantifiers of a theory can be removed by repeatedly applying the above theorem. Consider theory T_1 in our running example:

$$\forall x \in \Delta_x, \exists y \in \Delta_y : \neg R(x, y)$$

$$\forall x \in \Delta_x, y \in \Delta_y : R(x, y) \vee T(x, y)$$

The first sentence has an existential quantifier. Applying the conversion in Theorem 5.1, we convert T_1 into following theory T_3:

$$\forall x \in \Delta_x : Z(x)$$

$$\forall x \in \Delta_x, y \in \Delta_y : Z(x) \vee R(x, y)$$

$$\forall x \in \Delta_x : Z(x) \vee S(x)$$

$$\forall x \in \Delta_x, y \in \Delta_y : S(x) \vee R(x, y)$$

$$\forall x \in \Delta_x, y \in \Delta_y : R(x, y) \vee T(x, y)$$

We also construct Φ' and $\overline{\Phi'}$ by adding $\Phi'(Z) = \Phi'(S) = \overline{\Phi'}(Z) = 1$ and $\overline{\Phi'}(S) = -1$ to Φ and $\overline{\Phi}$. Then we know that $\mathrm{LWMC}(T_1, \Phi, \overline{\Phi}) = \mathrm{LWMC}(T_3, \Phi', \overline{\Phi'})$.

First-order Unit Propagation: (Lines 11-14 of Algorithm 2) Unit propagation can be used when a sentence in the theory contains a single atom (i.e. a sentence is a unit clause). In T_3, the first sentence is a unit clause. Applying unit propagation sets $Z(x)$ to *true* throughout the theory resulting in T_4:

$$\forall x \in \Delta_x, y \in \Delta_y : S(x) \vee R(x, y)$$

$$\forall x \in \Delta_x, y \in \Delta_y : R(x, y) \vee T(x, y)$$

$\text{LWMC}(T_3, \Phi', \overline{\Phi'}) = \Phi'(Z)^{|\Delta_x|} * \text{LWMC}(T_4, \Phi', \overline{\Phi'})$ as unit propagation has set Z to *true* for $|\Delta_x|$ ground atoms.

Decomposition: (Lines 15-16 of Algorithm 2) As mentioned earlier, all the rules we introduced for calculating the WMC of a propositional theory can be easily extended and used for first-order theories. As an example, in T, the first two sentences mention a totally separate set of (ground) atoms (or random variables) than the second two lines. Therefore, $\text{LWMC}(T, \Phi, \overline{\Phi}) = \text{LWMC}(T_1, \Phi, \Phi) * \text{LWMC}(T_2, \Phi, \overline{\Phi})$ where T_1 is a theory having the first two sentences in T and T_2 is a theory having the second two sentences in T. Decomposition corresponds to the *9th if statement* in Algorithm 2.

Lifted Decomposition: (Lines 17-18 of Algorithm 2) Assume we ground x in T_4. Considering $|\Delta_x| = n$, we get the following sentences:

$$\forall y \in \Delta_y : S(X_1) \vee R(X_1, y)$$
$$\forall y \in \Delta_y : R(X_1, y) \vee T(X_1, y)$$
$$\cdots$$
$$\forall y \in \Delta_y : S(X_i) \vee R(X_i, y)$$
$$\forall y \in \Delta_y : R(X_i, y) \vee T(X_i, y)$$
$$\cdots$$
$$\forall y \in \Delta_y : S(X_n) \vee R(X_n, y)$$
$$\forall y \in \Delta_y : R(X_n, y) \vee T(X_n, y)$$

It can be viewed that the clauses mentioning an arbitrary $X_i \in \Delta_x$ are totally disconnected from clauses mentioning $X_j \in \Delta_x$ ($j \neq i$), and are the same up to renaming X_i to X_j. Given the symmetry of the objects, we know for every i and j, the WMC of the clauses mentioning X_i is the same as the WMC of the clauses mentioning X_j. Thus, we can calculate the WMC of only the clauses mentioning some X_i and raise the result to the power of the number of connected components ($|\Delta_x|$). x is called a *decomposer* of the theory T_4. Assuming T_5 is the theory that results from substituting x with X_i (aka decomposing T_4 on x) as follows:

$$\forall y \subset \Delta_y : S(X_i) \vee R(X_i, y)$$
$$\forall y \in \Delta_y : R(X_i, y) \vee T(X_i, y)$$

$\text{LWMC}(T_4, \Phi', \overline{\Phi'}) = \text{LWMC}(T_5, \Phi', \overline{\Phi'})^{|\Delta_x|}$.

Lifted decomposition corresponds to the lines 15-16 of Algorithm 2. Lifted decomposition was first introduced by Poole (2003) in his first-order variable elimination algorithm and was used in the later extensions of this algorithm (de Salvo Braz et al., 2005; Milch et al., 2008; Taghipour et al., 2013c). With the advent of search-based lifted inference algorithms (Poole et al., 2011; Jha et al., 2010; Gogate and Domingos, 2010; Van den

Broeck et al., 2011), this rule was adapted and used in all these works. Variants of lifted decomposition rule have been used for MAP inference (de Salvo Braz et al., 2006; Sarkhel et al., 2014) and for asymmetric lifted inference in probabilistic databases (Dalvi and Suciu, 2012; Suciu et al., 2011). More recently, a variant of lifted decomposition has been also used for query answering in open-world probabilistic databases (Ceylan et al., 2016; Friedman and Van den Broeck, 2019). Examining the applicability of this rule to lifted marginal MAP inference is currently an open problem.

Lifted Case Analysis: (Lines 23-28 of Algorithm 2) Case analysis can be done for atoms having one logvar in a lifted way. Consider the $R(X_i, y)$ in T_5. Applying case analysis naively on this atom makes $2^{|\Delta_y|}$ calls to the LWMC function. Assuming $|\Delta_y| = m$, the first call receives the following theory $T_{5,1}$ as input:

$$\neg R(X_i, Y_1)$$
$$\ldots$$
$$\neg R(X_i, Y_m)$$
The sentences in T_5

The second call receives the following theory $T_{5,2}$ as input:

$$R(X_i, Y_1)$$
$$\neg R(X_i, Y_2)$$
$$\ldots$$
$$\neg R(X_i, Y_m)$$
The sentences in T_5

The third call receives the following theory $T_{5,3}$ as input:

$$\neg R(X_i, Y_1)$$
$$R(X_i, Y_2)$$
$$\neg R(X_i, Y_3)$$
$$\ldots$$
$$\neg R(X_i, Y_m)$$
The sentences in T_5

And the last call receives the following theory $T_{5,2^m}$ as input:

$$R(X_i, Y_1)$$
$$\ldots$$
$$R(X_i, Y_m)$$
The sentences in T_5

Consider the theories $T_{5,2}$ and $T_{5,3}$. The only difference in these two theories is that in $T_{5,2}$, $R(X_i, Y_1)$ is *true* and in $T_{5,3}$, $R(X_i, Y_2)$ is *true*. Due to the symmetry of the objects, we know that renaming two objects does not change the WMC. In $T_{5,2}$, renaming Y_1 to Y_2 and Y_2 to Y_1 gives us $T_{5,3}$, therefore $T_{5,2}$ and $T_{5,3}$ have the same WMC. We can extend this example and see that for two theories $T_{5,i}$ and $T_{5,j}$ in which for the same number of objects $Y \in \mathsf{y}$ we have $\mathsf{R}(X_i, Y)$ being assigned *true* and the same number assigned *false*, the WMC is the same.

According to this observation, we do not have to consider all $2^{|\Delta_y|}$ possible truth assignments to all ground atoms of $\mathsf{R}(X_i, \mathsf{y})$, but only the ones where the number of objects $Y \in \Delta_y$ for which $\mathsf{R}(X_i, Y)$ is *true* (or equivalently *false*) is different. This means considering $|\Lambda_y| + 1$ cases suffices, the jth case corresponding to $\mathsf{R}(X_i, Y)$ being *true* for exactly j out of $|\Delta_y|$ objects. Considering $|\Delta_y| + 1$ instead of $2^{|\Delta_y|}$ cases as described is in accordance with WMC being a symmetric function. Note that we must multiply by $\binom{|\Delta_y|}{j}$ to account for the number of ways one can select j out of $|\Delta_y|$ objects. Let Δ_{y_T} and Δ_{y_F} be the population of objects $Y \in \Delta_y$ for which $\mathsf{R}(X_i, Y)$ is assigned *true* and *false* respectively in lifted case analysis, and let $T_{5,j}$ be as follows:

$$\forall \mathsf{y_T} \in \Delta_{y_T} : \mathsf{R}(X_i, \mathsf{y_T})$$
$$\forall \mathsf{y_F} \in \Delta_{y_F} : \mathsf{R}(X_i, \mathsf{y_F})$$
$$\forall \mathsf{y} \in \Delta_y : \mathsf{S}(X_i) \vee \mathsf{R}(X_i, \mathsf{y})$$
$$\forall \mathsf{y} \in \Delta_y : \mathsf{R}(X_i, \mathsf{y}) \vee \mathsf{T}(X_i, \mathsf{y})$$

Then:

$$\mathsf{LWMC}(T_5, \Phi', \overline{\Phi'}) = \Phi(\mathsf{R})^j * \overline{\Phi}(\mathsf{R})^{|\Delta_x|-j} * \sum_{j=0}^{|\Delta_y|} \binom{|\Delta_y|}{j} \mathsf{LWMC}(T_{5,j}, \Phi', \overline{\Phi'})$$

Lifted case-analysis corresponds to the last *if statement* of Algorithm 2. It was first identified by de Salvo Braz et al. (2005), then extended by Milch et al. (2008), and then used in search-based lifted inference algorithms. Similar to lifted decomposition, variants of this rule have been used for MAP inference. Extending this rule for lifted marginal MAP inference is an open problem.

Shattering: In $T_{5,j}$, the objects in Δ_y are no longer symmetric: we know different things about those in Δ_{y_T} and those in Δ_{y_F}. We need to shatter the population Δ_y (and the clauses correspondingly) to the symmetric sub-populations. Shattering $T_{5,j}$ gives the following theory:

$$\forall y_T \in \Delta_{y_T} : R(X_i, y_T)$$
$$\forall y_F \in \Delta_{y_F} : R(X_i, y_F)$$
$$\forall y_T \in \Delta_{y_T} : S(X_i) \vee R(X_i, y_T)$$
$$\forall y_F \in \Delta_{y_F} : S(X_i) \vee R(X_i, y_F)$$
$$\forall y_T \in \Delta_{y_T} : R(X_i, y_T) \vee T(X_i, y_T)$$
$$\forall y_F \in \Delta_{y_F} : R(X_i, y_F) \vee T(X_i, y_F)$$

In the above theory, all populations are symmetric. The WMC of the above theory can be computed by a unit propagation on $R(X_i, y_T)$ and $R(X_i, y_F)$, applying decomposition, and then another unit propagation on each of the connected components.

Shattering can be done in two ways: 1- **shattering up-front:** shattering a theory as soon as a population is no longer symmetric, 2- **shattering as needed:** postponing the shattering until we can no longer apply a rule to the theory without shattering it. It has been shown that shattering as needed may perform better than shattering up-front (Kisynski and Poole, 2009).

Domain Recursion: (Lines 29-29 of Algorithm 2) Let us now consider theory T_2, the second connected component when we applied decomposition, which has the following sentences:

$$\forall x_1, x_2 \in \Delta_x, y \in \Delta_y : \neg Q(x_1, y) \vee \neg Q(x_2, y)$$
$$\forall x \in \Delta_x, y_1, y_2 \in \Delta_y : \neg Q(x, y_1) \vee \neg Q(x, y_2)$$

None of the rules we introduced so far can be applied to the above theory. In such cases, domain recursion grounds one object X from a population and applies the rules introduced so far to simplify the resulting theory and remove X. If after simplifying the theory, we get into the initial theory again (but with one or more populations reduced in size), domain recursion is *bounded* and WMC of the theory can be computed efficiently.

For T_2, let us ground one object $X \in \Delta_x$. Let $\Delta_{x'} = \Delta_x - \{X\}$. Then we get the following theory:

$$\forall x_2' \in \Delta_{x'}, y \in \Delta_y : \neg Q(X, y) \vee \neg Q(x_2', y)$$
$$\forall x_1' \in \Delta_{x'}, y \in \Delta_y : \neg Q(x_1', y) \vee \neg Q(X, y)$$
$$\forall x_1', x_2' \in \Delta_{x'}, y \in \Delta_y : \neg Q(x_1', y) \vee \neg Q(x_2', y)$$
$$\forall y_1, y_2 \in \Delta_y : \neg Q(X, y_1) \vee \neg Q(X, y_2)$$
$$\forall x' \in \Delta_{x'}, y_1, y_2 \in \Delta_y : \neg Q(x', y_1) \vee \neg Q(x', y_2)$$

Now we can apply lifted case analysis on $Q(X, y)$. We consider only the *j*th case here. Let Δ_{y_T} and Δ_{y_F} represent the population of objects $Y \in \Delta_y$ for which $Q(X, Y)$ is *true* and *false* respectively. Then we need to find the WMC of the following theory:

$$\forall y \in \Delta_{y_T} : Q(X, y_T)$$
$$\forall y \in \Delta_{y_F} : \neg Q(X, y_F)$$
$$\forall x_2' \in \Delta_{x'}, y \in \Delta_y : \neg Q(X, y) \vee \neg Q(x_2, y)$$
$$\forall x_1' \in \Delta_{x'}, y \in \Delta_y : \neg Q(x_1', y) \vee \neg Q(X, y)$$
$$\forall x_1', x_2' \in \Delta_{x'}, y \in \Delta_y : \neg Q(x_1', y) \vee \neg Q(x_2', y)$$
$$\forall y_1, y_2 \in \Delta_y : \neg Q(X, y_1) \vee \neg Q(X, y_2)$$
$$\forall x' \in \Delta_{x'}, y_1, y_2 \in \Delta_y : \neg Q(x', y_1) \vee \neg Q(x', y_2)$$

The third and the fourth clauses of the above theory are identical and we can remove one of them. Shatter the population of Δ_y and applying unit propagation on the first two sentences gives:

$$\forall x_2' \in \Delta_{x'}, y_T \in \Delta_{y_T} : \neg Q(x_2', y_T)$$
$$\forall x_1', x_2' \in \Delta_{x'}, y_T \in \Delta_{y_T} : \neg Q(x_1', y_T) \vee \neg Q(x_2', y_T)$$
$$\forall x_1', x_2' \in \Delta_{x'}, y_F \in \Delta_{y_F} : \neg Q(x_1', y_F) \vee \neg Q(x_2', y_F)$$
$$\forall x' \in \Delta_{x'}, y_{1T}, y_{2T} \in \Delta_{y_T} : \neg Q(x', y_{1T}) \vee \neg Q(x', y_{2T})$$
$$\forall x' \in \Delta_{x'}, y_{1T} \in \Delta_{y_T}, y_{2F} \in \Delta_{y_F} : \neg Q(x', y_{1T}) \vee \neg Q(x', y_{2F})$$
$$\forall x' \in \Delta_{x'}, y_{1F} \in \Delta_{y_F}, y_{2T} \in \Delta_{y_T} : \neg Q(x', y_{1F}) \vee \neg Q(x', y_{2T})$$
$$\forall x' \in \Delta_{x'}, y_{1F}, y_{2F} \in \Delta_{y_F} : \neg Q(x', y_{1F}) \vee \neg Q(x', y_{2F})$$

Applying unit propagation on the first clause gives the following theory:

$$\forall x_1', x_2' \in \Delta_{x'}, y_F \in \Delta_{y_F} : \neg Q(x_1', y_F) \vee \neg Q(x_2', y_F)$$
$$\forall x' \in \Delta_{x'}, y_{1F}, y_{2F} \in \Delta_{y_F} : \neg Q(x', y_{1F}) \vee \neg Q(x', y_{2F})$$

The above theory is identical to the initial theory T_2 that we started with, but with reduced population sizes. Therefore, domain recursion is bounded and by keeping the WMCs of sub-theories in a cache, the WMC of the above theory can be computed efficiently.

Domain recursion corresponds to the last four lines of Algorithm 2. It was first introduced by Van den Broeck (2011) and was central for efficient lifted inference in a class of models. However, later work showed that for all those models, simpler rules suffice (Taghipour et al., 2013c). This result caused the domain recursion rule to be ignored for several years, until Kazemi et al. (2016) revived it again by showing that it is actually more powerful than expected. Kazemi et al. (2016) identified several classes of theories for which lifted inference was only efficient when domain recursion was used. They also showed that while S4 (a theory left as an open problem by Beame et al. (2015), asking for a new inference rule) and symmetric transitivity theories fell outside their identified classes, domain recursion is bounded for them and inference for both of them becomes efficient when applying domain recursion on them. Identifying whether domain recursion is bounded for a theory or not without applying/simulating it on the theory is still an open problem.

Recently, domain recursion has been also considered as a substitution for Skolemization to deal with existential quantifiers. Kazemi et al. (2017) identify two theories with existential quantifier to which bounded domain recursion can be applied directly without Skolemization. They identify two advantages for applying bounded domain recursion directly on these theories without Skolemization: (i) potentially reducing the time complexity of inference, (ii) obviating the need for dealing with negative numbers in the weight functions that may be inconvenient for implementation purposes. Characterizing the classes of theories with existential quantifiers for which domain recursion is bounded is still an open problem.

5.3.4   Order of Applying Rules

Similar to WMC for propositional theories, the cost of WMC for first-order theories highly depends on the order of atoms on which (lifted) case-analysis is applied. Kazemi and Poole (2014) extended several heuristics that were proposed for finding a good order for propositional theories to the first-order theories. They showed that for first-order theories, an extension of min-size (aka min-neighbour) heuristic outperforms several other heuristics. A problem with Kazemi and Poole (2014)'s heuristics is that they are designed for finding an order for lifted inference algorithms based on variable elimination. Then the inverse of these orders is used for search-based approaches. Designing heuristics that find orderings directly for search-based approaches is an interesting direction for future research.

5.3.5   Bounding the Complexity of Lifted Inference

It is theoretically interesting and practically useful to know the upper-bound, lower-bound, or the exact cost of lifted inference for a theory before running lifted inference on it. Such bounds exist for propositional theories. Smith and Gogate (2015) take the first steps towards finding an upper-bound on the cost of lifted inference for a theory. With the new advancements in lifted inference (e.g., the new rules that have been identified) one direction for future research would be to extend Smith and Gogate (2015)'s work by considering these new advancements.

Algorithm 2 LWMC($T, \Phi, \overline{\Phi}$)

Input: A clausal first-order theory T and two weight functions Φ and $\overline{\Phi}$.
Output: LWMC($T, \Phi, \overline{\Phi}$).

 1: **if** $T = \{\}$ **then return** 1
 2: **if** $\{false\} \in T$ **then return** 0
 3: **if** $T \in Cache$ **then return** $Cache[T]$
 4: **if** $\exists \delta \in T$ such that $\delta = \cdots \vee true \vee \ldots$ **then**
 5:     $FAs = atoms(T) \setminus atoms(T \setminus \{\delta\})$
 6:     **return** LWMC($T - \{\delta\}, \Phi, \overline{\Phi}$) $* \prod_{V(\mathbf{x}) \in FAs}(\Phi(V) + \overline{\Phi}(V))^{\prod_{\mathsf{x} \in \mathbf{x}} |\Delta_\mathsf{x}|}$
 7: **if** $\exists \delta \in T$ such that $\delta = \cdots \vee false \vee \ldots$ **then**
 8:     **return** LWMC($T \setminus \{\delta\} \cup \{\delta \setminus \{false\}\}, \Phi, \overline{\Phi}$)
 9: **if** $\exists \delta \in T$ such that δ has a subexpression of the form $\exists \mathsf{x}, \varphi(\mathbf{y}_1, \mathsf{x}, \mathbf{y}_2)$ **then**
10:     **return** Skolemize(T, δ, x) // See Algorithm 3
11: **if** $\exists \{\neg V(\mathbf{x})\} \in T$ **then**
12:     **return** $\overline{\Phi}(V)^{\prod_{\mathsf{x} \in \mathbf{x}} |\Delta_\mathsf{x}|} *$ LWMC($T_{V(\mathbf{x})/false}, \Phi, \overline{\Phi}$)
13: **if** $\exists \{V(\mathbf{x})\} \in T$ **then**
14:     **return** $\Phi(V)^{\prod_{\mathsf{x} \in \mathbf{x}} |\Delta_\mathsf{x}|} *$ LWMC($T_{V(\mathbf{x})/true}, \Phi, \overline{\Phi}$).
15: **if** $|CC = connected\_components(T)| > 1$ **then**
16:     **return** $\prod_{cc \in CC}$ LWMC($cc, \Phi, \overline{\Phi}$)
17: **if** $\exists \mathsf{x}$ such that x is a decomposer of T **then**
18:     **return** LWMC($decompose(\mathrm{T}, \mathsf{x}), \Phi, \overline{\Phi}$)$^{|\Delta_\mathsf{x}|}$
19: **if** there are predicates in T that have no logvars **then**
20:     Select V from the predicates in T such that V has no logvars.
21:     $Cache[T] = \Phi(V) *$ LWMC($T_{V/true}, \Phi, \overline{\Phi}$) $+ \overline{\Phi}(V) *$ LWMC($T_{V/false}, \Phi, \overline{\Phi}$)
22:     **return** $Cache[T]$
23: **if** there are predicates in T that have exactly one logvar **then**
24:     Select V from the predicates in T such that V only has one logvar x.
25:     Let $\Delta_x = \Delta_{x_T^j} \cup \Delta_{x_F^j}$ such that $\Delta_{x_T^j} \cap \Delta_{x_F^j} = \emptyset$, $|\Delta_{x_T^j}| = j$ and $|\Delta_{x_F^j}| = \Delta_x - j$
26:     Let T_j be T with two more sentences: $\forall x_T^j \in \Delta_{x_T^j} : V(x_T^j)$ and $\forall x_F^j \in \Delta_{x_F^j} : \neg V(x_F^j)$
27:     $Cache[T] = \sum_{j=0}^{|\Delta_x|} \Phi(V)^j * \overline{\Phi}(V)^{|\Delta_x|-j} * \binom{|\Delta_x|}{j} *$ LWMC($T_j, \Phi, \overline{\Phi}$)
28:     **return** $Cache[T]$
29: Select logvar x in T and object X from Δ_x, and let $\Delta_{x'} = \Delta_x - \{X\}$.
30: Let T' be equivalent to T except that Δ_x is broken into two sub-populations $\{X\}$ and $\Delta_{x'}$.
31: **return** LWMC($T', \Phi, \overline{\Phi}$)

Algorithm 3 Skolemize(T, δ, x)

Input: A clausal first-order theory T having a sentence δ containing an existentially quantified logvar x.

Output: *Skolemize*(T, δ, x).

1: Create auxiliary atoms $\mathsf{Z}(\mathbf{y_1}, \mathbf{y_2})$ and $\mathsf{S}(\mathbf{y_1}, \mathbf{y_2})$.
2: Let T' be a theory having all sentences in T except δ.
3: Replace $\exists \mathsf{x}, \varphi(\mathbf{y_1}, \mathsf{x}, \mathbf{y_2})$ in δ with $\mathsf{Z}(\mathbf{y_1}, \mathbf{y_2})$ and add it to T'.
4: Add $\forall \mathbf{y_1} \in \Delta_{\mathbf{y_1}}, \mathsf{x} \in \Delta_x, \mathbf{y_2} \in \Delta_{\mathbf{y_2}} : \mathsf{Z}(\mathbf{y_1}, \mathbf{y_2}) \vee \neg\varphi(\mathbf{y_1}, \mathsf{x}, \mathbf{y_2})$ to T'
5: Add $\forall \mathbf{y_1} \in \Delta_{\mathbf{y_1}}, \mathbf{y_2} \in \Delta_{\mathbf{y_2}} : \mathsf{Z}(\mathbf{y_1}, \mathbf{y_2}) \vee \mathsf{S}(\mathbf{y_1}, \mathbf{y_2})$ to T'
6: Add $\forall \mathbf{y_1} \in \Delta_{\mathbf{y_1}}, \mathsf{x} \in \Delta_x, \mathbf{y_2} \in \Delta_{\mathbf{y_2}} : \mathsf{S}(\mathbf{y_1}, \mathbf{y_2}) \vee \neg\varphi(\mathbf{y_1}, \mathsf{x}, \mathbf{y_2})$ to T'
7: Let Φ' be equivalent to Φ except that $\Phi(\mathsf{Z}) = \Phi(\mathsf{S}) = 1$.
8: Let $\overline{\Phi'}$ be equivalent to $\overline{\Phi}$ except than $\overline{\Phi'}(\mathsf{Z}) = 1$ and $\overline{\Phi'}(\mathsf{S}) = -1$.
9: **return** LWMC($T', Phi', \overline{\Phi'}$)

---

6 Lifted Aggregation and Skolemization for Directed Models

Wannes Meert, Jaesik Choi, Jacek Kisyński, Hung Bui, Guy Van den Broeck,
Adnan Darwiche, Rodrigo de Salvo Braz, and David Poole

6.1 Introduction

The idea of *lifted inference* is to carry out as much inference as possible without propositionalizing. While original work considered directed parfactor models (Poole, 2003), most later work focused on undirected parfactor models (de Salvo Braz et al., 2007; Milch et al., 2008). Recently, methods based on weighted model counting for first-order logic became a successful alternative to parfactor representations (Van den Broeck, 2013). One aspect that arises in directed models that is absent in undirected graphical models is the need for **aggregation** that occurs when a parent random variable is parameterized by logical variables that are not present in a child random variable. First-order inference algorithms described in previous chapters (e.g. C-FOVE) do not allow a description of aggregation in first-order models that is independent of the sizes of the populations. To express aggregation patterns in first-order logic both universal and existential quantification is required whereas only universal quantification is required when considering undirected models. For similar reasons that parfactor-based inference algorithms ignore aggregation, weighted first-order model counting techniques do not support existential quantification. This chapter shows how aggregation and quantifiers relate to each other and introduces approaches to lift aggregate factors (Sec. 6.2 and 6.3) and techniques to handle existential quantifiers in weighted first-order model counting (Sec. 6.4).

Aggregate factors in probabilistic relational models can compactly represent dependencies among a large number of relational random variables. Problematic is that propositional inference on a factor aggregating n k-valued random variables into an r-valued result random variable is $O(rk2^n)$, that is, those based on aggregate functions such as SUM, AVERAGE, AND, etc. Lifted methods can ameliorate this to $O(rn^k)$ in general and $O(rk \log n)$ for commutative associative aggregators.

For parfactors, we show how to extend the C-FOVE algorithm to perform lifted inference in the presence of aggregation parfactors. We also show that there are cases where the polynomial time complexity (in the domain size of logical variables) of the C-FOVE algorithm can be reduced to logarithmic time complexity using aggregation parfactors. Next, we show further optimizations that offer an exact solution constant in n when $k = 2$ for certain aggregate operations such as AND, OR and SUM, and a close approximation for inference with aggregate factors with time complexity constant in n.

For weighted first-order model counting, we introduce a Skolemization procedure to deal with existential quantification. Lifted model counters apply only to Skolem normal form theories (i.e., no existential quantifiers). Since textbook Skolemization is not sound for model counting, this restriction precludes efficient model counting for directed models with aggregation. We present a Skolemization procedure to extend the applicability of first-order model counters to representations with existential quantifiers such as probabilistic logic programming.

By providing support for aggregation on the one hand and existential quantification on the other hand we now have inference procedures that are not limited to a subset of what can be expressed in StarAI models . This increases the practical potential of lifted inference.

6.2   Lifted Aggregation in Directed First-order Probabilistic Models

One aspect that arises in directed models is the need for aggregation that occurs when a parent random variable is parameterized by logical variables that are not present in a child random variable. Currently available first-order inference algorithms do not allow a description of aggregation in first-order models that is independent of the sizes of the populations. In this section we introduce a new data structure, *aggregation parfactors*, describe how to use it to represent aggregation in first-order models, and show how to perform lifted inference in its presence.

6.2.1   Parameterized Random Variables and Parametric Factors

Like previous chapters, this chapter is not tied to any particular first-order probabilistic language. We reason at the level of data structures and assume that various first-order languages (or their subsets) will compile to these data structures. First-order probabilistic languages share a concept of a *parameterized random variable*.

Parameterized Random Variables.   If S is a set, we denote by $|S|$ the size of the set S.

A *population* is a set of *individuals*. A population corresponds to a domain in logic. For example, a population may be a set of all soccer players involved in a soccer game, where *Rossi* is one of the individuals and the population size is 22.

A *parameter* corresponds to a logical variable and is typed with a population. For example, parameter *player* may be typed with the population of all players involved in a soccer game. Given parameter A, we denote its population by Δ_A. Given a set of constraints C, we denote a set of individuals from Δ_A that satisfy constraints in C by $\Delta_A : C$.

A *substitution* is of the form $\{X_1/t_1, \ldots, X_k/t_k\}$, where the X_i are distinct parameters, and each *term* t_i is a parameter typed with a population or a constant denoting an individual from a population. A *ground substitution* is a substitution where each t_i is a constant.

A *parameterized random variable* is of the form $f(t_1, \ldots, t_k)$, where f is a functor (either a function symbol or a predicate symbol) and t_i are terms. We denote a set of parameters of the parameterized random variable $f(t_1, \ldots, t_k)$ by $\mathbb{P}(f(t_1, \ldots, t_k))$. Each functor has a set of values called the *range* of the functor. We denote the range of the functor f by *range(f)*. Examples of parameterized random variables are *inj(player)* and *inj(Rossi)*. We have $\mathbb{P}(inj(player)) = \{player\}$ and $\mathbb{P}(inj(Rossi)) = \emptyset$.

A parameterized random variable $f(t_1, \ldots, t_k)$ represents a set of random variables, one for each possible ground substitution to all of its parameters. We denote this set by *ground*$(f(t_1, \ldots, t_k))$. For example, if *player* is typed with a population consisting of all 22 individuals playing the game, then *inj(player)* represents 22 random variables: *inj(Rossi)*, *inj(Panucci)*, ..., *inj(Desailly)* corresponding to the ground substitutions $\{player/Rossi\}$, $\{player/Panucci\}$, ..., $\{player/Desailly\}$, respectively. The range of the functor of the parameterized random variable is the domain of random variables represented by the parameterized random variable.

Let **v** denote an *assignment of values* to random variables; **v** is a function that takes a random variable and returns its value. We extend **v** to also work on parameterized random variables, where we assume that free parameters are universally quantified. For example, if **v**(*inj(player)*) = *true*, then each of the random variables represented by *inj(player)*, namely *inj(Rossi)*, *inj(Panucci)*, ..., *inj(Desailly)*, is assigned the value *true* by **v**.

The idea of parameterized random variables is similar to the notion of plates (Buntine, 1994); we use plate notation in our figures.

Parametric Factors. A *factor* on a set of random variables represents a function that, given an assignment of a value to each random variable from the set, returns a real number. Factors are used in the variable elimination algorithm (Zhang and Poole, 1994) to store initial conditional probabilities and intermediate results of computation during probabilistic inference in graphical models. Operations on factors include multiplication of factors and summing out random variables from a factor.

Let **v** be an assignment of values to random variables and let $\mathcal{F}$ be a factor on a set of random variables $\mathcal{S}$. We extend **v** to factors and denote by **v**$(\mathcal{F})$ the value of the factor $\mathcal{F}$ given **v**. If **v** does not assign values to all of the variables in $\mathcal{S}$, then **v**$(\mathcal{F})$ denotes a factor on the unassigned variables of $\mathcal{F}$.

A *parametric factor* or *parfactor* is a triple $\langle C, \mathcal{V}, \mathcal{F} \rangle$ where C is a set of inequality constraints on parameters (between a parameter and a constant or between two parameters), $\mathcal{V}$ is a set of parameterized random variables and $\mathcal{F}$ is a factor from the Cartesian product of ranges of parameterized random variables in $\mathcal{V}$ to the reals.

A parfactor $\langle C, \mathcal{V}, \mathcal{F} \rangle$ represents a set of factors, one for each ground substitution G to all free parameters in $\mathcal{V}$ that satisfies constraints in C. Each such factor $\mathcal{F}_G$ is a factor on the set of random variables obtained by applying G to $\mathcal{V}$.

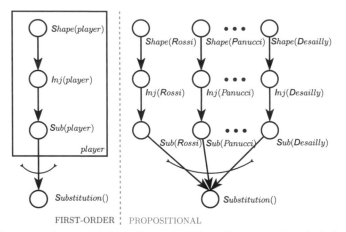

Figure 6.1: A first-order model from Example 6.1 and its equivalent belief network. Aggregation is denoted by curved arcs.

We use parfactors to represent probability distributions for parameterized random variables in first-order models and intermediate computation results during lifted inference.

6.2.2 Aggregation in Directed First-order Probabilistic Models

First-order probabilistic models describe probabilistic dependencies between parameterized random variables. A *grounding* of a first-order probabilistic model is a propositional probabilistic model obtained by replacing each parameterized random variable with the random variables it represents and replicating appropriate probability distributions.

Example 6.1 Consider the directed first-order probabilistic model and its grounding presented in Figure 6.1. The model is meant to represent that whether a player playing a soccer game is substituted or not during a single soccer game depends on whether he gets injured. The probability of an injury in turn depends on a physical condition of the player. The model has four nodes: a parameterized random variable *Shape(player)* with range {*Fit, Unfit*}, a parameterized random variable *Inj(player)* with range {*true, false*}, a parameterized random variable *Sub(player)* with range {*true, false*}, and a random variable *Substitution()* with range {*true, false*} that is *true* if a player was substituted during the game and *false* otherwise. We have $\Delta_{player} = \{Rossi, Panucci, \dots, Desailly\}$ and $|\Delta_{player}| = 22$.

A parameterized random variable *Shape(player)* represents the 22 random variables in the corresponding propositional model, as do variables *Inj(player)* and *Sub(player)*. Therefore, in the propositional model, the number of parent nodes influencing the node *Substitution()* is equal to 22. Their common effect aggregates in the child variable. In the discussed model we use the logical OR as an aggregation operator to describe the (deterministic) conditional probability distribution $\mathcal{P}(Substitution() \mid Sub(player))$. Note that *Substitution*() is a noisy-OR of *Inj(player)*.

In a directed first-order model, when a child variable has a parent variable with extra parameters, in the grounding the child variable has number of parents that varies with the size of the domain of those extra parameters. We need some **aggregation** operator to specify how the child variable depends on the parent variables no matter how many of them there are. Following Zhang and Poole (1996), we assume that the range of the parent variable is a subset of the range of the child variable, and use a commutative and associative deterministic binary operator over the range of the child variable as an aggregation operator $\otimes$.

Given probabilistic input to the parent variable, we can construct any causal independence model covered by the definition of causal independence from Zhang and Poole (1996), which in turn covers common causal independence models such as noisy-OR (Pearl, 1986) and noisy-MAX (Diez, 1993) as special cases. In other words, this allows any causal independence model to act as underlying mechanism for aggregation in directed first-order models. For other types of aggregation in first-order models, see Jaeger (2002).

In this chapter we require that the directed first-order probabilistic models satisfy the following conditions:

1. for each parameterized random variable, its parent has at most one extra parameter
2. if a parameterized random variable $C(\ldots)$ has a parent $P(\ldots, a, \ldots)$ with an extra parameter A, then:

   (a) $P(\ldots, a, \ldots)$ is the only parent of $C(\ldots)$

   (b) the range of P is a subset of the range of C

   (c) $C(\ldots)$ is a deterministic function of the parent: $C(\ldots) = P(\ldots, A_1, \ldots) \otimes \ldots \otimes p(\ldots, A_n, \ldots) = \bigotimes_{a \in \Delta_a} p(\ldots, A, \ldots)$, where $\otimes$ is a commutative and associative deterministic binary operator over the range of C.

At first the above conditions seem to be very restrictive, but they in fact are not. There is no need to define the aggregation over more than one parameter due to the associativity and commutativity of the $\otimes$ operator. We can obtain more complicated distributions by introducing auxiliary variables and combining multiple aggregations.

Example 6.2 Consider a parent variable $P(a, b, c)$ and a child variable $C(c)$. We can describe a $\otimes$-based aggregation over a and b, $C(c) = \bigotimes_{(A,B) \in \Delta_a \times \Delta_b} P(a, b, c)$ using an auxiliary parameterized random variable $C'(b, c)$ such that C' has the same range as C. Let $C'(b, c) = \bigotimes_{A \in \Delta_a} P(a, b, c)$, then $C(c) = \bigotimes_{B \in \Delta_b} C'(b, c)$.

Similarly, with the use of auxiliary nodes, we can construct a distribution that combines an aggregation with influence from other parent nodes or even combines multiple aggregations generated with different operators.

In the rest of the chapter, we assume that the discussed models satisfy conditions (1) and (2), for ease of presentation and with no loss of generality.

6.2.3   Existing Algorithm

Counting Formulas.   A *counting formula* is $\#_{a:C}[F(\ldots, a, \ldots)]$, where a is a parameter that is *bound* by the # sign, C is a set of inequality constraints involving a and $F(\ldots, a, \ldots)$ is a parameterized random variable.

The value of $\#_{a:C}[F(\ldots, a, \ldots)]$, given an assignment of values to random variables $\mathbf{v}$, is the *histogram H* that maps the range of F to natural numbers such that

$$H(X) = |\{A \in (\Delta_a : C) : \mathbf{v}(F(\ldots, A, \ldots)) = X\}|.$$

The range of such a counting formula is the set of histograms having a bucket for each element X in the range of F with entries adding up to $|\Delta_a : C|$. The number of such histograms is $\binom{|\Delta_a:C| + |range(F)| - 1}{|range(F)| - 1}$, which for small values of $|range(F)|$ is $O(|\Delta_a : C|^{|range(F)| - 1})$. Thus, any extensional representation of a function on a counting formula $\#_{a:C}[F(\ldots, a, \ldots)]$ requires amount of space at least linear in $|\Delta_a : C|$.

Counting formulas allow us to exploit interchangeability within factors. They were inspired by work on *cardinality potentials* (Gupta et al., 2007) and *counting elimination* (de Salvo Braz et al., 2007). They are a new form of parameterized random variables. Unless otherwise stated, by parameterized random variables we understand both forms: the standard, defined in Section 6.2.1, and counting formulas.

Normal-Form Constraints.   Let x be a parameter in $\mathcal{V}$ from a parfactor $\langle C, \mathcal{V}, \mathcal{F} \rangle$. In general, the size of the set $\Delta_x : C$ depends on other parameters in $\mathcal{V}$ (see discussions on *uniform solution counting partitions* in de Salvo Braz et al. (2007) and *normal form constraints* in Milch et al. (2008)).

Milch et al. (2008) introduced a special class of sets of inequality constraints. Let C be a set of inequality constraints on parameters and x be a parameter. We denote by $\mathcal{E}_x^C$ the set of terms t such that $(x \neq t) \in C$. Set C is in *normal form* if for each inequality $(x \neq y) \in C$, where x and y are parameters, $\mathcal{E}_x^C \setminus \{y\} = \mathcal{E}_y^C \setminus \{x\}$.

Consider a parfactor $\langle C, \mathcal{V}, \mathcal{F} \rangle$, where C is in normal form. For all parameters x in $\mathcal{V}$, $|\Delta_x : C| = |\Delta_x| - |\mathcal{E}_x^C|$.

We require that for a parfactor $\langle C, \mathcal{V}, \mathcal{F} \rangle$ involving counting formulas, the union of C and the constraints in all the counting formulas in $\mathcal{V}$ is in normal form. Other parfactors do not need to be in normal form.

C-FOVE.   Let Φ be a set of parfactors. Let $\mathcal{J}(\Phi)$ denote a factor equal to the product of all factors represented by elements of Φ. Let $\mathbf{u}$ be the set of all random variables represented by parameterized random variables present in parfactors in Φ. Let $\mathbf{q}$ be a subset of $\mathbf{u}$. The *marginal of $\mathcal{J}(\Phi)$ on* $\mathbf{q}$, denoted $\mathcal{J}_\mathbf{q}(\Phi)$, is defined as $\mathcal{J}_\mathbf{q}(\Phi) = \sum_{\mathbf{u} \setminus \mathbf{q}} \mathcal{J}(\Phi)$.

Given Φ and $\mathbf{q}$, the C-FOVE algorithm computes the marginal $\mathcal{J}_{\mathbf{q}}(\Phi)$ by summing out random variables from $\mathbf{u} \setminus \mathbf{q}$, in a lifted manner when possible. Evidence can be handled by adding to Φ additional parfactors on observed random variables.

As lifted summing out is only possible under certain conditions, the C-FOVE algorithm uses elimination enabling operations, such as applying substitutions to parfactors and multiplication. Below we show when and how these operations can be applied to aggregation parfactors. We refer the reader to Chapter 4 for more details on C-FOVE.

6.2.4 Incorporating Aggregation in C-FOVE

In Section 6.2.4.1, we show how to represent aggregation in first-order models using a simple form of **aggregation parfactors**. In Section 6.2.4.2, we show how these aggregation parfactors can be converted to parfactors that in turn can be used during inference with C-FOVE. In Section 6.2.4.3, we describe when and how reasoning directly in terms of these aggregation parfactors can achieve improved efficiency. In Section 6.2.4.4 we outline how a generalized version of aggregation parfactors increases the number of cases for which efficiency is improved.

6.2.4.1 Aggregation Parfactors

Example 6.3 Consider the model presented in Figure 6.1. We cannot represent the conditional probability distribution $\mathcal{P}(Substitution()|Sub(player))$ with a parfactor $\langle \emptyset, \{Sub(player), Substitution()\}, \mathcal{F} \rangle$ as even simple noisy-OR cannot be represented as a product. A parfactor $\langle \emptyset, \{Sub(Rossi), \ldots, Sub(Desailly), Substitution()\}, \mathcal{F} \rangle$ is not an adequate input representation of this distribution because its size would depend on $|\Delta_{player}|$. The same applies to a parfactor $\langle \emptyset, \{\#_{player:\emptyset}[Sub(player)], Substitution()\}, \mathcal{F} \rangle$ as the size of the range of $\#_{player:\emptyset}[Sub(player)]$ depends on $|\Delta_{player}|$.

Definition 6.1 An **aggregation parfactor** is a hex-tuple $\langle C, P(\ldots, a, \ldots), C(\ldots), \mathcal{F}_P, \otimes, C_a \rangle$, where

- $P(\ldots, a, \ldots)$ and $C(\ldots)$ are parameterized random variables
- the range of P is a subset of the range of C
- a is the only parameter in $P(\ldots, a, \ldots)$ that is not in $C(\ldots)$
- C is a set of inequality constraints not involving a
- $\mathcal{F}_P$ is a factor from the range of P to real numbers
- $\otimes$ is a commutative and associative deterministic binary operator over the range of C
- C_a is a set of inequality constraints involving a.

An aggregation parfactor $\langle C, P(\ldots, a, \ldots), C(\ldots), \mathcal{F}_P, \otimes, C_a \rangle$ represents a set of factors, one for each ground substitution $\mathcal{G}$ to parameters $\mathbb{P}(P(\ldots, a, \ldots)) \cup \mathbb{P}(C(\ldots)) \setminus \{a\}$ that satisfies constraints in C. Each factor $\mathcal{F}_{\mathcal{G}}$ is a mapping from the Cartesian product $(\times_{A \in \Delta_a : C_a} range(P)) \times range(C)$ to the reals, which, given an assignment of values to random variables $\mathbf{v}$, is defined as follows:

$$\mathbf{v}(\mathcal{F}_G) = \begin{cases} \prod_{A \in \Delta_a:C_a} \mathcal{F}_P(\mathbf{v}(P(\ldots,a,\ldots))), & \text{if } \bigotimes_{A \in \Delta_a:C_a} \mathbf{v}(P(\ldots,A,\ldots)) = \mathbf{v}(C(\ldots)); \\ 0, & \text{otherwise.} \end{cases}$$

It is important to notice that $\Delta_a : C_a$ might vary for different ground substitutions G if the set $C \cup C_a$ is not in normal form (see Section 6.2.3). The space required to represent an aggregation parfactor does not depend on the size of the set $\Delta_a : C_a$. It is also at most quadratic in the size of $range(C)$, as the operator $\otimes$ can be represented as a factor from $range(C) \times range(C)$ to $range(C)$.

When an aggregation parfactor $\langle C, P(\ldots,a,\ldots), C(\ldots), \mathcal{F}_P, \otimes, C_a \rangle$ is used to describe aggregation in a first-order model, the factor $\mathcal{F}_P$ will be a constant function with the value 1. An aggregation parfactor created during inference may have a non-trivial $\mathcal{F}_P$ component (see Section 6.2.4.3).

Example 6.4 Consider the first-order model and its grounding presented in Figure 6.1. We can represent the conditional probability distribution $P(Substitution()|Sub(player))$ with an aggregation parfactor $\langle \emptyset, Sub(player), Substitution(), \mathcal{F}_{Sub}, \text{OR}, \emptyset \rangle$, where $\mathcal{F}_{Sub}$ is a constant function with the value 1. The size of the representation does not depend on the population size of the parameter *player*.

In the rest of the chapter, Φ denotes a set of parfactors and aggregation parfactors. The notation introduced in Section 6.2.3 remains valid under the new meaning of Φ.

6.2.4.2   Conversion to Parfactors

Conversion using counting formulas.   Consider an aggregation parfactor $\langle C, P(\ldots,a,\ldots), C(\ldots), \mathcal{F}_P, \otimes, C_a \rangle$. Since $\otimes$ is an associative and commutative operator, given an assignment of values to random variables $\mathbf{v}$, it does not matter which of the variables $P(\ldots,A,\ldots)$, $A \in \Delta_a : C_a$ are assigned each value from $range(P)$, but only how many of them are assigned each value. This property was a motivation for the *counting elimination* algorithm (de Salvo Braz et al., 2007) and counting formulas (Milch et al., 2008), and allows us to convert aggregation parfactors to a product of two parfactors, where one of the parfactors involves a counting formula.

Proposition 6.1 *Let* $g_A = \langle C, P(\ldots,a,\ldots), C(\ldots), \mathcal{F}_P, \otimes, C_a \rangle$ *be an aggregation parfactor from* Φ *such that set* $C \cup C_a$ *is in normal form. Let* $\mathcal{F}_{\#}$ *be a factor from the Cartesian product* $range(\#_{a:C_a}[P(\ldots,a,\ldots)]) \times range(C)$ *to* $\{0,1\}$. *Given an assignment of values to random variables* $\mathbf{v}$, *the function is defined as follows:*

$$\mathcal{F}_{\#}(H, \mathbf{v}(C(\ldots))) = \begin{cases} 1, & \text{if } \bigotimes_{x \in range(P)} \bigotimes_{i=1}^{H(x)} x = \mathbf{v}(C(\ldots)); \\ 0, & \text{otherwise,} \end{cases}$$

where H is a histogram from $range(\#_{a:C_a}[P(\dots, a, \dots)])$. Then $\mathcal{J}(\Phi) = \mathcal{J}(\Phi \setminus \{g_a\} \cup \{\langle C \cup C_a, \{P(\dots, a, \dots)\}, \mathcal{F}_P\rangle, \langle C, \{\#_{a:C_a}[P(\dots, a, \dots)], C(\dots)\}, \mathcal{F}_\#\rangle\})$.

If the set $C \cup C_a$ is not in normal form we will need to use splitting operation described in Section 6.2.4.3 to convert the aggregation parfactor to a set of aggregation parfactors with constraint sets in normal form.

Conversion for MAX and MIN operators.   If in an aggregation parfactor $\otimes$ is the MAX operator (which includes the OR operator as a special case), we can use a factorization presented by Díez and Galán (2003) to convert the aggregation parfactor to parfactors without counting formulas. The factorization is an example of the *tensor rank-one decomposition* of a conditional probability distribution (Savicky and Vomlel, 2007).

Proposition 6.2 *Let $g_a = \langle C, P(\dots, a, \dots), C(\dots), \mathcal{F}_P, \text{MAX}, C_a\rangle$ be an aggregation parfactor from Φ, where MAX operator is induced by a total ordering $\prec$ of $range(C)$. Let $\mathbf{S}$ be a successor function induced by $\prec$. Let $C'(\dots)$ be an auxiliary parameterized random variable with the same parameterization and the same range as C. Let $\mathcal{F}_C$ be a factor from the Cartesian product $range(P) \times range(C)$ to real numbers that, given an assignment of values to random variables $\mathbf{v}$, is defined as follows:*

$$\mathcal{F}_C(\mathbf{v}(P(\dots, a, \dots)), \mathbf{v}(C'(\dots))) = \begin{cases} \mathcal{F}_P(\mathbf{v}(P(\dots, a, \dots))), & \text{if } \mathbf{v}(P(\dots, a, \dots)) \preccurlyeq \mathbf{v}(C'(\dots)); \\ 0, & \text{otherwise.} \end{cases}$$

Let $\mathcal{F}_\Delta$ be a factor from the Cartesian product $range(P) \times range(C)$ to real numbers that, given $\mathbf{v}$, is defined as follows:

$$\mathcal{F}_\Delta(\mathbf{v}(C(\dots)), \mathbf{v}(C'(\dots))) = \begin{cases} 1, & \text{if } \mathbf{v}(C(\dots)) = \mathbf{v}(C'(\dots)); \\ -1, & \text{if } \mathbf{v}(C(\dots)) = \mathbf{S}(\mathbf{v}(C'(\dots))); \\ 0, & \text{otherwise.} \end{cases}$$

Then $\mathcal{J}(\Phi) = \sum_{ground(C'(\dots))} \mathcal{J}(\Phi \setminus \{g_a\} \cup \{\langle C \cup C_a, \{P(\dots, a, \dots), C'(\dots)\}, \mathcal{F}_C\rangle, \langle C, \{C(\dots), C'(\dots)\}, \mathcal{F}_\Delta\rangle\})$.

An analogous proposition holds for the MIN operator. In both cases, as shown in Section 6.2.5, the above conversion is advantageous to the conversion described in Proposition 6.1, which uses counting formulas.

6.2.4.3   Operations on Aggregation Parfactors   In the previous section we showed that aggregation parfactors can be used during a modeling phase and then, during inference with the C-FOVE algorithm, once populations are known, aggregation parfactors can be translated to parfactors. Such a solution allows us to take advantage of the modeling properties of aggregation parfactors and C-FOVE inference capabilities. It is also possible to exploit aggregation parfactors during inference. In this section we describe operations on aggregation parfactors that can be added to the C-FOVE algorithm. These operations can delay or even avoid translation of aggregation parfactors to parfactors involving counting formulas. This in turn can result in more efficient inference, as we will see in Section 6.2.5.

Splitting.   The C-FOVE algorithm applies substitutions to parfactors to handle observations and queries and to enable the multiplication of parfactors. As this operation results in the creation of a residual parfactor, it is called splitting. Below we present how aggregation parfactors can be split on substitutions.

Proposition 6.3 *Let $g_a = \langle C, P(\ldots, a, \ldots), C(\ldots), \mathcal{F}_P, \otimes, C_a \rangle$ be an aggregation parfactor from Φ. Let $\{x/T\}$ be a substitution such that $(x \neq T) \notin C$ and $x \in \mathbb{P}(P(\ldots, a, \ldots)) \setminus \{a\}$ and term T is a constant such that $T \in \Delta_x$, or a parameter such that $T \in \mathbb{P}(P(\ldots, a, \ldots)) \setminus \{a\}$. Let $g_a[x/T]$ be a parfactor g_a with all occurrences of x replaced by term T. Then $\mathcal{J}(\Phi) = \mathcal{J}(\Phi \setminus \{g_a\} \cup \{g_a[x/T], \langle C \cup \{x \neq T\}, P(\ldots, a, \ldots), C(\ldots), \mathcal{F}_P, \otimes, C_a \rangle\})$.*

Proposition 6.3 allows us to split an aggregation parfactor on a substitution that does not involve the aggregation parameter. Below we show how to split on a substitution that involves the aggregation parameter a and a constant. Such an operation divides the individuals from $\Delta_a : C$ in two data structures: an aggregation parfactor and a standard parfactor. We have to make sure that, after splitting, $C(\ldots)$ is still equal to a $\otimes$-based aggregation over the whole $\Delta_a : C$.

Proposition 6.4 *Let $g_a = \langle C, P(\ldots, a, \ldots), C(\ldots), \mathcal{F}_P, \otimes, C_a \rangle$ be an aggregation parfactor from Φ. Let $\{a/T\}$ be a substitution such that $(a \neq T) \notin C_a$ and term T is a constant such that $T \in \Delta_a$, or a parameter such that $T \in \mathbb{P}(P(\ldots, a, \ldots)) \setminus \{a\}$. Let $C'(\ldots)$ be an auxiliary parameterized random variable with the same parameterization and range as $C(\ldots)$. Let $C_a[a/T]$ be a set of constraints C_a with all occurrences of a replaced by term T. Let $\mathcal{F}_1$ be a factor from the Cartesian product $range(P) \times range(C') \times range(C)$ to real numbers. Given an assignment of values to random variables $\mathbf{v}$, the function is defined as follows:*

$$\mathcal{F}_1(\mathbf{v}(P(\ldots, a, \ldots)), \mathbf{v}(C'(\ldots)), \mathbf{v}(C(\ldots))) =$$
$$\begin{cases} \mathcal{F}_P(P(\ldots, T, \ldots)), & \text{if } \mathbf{v}(P(\ldots, T, \ldots)) \otimes \mathbf{v}(C'(\ldots)) = \mathbf{v}(C(\ldots)); \\ 0, & \text{otherwise.} \end{cases}$$

Then $\mathcal{J}(\Phi) = \sum_{ground(C'(\ldots))} \mathcal{J}(\Phi \setminus \{g_a\} \cup \{\langle C, P(\ldots, a, \ldots), C'(\ldots), \mathcal{F}_P, \otimes, C_a \cup \{a \neq T\} \rangle,$ $\langle C \cup C_a[a/T], \{P(\ldots, T, \ldots), C'(\ldots), C(\ldots)\}, \mathcal{F}_1 \rangle\})$.

Splitting presented in Proposition 6.4 corresponds to the expansion of a counting formula in C-FOVE. The case where a substitution is of the form $\{x/a\}$ can be handled in a similar fashion as described in Proposition 6.4. If a substitution has more than one element, then we split recursively on its elements using the above propositions.

Multiplication.   The C-FOVE algorithm multiplies parfactors to enable elimination of parameterized random variables. An aggregation parfactor can be multiplied by a parfactor on $P(\ldots, a, \ldots)$.

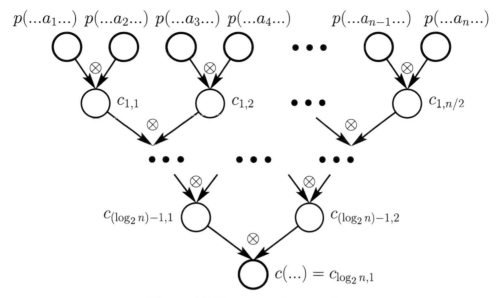

Figure 6.2: Decomposed aggregation.

Proposition 6.5 *Let $g_a = \langle C, P(\ldots, a, \ldots), C(\ldots), \mathcal{F}_P, \otimes, C_a \rangle$ be an aggregation parfactor from Φ and $g_1 = \langle C_1, \{P(\ldots, a, \ldots)\}, \mathcal{F}_1 \rangle$ be a parfactor from Φ such that $C_1 = C \cup C_a$. Let $g_2 = \langle C, P(\ldots, a, \ldots), C(\ldots), \mathcal{F}_P \mathcal{F}_1, \otimes, C_a \rangle$. Then $\mathcal{J}(\Phi) = \mathcal{J}(\Phi \setminus \{g_a, g_1\} \cup \{g_2\})$.*

We call g_2 the product of g_a and g_1.

Summing out.   The C-FOVE algorithm sums out random variables to compute the marginal. Below we show how in some cases we can sum out $P(\ldots, a, \ldots)$ directly from an aggregation parfactor.

   When $P(\ldots, a, \ldots)$ represents a set of random variables that can be treated as independent, aggregation decomposes into a binary tree of applications of the aggregation operator. Figure 6.2 illustrates this for a case where $n = |\Delta_a : C_a|$ is a power of two. The results at each level of the tree are identical, therefore we need to compute them only once. In the general case, covered by Proposition 6.6, we use a *square-and-multiply* method (Pingala, 200 BC), whose time complexity is logarithmic in $|\Delta_a : C_a|$, to eliminate $P(\ldots, a, \ldots)$ from an aggregation parfactor.

Proposition 6.6 *Let $g_a = \langle C, P(\ldots, a, \ldots), C(\ldots), \mathcal{F}_P, \otimes, C_a \rangle$ be an aggregation parfactor from Φ. Assume that $\mathbb{P}(C(\ldots)) = \mathbb{P}(P(\ldots, a, \ldots)) \setminus \{a\}$ and that set $C \cup C_a$ is in normal form. Assume that no other parfactor or aggregation parfactor in Φ involves parameterized random variables that represent random variables from $ground(P(\ldots, a, \ldots))$. Let $m = \lfloor \log_2 |\Delta_a : C_a| \rfloor$ and $B_m \ldots B_0$ be the binary representation of $|\Delta_a : C_a|$. Let $(\mathcal{F}_0, \ldots, \mathcal{F}_m)$ be* *a sequence of factors from range of C to the reals, defined recursively as follows:*

$$\mathcal{F}_0(x) = \begin{cases} \mathcal{F}_P(x), & \text{if } x \in range(P); \\ 0, & \text{otherwise,} \end{cases}$$

$$\mathcal{F}_k(x) = \begin{cases} \sum_{\substack{y,z \in range(C) \\ y \otimes z = x}} \mathcal{F}_{k-1}(y)\,\mathcal{F}_{k-1}(z), & \text{if } B_{m-k} = 0; \\ \sum_{\substack{w,y,z \in range(C) \\ w \otimes y \otimes z = x}} \mathcal{F}_P(w)\,\mathcal{F}_{k-1}(y)\,\mathcal{F}_{k-1}(z), & \text{otherwise.} \end{cases}$$

Then $\sum_{ground(P(\ldots,a,\ldots))} \mathcal{J}(\Phi) = \mathcal{J}(\Phi \setminus \{g_A\} \cup \{\langle C, \{C(\ldots)\}, \mathcal{F}_m\rangle\}).$

Proposition 6.6 does not allow variable $C(\ldots)$ in an aggregation parfactor to have extra parameters that are not present in variable $P(\ldots, a, \ldots)$. The C-FOVE algorithm handles extra parameters by introducing counting formulas on these parameters. Then it can proceed with standard summation. We cannot apply the same approach to aggregation parfactors as newly created counting formulas could have ranges incompatible with the range of the aggregation operator. We need a special summation procedure, described below in Proposition 6.7.

Proposition 6.7 *Let* $g_a = \langle C, P(\ldots, a, \ldots), C(\ldots, e, \ldots), \mathcal{F}_P, \otimes, C_a\rangle$ *be an aggregation parfactor from* Φ. *Assume that* $\mathbb{P}(C(\ldots)) \setminus \{e\} = \mathbb{P}(P(\ldots, a, \ldots)) \setminus \{a\}$ *and that set* $C \cup C_a$ *is in normal form. Assume that no other parfactor or aggregation parfactor in* Φ *involves parameterized random variables that represent random variables from* $ground(P(\ldots, a, \ldots))$. *Let* $m = \lfloor \log_2 |\Delta_a : C_a| \rfloor$ *and* $B_m \ldots B_0$ *be the binary representation of* $|\Delta_a : C_a|$. *Let* $(\mathcal{F}_0, \ldots, \mathcal{F}_m)$ *be a sequence of factors from range of C to real numbers, defined recursively as follows:*

$$\mathcal{F}_0(x) = \begin{cases} \mathcal{F}_P(x), & \text{if } x \in range(P); \\ 0, & \text{otherwise,} \end{cases}$$

$$\mathcal{F}_k(x) = \begin{cases} \sum_{\substack{y,z \in range(C) \\ y \otimes z = x}} \mathcal{F}_{k-1}(y)\,\mathcal{F}_{k-1}(z), & \text{if } B_{m-k} = 0; \\ \sum_{\substack{w,y,z \in range(C) \\ w \otimes y \otimes z = x}} \mathcal{F}_P(w)\,\mathcal{F}_{k-1}(y)\,\mathcal{F}_{k-1}(z), & \text{otherwise.} \end{cases}$$

Let C_e *be a set of constraints from C that involve e. Let* $\mathcal{F}_\#$ *be a factor from the range of counting formula* $\#_{e:C_e}[C(\ldots, e, \ldots)]$ *to real numbers defined as follows:*

$$\mathcal{F}_\#(H) = \begin{cases} \mathcal{F}_m(x), & \text{if } \exists x \in range(C) \ H(x) = |\Delta_e : C_e|; \\ 0, & \text{otherwise.} \end{cases}$$

Then $\sum_{ground(P(\ldots,a,\ldots))} \mathcal{J}(\Phi) = \mathcal{J}(\Phi \setminus \{g_A\} \cup \{\langle C \setminus C_e, \{\#_{e:C_e}[c(\ldots, e, \ldots)]\}, \mathcal{F}_\#\rangle\}).$

The above proposition can be generalized to the cases where $C(\ldots)$ has more than one extra parameter.

If set $C \cup C_a$ is not in normal form, then $|\Delta_a : C_a|$ might vary for different ground substitutions to parameters in $P(\ldots, a, \ldots)$ and we will not be able to apply Propositions 6.6 and 6.7. We can follow Milch et al. (2008) and bring constraints in the aggregation parfactor to a normal form by splitting it on appropriate substitutions. Once the constraints are in normal form, $|\Delta_a : C_a|$ does not change for different ground substitutions. The other approach is to compute *uniform solution counting partitions* (de Salvo Braz et al., 2007) using a constraint solver and use this information when summing out $P(\ldots, a, \ldots)$.

6.2.4.4 Generalized Aggregation Parfactors Propositions 6.6 and 6.7 require that random variables represented by $P(\ldots, a, \ldots)$ are independent. They are only dependent if they either have a common ancestor in the grounding or a common observed descendant. If during inference we eliminate the common ancestor or condition on the observed descendant before we eliminate $P(\ldots, a, \ldots)$ through aggregation, we may introduce a counting formula on $P(\ldots, a, \ldots)$. This would prevent us from applying results of Propositions 6.6 and 6.7 and performing efficient lifted aggregation.

We need to delay such conditioning and summing out until we eliminate $P(\ldots, a, \ldots)$. It requires a generalized version of the aggregation parfactor data structure, a septuple $\langle C, P(\ldots, a, \ldots), C(\ldots, e, \ldots), \mathcal{V}, \mathcal{F}_{P \cup \mathcal{V}}, \otimes, C_a \rangle$ where $\mathcal{V}$ is a set of *context parameterized random variables* and $\mathcal{F}_{P \cup \mathcal{V}}$ is a factor from the Cartesian product of ranges of parameterized random variables in $\{P(\ldots, a, \ldots)\} \cup \mathcal{V}$ to the reals. The factor $\mathcal{F}_{P \cup \mathcal{V}}$ stores the dependency between $P(\ldots, a, \ldots)$ and context variables.

Generalization of propositions from Sections 6.2.4.2 and 6.2.4.3 is straightforward. Proposition 6.5 has to be generalized so aggregation parfactors can be multiplied by parfactors on variables other than $P(\ldots, a, \ldots)$, and generalized versions of Propositions 6.6 and 6.7 have to manipulate larger factors.

The third experiment from Section 6.2.5 involves inference with generalized aggregation parfactors.

6.2.5 Experiments

In our experiments, we investigated how the population size of parameters and the size of the range of parameterized random variables affect inference in the presence of aggregation.

We compared the performance of variable elimination (VE), variable elimination with the noisy-MAX factorization (Díez and Galán, 2003) (VE-FCT), C-FOVE, C-FOVE with the lifted noisy-MAX factorization described in Section 6.2.4.2 (C-FOVE-FCT), and C-FOVE with aggregation parfactors (AC-FOVE). We used Java implementations of the above algorithms on an Intel Core 2 Duo 2.66 GHz processor with 1GB of memory made available to the JVM.

In the first experiment, we tested the above algorithms on the model introduced in Example 6.1 and depicted in Figure 6.1. In the second experiment, we used a modified version

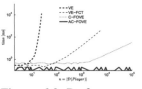

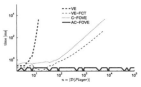

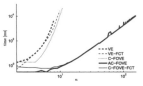

Figure 6.3: Performance on the model with noisy-OR aggregation.

Figure 6.4: Performance on the model with noisy-MAX aggregation.

Figure 6.5: Performance on the smoking-friendship model.

of this model, in which $range(Inj) = range(Sub) = range(Substitution) = \{0, 1, 2\}$, and *Substitution*() is a noisy-MAX of *Inj*(*player*). In both experiments, we measured the time necessary to compute the marginal of the variable *Substitution*() using a top-down elimination ordering. We varied the population size n of the parameter *players* from 1 to 100, 000.

Figures 6.3 and 6.4 show the results of the experiments. The time complexity for VE is exponential in n, and the algorithm did not scale as n increased. The time complexity for VE-FCT is linear in n. In the model with noisy-OR, the time complexity for C-FOVE is also linear in n, but C-FOVE does lifted inference and achieved better results than VE-FCT, which performs inference at the propositional level. In the model with noisy-MAX, the time complexity for C-FOVE is quadratic in n, and C-FOVE was outperformed by VE-FCT. C-FOVE-FCT and AC-FOVE, for which the time complexity is logarithmic in n, performed best in both cases (for clarity we do not show the C-FOVE-FCT performance in Figures 6.3 and 6.4). The difference between their performance and the performance of C-FOVE was apparent even for small populations in the second experiment, which involved aggregation over non-binary random variables.

For the third experiment we used an ICL theory (Poole, 2003) from Carbonetto et al. (2009) that explains how people alter their smoking habits within their social network. Parameters of the model were learned from data of smoking and drug habits among teenagers attending a school in Scotland (Pearson and Michell, 2000) using methods described by Carbonetto et al. (2009). Given the population size n, the equivalent propositional graphical model has $3n^2 + n$ nodes and $12n^2 - 9n$ arcs. We varied n from 2 to 140 and for each value, we computed a marginal probability of a single individual being a smoker. Figure 6.5 shows the results of the experiment. VE, VE-FCT and C-FOVE algorithms failed to solve instances with a population size greater than 8, 10, and 11, respectively. AC-FOVE was able to handle efficiently much larger instances and it ran out of memory for a population size of 159. The AC-FOVE algorithm performed equally to the C-FOVE-FCT algorithm except for small populations. It is important to remember that the C-FOVE-FCT algorithm, unlike AC-FOVE, can only be applied to MAX and MIN-based aggregation.

6.3 Efficient Methods for Lifted Inference with Aggregation Parfactors

In this section, we present efficient lifted inference algorithms for models with aggregate operators. Aggregation factors (that is, those based on aggregate functions such as *SUM, AVERAGE, AND* etc) in probabilistic relational models can compactly represent dependencies among a large number of relational random variables. However, propositional inference on a factor aggregating n k-valued random variables into an r-valued result random variable is $O(rk2^n)$. Lifted methods can ameliorate this to $O(rn^k)$ in general and $O(rk \log n)$ for commutative associative aggregators.

We present (a) an *exact* solution *constant* in n when $k = 2$ for certain aggregate operations such as *AND, OR* and *SUM*, and (b) a close approximation for inference with aggregation factors with time complexity *constant* in n. This approximate inference involves an analytical solution for some operations when $k > 2$. The approximation is based on the fact that the typically used aggregate functions can be represented by linear constraints in the standard $(k - 1)$-simplex in $\mathbb{R}^k$ where k is the number of possible values for random variables. This includes even aggregate functions that are commutative but not associative (e.g., the *MODE* operator that chooses the most frequent value). Our algorithm takes polynomial time in k (which is only 2 for binary variables) regardless of r and n, and the error decreases as n increases. Therefore, for most applications (in which a close approximation suffices) our algorithm is a much more efficient solution than previous ones, as shown in Section 6.3.5. We also describe a (c) third method which further optimizes aggregations over multiple groups of random variables with distinct distributions.

This section only considers *deterministic* aggregation factors based on functions *OR, MAX, AND, XOR, SUM, AVERAGE, MODE* and *MEDIAN*, represented in the same formalism used in Section 6.2. It does not address noisy versions of the aggregations (such as *Noisy-OR*) since these can be further normalized into a model involving the respective deterministic aggregation plus an extra parfactor for the noise on x_i (corresponding to $\mathcal{F}_P$ in Section 6.2).

Example 6.5 The dependence between political ads and votes in the example in Figure 6.6 can be compactly represented by an aggregation parfactor with a domain formed by the set of voters. The figure uses the more traditional notation V_i, equivalent to $V(i)$.

Aggregation Factor Marginalization (AFM) Problems   This section is concerned about the inference problem of marginalizing a set of random variables in an FOPM with aggregation factors to determine the marginal density of others. As shown in Section 6.2, this can be done by using C-FOVE (Milch et al., 2008) extended with a lifted operation for summing random variables out of an aggregation parfactor. These summations can be

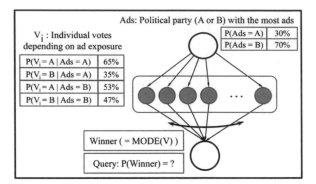

Figure 6.6: Graphical model on the domain of the election of one of two parties A and B. The random variable *Ads* indicates which party has the most ads in the media. The variables V_i indicate the vote of each person in a population, modeled as a dependence of ad exposure. The *Winner* variable indicates the winner and it is determined by the majority (*MODE*) of votes. One would like to estimate the probability of each party winning the election given this model.

reduced to the **Aggregation Factor Marginalization** (AFM) calculation:

$$\phi'_{\mathbf{y}}(y) = \sum_{x_1,,x_n} \left(\phi_{\otimes}(y, x_1, \ldots, x_n) \prod_{1 \le i \le n} \phi_{\mathbf{x}}(x_i) \right).$$

where $\phi_{\mathbf{x}}$ is the (same for all i) potential product of all other factors in the model that have X_i as an argument, and $\phi'_{\mathbf{y}}$ is the resulting potential on y alone. This subproblem is also one that needs to be solved in extending Lifted Belief Propagation (Singla and Domingos, 2008) to deal with aggregation factors.

Section 6.2 shows how, when different x_i have different potential functions on them, the problem can be normalized (by splitting and using auxiliary variables) to multiple such sums in which this uniformity holds. Similarly, one can separate the case in which only *some* x_i need to be summed out into two different aggregation parfactors, one for all aggregate random variables being summed out, and another for the remaining ones.

Inference Problems with Inequality   We also define aggregation factors with inequality constraints as follows:

$$\phi_{\otimes_{\le}}(y, x_1, \ldots, x_n) = \begin{cases} 1, & \text{if } y \le x_1 \otimes \cdots \otimes x_n; \\ 0, & \text{otherwise.} \end{cases}$$

with the corresponding problem **AFM[$\le$]** defined as

$$\sum_{x_1, \ldots, x_n} \left(\phi_{\otimes_{\le}}(y, x_1, \ldots, x_n) \cdot \prod_{1 \le i \le n} \phi_{\mathbf{x}}(x_i) \right).$$

$\phi_{\otimes_2}$ and **AFM**[$\geq$] are defined analogously.

6.3.1   Efficient Methods for AFM Problems

The exact solutions for **AFM** problems presented in Section 6.2 are efficient, but their applicability is limited to some operations (Díez and Galán, 2003), or their computational complexity still depends on the number of random variables. In this section, we present efficient constant-time exact and approximate marginalizations that are applicable to more aggregate functions.

Normal Distribution with Linear Constraints   Section 6.2 shows how the potential of an aggregation parfactor depends only on the value histogram (also described as counting formulas) of its aggregated random variables.

Given values $x_1, \ldots, x_n$ for n random variables with the same range, the value histogram of x is a vector h with $h_u = |\{i : x_i = u\}|$ for each u in the random variables' range. When a potential function on $x_1, \ldots, x_n$ depends on the histogram alone, as in the case of aggregation factors, then there is a function $\phi_\mathbf{h}$ on histograms such that $\phi(y, x_1, \ldots, x_n) = \phi_\mathbf{h}(y, h)$ and $\phi_\bigotimes(y, x_1, \ldots, x_n) = \phi_{\bigotimes \mathbf{h}}(y, h)$. In what follows, this section describes the binomial case (range of x_i equal to 2) for clarity, but it applies to the multinomial case as well. One can write

$$\sum_{x_1, \ldots, x_n} \phi(y, x_1, \ldots, x_n) \prod_i \phi_\mathbf{x}(x_i) = \sum_h \binom{n}{h_1} \phi_\mathbf{h}(y, h) p_1^{h_1} p_0^{n-h_1}, \tag{6.1}$$

where p_0 and p_1 are the normalizations of $\phi_\mathbf{x}$. This corresponds to grouping assignments on x into their corresponding histograms h, and iterating over the histograms (which are exponentially less many), taking into account that each histogram corresponds to $\binom{n}{h_1}$ assignments.

One now observes that functions $\phi_\mathbf{h}(y, h)$ coming from aggregation factors always evaluate to 0 or 1. Moreover, the set of histograms for which they evaluate to 1 can be described by linear constraints on the histogram components. For example, $\phi_{MODE}(y, h)$ will only be 1 if $h_y \geq h_{y'}$ for all $y' \neq y$. Given $\phi_\mathbf{h}$ and y, let C_y be the set of histograms h such that $\phi_\mathbf{h}(y, h) = 1$. Then (6.1) can be rewritten as

$$\sum_{h \in C_y} \binom{n}{h_1} p_1^{h_1} p_0^{n-h_1},$$

which is the probability of a set of h_1 values under a binomial distribution. For large n, according to the Central Limit Theorem (Rice, 2006), the binomial distribution is approximated by the normal distribution $N(np_1, np_1 p_0)$ with density function f. Then

$$\sum_{h \in C_y} \binom{n}{h_1} p_1^{h_1} p_0^{n-h_1} \approx \int_{h' \in C'_y} f(h') \, dh',$$

where C_y' is a continuous region in the $(k-1)$-simplex corresponding to C_y (which is defined in discrete space). Figure 6.8 lists C_y and an appropriate C_y' for the several aggregation factor potentials, for both **AFM** and **AFM**[≥].

Let us see two examples. In the first one, for **AFM** on *MODE* on binary variables, $y = 1$, and histograms with $h(1) = t$, C_y is $h_1 \geq h_0$ and C_y' is $t \in \left[\left\lfloor \frac{n}{2} \right\rfloor + 0.5, n + 0.5\right]^{11}$, so one computes

$$\int_{t=\left\lfloor \frac{n}{2} \right\rfloor + 0.5}^{n+0.5} f(t)\, dt,$$

which can be done in constant time.

In the second example we consider **AFM** and **AFM**[≥] on *SUM* with $n = 100$ random variables representing ratings of 100 people who watch a movie. Each person gives ratings of either 0 (negative) or 1 (positive), with probabilities 0.55 and 0.45, respectively ($p_0 = 0.55$). One may be interested in the summation of those votes ($r = 100$).

Figure 6.7 shows the probability density of the number of positive ratings. The bars in red in (a) and (b) panels show the area corresponding to the result for **AFM** and **AFM**[≥], respectively, for $y = 50$. The former can have the exact binomial distribution form computed in constant time, while the latter can have the normal distribution approximation computed in constant time. Therefore, the marginal on Y can be approximated in $O(r)$. The methods in Section 6.2, on the other hand, take $O(r \log n)$, and (Díez and Galán, 2003) is not applicable.

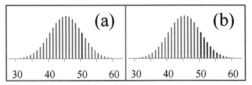

Figure 6.7: Histogram with a binomial distribution with (a) equality and (b) inequality constraints.

This section now explains the method in more detail for two different cases: aggregated binary random variables ($k = 2$), which can be dealt with analytically, and aggregated multivalued random variables ($k > 2$).

Binary Variable Case AFM Problem   For *AND, OR, MIN, MAX* and *SUM*, an exact solution with time constant in n for **AFM** for the binary case can be computed, for the

---

[11] Here, +0.5 and −0.5 are continuity corrections for accurate approximations.

Operator	Problem	y	C_y	C'_y
AND	**AFM**	*TRUE*	$h_{TRUE} = n$	not needed (cheap exact solution)
OR	**AFM**	*FALSE*	$h_{FALSE} = n$	not needed (cheap exact solution)
SUM	**AFM**	y	$\sum_i i \times h_i = y$	$y - 0.5 \le \sum_i i \times h_i \le y + 0.5$
SUM	**AFM[$\ge$]**	y	$\sum_i i \times h_i \le y$	$\sum_i i \times h_i \le y - 0.5$
MAX	**AFM**	y	$h_y > 0,$ $\forall i > y \ \ h_i = 0$	$h_y > 0.5,$ $\forall i > y \ \ -0.5 \le h_i \le 0.5$
MAX	**AFM[$\ge$]**	y	$\forall i > y \ \ h_i = 0$	$\forall i > y \ \ -0.5 \le h_i \le 0.5$
MODE	**AFM**	y	$\forall i \ne y \ \ h_y > h_i$	$\forall i \ne y \ \ h_y > h_i$
MEDIAN	**AFM**	y	$\sum_{i=1}^{y-1} h(i) < \frac{n}{2},$ $\sum_{i=y}^{n} h(i) \ge \frac{n}{2}$	$\sum_{i=1}^{y-1} h(i) + 0.5 \le \lfloor \frac{n}{2} \rfloor \le \sum_{i=y}^{n} h(i) - 0.5$
MEDIAN	**AFM[$\ge$]**	y	$\sum_{i=1}^{y-1} h(i) \ge \frac{n}{2}$	$\sum_{i=1}^{y-1} h(i) - 0.5 \ge \lfloor \frac{n}{2} \rfloor$

Figure 6.8: Constraints to be used in binomial (multinomial) distribution exact calculations (C_y) and (multivariate) normal distribution approximations (C'_y). The table does not exhaust all combinations. However those omitted are easily obtained from the presented ones. For example, $\phi_{OR}(T, x) = 1 - \phi_{OR}(F, x)$, $\phi_{AVERAGE}(y, x) = \phi_{SUM}(y \times n, x)$, and $\phi_{MODE \ge}(y, x) = \sum_{y' \le y} \phi_{MODE}(y', x)$.

appropriate choices of p_0 and p_1, as

$$\phi'_y(y) = \binom{n}{y} p_0^{n-y} \cdot p_1^{y}.$$

AVERAGE can be solved by using ϕ'_y obtained from *SUM* on y/n. This solution follows from the fact that, for the above cases, one needs the potential of a single histogram.

For *MODE* and *MEDIAN*, exact solutions for **AFM** are of the following form, with time linear in n:

$$\phi'_y(TRUE) = \sum_{i=\lfloor \frac{n}{2} \rfloor + 1}^{n} \binom{n}{i} p_0^{n-i} \cdot p_1^{i}.$$

Such solutions are more expensive because they measure the density of a region of histograms. They can be approximated by the normal distribution in the following way:

$$\phi'_y(TRUE) \approx \int_{t=\lfloor \frac{n}{2} \rfloor + 0.5}^{n+0.5} \frac{exp\left(-\frac{(t - np_1)^2}{2 \cdot np_1(1 - p_1)}\right)}{\sqrt{2\pi \cdot np_1(1 - p_1)}} \, dt.$$

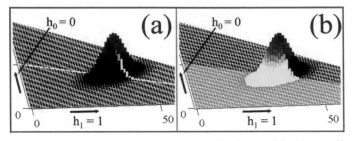

Figure 6.9: Histogram space for multinomial distributions with (a) equality and (b) inequality constraints.

Note that *MODE* is not solved by either (Díez and Galán, 2003)'s factorization or the logarithmic algorithm presented in Section 6.2, while this approach can compute an approximation in constant time.

Example 6.6 For n is 100, $p_1 = 0.45$, the exact solution is about 0.18272. Our approximate solution is about 0.18286. Thus, the error is less than 0.1% of the exact solution.

AFM[≤] and AFM[≥] Problems   For binary aggregated random variables, these problems are different from **AFM** only for the *SUM* (and thus, *AVERAGE*) case. For *SUM* one can use the approximation

$$\phi'_{\mathbf{y}}(y) = \sum_{i=y}^{n} \binom{n}{i} p_1^i (1 - p_1)^{n-i} \approx \int_{t=y-0.5}^{n+0.5} \frac{exp\left(-\frac{(t-np_1)^2}{2 \cdot np_1(1-p_1)}\right)}{\sqrt{2\pi \cdot np_1(1 - p_1)}} \, dt.$$

Multivalued Variable Case   In the multivalued ($k > 2$) case, there is a need to compute the probability of a linearly constrained region of histograms, which motivates us to consider approximate solutions with the multivariate normal distribution.

Example 6.7 Suppose that the aggregation function is *SUM*. There are 100 random variables representing ratings of 100 people who watch a movie. Each person gives ratings among 0, 1 and 2 (0 is lowest and 2 is highest). One may want to calculate the sum of ratings from 100 people when each person gives a rating 0 with 0.35 ($p(x_i = r0) = 0.35$), 1 with 0.35 ($p(x_i = r1) = 0.35$), and 2 with 0.3 ($p(x_i = r2) = 0.3$). The probability of histograms is provided by the multinomial distribution, as shown in Figure 6.9. The colored bars in (a) represent the probability of the ratings sum being exactly 100. If instead one wishes to determine the probability of the ratings sum exceeding 100, one may have an **AFM[≥]** instance, with a probability corresponding to the colored bars in the (b) panel. In both cases, we need to compute the volume of a histogram region.

Analytical Solution for Operators with a Single Linear Constraint   As in the previous sub-section, the multinomial distribution can be approximated by the multivariate normal distribution.

Suppose that each rv may have three values with probability p_0, p_1 and p_2 ($p_0 + p_1 + p_2 = 1$), respectively. Then the multinomial distribution of h_0, h_1 and h_2 chosen from n random variables is

$$\binom{n}{h_0 \, h_1 \, h_2} \cdot p_0^{h_0} \cdot p_1^{h_1} \cdot p_2^{h_2} = \frac{n!}{h_0! h_1! h_2!} \cdot p_0^{h_0} \cdot p_1^{h_1} \cdot p_2^{h_2}.$$

The corresponding bivariate (i.e. (3-1) multivariate) normal distribution of $\mathbb{X} = [h_0 \, h_1]$ chosen from n random variables is as follows (Note that $h_2 = n - h_1 - h_2$),

$$\frac{1}{(2\pi)^{2/2}|\Sigma|^{1/2}} \cdot exp\left(-\frac{1}{2}(\mathbb{X} - \mu)\Sigma^{-1}(\mathbb{X} - \mu)'\right),$$

when the μ and Σ are

$$\mu = [np_0 \; np_1], \; \Sigma = \begin{pmatrix} np_0(1 - p_0) & np_1 p_2 \\ np_2 p_1 & np_2(1 - p_2) \end{pmatrix}.$$

Example 6.8 Suppose that one sets p_0, p_1 and p_2 as 0.35, 0.35 and 0.3 respectively and y as 100. Any operator with a single linear constraint (e.g. **AFM**, **AFM**[$\leq$] and **AFM**[$\geq$] on *SUM*, and **AFM**[$\leq$] and **AFM**[$\geq$] on *MEDIAN*) allows an analytical solution because there is a linear transformation from $\mathbb{X} = [h_0 \, h_1]$ to y. Consider the following linear transform $y = 0 \cdot h_0 + 1 \cdot h_1 + 2 \cdot h_2 = 200 - 2 \cdot h_0 - h_1$. When one represents the transform as $y = A\mathbb{X} + B$, the new distribution of y is given by the 1-D normal distribution:

$$\frac{1}{\sqrt{2\pi\Sigma_y}} \cdot exp\left(-\frac{(y-\mu_y)^2}{2\Sigma_y}\right),$$

where $\mu_y = A\mu + B$ and $\Sigma_y = A\Sigma A^T$ are scalars. From the transformation the solution of **AFM** for $y = 100$ can be calculated in the following way:

$$\frac{1}{\sqrt{2\pi\Sigma_y}} \int_{y=100-0.5}^{100+0.5} exp\left(-\frac{(y-\mu_y)^2}{2\Sigma_y}\right) dy.$$

The solutions of **AFM**[$\leq$] and **AFM**[$\geq$] for $y = 100$ can be calculated in similar ways.

Sampling for Remaining Operators   In general, integration of a multivariate truncated normal does not allow an analytical solution. Fortunately, efficient **Gibbs sampling** methods (e.g. (Geweke, 1991; Damien and Walker, 2001)) are applicable to the truncated normal in straightforward ways, even with several linear constraints. This immediately feeds to an approximation with time complexity not depending on n, the number of random variables.

6.3.2   Aggregation Factors with Multiple Atoms

This sub-section now considers a generalized situation. Previous sub-sections assume that all random variables in a relational atom have the same distribution. Here, it deals with the issue of aggregating J distinct groups of random variables, each represented by a relational atom X_j with n_j groundings and a distinct potential $\phi_{\mathbf{x_j}}$, for $1 \leq j \leq J$.

$$y = \bigotimes_{\substack{1<j<J \\ 1<i<n_j}} x_{j,i}.$$

This problem, **AFM-M**, is an extension of the **AFM** and consists in calculating the following marginal:

$$\sum_{x_{1,1},\cdots,x_{J,n_J}} \phi_\otimes(y, x_{1,1}, \cdots, x_{J,n_J}) \prod_{j=1}^{J} \prod_{1 \leq i \leq n_j} \phi_{\mathbf{x_j}}(x_{j,i}).$$

One solution approach is to compute an aggregate y_j^0 per atom j, and then combine each pair y_j^i and y_{j+1}^i into $y_{\lfloor j/2 \rfloor}^{i+1}$ until they are all aggregated. This will have complexity $O(J \log J)$ but works only for associative operators. For non-associative operators, one may need to calculate the marginal for each X_j independently:

$$\sum_{h^1,\cdots,h^J} \phi_{\otimes \mathbf{h}}(y, \mathbf{h}) \left(\binom{n_1}{h_1^1} p_{1,0}^{h_0^1} p_{1,1}^{h_1^1} \cdots \binom{n_J}{h_1^J} p_{J,0}^{h_0^J} p_{J,1}^{h_1^J} \right),$$

where $p_{j,0}$ and $p_{j,1}$ are the normalization of $\phi_{\mathbf{x_j}}(0)$ and $\phi_{\mathbf{x_j}}(1)$; h^j is a histogram for atom j, and $\mathbf{h}$ is the combined histogram. The complexity of this approach is $O(exp(J))$.

Another approach is to make use of the representation of the aggregation operator as a set of linear constraints (Figure 6.8). Note that h^j is approximately normally distributed when n_j is large, and h^i and h^j are independent when $i \neq j$. Thus, the all-group histogram vector $\mathbf{h}$ is also approximately normally distributed because it is the normal sum ($\mathbf{h}_i = \sum_j h_i^j$).

Any linear constraint in Figure 6.8 can be re-expressed as a linear constraint using elements of $\mathbf{h}$, and the multinomial-normal approximation can be used to yield a similar approximate solution in time constant n, the total number of random variables.

For example, for binary random variables, the normal approximation of the all-group histogram is:

$$N\left(\sum_{j=1}^{J} n_j p_{j,1}, \sum_{j=1}^{J} n_j p_{j,1} p_{j,0} \right). \tag{6.2}$$

This way, the time complexity is only $O(J)$ instead of $O(J \log J)$ (or $O(exp(J))$ for non-associative operators).

6.3.3   Efficient Lifted Belief Propagation with Aggregation Parfactors

Here, we show that lifted belief propagation can be performed efficiently with these methods. Given an aggregate operator ϕ, messages of the form $\mu_{\phi \to c}$ can be directly calculated with **AFM** solutions. On the other hand, messages from the aggregation factor to one of its aggregated random variables, of the form $\mu_{\phi \to x_i}$, need to be calculated in the following way:

$$
\begin{aligned}
\mu_{\phi \to x_1}(x_1) &\propto \sum_{y, x_2, \cdots, x_n} \left(\phi_y(y) \phi_\otimes(y, x_1, \cdots, x_n) \prod_{2 \le i \le n} \phi_x(x_i) \right) \\
&= \sum_y \phi_y(y) \sum_{x_2, \cdots, x_n} \left(\phi_\otimes(y, x_1, \cdots, x_n) \prod_{2 \le i \le n} \phi_x(x_i) \right) \\
&= \sum_y \phi_y(y) \sum_{y = c' \otimes x_1} \phi_{y'}(c')
\end{aligned}
\tag{6.3}
$$

This consists in a *lifted* belief propagation step because the same calculation and value hold for all x_i since the aggregated random variables are symmetric (after shattering).

In case of the extended problems in Section 6.3.2, the calculation is very similar to Equation (6.3) given Equation (6.2).

6.3.4   Error Analysis

This section discusses error bounds for the multinomial-normal approximations. In general, the Berry-Esseen theorem (Esseen, 1942) gives an upper bound on the error. Suppose that $\phi_\mathbf{y}(y)$ and $\widetilde{\phi}_\mathbf{y}(y)$ represent the probability mass of a binomial distribution and density of its normal approximation, respectively. Furthermore, one may represent the cumulative probabilities as $\Phi_\mathbf{y}(y)$ and $\widetilde{\Phi}_\mathbf{y}(y)$[12]. Then, given any y, the error between the two cumulative probabilities is bounded (Esseen, 1942) as in

$$
\left| \Phi_\mathbf{y}(y) - \widetilde{\Phi}_\mathbf{y}(y) \right| < c \cdot \frac{p^2 + (1-p)^2}{\sqrt{np(1-p)}},
$$

where c is a small (< 1) constant. Thus, the asymptotic error bound is $O(1/\sqrt{n})$, and this extends to probability on any interval.

For k-valued multinomials, suppose that $\Phi_\mathbf{Y}(A)$ and $\widetilde{\Phi}_\mathbf{Y}(A)$ represent the probability of a multinomial distribution and its multivariate normal approximation over a measurable convex set A in R^k. Then, the approximation error is bounded (Gotze, 1991):

$$
sup_A \left| \Phi_\mathbf{Y}(A) - \widetilde{\Phi}_\mathbf{Y}(A) \right| < c \cdot \frac{k}{\sqrt{n}},
$$

---

[12] That is, $\Phi_\mathbf{y}(y) = \sum_{i=0}^{y} \phi_\mathbf{y}(i)$, and $\widetilde{\Phi}_\mathbf{y}(y) = \int_{t=-\infty}^{y} \widetilde{\phi}_\mathbf{y}(t)\, dt$.

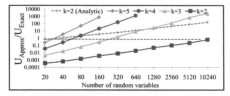

Figure 6.10: Ratios of utilities of approximate algorithms and exact method (histogram based counting).

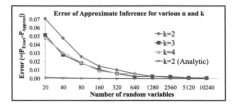

Figure 6.11: Error curves for different values of k and n.

where c depends only on the multinomial parameters and not on n. In our problem, A is determined by linear constraints, hence it is convex. Thus, the asymptotic error bound is $O(k/\sqrt{n})$.

6.3.5 Experiments

This section provides experimental results on the example in Figure 6.6 (which uses the *MODE* aggregate function) which provides insight on when to use the approximate algorithm instead of the generally applicable exact algorithm based on counting formulas (the logarithmic method in Section 6.2 does not apply to *MODE*).

One may compute the utility of any of the methods tested, approximate or exact alike, in the following manner. One assumes a typical application in which the utility of an error is an inverse quadratic function $U(err) = 1 - err^2$. The utility of a method obtaining error err is normalized by the time t it takes to run, so $U(err, t) = U(err)/t$. For sampling methods, t is the time to convergence. To compare our approximate methods to their exact counterparts, we plot the *ratio* between their utilities. Therefore, an approximation method is generally superior to its exact counterpart when this ratio is greater than 1.

We apply both the approximate and the exact inference methods to the model in Figure 6.6. For $k = 2$, both the analytical and the sampling methods are applied. Given k and n, model potentials are randomly chosen, and the error and convergence time are recorded. This is averaged over 100 trials to calculate the utilities U_{Approx} and U_{Exact}.

As shown in Figure 6.10, the suggested approximate algorithm has much higher utility than the exact method for larger k and n. However, when $k = 2$ (binary variables), the exact method has higher utility than sampling for relatively large n (e.g. $n = 10,240$). In this case, one can use the efficient analytic integration which applies for $k = 2$. We also show in Figure 6.11 how the error decreases for different values of k and n.

In addition, the results indicate that convergence time stays flat for various k and n. However, the error of sampling method is noticeable for small n. For example, when $k = 4$, the error is 3.07% with $n = 40$ and 1.82% with $n = 80$. For larger n, this issue is resolved. The error becomes less than 1% when $n = 320$ and negligible when $n > 5120$. These observations are consistent for various k from 2 to 6.

6.4 Lifted Aggregation through Skolemization for Weighted First-order Model Counting

In this section we focus on lifted inference algorithms that are based on weighted model counting (WMC). More specifically, *weighted first-order model counting* (WFOMC) (see Chapter 5). WFOMC assigns a weight to interpretations in finite-domain, function-free, first-order logic, and computes the sum of the weights of all models.

One major limitation of first-order model counters, however, is that they require input in *Skolem normal form* (i.e., without existential quantifiers). This is a common requirement for first-order automated reasoning algorithms, such as theorem provers. It is usually dealt with by Skolemization, which introduces Skolem constants and functions. However, the introduction of functions is problematic for first-order model counters as they expect a function-free input.

In this section we introduce a *Skolemization procedure* that is specific for weighted first-order model counting. The procedure maps a logical input theory to an output theory that is devoid of existential quantifiers and functions, yet has an identical weighted first-order model count. The procedure is modular, in that it remains sound when extending the input and output theories with a new sentence. Furthermore, it is purely first-order as it is independent of the domain of discourse.

The proposed Skolemization algorithm has a range of implications. The WFOMC representation of various forms of directed models such as Probabilistic Logic Programs generally contains existential quantifiers to express aggregation over multiple parents (De Raedt et al., 2007). Also Markov Logic Networks can contain quantifiers (Richardson and Domingos, 2006). The proposed Skolemization algorithm allows us to perform lifted inference on these representations.

6.4.1 Normal Forms

It is common for logical reasoning algorithms to operate on *normal form* representations instead of arbitrary sentences. For example, propositional SAT solvers and weighted model counters often expect CNF inputs. We distinguish the following first-order normal forms: A theory in *prenex normal form* consists of formulas $Q_1 x_1, \ldots, Q_n x_n, \phi$, where each Q_i is either a universal or existential quantifier, and ϕ is quantifier-free. A theory in *prenex clausal form* is a theory in prenex normal form where ϕ is a clause. A theory in *Skolem normal form* is a theory in prenex normal form where all Q_i are universal quantifiers. A *first-order CNF* is a theory in Skolem and prenex clausal form. Thus, all sentences take the form $\forall x_1, \ldots, \forall x_n, l_1 \vee \cdots \vee l_m$. WFOMC algorithms require a theory to be in *first-order CNF*.

6.4.2 Skolemization for WFOMC

It is well known that one can take any arbitrary formula and convert it to prenex clausal form. This involves pushing negations inside, pushing quantifiers to the front, and distributing disjunctions over conjunctions. The situation for Skolem normal form is different. Not every formula can be transformed into an equivalent Skolem normal form. This problem is typically dealt with by *Skolemization*, which eliminates existential quantifiers from a prenex normal form. This is done by replacing existentially quantified variables by Skolem constants and functions. The result is not logically equivalent to the original formula, but only *equisatisfiable* (i.e., satisfiable precisely when the original formula is satisfiable).

The standard Skolemization algorithm is specific to the satisfiability task and may be unsuitable for other tasks. It is particularly unsuitable for WFOMC as it may produce a result with functions, which are not permitted in the WFOMC task. For example, standard Skolemization would transform the formula

$$\forall x, \exists y, \ \mathsf{WorksFor}(x, y) \vee \mathsf{Boss}(x) \tag{6.4}$$

into the following formula with the Skolem function Sk().

$$\forall x, \ \mathsf{WorksFor}(x, \mathsf{Sk}(x)) \vee \mathsf{Boss}(x).$$

As soon as we allow functions, the Herbrand base becomes infinite, which makes the model counting task ill-defined, therefore, ruling out standard Skolemization for WFOMC.[13]

6.4.2.1   Algorithm   This section introduces a Skolemization technique for WFOMC. It takes as input a triple $(\Delta, \mathrm{w}, \bar{\mathrm{w}})$ whose Δ is an arbitrary sentence and returns a triple $(\Delta', \mathrm{w}', \bar{\mathrm{w}}')$ whose Δ' is in Skolem normal form (i.e., no existential quantifiers). Such a Δ' can then be turned into first-order CNF using standard transformations. The proposed technique does not introduce functions. It satisfies two properties: one is essential and the other expands the applications of the technique.

The essential property is soundness.

Property 6.1 (Soundness)   *Skolemization of* $(\Delta, \mathrm{w}, \bar{\mathrm{w}})$ *to* $(\Delta', \mathrm{w}', \bar{\mathrm{w}}')$ *is sound iff for any* **D**, *we have that*

$$WFOMC(\Delta, \mathbf{D}, \mathrm{w}, \bar{\mathrm{w}}) = WFOMC(\Delta', \mathbf{D}, \mathrm{w}', \bar{\mathrm{w}}').$$

To motivate the second property, modularity, we note that one may be interested in queries of the form $WFOMC(\Delta \wedge \phi, \mathbf{D}, \mathrm{w}, \bar{\mathrm{w}})$, where Δ, w and $\bar{\mathrm{w}}$ are fixed, but where ϕ is changing. For example, we will see in Section 6.4.3 that probabilistic inference can

---

[13] One could obtain a Skolem normal form by grounding existential quantifiers, replacing them by large, but finite disjunctions. While this may still permit limited runtime improvements on vacuous formulas, it is for all practical purposes equivalent to reducing the WFOMC problem to a WMC problem. Moreover, that transformation is dependent on the domain and leads to large formulas whose conversion to CNF blows up (e.g., when grounding $\exists x \forall y$).

be reduced to these types of queries. Therefore, we want to achieve a stronger form of soundness.

Property 6.2 (Modularity) *Skolemization of* $(\Delta, w, \bar{w})$ *to* $(\Delta', w', \bar{w}')$ *is modular iff for any* **D** *and any sentence* ϕ,

$$WFOMC(\Delta \wedge \phi, \mathbf{D}, w, \bar{w}) = WFOMC(\Delta' \wedge \phi, \mathbf{D}, w', \bar{w}').$$

That is, by replacing ϕ, one does not invalidate the Skolemization obtained under a different ϕ.

The proposed Skolemization algorithm eliminates existential quantifiers one by one. Its basic building block is the following transformation.

Definition 6.2 *Suppose that* Δ *contains a subexpression of the form* $\exists x, \phi(x, \mathbf{y})$, *where* $\phi(x, \mathbf{y})$ *is an arbitrary sentence containing the free logical variables* x *and* $\mathbf{y}$. *Let* n *be the number of variables in* $\mathbf{y}$. *First, we introduce two new predicates: the* Tseitin *predicate* Z/n *and the* Skolem *predicate* S/n. *Second, we replace the expression* $\exists x, \phi(x, \mathbf{y})$ *in* Δ *by the atom* $Z(\mathbf{y})$, *and append the formulas*

$$\forall \mathbf{y}, \forall x, \ Z(\mathbf{y}) \vee \neg \phi(x, \mathbf{y})$$

$$\forall \mathbf{y}, \ S(\mathbf{y}) \vee Z(\mathbf{y})$$

$$\forall \mathbf{y}, \forall x, \ S(\mathbf{y}) \vee \neg \phi(x, \mathbf{y}).$$

The functions w' *and* $\bar{w}'$ *are equal to* w *and* $\bar{w}$, *except that* $w'(Z) = \bar{w}'(Z) = w'(S) = 1$ *and* $\bar{w}'(S) = -1$.

In the resulting theory Δ', a single existential quantifier is now eliminated. This building block can eliminate single universal quantifiers as well. When Δ contains a subexpression $\forall x, \phi(x, \mathbf{y})$, we replace it by $\neg \exists x, \neg \phi(x, \mathbf{y})$, whose existential quantifier can be eliminated with the Skolemization step. Repeated application of the Skolemization step comprises a modular Skolemization algorithm. The proof and detailed intuition can be found in Van den Broeck et al. (2014). Additionally, repeated application of the Skolemization step will terminate with a sentence in Skolem normal form. Moreover, this can be achieved in time polynomial in the size of Δ.

6.4.3 Skolemization of Markov Logic Networks

We will show in this section how the proposed Skolemization technique can extend the scope of first-order model counters to new situations. We will consider in particular an undirected first-order probabilistic language, Markov Logic Networks (MLN) (Richardson and Domingos, 2006), for which we will now introduce a WFOMC encoding.

Representation. An MLN is a set of tuples (w, ψ), where w is a real number representing a weight and ψ is a formula in first-order logic. When w is infinite, ψ represents a first-order logic constraint, also called a *hard formula*. Building further on the example given before, consider the following MLN

$$1.3 \quad \exists y, \; \mathsf{WorksFor}(x, y) \lor \mathsf{Boss}(x). \tag{6.5}$$

This statement softens the logical sentence we saw earlier. Instead of saying that every person either has a boss, or is a boss, it states that worlds with many employed people are more likely. That is, it is now possible to have a world with unemployed people, but the more unemployed people there are, the lower the probability of that world.

The semantics of a first-order MLN Φ is defined in terms of its *grounding* for a given domain of constants **D**. The grounding of Φ is the MLN obtained by first grounding all its quantifiers and then replacing each formula in Φ with all its groundings (using the same weight). With the domain **D** = {A, B} (e.g., two people, Alice and Bob), the above first-order MLN represents the following grounding.

$$1.3 \quad \mathsf{WorksFor}(A, A) \lor \mathsf{WorksFor}(A, B) \lor \mathsf{Boss}(A)$$

$$1.3 \quad \mathsf{WorksFor}(B, A) \lor \mathsf{WorksFor}(B, B) \lor \mathsf{Boss}(B)$$

This ground MLN contains six different random variables, which correspond to all groundings of atoms $\mathsf{WorksFor}(x, y)$ and $\mathsf{Boss}(x)$. This leads to a distribution over 2^6 possible worlds (i.e., interpretation). The weight of each world is simply the product of all weights e^w, where (w, γ) is a ground MLN formula and γ is satisfied by the world. The weights of worlds that do not satisfy a hard formula are set to zero. The probabilities of worlds are obtained by normalizing their weights.

Encoding a Markov Logic Network. The WFOMC encoding $(\Delta, w, \bar{w})$ of an MLN is constructed as follows. For each MLN formula $(w_i, \phi_i(\boldsymbol{x}_i))$, where $\boldsymbol{x}_i$ denotes the free logical variables in ϕ_i, we introduce a parameter predicate $\mathsf{P}_i/|\boldsymbol{x}_i|$. For each MLN formula, Δ contains the sentence $\forall \boldsymbol{x}_i, \; \mathsf{P}_i(\boldsymbol{x}_i) \Leftrightarrow \phi_i(\boldsymbol{x}_i)$. The weight function sets $w(\mathsf{P}_i) = e^{w_i}$, $\bar{w}(\mathsf{P}_i) = 1$, and $w(\mathsf{Q}) = \bar{w}(\mathsf{Q}) = 1$ for all other predicates Q. Each P_i captures the truth value of ϕ_i and carries its weight. Hard formulas can directly be encoded as constraints.

The encoding of Formula 6.5 has Δ equal to

$$\forall x, \; \mathsf{P}(x) \Leftrightarrow \exists y, \; \mathsf{WorksFor}(x, y) \lor \mathsf{Boss}(x).$$

Its w maps P to $e^{1.3}$ and all other predicates to 1. Its $\bar{w}$ maps all predicates to 1.

As discussed in Section 6.4.1, WFOMC algorithms require first-order CNF input. Applying the WFOMC encoding for an MLN will only yield a Δ in Skolem normal form (and thus rewritable into CNF) if the MLN formulas are quantifier-free. Then, the only quantifiers in Δ are the universal ones introduced by the encoding itself. Therefore, Van den Broeck et al. (2011) and Gogate and Domingos (2010) resort to *grounding all quantifiers* in the MLN formulas so as to obtain a CNF. This makes the WFOMC encoding specific to the domain **D**, and partly removes first-order structure from the problem.

Applying Skolemization. We can now perform WFOMC inference in MLNs with quantifiers. Skolemization and CNF conversion for the example above results in a Δ' equal to

$$\forall x, \; \mathsf{P}(x) \vee \neg\mathsf{Z}(x)$$
$$\forall x, \; \neg\mathsf{P}(x) \vee \mathsf{Z}(x)$$
$$\forall x, \forall y, \; \mathsf{Z}(x) \vee \neg\mathsf{WorksFor}(x, y)$$
$$\forall x, \forall y, \; \mathsf{Z}(x) \vee \neg\mathsf{Boss}(x)$$

$$\forall x, \forall y, \; \mathsf{S}(x) \vee \neg\mathsf{WorksFor}(y, x)$$
$$\forall x, \; \mathsf{S}(x) \vee \neg\mathsf{Boss}(x)$$
$$\forall x, \; \mathsf{S}(x) \vee \mathsf{Z}(x)$$

This theory can be used for WFOMC inference.

6.4.4   Skolemization of Probabilistic Logic Programs

We now show a WFOMC encoding for a directed first-order probabilistic language. The encoding is explained for the *ProbLog* language (De Raedt et al., 2007; Fierens et al., 2015).

Representation. ProbLog extends logic programs with facts that are annotated with probabilities. A ProbLog program Φ is a set of probabilistic facts F and a regular logic program L. A probabilistic fact $p :: a$ consists of a probability p and an atom a. A logic program is a set of rules, with the form Head : - Body, where the head is an atom and the body is a conjunction of literals. For example,

$$0.1 :: \mathsf{Attends}(x).$$
$$0.3 :: \mathsf{ToSeries}(x).$$
$$\mathsf{Series} : \text{-} \mathsf{Attends}(x), \mathsf{ToSeries}(x).$$

This program expresses that if more people attend a workshop, it more likely turns into a series of workshops.

The semantics of a ProbLog program Φ are defined by a distribution over the *groundings* of the probabilistic facts for a given domain of constants **D** (Sato, 1995).[14] The probabilistic facts $p_i :: a_i$ induce a set of possible worlds, one for each possible partition of a_i in positive and negative literals. The set of true a_i literals with the logic program L defines a well-founded model (Van Gelder et al., 1991). The probability of such a model is the product of p_i for all true a_i literals and $1 - p_i$ for all false a_i literals.

---

[14] Our treatment assumes a function-free and finite-domain fragment of ProbLog. Starting from classical ProbLog semantics, one can obtain the a finite function-free domain for a given query by exhaustively executing the Prolog program and keeping track of the goals that are called during resolution.

For the domain $\mathbf{D} = \{A, B\}$ (two people), the above first-order ProbLog program represents the following grounding:

$$0.1 :: \mathsf{Attends}(A). \qquad 0.1 :: \mathsf{Attends}(B).$$
$$0.3 :: \mathsf{ToSeries}(A). \qquad 0.3 :: \mathsf{ToSeries}(B).$$
$$\mathsf{Series} : \text{-} \, \mathsf{Attends}(A), \mathsf{ToSeries}(A).$$
$$\mathsf{Series} : \text{-} \, \mathsf{Attends}(B), \mathsf{ToSeries}(B).$$

This ground ProbLog program contains 4 probabilistic facts which correspond to 2^4 possible worlds. The weight of, for example, the world in which $\mathsf{Attends}(A)$ and $\mathsf{ToSeries}(A)$ are true would be $0.1 \cdot (1 - 0.1) \cdot 0.3 \cdot (1 - 0.3) = 0.0189$ and the model would be $\{\mathsf{Attends}(A), \mathsf{ToSeries}(A), \mathsf{Series}\}$.

Encoding a ProbLog Program. The transformation from a ProbLog program to a first-order logic theory is based on Clark's completion (Clark, 1978). This is a transformation from logic programs to first-order logic. For certain classes of programs, called *tight* logic programs (Fages, 1994), it is correct, in the sense that every model of the logic program is a model of the completion, and vice versa. Intuitively, for each predicate P, the completion contains a single sentence encoding all its rules. These rules have the form $\mathsf{P}(\mathbf{x}) : \text{-} \, b_i(\mathbf{x}, \mathbf{y}_i)$, where b_i is a body and $\mathbf{y}_i$ are the variables that appear in the body b_i but not in the head. The sentence encoding these rules in the completion is $\forall \mathbf{x}, \ \mathsf{P}(\mathbf{x}) \Leftrightarrow \bigvee_i \exists \mathbf{y}_i, \ b_i(\mathbf{x}, \mathbf{y}_i)$. If the program contains cyclic rules, the completion is not sound, and, it is necessary to first apply a conversion to remove positive loops (Janhunen, 2004).

The WFOMC encoding $(\Delta, \mathrm{w}, \bar{\mathrm{w}})$ of a tight ProbLog program has Δ equal to Clark's completion of L. For each probabilistic fact[15] $p :: a$ we set the weight function to $\mathrm{w}(pred(a)) = p$ and $\bar{\mathrm{w}}(pred(a)) = 1 - p$.

Again, a Skolem normal form is required to use WFOMC. However, we get this form only when the variables that appear in the body of a rule also appear in the head of a rule. This is not the case for most Prolog programs though. For example, if we apply the encoding to the example above, an existential quantifier appears in the sentence:

$$\mathsf{Series} \Leftrightarrow \exists x, \mathsf{Attends}(x) \wedge \mathsf{ToSeries}(x).$$

Furthermore, w maps $\mathsf{Attends}$ to 0.1 and $\mathsf{ToSeries}$ to 0.3, and $\bar{\mathrm{w}}$ maps $\mathsf{Attends}$ to 0.9 and $\mathsf{ToSeries}$ to 0.7. Both w and $\bar{\mathrm{w}}$ are 1 for all other predicates. This example is not in Skolem normal form and requires Skolemization before it can be processed by WFOMC algorithms.

Applying Skolemization. Skolemization and CNF conversion returns following Δ':

---

[15] If multiple probabilistic facts are defined for the same predicate, auxiliary predicates need to be introduced.

$$\forall x, \ Z \lor \neg Attends(x) \lor \neg ToSeries(x)$$
$$Z \lor S$$
$$\forall x, \ S \lor \neg Attends(x) \lor \neg ToSeries(x)$$

$$Series \lor \neg Z$$
$$\neg Series \lor Z$$

Sentence Δ' is in Skolem normal form and is now processable by WFOMC algorithms. A simple ProbLog program as the one above is identical to a noisy-or structure (Cozman, 2004), popular in Bayesian network modeling.

6.4.5   Negative Probabilities

In the encodings for MLNs and probabilistic logics, the weight functions (indirectly) represent probabilities and are therefore always positive. Our Skolemization algorithm introduces negative weights. This might appear odd when interpreting the weights as negative probabilities. This issue has been discussed before. For example, Feynman (1987) writes *"Negative probabilities allow an abstract calculation which permits freedom to do mathematical calculations in any order simplifying the analysis enormously"*.

The potential of negative probabilities was already observed by Jha and Suciu (2012) for answering queries in probabilistic databases and served as inspiration for our approach. Probabilistic databases (Suciu et al., 2011) are fundamentally a type of first-order probabilistic model. It can be viewed as a special type of weighted model counting problem (Δ, w), where the weight function encodes the probability $w(t)$ with which a tuple t can be found in the database. A query on such a database is typically a union of conjunctive queries (UCQ), which corresponds to a monotone DNF sentence Δ. A noticeable difference with most WMC solvers (and WFOMC) is that the solvers for probabilistic databases expect the theory Δ to be in DNF instead of CNF. Different from WFOMC is that although the query (i.e., Δ) is first-order, the weight function is defined on the propositional level like in WMC. Weights are thus assigned to ground literals (the tuples) whereas for WFOMC weights are assigned to predicates (the tables). This allows WFOMC to exploit more types of symmetries.

Jha and Suciu (2012) propose to extend probabilistic databases with MarkoViews, a representation similar to MLNs, in which each weighted formula is again a UCQ query, that is, a monotone DNF. To compute the probability of a query, they introduce negative tuple probabilities.

The use of negative probabilities has also come up for optimizing calculations for weighted model counting (Meert et al., 2016) and for specific structures in probabilistic graphical models like noisy-or (Díez and Galán, 2003). This particular case can also be expressed compactly using aggregation parfactors as shown in Section 6.2 and can be seen as a special case of the Skolemization algorithm applied to a noisy-or model.

Jaeger and Van den Broeck (2012) shows a negative liftability proof that uses *relational Skolemization*. Similar to our approach, subexpressions containing an existential quantifier

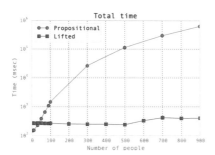

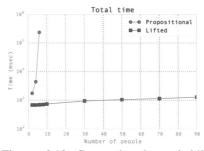

Figure 6.12: Computing the probability of Series for the workshop example.

Figure 6.13: Computing the probability of Series for the extended workshop example.

are transformed and relaxed to eliminate the quantifier. Relational Skolemization, however, does not guarantee a correct model count. It rather guarantees that if the weight of a model is non-zero it will also be non-zero in the Skolemized version.

6.4.6 Experiments

Given the resulting CNF from the workshop example described in the previous section, we computed the probability of Series using both propositional WMC and lifted WFOMC. The results are depicted in Fig. 6.12 and show a significant speedup for the lifted approach with respect to propositional inference.

If we extend the example to express that we are mainly interested in attendees that have joint publications, we obtain:

$$0.1 :: \text{Attends}(x).$$

$$0.3 :: \text{ToSeries}(x).$$

$$\text{Series} : \text{-} \text{Attends}(x), \text{Coauthor}(x, y), \text{Attends}(y), \text{ToSeries}(x, y).$$

Because of the extra logic variable, propositional inference becomes exponential in the size of the domain (see Fig. 6.13). Lifted inference on the other hand is polynomial in the size of the domain resulting in an exponential speedup.

6.5 Conclusions

Aggregate factors in probabilistic relational models can compactly represent dependencies among a large number of relational random variables. In general, propositional inference on a factor aggregating n k-valued random variables into an r-valued result random variable is $O(rk2^n)$ (that is, those based on aggregate functions such as SUM, AVERAGE, AND, etc).

In this chapter, we present three approaches to compute conditional probabilities in directed and undirected first-order models with aggregate operators. The first is by means of *aggregation parfactors*, a specialized data structure that can be incorporated into lifted variable elimination. The second is efficient lifted inference algorithms which compute conditional probabilities exactly or approximately with aggregate factors with time complexity constant in n. The third is by representing aggregation as a quantifier and performing a specialized skolemization operation that allows for weighted first-order model counting.

Acknowledgments

Authors wish to thank Peter Carbonetto, Michael Chiang, Mark Crowley, Craig Wilson, James Wright for valuable comments. This work was supported by NSERC grant to David Poole, ONR grant #N00014-12-1-0423, NSF grant #IIS-1118122, NSF grant #IIS-0916161, and the Research Foundation-Flanders (FWO-Vlaanderen).

7 First-order Knowledge Compilation

Seyed Mehran Kazemi, Guy Van den Broeck, and David Poole

Knowledge compilation approaches aim at compiling a theory into a secondary structure which enables processing desired queries more easily. Examples of the secondary structures for propositional theories include junction trees (Lauritzen and Spiegelhalter, 1988; Shafer and Shenoy, 1990; Jensen et al., 1990), arithmetic circuits (Darwiche, 2003), and probabilistic sentential decision diagrams (Kisa et al., 2014).

The advent of search-based lifted inference algorithms inspired leveraging knowledge compilation techniques (Darwiche and Marquis, 2002; Darwiche, 2003) for lifted inference: lifted operations were first compiled into a secondary structure, and then the secondary structure was executed to produce the results. Knowledge compilation approaches generally offer two advantages: 1- speeding up inference by compiling the high-level lifted rules to a secondary structure composed of lower-level operations and then executing the low-level operations, 2- breaking the inference into an offline (compilation) and an online (executing the secondary structure) phase thus providing the grounds for most of the computations to be shifted to the offline phase thus speeding up the online phase, and for reusing the secondary structure for answering several queries. While the initial knowledge compilation works for lifted inference used a data structure as the secondary structure (Van den Broeck, 2013), it was shown later on that using programs in low-level languages (e.g., C++ programs) as the secondary structure is more efficient Kazemi et al. (2016).

In a parallel line of work, Braun and Möller (2016, 2017a) generalized the secondary structure of junction trees to the first-order setting. Gehrke et al. (2018) define lifted dynamic junction trees for temporal distributions, and Braun and Möller (2017b) include support for incorporating evidence more efficiently.

In this chapter, we describe two existing knowledge compilation approaches for lifted inference that are based on search: **weighted first-order model counting** (WFOMC)[16] (Van den Broeck et al., 2011; Van den Broeck, 2013) and L2C[17] (Kazemi et al., 2016). We com-

---

[16] aka Forclift: `https://github.com/UCLA-StarAI/Forclift`
[17] aka LRC2CPP: `https://github.com/Mehran-k/L2C`

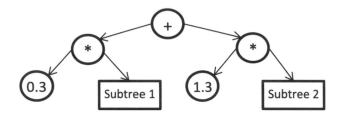

Figure 7.1: The trace of case analysis rule for the theory and weight functions in Example 7.1.

pare the two approaches together and with probabilistic theorem proving[18] (Gogate and Domingos, 2010), a searched-based approach to lifted inference. We start with describing how knowledge compilation can be used to find the WMC of a propositional theories.

7.1   Calculating WMC of Propositional Theories through Knowledge Compilation

The idea behind knowledge compilation (Darwiche and Marquis, 2002) approaches is to take a theory and compile it into a secondary structure on which several desired queries can be answered efficiently. The compilation phase is offline and is performed only once, whereas the query answering phase is online and may be performed multiple times. The aim of knowledge compilation is to push the operations to the offline phase as much as possible, thus speeding up the query answering in online phase.

As described in Section 5.2, the WMC of a theory can be calculated through a search-based algorithm as in Algorithm 1 by recursively applying a set of rules to the theory. Huang and Darwiche (2007) realized that any exhaustive search-based algorithm can be turned into a knowledge compiler by running the search-based algorithm symbolically and keeping the trace of the search as a data structure. We describe how Algorithm 1 can be turned into a knowledge compiler using an example.

Example 7.1 Consider a theory T as follows:

$$A \vee B \vee C$$
$$\neg A \vee B \vee \neg C$$
$$D \vee E \vee C$$
$$\neg D \vee E \vee \neg C$$

and two weight functions Φ and $\overline{\Phi}$ as follows:

$$\Phi(A) = 0.1, \Phi(B) = 0.2, \Phi(C) = 0.3, \Phi(D) = 0.4, \Phi(E) = 0.5$$
$$\overline{\Phi}(A) = 1.1, \overline{\Phi}(B) = 1.2, \overline{\Phi}(C) = 1.3, \overline{\Phi}(D) = 1.4, \overline{\Phi}(E) = 1.5$$

---

[18] Available on Alchemy-2: `https://code.google.com/archive/p/alchemy-2/`

Suppose Algorithm 1 selects C for case analysis. We know from Section 5.2 that Algorithm 1 will make two recursive calls: one where C has been replaced with *true*, and one where C has been replaced with *false*. The result of the former call is multiplied by $\Phi(C) = 0.3$, the result of the latter call is multiplied by $\overline{\Phi}(C) = 1.3$, and the sum of these two products are returned. The trace of this rule can be recorded as a data structure as in Fig 7.1, where subtree 1 and subtree 2 represent the subtrees returned by the two recursive calls respectively. The trace of the other rules can be recorded similarly to create a data structure.

Once the data structure is fully constructed, the WMC of the input theory for the two weight functions can be computed by executing the data structure. If the weight functions change, we do not need to do the compilation again: we simply change the weights (e.g., 0.3 and 1.3 in Fig 7.1) in the leaves of the data structure. Furthermore, if we observe the values of some random variables, (under mild conditions) we do not need to do the compilation again: we only consider the branches corresponding to the observed values in the tree (for instance if we observe *C* is *true*, then we only traverse the left branch of the tree in Fig 7.1).

While semantically equivalent, the data structure generated as a result of tracing the search comes under several names and shapes. The most well-known ones include dDNNFs (Darwiche and Marquis, 2002), arithmetic circuits (Darwiche, 2003), AND/OR trees (Dechter and Mateescu, 2007), and sum-product networks (Poon and Domingos, 2011).

7.2 Knowledge Compilation for First-order Theories

Van den Broeck et al. (2011) extended the idea of recording the search trace for generating knowledge compilers to first-order theories. They proposed a search-based algorithm to calculate the WMC of a first-order theory and then showed how a knowledge compiler can be constructed for these theories by recording the trace of the search. They recorded the trace of the search as a data structure corresponding to a first-order variant of the dDNNFs called FO-dDNNFs. Similar to the dDNNFs, if the weight functions change, one does not need to compile the input theory into a FO-dDNNF again, but can simply change the numbers in the data structure. Furthermore, Van den Broeck and Davis (2012) show how the compilation can be slightly modified so that for certain types of observations, the compilation needs not be done again.

Kazemi et al. (2016) also recorded the trace of a search-based lifted algorithm to construct a knowledge compiler, but instead of data structures they used low-level programs (C++ programs in particular) as their secondary structure. We describe how a search-based lifted inference algorithm can be executed symbolically to extract a secondary structure in the form a C++ program.

Compiling case-analysis to C++:   Let T contain an atom S with no logvars, T_1 represent T when S is replaced with *true*, and T_2 represent T when S is replaced by *false*. Then case-analysis on S generates the following C++ code:

```
{Code for storing WMC(T₁) in v1}
{Code for storing WMC(T₂) in v2}
v3 = posw[S] * v1 + negw[S] * v2;
```

Compiling lifted case-analysis to C++:   Let T be a theory and $S(x)$ be a unary atom in T. Let T_i be the theory resulting from branching on $S(x)$ being *true* for i out of $|\Delta_x|$ entities, and *false* for the rest. Then lifted case-analysis on $S(x)$ generates the following C++ code:

```
v1 = 0;
for(i=0; i <= popSize[x]; i++){
   {Code for storing WMC(Tᵢ) in v2}
   v1 += choose(popSize[x], i) * pow(posw[S], i)
         * pow(negw[S], popSize[x] - i) * v2;
}
```

Compiling decomposition to C++:   Let T be a theory which consists of two connected components T_1 and T_2 (can be easily extended to more than two connected components). Then decomposition on theory T generates the following C++ code:

```
{Code for storing WMC(T₁) in v1}
{Code for storing WMC(T₂) in v2}
v3 = v1 * v2;
```

Compiling lifted decomposition to C++:   Let T be a theory, x be a logvar in T on which lifted decomposition can be applied, and T' be the theory resulting from replacing x in T by some entity $X \in \Delta_x$. The lifted decomposition generates the following C++ code:

```
{Code for storing WMC(T') in v2}
v1 = pow(v2,popSize[x]);
```

Compiling unit propagation to C++:   Let T be a theory containing a unit clause $\forall x : S(x)$ (the cases where S is negated or has more or less logvars similar). Let T' be the theory resulting from propagating $S(x)$. Then unit propagation on $S(x)$ for theory T generates the following C++ code:

```
{Code for storing WMC(T') in v2}
v1 = pow(posw[S], popSize[x]) * v2;
```

Compiling shattering and caching to C++:   Shattering does not produce any C++ code, but prepares the theory for applying other rules on it. Before applying any rules, we check if the cache contains the theory T. If it does, then we generate the following C++ code:

```
v1 = cache[T];
```

In the compilation stage, we keep record of the cache entries that are used in future and remove from the C++ program all other cache inserts.

7.2.1 Compilation Example

Example 7.2 Consider compiling a theory T with the following sentences to C++:

$$\forall x \in \Delta_r, m \in \Delta_m : \mathsf{Y}(x, m) \Leftrightarrow \mathsf{R}(x, m) \wedge \mathsf{S}(x, m)$$
$$\forall x \in \Delta_x, m \in \Delta_m : \mathsf{Z}(x, m) \Leftrightarrow \mathsf{S}(x, m) \wedge \mathsf{V}(x)$$

where $\Delta_x = \{X_1, X_2, X_3, X_4, X_5\}$ and $\Delta_m = \{M_1, M_2\}$. Let $\Phi(Y) = 1.2$, $\Phi(Z) = 0.2$, and the rest of the weights be 1. Lifted decomposition can be applied on x. Let T' represent T when lifted decomposition has been applied on x. *L2C* generates the following C++ code:

{Code for storing WMC(T') in $v2$ }
$v1 = \text{pow}(v2, 5)$;

where 5 represents $|\Delta_x|$. T' has the following sentences:

$$\forall m \in \Delta_m : \mathsf{Y}(X_1, m) \Leftrightarrow \mathsf{R}(X_1, m) \wedge \mathsf{S}(X_1, m)$$
$$\forall m \in \Delta_m : \mathsf{Z}(X_1, m) \Leftrightarrow \mathsf{S}(X_1, m) \wedge \mathsf{V}(X_1)$$

Suppose we choose to do a case analysis on $\mathsf{S}(X_1, m)$. Assuming T'_i represents the theory when $\mathsf{S}(X_1, m)$ is *true* for i out of $|\Delta_m|$ entities and *false* for the rest, *L2C* generates the following code:

$v2 = 0$;
for(int $i = 0$; $i <= \text{popSize}[m]$; i++){
  {Code for storing WMC(T'_i) in $v3$ }
  $v2 \mathrel{+}= \text{choose}(\text{popSize}[m], i) * v3$;
$v1 = \text{pow}(v2, 5)$;

Note that we ignored the weights of assigning S to *true* and *false* in the code as both weights are 1. Continuing this process and applying the other rules may generate a code such as the one in Fig 7.2(a).

7.2.2 Pruning

Since the program obtained from *L2C* is generated automatically (not by a developer), a post-pruning step might be required to reduce the size of the program. For instance, one can remove lines 11 and 12 of the program in Fig 7.2(a) and replace line 13 with "$v6 = v9+1$;". The same can be done for lines 5 and 9. One may also notice that some variables are set to some values and are then being used only once. For example in the program of Fig 7.2(a), $v7$ and $v9$ are two such variables. The program can be pruned by removing these lines and replacing them with their values whenever they are being used. One can obtain the program in Fig 7.2(b) by pruning the program in Fig 7.2(a). Pruning can potentially save time and memory at run-time, but the pruning itself may be time-consuming.

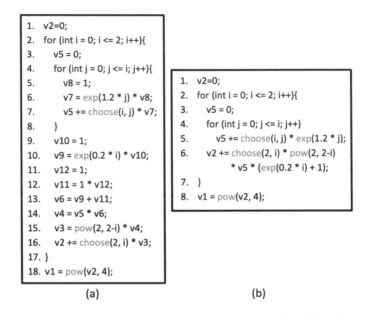

Figure 7.2: (a) The C++ program for the theory in Example 7.2. The WMC is stored in $v1$. (b) The C++ program in part (a) after pruning.

When the target program is C++, one can take advantage of the available packages developed for pruning/optimizing C++ programs. In particular, one can use the $-O3$ flag when compiling our programs which optimizes the code at compile time. Using $-O3$ slightly increases the compile time, but substantially reduces the run time when the program and the population sizes are large.

7.3   Why Compile to Low-level Programs?

Kazemi et al. (2016) reported an average of $175x$ speedup on their benchmarks for compiling to C++ programs instead of FO-dDNNFs. They also report that compiling to FO-dDNNFs offers speedup compared to not compiling. Fig 7.3 represents the performance of L2C, WFOMC, and probabilistic theorem proving (PTP), which is a search-based algorithm without compilation, on a benchmark theory.

To understand the reason why compiling to low-level programs can be more efficient than compiling to data structures, Kazemi et al. (2016) compare PTP, WFOMC, and L2C. Before getting into the explanation, we need to review the difference between program interpreters and program compilers. When we write a program in an interpreter language (e.g., Python, Ruby, etc.), at the execution time the commands in our code are considered and executed one by one. However, when we write a program in a compiler language

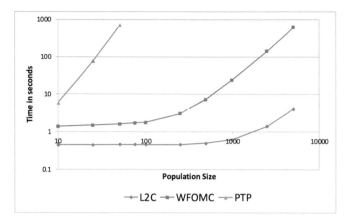

Figure 7.3: A comparison of L2C, WFOMC, and PTP on a benchmark theory taken from (Kazemi et al., 2016).

(e.g., C/C++, Java, etc.), the program is first compiled into lower-level operations, and then the lower-level operations are executed one by one. Compilers are generally orders of magnitude faster than interpreters as they have a lower overhead.

One way of viewing the process of calculating the WMC of a theory in Algorithm 2 is as follows: until a final criteria is met, we select the next rule that can be applied and execute it. Repeatedly selecting the next rule and executing it is similar to what interpreter languages do. Therefore, the PTP algorithm (or other search-based algorithms) can be viewed as running the lifted inference rules on a theory using an interpreter. WFOMC first compiles these rules into a data structure containing lower-level operations, and then executes those lower-level operations one by one. Therefore, WFOMC can be viewed as a compilation step, followed by an interpreting step. L2C, similar to WFOMC, first compiles the rules into a low-level program containing lower-level operations. Since the lower-level operations are stored as a programming language (e.g., C++), they can compile these lower-level operations and generate even lower-level operations, and then execute these operations one by one. Thus L2C can be viewed as two compilations steps followed by an interpreting step. Kazemi et al. (2016) show empirically that the extra compilation step in L2C compared to WFOMC is the reason why L2C outperforms WFOMC, and the extra compilation in WFOMC compared to PTP is the reason why WFOMC outperforms PTP. Note that all these algorithms result in the same time complexity and the boost in the performance comes from reducing the overhead. See Fig 7.4 for an illustrative explanation.

PTP, WFOMC, and L2C were all proposed before domain recursion rule was revived. While using the domain recursion rule for PTP and WFOMC is straight-forward, determining if/how it can be wired into L2C is currently an open problem. Furthermore, the

Figure 7.4: A comparison of search, compiling to data structures, and compiling to low-level programs for lifted inference.

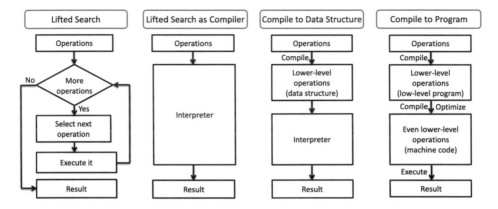

idea of compiling to low-level programs instead of data structures for other knowledge compilation problems is an interesting future direction.

8 Domain Liftability

Seyed Mehran Kazemi, Angelika Kimmig, Guy Van den Broeck, and David Poole

With the introduction of several lifting rules, a theoretical study of the power of these rules gained popularity. The theoretical study of lifted inference began with the complexity notion of *domain-lifted* inference and *domain-liftable* classes of models (Van den Broeck, 2011). Domain-liftability studies the classes of theories for which efficient lifted inference is or is not possible. Inference is domain-lifted when it runs in time polynomial in the number of objects in the domain. A class of models is domain-liftable when inference is domain-lifted for all models in that class. The theoretical power of lifting rules can be identified by characterizing the classes of models that become domain-liftable using those rules.

So far, several classes of domain-liftable models have been identified namely FO^2 (Van den Broeck, 2011; Van den Broeck et al., 2014), *recursively unary* (Poole et al., 2011), γ-acyclic (Beame et al., 2015), S^2FO^2 and S^2RU (Kazemi et al., 2016), and $BDR + RU$ (Kazemi et al., 2017). Characterizing such classes is not only theoretically interesting but also useful in practice, e.g., for lifted learning (Van Haaren et al., 2016) where the search space for theories is restricted to the ones for which lifted inference runs efficiently. One of the ultimate goals of domain-liftability is to fully characterize the class of theories for which lifted inference runs efficiently. While for several classes of models, it has been proved that lifted inference runs efficiently, none of these classes have been proved to be complete. Kazemi et al. (2017) conjecture that their $BDR + RU$ is the largest possible domain-liftable class. To this date, this conjecture has not been proved or disproved.

In this chapter, we formally define domain-liftability. Then we formally define the classes of theories identified so far for which lifted inference is or is not efficient.

8.1 Domain-liftable Classes

We start with the following the definition:

Definition 8.1 *A theory T is **domain-liftable** (Van den Broeck, 2011) if calculating its WMC is polynomial in population sizes* $|\Delta_{x_1}|, |\Delta_{x_2}|, \ldots, |\Delta_{x_k}|$ *where* $x_1, x_2, \ldots, x_k$ *represent the logvars in T. A class C of theories is domain-liftable if* $\forall T \in C$, *T is domain-liftable.*

So far, several classes of domain-liftable theories have been identified. We start with three of the first domain-liftable classes that have been identified so far: *FO²* (Van den Broeck, 2011; Van den Broeck et al., 2014), *recursively unary* (Poole et al., 2011), and *γ-acyclic* (Beame et al., 2015). Let $\mathcal{R}^{-\mathcal{DR}}$ denote the set of rules introduced in Section 5.3 except domain recursion. We define each of these classes below:

Definition 8.2 *A theory is in FOⁱ if all its clauses have up to i logvars.*

Hence, a theory is in *FO²* if all its clauses have up to 2 logvars.

Definition 8.3 *A theory T is* recursively unary (RU) *if for every theory T′ resulting from applying rules in $\mathcal{R}^{-\mathcal{DR}}$ except for lifted case analysis to T, until no more rules apply, either T′ has no sentences, or there exists some unary atom* $S(\ldots, x, \ldots)$ *in T′, such that for any* $\Delta_{x_T} \subset \Delta_x$ *and* $\Delta_{x_F} \subset \Delta_x$ *where* $\Delta_{x_T} \cup \Delta_{x_F} = \Delta_x$ *and* $\Delta_{x_T} \cap \Delta_{x_F} = \emptyset$, *the theory* $\forall x_T \in \Delta_{x_T} : S(\ldots, x_T, \ldots) \;\wedge\; \forall x_F \in \Delta_{x_F} : \neg S(\ldots, x_F, \ldots) \;\wedge\; T'$ *is* RU.

To define the *γ-acyclic* class, we need to define hypergraphs and *γ-cycles*. We use the definition in Duris (2012). A **hypergraph** is a set of vertices $\mathcal{V}$ and a set of hyperedges $\mathcal{E}$, where a hyperedge is a subset of vertices in $\mathcal{V}$. A **γ-cycle** is a sequence $E_1, x_1, E_2, x_2, \ldots, E_n, x_n$ $(n \geq 3)$ where E_is are distinct hyperedges and x_is distinct vertices such that for each $1 \leq i \leq n$, $x_i \in E_i$, $x_i \in E_{(i+1) \bmod n}$ and $x_i \notin E_j$ for any other E_j.

Definition 8.4 *A theory T containing only one clausal sentence is γ-acyclic if it is without self-joins and a hypergraph whose vertices are the logvars in T and whose hyperedges are the atoms in T has no γ-cycles.*

Note that every theory in one of the above classes is domain-liftable, and a theory that is not in any of the above classes may or may not be domain-liftable.

Beame et al. (2015) identified a theory that was not in any of the above classes and none of the rules in $\mathcal{R}^{-\mathcal{DR}}$ where applicable to it, yet its WMC could be computed in polynomial time. They asked for a new rule to solve their theory. The theory they identified (named *S4*) is as follows:

$$\forall x_1, x_2 \in \Delta_x, y_1, y_2 \in \Delta_y : S(x_1, y_1) \vee \neg S(x_2, y_1) \vee S(x_2, y_2) \vee \neg S(x_1, y_2)$$

The identification of the *S4* theory led into the revival of the domain recursion rule as Kazemi et al. (2016) showed that domain recursion is indeed the rule that solves the *S4* theory. Besides solving the *S4* theory, Kazemi et al. (2016) showed that domain recursion also makes domain-lifted inference (through general-purpose rules) possible for the *symmetric transitivity* theory:[19]

---

[19] Finding the WMC for the non-symmetric case is more difficult and is a long-lasting open problem in lifted probabilistic inference.

$$\forall x, y, z \in \Delta_p : \neg F(x, y) \lor \neg F(y, z) \lor F(x, z)$$
$$\forall x, y \in \Delta_p : \neg F(x, y) \lor F(y, x)$$

The revival of the domain recursion rule resulted in identification of three new domain-liftable classes: S^2FO^2, S^2RU (Kazemi et al., 2016), and $BDR + RU$ (Kazemi et al., 2017). Let $\mathcal{R}$ represent $\mathcal{R}^{-\mathcal{DR}}$ plus bounded domain recursion. We define these classes below:

Definition 8.5 *Let $\alpha(S)$ be a clausal theory that uses a single binary predicate S, such that each clause has exactly two different literals of S. Let $\alpha = \alpha(S_1) \land \alpha(S_2) \land \cdots \land \alpha(S_n)$ where the S_i are different binary predicates. Let β be a theory where all clauses contain at most one S_i literal, and the clauses that contain an S_i literal contain no other literals with more than one logvar. Then, S^2FO^2 and S^2RU are the classes of theories of the form $\alpha \land \beta$ where $\beta \in FO^2$ and $\beta \in RU$ respectively.*

In what follows, we provide an example of a theory that is not in FO^2 and not in RU, but is in S^2FO^2 and *StwoRU*.

Example 8.1 Suppose we have a set Δ_j of jobs and a set Δ_v of volunteers. Every volunteer must be assigned to at most one job, and every job requires no more than one person. If the job involves working with gasoline, the assigned volunteer must be a non-smoker. And we know that smokers are most probably friends with each other. Then we have the following first-order theory:

$$\forall v_1, v_2 \in \Delta_v, j \in \Delta_j : \neg\mathsf{Assigned}(v_1, j) \lor \neg\mathsf{Assigned}(v_2, j)$$
$$\forall v \in \Delta_v, j_1, j_2 \in \Delta_j : \neg\mathsf{Assigned}(v, j_1) \lor \neg\mathsf{Assigned}(v, j_2)$$
$$\forall v \in \Delta_v, j \in \Delta_j : \mathsf{InvolvesGas}(j) \land \mathsf{Assigned}(v, j) \Rightarrow \neg\mathsf{Smokes}(v)$$
$$\forall v_1, v_2 \in \Delta_v : \mathsf{Aux}(v_1, v_2) \Leftrightarrow (\mathsf{Smokes}(v_1) \land \mathsf{Friends}(v_1, v_2) \Rightarrow \mathsf{Smokes}(v_2))$$

Predicate Aux is added to capture the probability assigned to the last rule (as in MLNs). This theory is not in FO^2, not in RU, and is not domain-liftable using $\mathcal{R}^{-\mathcal{DR}}$. However, the first two clauses are of the form described in Definition 8.5, the third and fourth are in FO^2 (and also in RU), and the third clause, which contains $\mathsf{Assigned}(v, j)$, has no other atoms with more than one logvar. Therefore, this theory is in S^2FO^2 (and also in S^2RU) and domain-liftable.

In what follows, we define the *BDR+RU* class from (Kazemi et al., 2017).

Definition 8.6 *A theory T is in* BDR+RU *if for every theory T' resulting from applying rules in $\mathcal{R}$ except for lifted case analysis to T, until no more rules apply, either T' has no sentences, or there exists a unary atom $S(\ldots, x, \ldots)$ in T', such that for any $\Delta_{x_T} \subset$*

Δ_x *and* $\Delta_{x_F} \subset \Delta_x$ *where* $\Delta_{x_T} \cup \Delta_{x_F} = \Delta_x$ *and* $\Delta_{x_T} \cap \Delta_{x_F} = \emptyset$, *the theory* $\forall x_T \in \Delta_{x_T}$: $S(\ldots, x_T, \ldots) \ \wedge \ \forall x_F \in \Delta_{x_F} : \neg S(\ldots, x_F, \ldots) \ \wedge \ T'$ *is in* BDR+RU.

Note that *BDR+RU* is defined similarly to *RU* with the only difference being that $\mathcal{R}^{-\mathcal{DR}}$ is replaced with $\mathcal{R}$.

Kazemi et al. (2016) prove that $FO^2 \subset S^2FO^2$, $FO^2 \subset RU$, $RU \subset S^2RU$, and $S^2FO^2 \subset S^2RU$. *BDR+RU* is by definition a superset of S^2RU. Kazemi et al. (2017) conjecture that *BDR+RU* is complete (i.e. the largest possible domain-liftable class). The conjecture can be also viewed as $\mathcal{R}$ being a complete set of rules for lifted inference. Proving or disproving this conjecture is an open problem.

8.1.1   Membership Checking

As mentioned earlier, Kazemi et al. (2016) proved that S^2RU is a superclass of FO^2, S^2FO^2 and *RU*. However, this proof does not suggest that one may only consider the S^2RU class and forget about the other ones, as determining whether a theory is in each of the above classes (aka membership checking) may not have the same time complexity. Membership checking for FO^2 and S^2FO^2 may be easier than membership checking for *RU* or S^2RU as the latter requires symbolic derivation of the rules.

Based on membership checking, we separate between two types of domain-liftable classes:

Definition 8.7 *Let C be a domain-liftable class of theories. C is* syntactically domain-liftable (SynDL) *if for any given theory T, determining whether* $T \in C$ *can be done by just looking at the syntax of T. C is* symbolically domain-liftable (SymDL) *if for any given theory T, determining whether* $T \in C$ *requires symbolic derivation of the rules.*

Having the above definition, FO^2, and S^2FO^2 are SynDL classes and *RU*, S^2RU and *BDR + RU* are SymDL classes. γ-*acyclic* is also a SymDL class as determining if a hypergraph has no γ-*cycles* requires a symbolic derivation of rules until an empty hypergraph is obtained or the rules fail (Fagin, 1983; Beame et al., 2015). There is a trade-off between SynDL and SymDL classes: SynDL classes may be smaller than SymDL classes, but determining if a theory belongs to a SynDL class may take less time than determining if it belongs to a SymDL class. For both SynDL and SymDL classes, the time required to determine if a theory belongs to the class is independent of the size of the populations.

A richer study of these two classes and determining the time complexities of membership checking for each class is an interesting direction for future research.

8.2   Negative Results

With the discovery of large domain-liftable classes, the question was raised as to whether a limit exists on what could be lifted or not. In other words, the question was whether all theories in finite-domain, function-free subset of first-order logic were domain-liftable. Jaeger (2015) and Jaeger and Van den Broeck (2012) answered this question negatively. They

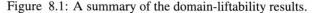

Figure 8.1: A summary of the domain-liftability results.

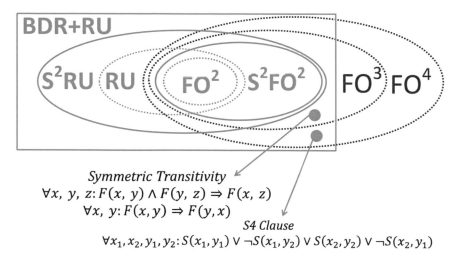

proved that there exists at least one theory for which lifted inference is not domain-liftable, unless *ETIME* = *NETIME*. The class *ETIME* is the class of problems that are solvable in time $O(2^{cn})$ for some constant c, and *NETIME* is the corresponding non-deterministic class (Leeuwen, 1990).

Beame et al. (2015) extended the negative result in (Jaeger, 2015; Jaeger and Van den Broeck, 2012) and showed that there exists at least one theory in FO^3 for which finding the WMC is $\#P_1$-*complete*. The class $\#P_1$ (Valiant, 1979) is the class of counting problems for NP computations over a single-letter input alphabet.

Fig 8.1 from Kazemi et al. (2017) puts together the positive and negative liftability results. This figure shows the hierarchy of the domain-liftable classes introduced earlier except for the *γ-acyclic* class. Finding the appropriate place for *γ-acyclic* theories in the hierarchy of Fig 8.1 is an open problem.

Finally, Kuzelka and Wang (2020) show that domain-liftability result for probabilistic inference can be generalized to apply to lifted parameter learning (Van den Broeck et al., 2013; Kuzelka and Kungurtsev, 2019), as well as the problem of constructing the relational marginal polytope.

9 Tractability through Exchangeability: The Statistics of Lifting

Mathias Niepert and Guy Van den Broeck

Abstract. Exchangeability is a central notion in statistics and probability theory. The assumption that an infinite sequence of data points is exchangeable is at the core of Bayesian statistics. However, *finite* exchangeability as a statistical property that renders probabilistic inference tractable is less well-understood. We develop a theory of finite exchangeability and its relation to tractable probabilistic inference. The theory is complementary to that of independence and conditional independence. We show that tractable inference in probabilistic models with high treewidth and millions of variables can be explained with the notion of finite (partial) exchangeability. We also show that existing lifted inference algorithms implicitly utilize a combination of conditional independence and partial exchangeability.

9.1 Introduction

Probabilistic graphical models such as Bayesian and Markov networks explicitly represent *conditional independencies* of a probability distribution with their structure (Pearl, 1988; Lauritzen, 1996; Koller and Friedman, 2009; Darwiche, 2009). Their wide-spread use in research and industry can largely be attributed to this structural property and their declarative nature, separating representation and inference algorithms. Conditional independencies often lead to a more concise representation and facilitate efficient algorithms for parameter estimation and probabilistic inference. It is well-known, for instance, that probabilistic graphical models with a tree structure admit efficient inference. In addition to conditional independencies, modern inference algorithms exploit *contextual independencies* (Boutilier et al., 1996) to speed up probabilistic inference.

The time complexity of classical probabilistic inference algorithms is exponential in the treewidth (Robertson and Seymour, 1986) of the graphical model. Independence and its various manifestations often reduce treewidth and treewidth has been used in the literature as the decisive factor for assessing the tractability of probabilistic inference (cf. Koller and Friedman (2009), Darwiche (2009)). However, recent algorithmic developments have shown that inference in probabilistic graphical models can be highly tractable, even in high-treewidth models without any conditional independencies. In particular, lifted probabilistic inference algorithms often perform efficient inference in densely connected graphical models with millions of random variables. With the success of lifted inference, understanding these algorithms and their tractability on a more fundamental level has become a central challenge. A pressing question concerns the underlying statistical principle that allows inference to be tractable in the absence of independence.

This chapter develops a deeper understanding of the statistical properties that render inference tractable. We consider an inference problem tractable when it is solved by an efficient algorithm, running in time polynomial in the number of random variables. The crucial contribution is a comprehensive theory that relates the notion of finite partial *exchangeability* (Diaconis and Freedman, 1980) to tractability. One instance is full exchangeability where the distribution is invariant under variable permutations. We develop a theory of exchangeable decompositions that results in novel tractability conditions. Similar to conditional independence, partial exchangeability decomposes a probabilistic model so as to facilitate efficient inference. Most importantly, the notions of conditional independence and partial exchangeability are complementary, and when combined, define a much larger class of tractable models than the class of models rendered tractable by conditional independence alone.

Conditional and contextual independence are such powerful concepts because they are *statistical* properties of the distribution, regardless of the representation used. Partial exchangeability is such a statistical property that is independent of any representation, be it a joint probability table, a Bayesian network, or a statistical relational model. We introduce novel forms of exchangeability, discuss their sufficient statistics, and efficient inference algorithms. The resulting exchangeability framework allows us to state known liftability results as corollaries, providing a first statistical characterization of exact lifted inference. As an additional contribution, we connect the semantic notion of exchangeability to *syntactic* notions of tractability by showing that liftable statistical relational models have the required exchangeability properties due to their syntactic symmetries. We thereby unify notions of lifting from the exact and approximate inference community into a common framework.

9.2   A Case Study: Markov Logic

The analysis of exchangeability and tractable inference is developed in the context of arbitrary discrete probability distributions, independent of a particular representational formalism. Nevertheless, for the sake of accessibility, we will provide examples and intuitions for Markov logic, a well-known statistical relational language that exhibits several forms of exchangeability. Hence, after the derivation of the theoretical results in each section, we apply the theory to the problem of inference in Markov logic networks. This also allows us to link the theory to existing results from the lifted probabilistic inference literature.

9.2.1   Markov Logic Networks Review

Recall some standard concepts from function-free first-order logic. An *atom* $P(t_1, \ldots, t_n)$ consists of a predicate P/n of arity n followed by n argument terms t_i, which are either *constants*, $\{A, B, \ldots\}$ or *logical variables* $\{x, y, \ldots\}$. A *unary atom* has one argument and a *binary atom* has two. A formula combines atoms with connectives (e.g., $\wedge$, $\Leftrightarrow$). A formula is *ground* if it contains no logical variables. The groundings of a formula are obtained by instantiating the variables with particular constants.

A *Markov logic network* (MLN) (Richardson and Domingos, 2006) is a set of tuples (w, f), where w is a real number representing a weight and f is a formula in first-order logic. Let us, for example, consider the following MLN

$$1.3 \quad \text{Smokes}(x) \Rightarrow \text{Cancer}(x) \tag{9.1}$$

$$1.5 \quad \text{Smokes}(x) \wedge \text{Friends}(x, y) \Rightarrow \text{Smokes}(y) \tag{9.2}$$

which states that smokers are more likely to (9.1) get cancer and (9.2) be friends with other smokers.

Given a domain of constants $\mathbf{D}$, a first-order MLN Δ induces a *grounding*, which is the MLN obtained by replacing each formula in Δ with all its groundings (using the same weight). Take for example the domain $\mathbf{D} = \{A, B\}$ (e.g., two people, Alice and Bob), the above first-order MLN represents the following grounding.

$$1.3 \quad \text{Smokes}(A) \Rightarrow \text{Cancer}(A)$$

$$1.3 \quad \text{Smokes}(B) \Rightarrow \text{Cancer}(B)$$

$$1.5 \quad \text{Smokes}(A) \wedge \text{Friends}(A, A) \Rightarrow \text{Smokes}(A)$$

$$1.5 \quad \text{Smokes}(A) \wedge \text{Friends}(A, B) \Rightarrow \text{Smokes}(B)$$

$$1.5 \quad \text{Smokes}(B) \wedge \text{Friends}(B, A) \Rightarrow \text{Smokes}(A)$$

$$1.5 \quad \text{Smokes}(B) \wedge \text{Friends}(B, B) \Rightarrow \text{Smokes}(B)$$

This ground MLN contains eight different random variables, which correspond to all groundings of atoms $\text{Smokes}(x)$, $\text{Cancer}(x)$ and $\text{Friends}(x, y)$. This leads to a distribution over 2^8 possible worlds. The weight of each world is the product of the expressions $\exp(w)$, where (w, f) is a ground MLN formula and f is satisfied by the world. The probabilities of worlds are obtained by normalizing their weights. Without loss of generality (Jha et al., 2010), we assume that first-order formulas contain no constants.

9.2.2   Inference without Independence

Surprisingly, lifted inference algorithms perform tractable inference even in the absence of conditional independencies. For example, when interpreting the above MLN as an (undirected) probabilistic graphical model, all pairs of variables in $\{\text{Smokes}(A), \text{Smokes}(B), \ldots\}$ are connected by an edge due to the groundings of Formula 9.2. The model has no con-

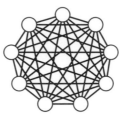

Figure 9.1: An undirected graphical model with 9 finitely exchangeable Bernoulli variables. There are no (conditional) independencies that hold among the variables.

ditional or contextual independencies between the Smokes variables. Nevertheless, lifted inference algorithms exactly compute its single marginal probabilities in time linear in the size of the corresponding graphical model (Van den Broeck, 2011), scaling up to millions of random variables.

As lifted inference research makes algorithmic progress, the quest for the source of tractability and its theoretical properties becomes increasingly important. For exact lifted inference, most theoretical results are based on the notion of domain-lifted inference (Van den Broeck, 2011).

Definition 9.1 (Domain-Lifted) *Domain-lifted inference algorithms run in time polynomial in* $|\mathbf{D}|$.

Note that domain-lifted algorithms can be exponential in other parameters, such as the number of formulas and predicates. Our current understanding of exact lifted inference is that syntactic properties of MLN formulas permit domain-lifted inference (Van den Broeck, 2011; Jaeger and Van den Broeck, 2012; Taghipour et al., 2013c; Van den Broeck et al., 2014; Beame et al., 2015; Kazemi et al., 2016). We will review these results where relevant. Moreover, the (fractional) automorphisms of the graphical model representation have been related to lifted inference (Niepert, 2012b; Bui et al., 2013b; Noessner et al., 2013; Mladenov et al., 2012). While there are deep connections between automorphisms and exchangeability (Niepert, 2012b, 2013; Bui et al., 2012), we refer these to future work.

9.3   Finite Exchangeability

This section provides some background on the concept of **finite partial exchangeability**. We proceed by showing that particular forms of finite exchangeability permit tractable inference. For the sake of simplicity and to provide links to statistical relational models such as MLNs, we present the theory for *finite* sets of (upper-case) *binary* random variables $\mathbf{X} = \{X_1, \ldots, X_n\}$. However, the theory applies to all distributions over finite valued discrete random variables. Lower-case $\mathbf{x}$ denote an assignments to $\mathbf{X}$.

We begin with the most basic form of exchangeability.

Definition 9.2 (full exchangeability) *A set of variables* $\mathbf{X} = \{X_1, ..., X_n\}$ *is fully exchangeable if and only if* $\Pr(X_1 = x_1, \ldots, X_n = x_n) = \Pr(X_1 = x_{\pi(1)}, \ldots, X_n = x_{\pi(n)})$ *for all permutations* π *of* $\{1, \ldots, n\}$.

Full exchangeability is best understood in the context of a finite sequence of binary random variables such as a number of coin tosses. Here, exchangeability means that it is only the number of heads that matters and not their particular order. Figure 9.1 depicts an undirected graphical model with 9 finitely exchangeable dependent Bernoulli variables.

9.3.1   Finite Partial Exchangeability

The assumption that all variables of a probabilistic model are exchangeable is often too strong. Fortunately, exchangeability can be generalized to the concept of partial exchangeability using the notion of a **sufficient statistic** (Diaconis and Freedman, 1980; Lauritzen et al., 1984; Lauritzen and Spiegelhalter, 1988). Particular instances of exchangeability such as full finite exchangeability correspond to particular statistics.

Definition 9.3 (Partial Exchangeability) *Let* $\mathcal{D}_i$ *be the domain of* X_i, *and let* $\mathcal{T}$ *be a finite set. A set of random variables* $\mathbf{X}$ *is* partially exchangeable *with respect to the statistic* $T : \mathcal{D}_1 \times \cdots \times \mathcal{D}_n \to \mathcal{T}$ *if and only if*

$$T(\mathbf{x}) = T(\mathbf{x}') \text{ implies } \Pr(\mathbf{x}) = \Pr(\mathbf{x}').$$

The following theorem states that the joint distribution of a set of random variables that is partially exchangeable with a statistic T is a unique mixture of uniform distributions.

Theorem 9.1 (Diaconis and Freedman (1980)) *Let* $\mathcal{T}$ *be a finite set. Let* $T : \{0, 1\}^n \to \mathcal{T}$ *be a statistic of a partially exchangeable set* $\mathbf{X}$. *Let* $S_t = \{\mathbf{x} \in \mathcal{D}_1 \times \cdots \times \mathcal{D}_n \mid T(\mathbf{x}) = t\}$, *let* U_t *be the uniform distribution over* S_t, *and let* $\Pr(S_t) = \Pr(T(\mathbf{x}) = t)$. *Then,*

$$\Pr(\mathbf{X}) = \sum_{t \in \mathcal{T}} \Pr(S_t) U_t(\mathbf{X}).$$

Hence, a distribution that is partially exchangeable with respect to a statistic T can be parameterized as a unique mixture of uniform distributions. We will see that several instances of partial exchangeability render probabilistic inference tractable. Indeed, the major theme of this chapter can be summarized as finding methods for constructing the above representation and exploiting it for tractable probabilistic inference for a given probabilistic model.

Let $[[\cdot]]$ be the indicator function. The uniform distribution of each equivalence class S_t is $U_t(\mathbf{X}) = [[T(\mathbf{X}) = t]]/|S_t|$; and the probability of S_t is $\Pr(S_t) = \Pr(\mathbf{x})|S_t|$ for every $\mathbf{x} \in S_t$. Hence, every value of the statistic T corresponds to one equivalence class S_t of joint assignments with identical probability. We will refer to these equivalence classes as **orbits**. We write $\mathbf{x} \sim e$ when assignments $\mathbf{x}$ and e agree on the values of their shared variables (Darwiche, 2009). The *suborbit* $S_{t,e} \subseteq S_t$ for some evidence state e is the set of those states in S_t that are compatible with e, that is, $S_{t,e} = \{\mathbf{x} \mid T(\mathbf{x}) = t \text{ and } \mathbf{x} \sim e\}$.

9.3.2   Partial Exchangeability and Probabilistic Inference

We are now in the position to relate finite partial exchangeability to tractable probabilistic inference, using notions from Theorem 9.1. The inference tasks we consider are

– *MPE inference*, i.e., finding $\text{argmax}_\mathbf{y}\,\text{Pr}(\mathbf{y}, e)$ for any given assignment e to variables $\mathbf{E} \subseteq \mathbf{X}$, and
– *marginal inference*, i.e., computing $\text{Pr}(e)$ for any given e.

For a set of variables $\mathbf{X}$, we say that $P(\mathbf{x})$ can be computed *efficiently* iff it can be computed in time polynomial in $|\mathbf{X}|$. We make the following complexity claims

Theorem 9.2 *Let* $\mathbf{X}$ *be partially exchangeable with statistic T. If we can efficiently*
– *for all* $\mathbf{x}$, *evaluate* $\text{Pr}(\mathbf{x})$, *and*
– *for all* e *and* $t \in \mathcal{T}$ *decide whether there exists an* $\mathbf{x} \in S_{t,e}$, *and if so, construct it,*

then the complexity of MPE *inference is polynomial in* $|\mathcal{T}|$. *If we can additionally compute* $|S_{t,e}|$ *efficiently, then the complexity of* marginal *inference is also polynomial in* $|\mathcal{T}|$.

Proof   For MPE inference, we construct an $\mathbf{x}_t \in S_{t,e}$ for each $t \in \mathcal{T}$, and return the one maximizing $\text{Pr}(\mathbf{x}_t)$. For marginal inference, we return $\sum_{t \in \mathcal{T}} \text{Pr}(\mathbf{x}_t)|S_{t,e}|$.

If the above conditions for tractable inference are fulfilled we say that a distribution is *tractably* partially exchangeable for MPE or marginal inference. We will present notions of exchangeability and related statistics T that make distributions tractably partially exchangeable. Please note that Theorem 9.2 generalizes to situations in which we can only efficiently compute $\text{Pr}(\mathbf{x})$ up to a constant factor Z, as is often the case in undirected probabilistic graphical models.

9.3.3   Markov Logic Case Study

Exchangeability and independence are not mutually exclusive. Independent and identically distributed (iid) random variables are also exchangeable. Take for example the MLN

$$1.5 \quad \text{Smokes}(x)$$

The random variables $\text{Smokes}(A), \text{Smokes}(B), \dots$ are independent. Hence, we can compute their marginal probabilities independently as

$$\text{Pr}(\text{Smokes}(A)) = \text{Pr}(\text{Smokes}(B)) = \frac{\exp(1.5)}{\exp(1.5) + 1}$$

The variables are also finitely exchangeable. For example, the probability that A smokes and B does not is equal to the probability that B smokes and A does not. The sufficient statistic $T(\mathbf{x})$ counts *how many* people smoke in the state $\mathbf{x}$ and the probability of a state in which n out of N people smoke is $\exp(1.5n)/(\exp(1.5) + 1)^N$.

Exchangeability can occur without independence, as in the following Markov logic network

$$1.5 \quad \text{Smokes}(x) \wedge \text{Smokes}(y)$$

This distribution has neither independencies nor conditional independencies. However, its variables are finitely exchangeable and the probability of a state $\mathbf{x}$ is only a function of the sufficient statistic $T(\mathbf{x})$ counting the number of smokers in $\mathbf{x}$. The probability of a state now increases by a factor of $\exp(1.5)$ with every *pair of smokers*. When n people smoke there are n^2 pairs and, hence, $\Pr(\mathbf{x}) = \exp\left(1.5n^2\right)/Z$, where Z is a normalization constant. Let Y consist of all $\text{Smokes}(x)$ variables except for $\text{Smokes}(A)$, and let $\mathbf{y}$ be an assignment to Y in which m people smoke. The probability that A smokes given $\mathbf{y}$ is

$$\Pr\left(\text{Smokes}(A) \mid \mathbf{y}\right) = \frac{\exp\left(1.5(m+1)^2\right)}{\exp\left(1.5(m+1)^2\right) + \exp\left(1.5m^2\right)},$$

which clearly depends on the number of smokers in $\mathbf{y}$. Hence, $\text{Smokes}(A)$ is not independent of Y but the random variables are exchangeable with sufficient statistic n. Figure 9.1 depicts the graphical representation of the corresponding ground Markov logic network.

9.4 Exchangeable Decompositions

We now present novel instances of partial exchangeability that render probabilistic inference tractable. These instances generalize exchangeability of single variables to exchangeability of sets of variables. We describe the notion of an **exchangeable decomposition** and prove that it fulfills the tractability requirements of Theorem 9.2. We proceed by demonstrating that these forms commonly occur in MLNs.

9.4.1 Variable Decompositions

The notions of independent and exchangeable decompositions are at the core of the developed theoretical results.

Definition 9.4 (Variable Decomposition) A variable decomposition $\mathcal{X} = \{\mathbf{X}_1, \ldots, \mathbf{X}_k\}$ *partitions* $\mathbf{X}$ *into subsets* $\mathbf{X}_i$. *We call* $w = \max_i |\mathbf{X}_i|$ *the* width *of the decomposition.*

Definition 9.5 (Independent Decomposition) *A variable decomposition is* independent *if and only if* $\Pr$ *factorizes as*

$$\Pr\left(\mathbf{X}\right) = \prod_{i=1}^{k} Q_i(\mathbf{X}_i).$$

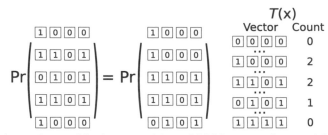

Figure 9.2: An exchangeable decomposition of 20 binary random variables (the boxes) into 5 components of width 4 (the rows). The statistic T counts the occurrences of each unique binary vector.

Definition 9.6 (Exchangeable Decomposition) *A variable decomposition is* exchangeable *iff for all permutations π,*

$$\Pr\left(\mathbf{X}_1 = \mathbf{x}_1, \ldots, \mathbf{X}_k = \mathbf{x}_k\right)$$
$$= \Pr\left(\mathbf{X}_1 = \mathbf{x}_{\pi(1)}, \ldots, \mathbf{X}_k = \mathbf{x}_{\pi(k)}\right).$$

Figure 9.2 depicts an example distribution Pr with 20 random variables and a decomposition into 5 subsets of width 4. The joint distribution is invariant under permutations of the 5 sequences. The corresponding sufficient statistic T counts the number of occurrences of each binary vector of length 4 and returns a tuple of counts.

Please note that the definition of full finite exchangeability (Definition 9.2) is the special case when the exchangeable decomposition has width 1. Also note that the size of all subsets in an exchangeable decomposition equal the width.

9.4.2   Tractable Variable Decompositions

The core observation of this chapter is that variable decompositions that are exchangeable and/or independent result in tractable probabilistic models. For independent decompositions, the following tractability guarantee is used in most existing inference algorithms.

Proposition 9.1 *Given an independent decomposition of $\mathbf{X}$ with bounded width, and a corresponding factorized representation of the distribution (cf. Definition 9.5), the complexity of MPE and marginal inference is polynomial in $|\mathbf{X}|$.*

While the decomposition into independent components is a well-understood concept, the combination with finite exchangeability has not been previously investigated as a statistical property that facilitates tractable probabilistic inference. We can now prove the following result.

Theorem 9.3 *Suppose we can compute* $\Pr(\mathbf{x})$ *in time polynomial in* $|\mathbf{X}|$. *Then, given an exchangeable decomposition of* $\mathbf{X}$ *with bounded width, the complexity of MPE and marginal inference is polynomial in* $|\mathbf{X}|$.

Proof Following Theorem 9.2, we have to show that there exists a statistic T so that (a) $|\mathcal{T}|$ is polynomial in $|\mathbf{X}|$; (b) we can efficiently decide whether an $\mathbf{x} \in S_{t,e}$ exists and if so, construct it; and (c) efficiently compute $|S_{t,e}|$ for all e and $t \in \mathcal{T}$. Statements (b) and (c) ensure that the assumptions of Theorem 9.2 hold for exchangeable decompositions, which combined with (a) proves the theorem.

To prove (a), let us first construct a sufficient statistic for exchangeable decompositions. A full joint assignment $\mathbf{X} = \mathbf{x}$ decomposes into assignments $\mathbf{X}_1 = \mathbf{x}_1, \ldots, \mathbf{X}_k = \mathbf{x}_k$ in accordance with the given variable decomposition. Each $\mathbf{x}_\ell$ is a bit string $b \in \{0,1\}^w$. Consider a statistic $T(\mathbf{x}) = (c_1, \ldots, c_{2^w})$, where each c_i has a corresponding unique bit string $b_i \in \{0,1\}^w$ and $c_i = \sum_{\ell=1}^{k}[[\mathbf{x}_\ell = b_i]]$. The value c_i of the statistic thus represents the number of components in the decomposition that are assigned bit string b_i. Hence, we have $\sum_{i=1}^{2^w} c_i = k$, and we prove (a) by observing that

$$|\mathcal{T}| = \binom{k + 2^w - 1}{2^w - 1} \le k^{2^w - 1} \le n^{2^w - 1}.$$

To prove statements (b) and (c) we have to find, for each partial assignment $E = e$, an algorithm that generates an $\mathbf{x} \in S_{t,e}$ and that computes $|S_{t,e}|$ in time polynomial in $|\mathbf{X}|$. To hint at the proof strategy, we give the formula for the orbit without evidence $|S_t|$:

$$|S_{(c_1,\ldots,c_{2^w})}| = \binom{k}{c_1}\binom{k - c_1}{c_2} \cdots \binom{k - \sum_{i=1}^{2^w-2} c_i}{c_{2^w-1}}.$$

The proof is very technical and deferred to the appendix.

9.4.3 Markov Logic Case Study

Let us consider the following MLN

$$1.3 \quad \mathsf{Smokes}(x) \Rightarrow \mathsf{Cancer}(x)$$
$$1.5 \quad \mathsf{Smokes}(x) \wedge \mathsf{Smokes}(y)$$

It models a distribution in which every non-smoker or smoker with cancer, that is, every x satisfying the first formula, increases the probability by a factor of $\exp(1.3)$. Each pair of smokers increases the probability by a factor of $\exp(1.5)$. This model is not fully exchangeable: swapping $\mathsf{Smokes}(A)$ and $\mathsf{Cancer}(A)$ in a state yields a different probability. There are no (conditional) independencies between the $\mathsf{Smokes}(x)$ atoms.

The variables in this MLN do have an exchangeable decomposition whose width is two, namely

$$X = \{\{\mathsf{Smokes(A), Cancer(A)}\},$$
$$\{\mathsf{Smokes(B), Cancer(B)}\},$$
$$\{\mathsf{Smokes(C), Cancer(C)}\}, \dots \}.$$

The sufficient statistic of this decomposition counts the number of people in each of four groups, depending on whether they smoke, and whether they have cancer. The probability of a state only depends on the number of people in each group and swapping people between groups does not change the probability of a state. For example,

$$\Pr(\mathsf{Smokes(A)} = 0, \mathsf{Cancer(A)} = 1,$$
$$\mathsf{Smokes(B)} = 1, \mathsf{Cancer(B)} = 0)$$
$$= \Pr(\mathsf{Smokes(A)} = 1, \mathsf{Cancer(A)} = 0,$$
$$\mathsf{Smokes(B)} = 0, \mathsf{Cancer(B)} = 1).$$

Theorem 9.3 says that this MLN permits tractable inference.

The fact that this MLN has an exchangeable decomposition is not a coincidence. In general, we can show this for MLNs of unary atoms, which are called *monadic* MLNs.

Theorem 9.4 *The variables in a monadic MLN have an exchangeable decomposition. The width of this decomposition is equal to the number of predicates.*

The proof builds on syntactic symmetries of the MLN, called renaming automorphisms (Bui et al., 2012; Niepert, 2012b). Please see the appendix for further details.

It now follows as a corollary of Theorems 9.3 and 9.4 that MPE and marginal inference in monadic MLNs is polynomial in $|\mathbf{X}|$, and therefore also in the domain size $|\mathbf{D}|$.

Corollary 9.4.1 *Inference in monadic MLNs is domain-lifted.*

9.5   Marginal and Conditional Exchangeability

Many distributions are not decomposable into independent or exchangeable decompositions. Similar to conditional independence, the notion of exchangeability can be extended to conditional exchangeability. We generalize exchangeability to conditional distributions, and state the corresponding tractability guarantees.

9.5.1   Marginal and Conditional Decomposition

Tractability results for exchangeable decompositions on all variables under consideration also extend to subsets.

Definition 9.7 (marginal exchangeability) *When a subset $\mathbf{Y}$ of the variables under consideration has an exchangeable decomposition $\mathcal{Y}$, we say that $\mathbf{Y}$ is marginally exchangeable.*

This means that $\mathcal{Y}$ is still an exchangeable decomposition when considering the distribution $\Pr(\mathbf{Y}) = \sum_{\mathbf{X}\setminus\mathbf{Y}} \Pr(\mathbf{X})$.

Theorem 9.5 *Suppose we are given a marginally exchangeable decomposition of $\mathbf{Y}$ with bounded width and let $\mathbf{Z} = \mathbf{X} \setminus \mathbf{Y}$. If computing $\Pr(\mathbf{y}) = \sum_{\mathbf{z}} \Pr(\mathbf{y}, \mathbf{z})$ is polynomial in $|\mathbf{X}|$ for all $\mathbf{y}$, then the complexity of MPE and marginal inference over variables $\mathbf{Y}$ is polynomial in $|\mathbf{X}|$.*

Proof   Let T be the statistic associated with the given decomposition, and let e be evidence given on $\mathbf{E} \subseteq \mathbf{Y}$. Then,

$$P(e) = \sum_{t \in \mathcal{T}} |S_{t,e}| \cdot \sum_{\mathbf{z}} \Pr(\mathbf{y}_{t,e}, \mathbf{z}), \quad \text{where } \mathbf{y}_{t,e} \in S_{t,e}.$$

By the assumption that $\mathbf{Y}$ is marginally exchangeable and the proof of Theorem 9.3, we can compute $|S_{t,e}|$ and an $\mathbf{y}_{t,e} \in S_{t,e}$ in time polynomial in $|\mathbf{X}|$. An analogous argument holds for MPE inference on $\Pr(\mathbf{Y})$.

We now need to identify distributions $\Pr(\mathbf{Y}, \mathbf{Z})$ for which we can compute $\Pr(\mathbf{y})$ efficiently. This implies tractable probabilistic inference over $\mathbf{Y}$. Given a particular $\mathbf{y}$, we have already seen sufficient conditions: when $\mathbf{Z}$ decomposes exchangeably or independently conditioned on $\mathbf{y}$, Proposition 9.1 and Theorem 9.3 guarantee that computing $\Pr(\mathbf{y})$ is tractable. This suggests the following general notion.

Definition 9.8 (conditional decomposability) *Let $\mathbf{X}$ be a set of variables with $\mathbf{Y} \subseteq \mathbf{X}$ and $\mathbf{Z} = \mathbf{X} \setminus \mathbf{Y}$. We say that $\mathbf{Z}$ is* exchangeably (independently) decomposable given $\mathbf{Y}$ *if and only if for each assignment $\mathbf{y}$ to $\mathbf{Y}$, there exists an exchangeable (independent) decomposition $\mathcal{Z}_{\mathbf{y}}$ of $\mathbf{Z}$.*

Furthermore, we say that $\mathbf{Z}$ is decomposable with *bounded width* iff the width w of each $\mathcal{Z}_{\mathbf{y}}$ is bounded. When the decomposition can be computed in time polynomial in $|\mathbf{X}|$ for all $\mathbf{y}$, we say that $\mathbf{Z}$ is *efficiently decomposable*.

Theorem 9.6 *Let $\mathbf{X}$ be a set of variables with $\mathbf{Y} \subseteq \mathbf{X}$ and $\mathbf{Z} = \mathbf{X} \setminus \mathbf{Y}$. Suppose we are given a marginally exchangeable decomposition of $\mathbf{Y}$ with bounded width. Suppose further that $\mathbf{Z}$ is efficiently (exchangeably or independently) decomposable given $\mathbf{Y}$ with bounded width. If we can compute $\Pr(\mathbf{x})$ efficiently, then the complexity of MPE and marginal inference over variables $\mathbf{Y}$ is polynomial in $|\mathbf{X}|$.*

Proof   Following Theorem 9.5, we only need to show that we can compute $\Pr(\mathbf{y}) = \sum_{\mathbf{z}} \Pr(\mathbf{y}, \mathbf{z})$ in time polynomial in $|\mathbf{X}|$ for all $\mathbf{y}$. When $\mathbf{Z}$ is exchangeably decomposable given Y, this follows from constructing $\mathcal{Z}_\mathbf{y}$ and employing the arguments made in the proof of Theorem 9.3. The case when $\mathbf{Z}$ is independently decomposable is analogous.

Theorems 9.5 and 9.6 are powerful results and allow us to identify numerous probabilistic models for which inference is tractable. For instance, we will prove liftability results for Markov logic networks. However, we are only at the beginning of leveraging these tractability results to their fullest extent. Especially Theorem 9.5 is widely applicable because the computation of $\sum_\mathbf{z} \Pr(\mathbf{y}, \mathbf{z})$ can be tractable for many reasons. For instance, conditioned on the variables Y, the distributions $\Pr(\mathbf{y}, \mathbf{Z})$ could be bounded treewidth graphical models, such as tree-structured Markov networks. Tractability for $\Pr(Y)$ follows immediately from Theorem 9.5.

Since Theorem 9.5 only speaks to the tractability of querying variables in Y, there is the question of when we can also efficiently query the variables $\mathbf{Z}$. Results from the lifted inference literature may provide a solution by bounding or approximating queries and evidence that includes $\mathbf{Z}$ to maintain marginal exchangeability (Van den Broeck and Darwiche, 2013). The next section shows that certain restricted situations permit tractable inference on the variables in $\mathbf{Z}$.

9.5.2   Markov Logic Case Study

Let us again consider the MLN

> 1.3   Smokes(x) $\Rightarrow$ Cancer(x)
>
> 1.5   Smokes(x) $\wedge$ Friends(x, y) $\Rightarrow$ Smokes(y)

having the marginally exchangeable decomposition

$$\mathcal{Y} = \{\{\text{Smokes(A)}\}, \{\text{Smokes(B)}\}, \{\text{Smokes(C)}\}, \dots \}$$

whose width is one. To intuitively see why this decomposition is marginally exchangeable, let us consider two states $\mathbf{y}$ and $\mathbf{y}'$ of the Smokes(x) atoms in which *only* the values of two atoms, for example Smokes(A) and Smokes(B), are swapped. There is a symmetry of the MLNs joint distribution that swaps these atoms: the renaming automorphism (Bui et al., 2013a; Niepert, 2012a) that swaps constants A and B in all atoms. For marginal exchangeability, we need that $\sum_\mathbf{z} \Pr(\mathbf{y}, \mathbf{z}) = \sum_\mathbf{z} \Pr(\mathbf{y}', \mathbf{z})$. But this holds since the renaming automorphism is an automorphism of the set of states $\{\mathbf{yz} \mid \mathbf{z} \in \mathcal{D}_\mathbf{z}\}$ – for every $\mathbf{y}$, $\mathbf{y}'$, and $\mathbf{z}$ there exists an automorphism that maps $\mathbf{yz}$ to $\mathbf{y}'\mathbf{z}'$ with $\Pr(\mathbf{y}, \mathbf{z}) = \Pr(\mathbf{y}', \mathbf{z}')$.

The given MLN has several marginally exchangeable decompositions, with the most general one being

$$\mathcal{Y} = \{\{\text{Smokes(A), Cancer(A), Friends(A, A)}\},$$
$$\{\text{Smokes(B), Cancer(B), Friends(B, B)}\},$$
$$\{\text{Smokes(C), Cancer(C), Friends(C, C)}\}, \dots \}.$$

For that decomposition, the remaining $\mathbf{Z}$ variables

$$\{\mathsf{Friends}(A, B), \mathsf{Friends}(B, A), \mathsf{Friends}(A, C), \dots\}$$

are independently decomposable given $\mathbf{Y}$. The $\mathbf{Z}$ variables appear at most once in any formula. In a probabilistic graphical model representation, evidence on the $\mathbf{Y}$ variables would therefore decompose the graph into independent components. Thus, it follows from Theorem 9.6 that we can efficiently answer any query over the variables in $\mathbf{Y}$.

This insight generalizes to a large class of MLNs, called the *two-variable fragment*. It consists of all MLNs whose formulas contain at most two logical variables.

Theorem 9.7 *In a two-variable fragment MLN, let $\mathbf{Y}$ and $\mathbf{Z}$ be the ground atoms with one and two distinct arguments respectively. Then there exists a marginally exchangeable decomposition of $\mathbf{Y}$, and $\mathbf{Z}$ is efficiently independently decomposable given $\mathbf{Y}$. Each decomposition's width is at most twice the number of predicates.*

The proof of Theorem 9.7 is rather technical and we refer the reader to the appendix for a detailed proof. It now follows from Theorems 9.6 and 9.7 that the complexity of inference over the unary atoms in the two-variable fragment is polynomial in the domain size $|\mathbf{D}|$.

What happens if our query involves variables from $\mathbf{Z}$ – the binary atoms? It is known in the lifted inference literature that we cannot expect efficient inference of general queries that involve the binary atoms. Assignments to the $\mathbf{Z}$ variables break symmetries and therefore break marginal exchangeability. This causes inference to become #P-hard as a function of the query (Van den Broeck and Davis, 2012). Nevertheless, if we bound the number of binary atoms involved in the query, we can use the developed theory to show a general liftability result.

Theorem 9.8 *For any MLN in the two-variable fragment, MPE and marginal inference over the unary atoms and a bounded number of binary atoms is domain-lifted.*

This theorem corresponds to a theoretical results in the lifted inference literature (Jaeger and Van den Broeck, 2012). We refer the interested reader to the appendix for the proof. A consequence of Theorem 9.8 is that we can efficiently compute *all single marginals* in the two-variable fragment of quantifier-free Markov logic, given arbitrary evidence on the unary atoms.

9.6 Discussion and Conclusion

We conjecture that the concept of (partial) exchangeability has potential to contribute to a deeper understanding of tractable probabilistic models. The important role conditional independence plays in the research field of graphical models is evidence for this hypothesis. Similar to conditional independence, it might be possible to develop a theory of exchangeability that mirrors that of independence. For instance, there might be a (graphical) structural representations of particular types of partial exchangeability and corresponding logical axiomatizations (Pearl, 1988). Moreover, it would be interesting to develop graphical models with exchangeability and independence, and notions like d-separation to detect marginal exchangeability and conditional decomposability from a structural representation. The first author has taken steps in this direction by introducing exchangeable variable models, a class of (non-relational) probabilistic models based on finite partial exchangeability (Niepert and Domingos, 2014).

Recently, there has been considerable interest in computing and exploiting the automorphisms of graphical models (Niepert, 2013; Bui et al., 2012). There are several interesting connections between automorphisms, exchangeability, and lifted inference (Niepert, 2012b). Moreover, there are several group theoretical algorithms that one could apply to the automorphism groups to discover the structure of exchangeable variable decompositions from the structure of the graphical models. Since we presently only exploit renaming automorphisms, there is a potential for tractable inference in MLNs that goes beyond what is known in the lifted inference literature.

Partial exchangeability is related to collective graphical models (Sheldon and Dietterich, 2011) (CGMs) and cardinality-based potentials (Gupta et al., 2007) as these models also operate on sufficient statistics. However, probabilistic inference for CGMs is not tractable and there are no theoretical results that identify tractable CGMs models. The presented work may help to identify such situations. The presented theory generalizes the statistics of cardinality-based potentials.

Lifted Inference and Exchangeability   Our case studies identified a deep connection between lifted probabilistic inference and the concepts of partial, marginal and conditional exchangeability. In this new context, it appears that exact lifted inference algorithms (de Salvo Braz et al., 2005; Milch et al., 2008; Jha et al., 2010; Van den Broeck et al., 2011; Gogate et al., 2012; Taghipour and Davis, 2012) can all be understood as performing essentially three steps: (i) construct a sufficient statistic $T(\mathbf{x})$, (ii) generate all possible values of the sufficient statistic, and (iii) count suborbit sizes for a given statistic. For an example of (i), we can show that a compiled first-order circuit (Van den Broeck et al., 2011) or the trace of probabilistic theorem proving (Gogate et al., 2012) encode a sufficient statistic in their existential quantifier and splitting nodes. Steps (ii) and (iii) are manifested in all these algorithms through summations and binomial coefficients.

Between Corollary 9.4.1 and Theorem 9.8, we have re-proven a sequence of initial liftability results from the lifted inference literature (Jaeger and Van den Broeck, 2012) within the exchangeability framework, and extended these to MPE inference. There is an essential difference though: liftability results make assumptions about the syntax (e.g., MLNs), whereas our exchangeability theorems apply to all distributions. We expect Theorem 9.6 to be used to show liftability, and more general tractability results for many other representation languages, including but not limited to the large number of statistical relational languages that have been proposed (Getoor and Taskar, 2007; De Raedt et al., 2008).

Acknowledgments

This work initially appeared as Niepert and Van den Broeck (2014) and was partially supported by ONR grants #N00014-12-1-0423, #N00014-13-1-0720, and #N00014-12-1-0312; NSF grants #IIS-1118122 and #IIS-0916161; ARO grant #W911NF-08-1-0242; AFRL contract #FA8750-13-2-0019; the Research Foundation-Flanders (FWO-Vlaanderen); and a Google research award to MN.

Appendix: Continued Proof of Theorem 9.3

Proof   To prove statements (b) and (c), we need to represent partial assignments $E = e$ with $E \subseteq X$. The partial assignments decompose into partial assignments $E_1 = e_1, \ldots, E_k = e_k$ in accordance with X. Each e_ℓ corresponds to a string $m \in \{0, 1, *\}^w$ where characters 0 and 1 encode assignments to variables in E_ℓ and $*$ encodes an unassigned variable in $X_\ell - E_\ell$. In this case, we say that e_ℓ is of type m. Please note that there are 2^w distinct b and 3^w distinct m. We say that x *agrees with* e, denoted by $x \sim e$, if and only if their shared variables have identical assignments.

A *completion* c *of* e *to* x is a bijection $c : \{e_1, \ldots, e_k\} \rightarrow \{x_1, \ldots, x_k\}$ such that $c(e_i) = x_j$ implies $e_i \sim x_j$. Every completion corresponds to a unique way to assign elements in $\{0, 1\}$ to unassigned variables so as to turn the partial assignment e into the full assignment x.

Let $t = (c_1, \ldots, c_{2^w}) \in \mathcal{T}$, let $E \subseteq X$, and let $e \in \mathcal{D}_E$. Moreover, let $d_j = \sum_{\ell=1}^{k}[[e_\ell = m_j]]$ for each $m_j \in \{0, 1, *\}^w$. Consider the set of matrices

$$\mathcal{A}_{t,e} = \Big\{ A \in \mathbb{M}(2^w, 3^w) \,\Big|\, \sum_i a_{i,j} = d_j, \sum_j a_{i,j} = c_i$$

$$\text{and } a_{i,j} = 0 \text{ if } b_i \not\sim m_j \Big\}.$$

Every $A \in \mathcal{A}_{t,e}$ represents a set of completions from e to x for which $T(x) = t$. The value $a_{i,j}$ indicates that each completion represented by A maps $a_{i,j}$ elements in $\{e_1, \ldots, e_k\}$ of type m_j to $a_{i,j}$ elements in $\{x_1, \ldots, x_k\}$ of type b_i. We write $\gamma(A)$ for the set of completions represented by A.

We have to prove the following statements

1. For every $A \in \mathcal{A}_{t,e}$ and every $c \in \gamma(A)$ there exists an $x \in \mathcal{D}_X$ with $T(x) = t$ and c is a completion of e to x;
2. For every $x \in \mathcal{D}_X$ with $T(x) = t$ and every completion c of e to x there exists an $A \in \mathcal{A}_{t,e}$ such that $c \in \gamma(A)$;
3. For all $A, A' \in \mathcal{A}_{t,e}$ with $A \neq A'$ we have that $\gamma(A) \cap \gamma(A') = \emptyset$;
4. For every $A \in \mathcal{A}_{t,e}$, we can efficiently compute $|\gamma(A)|$, the size of the set of completions represented by A.

To prove statement (1), let $A \in \mathcal{A}_{t,e}$ and $c \in \gamma(A)$. By the definition of $\mathcal{A}_{t,e}$, c maps $a_{i,j}$ elements in $\{e_1, \ldots, e_k\}$ of type m_j to $a_{i,j}$ elements in $\{x_1, \ldots, x_k\}$ of type b_i. By the conditions

$\sum_i a_{i,j} = d_j$ and $\sum_j a_{i,j} = c_i$ of the definition of $\mathcal{A}_{t,e}$ we have that c is a bijection. By the condition $a_{i,j} = 0$ if $b_i \nsim m_j$ of the definition of $\mathcal{A}_{t,e}$ we have that $\mathsf{c}(e_i) = \mathbf{x}_j$ implies $e_i \sim \mathbf{x}_j$ and, therefore, $e \sim \mathbf{x}$. Hence, c is a completion. Moreover, c completes e to an $\mathbf{x}$ with $\sum_{\ell=1}^k [[\mathbf{x}_\ell = b_i]] = \sum_j a_{i,j} = c_i$ by the definition of $\mathcal{A}_{t,e}$. Hence, $T(\mathbf{x}) = t$.

To prove statement (2), let $\mathbf{x} \in \mathcal{D}_\mathbf{X}$ with $T(\mathbf{x}) = t$ and let c be a completion of e to $\mathbf{x}$. We construct an A with $\mathsf{c} \in \gamma(A)$ as follows. Since c is a completion we have that $\mathsf{c}(e_i) = \mathbf{x}_j$ implies $e_i \sim \mathbf{x}_j$ and, hence, we set $a_{i,j} = 0$ if $b_i \nsim m_j$. For all other entries in A we set $a_{i,j} = |\{e_j \mid \mathsf{c}(e_j) = \mathbf{x}_i\}|$. Since c is surjective, we have that $\sum_j a_{i,j} = c_i$ and since c is injective, we have that $\sum_i a_{i,j} = d_j$. Hence, $A \in \mathcal{A}_{t,e}$.

To prove statement (3), let $A, A' \in \mathcal{A}_{t,e}$ with $A \neq A'$. Since $A \neq A'$ we have that there exist i, j such that, without loss of generality, $a_{i,j} < a'_{i,j}$. Hence, every $\mathsf{c} \in A$ maps fewer elements of type m_j to elements of type b_i than every $\mathsf{c}' \in A'$. Hence, $\mathsf{c} \neq \mathsf{c}'$ for every $\mathsf{c} \in A$ and every $\mathsf{c}' \in A'$.

To prove statement (4), let $A \in \mathcal{A}_{t,e}$. Every $\mathsf{c} \in A$ maps $a_{i,j}$ elements in $\{e_1, ..., e_k\}$ of type m_j to $a_{i,j}$ elements in $\{\mathbf{x}_1, ..., \mathbf{x}_k\}$ of type b_i. Hence, the size of the set of completions represented by A is, for each $1 \leq j \leq 3^w$, the number of different ways to place $a_{i,j}$ balls of color i, $1 \leq i \leq 2^w$, into d_j urns. Hence,

$$|\gamma(A)| = \prod_{j=1}^{3^w} \prod_{i=1}^{2^w} \binom{d_j - \sum_{q=1}^{i-1} a_{i,q}}{a_{i,j}}.$$

From the statements (1)-(4) we can conclude that

$$|S_{t,e}| = \sum_{A \in \mathcal{A}_{t,e}} |\gamma(A)|.$$

This allows us to prove (b) and (c). We can construct $\mathcal{A}_{t,e}$ in time polynomial in n as follows. There are $2^w 3^w$ entries in a $2^w \times 3^w$ matrix and each entry has at most k different values. Hence, we can enumerate all $k^{6^w} \leq n^{6^w}$ possible matrices $A \in \mathbb{M}(2^w, 3^w)$. We simply select those A for which the conditions in the definition of $\mathcal{A}_{t,e}$ hold. For one $A \in \mathcal{A}_{t,e}$ we can efficiently construct one c and the $\mathbf{x}$ that it completes e to. This proves (b). Finally, we compute $|S_{t,e}|$. This proves (c).

Appendix: Proof of Theorem 9.4

Proof   Let $P_1, ..., P_N$ be the N unary predicates of a given MLN and let $\mathbf{D} = \{1, ..., k\}$ be the domain. After grounding, there are k ground atoms per predicate. We write $P_j(i)$ to denote the ground atom that resulted from instantiating predicate P_j with domain element i. Let $\mathcal{X} = \{\mathbf{X}_1, ..., \mathbf{X}_k\}$ be a decomposition of the set of ground atoms with $\mathbf{X}_i = \{P_1(i), P_2(i), ..., P_N(i)\}$ for every $1 \leq i \leq k$. A renaming automorphism (Bui et al., 2013a; Niepert, 2012b) is a permutation of the ground atoms that results from a permutation of the domain elements. The joint distribution over all ground atoms remains invariant under these permutations. Consider the permutation of ground atoms that results from swapping two domain elements $i \leftrightarrow i'$. This permutation acting on the set of ground atoms permutes the components $\mathbf{X}_i$ and $\mathbf{X}_{i'}$ and leaves all other components invariant. Since this is possible for each pair $i, i' \in \{1, ..., k\}$ it follows that the decomposition X is exchangeable.

Appendix: Proof of Theorem 9.7

Proof   Let $P_1, ..., P_M$ be the M unary predicates and let $Q_1, ..., Q_N$ be the N binary predicates of a given MLN and let $\mathbf{D} = \{1, ..., k\}$ be the domain. After grounding, there are k ground atoms per unary and k^2 ground atoms per binary predicate. We write $P_\ell(i)$ to denote the ground atom that resulted from instantiating unary predicate P_ℓ with domain element i and $Q_\ell(i, j)$ to denote the ground atom that resulted from instantiating binary predicate Q_ℓ with domain elements i and j.

Let $\mathbf{X}$ be the set of all ground atoms, let $Y = \{P_\ell(i) \mid 1 \leq i \leq k, 1 \leq \ell \leq M\} \cup \{Q_\ell(i, i) \mid 1 \leq i \leq k, 1 \leq \ell \leq N\}$ and let $\mathbf{Z} = \mathbf{X} - Y$. Moreover, let $\mathcal{Y} = \{Y_1, ..., Y_k\}$ with $Y_i = \{P_\ell(i) \mid 1 \leq \ell \leq M\} \cup \{Q_\ell(i, i) \mid 1 \leq \ell \leq N\}$. We can make the same arguments as in the proof of Theorem 9.4 to show that $\mathcal{Y}$ is exchangeable.

Now, we prove that the variables $\mathbf{Z}$ are independently decomposable given Y. Let $\mathbf{Z}_{i,j} = \{Q_1(i, j), ..., Q_N(i, j), Q_1(j, i), ..., Q_N(j, i)\}$ for all $1 \leq i < j \leq k$. Now, let f be any ground formula and let G be the set of ground atoms occurring in both f and $\mathbf{Z}$. Then, either $G \subseteq \mathbf{Z}_{i,j}$ or $G \cap \mathbf{Z}_{i,j} = \emptyset$ since every formula in the MLN has at most two variables. Hence, $\{\mathbf{Z}_{i,j} \mid 1 \leq i < j \leq k\}$ is a decomposition of $\mathbf{Z}$ with $\binom{k}{2}$ components, width $2N$, and $\Pr(\mathbf{Z}, \mathbf{y})$ factorizes as

$$\Pr(\mathbf{Z}, \mathbf{y}) = \prod_{\substack{i,j \\ i<j}} Q_{i,j}(\mathbf{Z}_{i,j}, \mathbf{y}).$$

By the properties of MLNs, we have that the $Q_{i,j}(\mathbf{Z}_{i,j}, \mathbf{y})$ are computable in time exponential in the width of the decomposition but polynomial in k.

Appendix: Proof of Theorem 9.8

Proof   Suppose that the query e contains a bounded number of binary atoms whose arguments are constants from the set K. Consider the set of variables $\mathbf{Q}$ consisting of all unary atoms whose argument comes from K, and all binary atoms whose arguments both come from K. The unary atoms in $\mathbf{Q}$ are no longer marginally exchangeable, because their arguments can appear asymmetrically in e. We can now answer the query by simply enumerating all states $\mathbf{q}$ of $\mathbf{Q}$ and performing inference in each $\Pr(Y, \mathbf{Z}, \mathbf{q})$ separately, where all variables Y have again become marginally exchangeable, and all variables $\mathbf{Z}$ have become independently decomposable given Y. The construction of Y and $\mathbf{Z}$ is similar to the proof of Theorem 9.7, except that some additional binary atoms are now in Y instead of $\mathbf{Z}$. These atoms have one argument in K, and one not in K, and are treated as unary. When we bound the number of binary atoms in the query, the size of $\mathbf{Q}$ will not be a function of the domain size, and enumerating over all states $\mathbf{q}$ is domain-lifted.

III APPROXIMATE INFERENCE

10 Lifted Markov Chain Monte Carlo

Mathias Niepert and Guy Van den Broeck

Abstract.

This chapter presents an approach to utilize exact and approximate symmetries in probabilistic graphical models during Markov chain Monte Carlo (MCMC) inference. We discuss permutation groups representing the symmetries of graphical models and how to compute them. Next, we introduce orbital Markov chains, a family of lifted Markov chains leveraging model symmetries to reduce mixing times. Unfortunately, the majority of real-world graphical models is asymmetric. This is even the case for relational representations when evidence is given. Therefore, we extend lifted MCMC to instead utilize approximate symmetries. Lifted MCMC leads to improved probability estimates while remaining unbiased. Experiments demonstrate that the approach outperforms existing MCMC algorithms.

10.1 Introduction

This chapter describes the use of group theoretical concepts and algorithms to perform Markov chain Monte Carlo sampling in probabilistic models. Since relational models often exhibit strong topological symmetries, permutation groups offer a compact and well-understood representation. Moreover, numerous efficient group theoretical algorithms are implemented in comprehensive open-source group algebra frameworks such as GAP (GAP).

Symmetries on different syntactical levels of statistical relational formalism ultimately lead to symmetries in the space of joint variable assignments. This space of possible assignments corresponds to the state space of Monte Carlo Markov chains such as the Gibbs sampler that are often used for approximate probabilistic inference. Since the permutation group modeling the symmetries induces a partition (the so-called orbit partition) on the state space of these Markov chains, we investigate whether this can be exploited for more efficient MCMC approaches to probabilistic inference. The basic idea is that *lifted* Markov chains implicitly or explicitly operate on the partition of the state space instead of the space of individual assignments. We describe orbital Markov chains, which are derived from an existing Markov chain so as to leverage the symmetries in the underlying model. Under mild conditions, orbital Markov chains have the same convergence properties as chains operating on the state space partition without the need to explicitly compute this partition.

While lifted inference algorithms perform well for highly symmetric graphical models, they depend heavily on the presence of symmetries and perform worse for asymmetric models due to their computational overhead. This is especially unfortunate as numerous real-world graphical models are not symmetric. To bring the achievements of the lifted inference community to the mainstream of machine learning and uncertain reasoning it is crucial to explore ways to apply ideas from the lifted inference literature to inference problems in asymmetric graphical models. This chapter further describes a lifted inference algorithm for asymmetric graphical models. It uses a symmetric approximation of the original model to compute a proposal distribution for a Metropolis-Hastings chain. The approach combines a base MCMC algorithm such as the Gibbs sampler with the Metropolis chain that performs jumps in the approximate symmetric model, while producing unbiased probability estimates.

We conducted several experiments verifying that orbital Markov chains converge faster to the true distribution than state of the art Markov chains. We also conduct experiments where lifted inference is applied to graphical models with no exact symmetries and no color-passing symmetries, and where every random variable has distinct soft evidence. Yet, we are able to show improved probability estimates while remaining unbiased.

10.2   Background

We first recall basic concepts of group theory and finite Markov chains both of which are crucial for understanding this chapter.

10.2.1   Group Theory

A **symmetry** of a discrete object is a structure-preserving bijection on its components. A natural way to represent symmetries are **permutation groups**. A group is an algebraic structure $(\mathfrak{G}, \circ)$, where $\mathfrak{G}$ is a set closed under a binary associative operation $\circ$ with an identity element and a unique inverse for each element. We often write $\mathfrak{G}$ rather than $(\mathfrak{G}, \circ)$.

A permutation group $\mathfrak{G}$ *acting on* a finite set Ω is a finite set of bijections $\mathfrak{g} : \Omega \to \Omega$ that form a group. Let Ω be a finite set and let $\mathfrak{G}$ be a permutation group acting on Ω. If $\alpha \in \Omega$ and $\mathfrak{g} \in \mathfrak{G}$ we write $\alpha^{\mathfrak{g}}$ to denote the image of α under $\mathfrak{g}$. A cycle $(\alpha_1 \ \alpha_2 \ ... \ \alpha_n)$ represents the permutation that maps α_1 to α_2, α_2 to α_3,..., and α_n to α_1. Every permutation can be written as a product of disjoint cycles where each element that does not occur in a cycle is understood as being mapped to itself. A generating set R of a group is a subset of the group's elements such that every element of the group can be written as a product of finitely many elements of R and their inverses.

We define a relation $\sim$ on Ω with $\alpha \sim \beta$ if and only if there is a permutation $\mathfrak{g} \in \mathfrak{G}$ such that $\alpha^{\mathfrak{g}} = \beta$. The relation partitions Ω into equivalence classes which we call **orbits**. We call this partition of Ω the orbit partition induced by $\mathfrak{G}$. We use the notation $\alpha^{\mathfrak{G}}$ to denote the orbit $\{\alpha^g \mid g \in \mathfrak{G}\}$ containing α. Let $f : \Omega \to \mathbb{R}$ be a function from Ω into the real numbers and let $\mathfrak{G}$ be a permutation group acting on Ω. We say that $\mathfrak{G}$ is an **automorphism group** for (Ω, f) if and only if for all $\omega \in \Omega$ and all $g \in \mathfrak{G}, f(\omega) = f(\omega^g)$.

10.2.2 Finite Markov Chains

Given a finite set Ω a **Markov chain** defines a random walk $(\mathbf{x}_0, \mathbf{x}_1, ...)$ on elements of Ω with the property that the conditional distribution of $\mathbf{x}_{n+1}$ given $(\mathbf{x}_0, \mathbf{x}_1, ..., \mathbf{x}_n)$ depends only on $\mathbf{x}_n$. For all $\mathbf{x}, \mathbf{y} \in \Omega$, $P(\mathbf{x} \to \mathbf{y})$ is the chain's probability to transition from $\mathbf{x}$ to $\mathbf{y}$, and $P^t(\mathbf{x} \to \mathbf{y}) = P^t_{\mathbf{x}}(\mathbf{y})$ the probability of being in state $\mathbf{y}$ after t steps if the chain starts at state $\mathbf{x}$. We often refer to the conditional probability matrix P as the *kernel* of the Markov chain. A Markov chain is *irreducible* if for all $\mathbf{x}, \mathbf{y} \in \Omega$ there exists a t such that $P^t(\mathbf{x} \to \mathbf{y}) > 0$ and *aperiodic* if for all $\mathbf{x} \in \Omega$, $\gcd\{t \geq 1 \mid P^t(\mathbf{x} \to \mathbf{x}) > 0\} = 1$.

Theorem 10.1 *Any irreducible and aperiodic Markov chain has exactly one* ***stationary distribution****.*

A distribution π on Ω is reversible for a Markov chain with state space Ω and transition probabilities P, if for every $\mathbf{x}, \mathbf{y} \in \Omega$

$$\pi(\mathbf{x})P(\mathbf{x} \to \mathbf{y}) = \pi(\mathbf{y})P(\mathbf{y} \to \mathbf{x}).$$

We say that a Markov chain is reversible if there exists a reversible distribution for it. The AI literature often refers to reversible Markov chains as Markov chains satisfying the detailed balance property.

Theorem 10.2 *Every reversible distribution for a Markov chain is also a stationary distribution for the chain.*

10.2.3 Markov Chain Monte Carlo

Numerous approximate inference algorithms for probabilistic graphical models draw sample points from a Markov chain whose stationary distribution is that of the probabilistic model, and use the sample points to estimate marginal probabilities. Sampling approaches of this kind are referred to as **Markov chain Monte Carlo** methods. We discuss the **Gibbs sampler**, a sampling algorithm often used in practice.

Let $\mathbf{X}$ be a finite set of random variables with probability distribution π. The Markov chain for the *Gibbs sampler* is a Markov chain $\mathcal{M} = (\mathbf{x}_0, \mathbf{x}_1, ...)$ which, being in state $\mathbf{x}_t$, performs the following steps at time $t + 1$:

1. Select a variable $X \in \mathbf{X}$ uniformly at random;
2. Sample $\mathbf{x}'_{t+1}(X)$, the value of X in the state $\mathbf{x}'_{t+1}$, according to the conditional π-distribution of X given that all other variables take their values according to $\mathbf{x}_t$; and
3. Let $\mathbf{x}'_{t+1}(Y) = \mathbf{x}_t(Y)$ for all variables $Y \in \mathbf{X} \setminus \{X\}$.

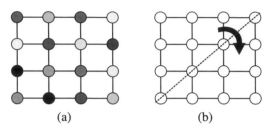

(a)   (b)

Figure 10.1: A ferromagnetic Ising model with constant interaction strength. In the presence of an external field, that is, when the variables have different unary potentials, the probabilistic model is asymmetric (a). However, the model is rendered symmetric by assuming a constant external field (b). In this case, the symmetries of the model are generated by the reflection and rotation automorphisms.

The Gibbs chain is aperiodic and has π as a stationary distribution. If the chain is irreducible, then the marginal estimates based on sample points drawn from the chain are unbiased once the chain reaches the stationary distribution.

10.3   Symmetries of Probabilistic Models

This section describes notions of symmetry in the context of probabilistic graphical models, as well as their relational extensions.

10.3.1   Graphical Model Symmetries

Symmetries of a set of random variables and graphical models have been formally defined in the lifted and symmetry-aware probabilistic inference literature with concepts from group theory (Niepert, 2013; Bui et al., 2012).

Definition 10.1 *Let* **X** *be a set of discrete random variables with distribution π and let Ω be the set of states (configurations) of* **X**. *We say that a permutation group $\mathfrak{G}$ acting on Ω is an automorphism group for* **X** *if and only if for all* **x** $\in \Omega$ *and all* $g \in \mathfrak{G}$ *we have that* $\pi(\mathbf{x}) = \pi(\mathbf{x}^g)$.

Note that the definition of an automorphism group is independent of the particular representation of the probabilistic model. For particular representations, there are efficient algorithms for computing the automorphism group. Typically, one computes the generators of the automorphism group with algorithms that derive permutation groups for colored undirected graphs such as SAUCY and NAUTY (Niepert, 2012b). Note that we do not require the automorphism group to be maximal, that is, it can be a subgroup of a different automorphism group for the same set of random variables.

Most probabilistic models are asymmetric. For instance, the Ising model which is used in numerous applications, is asymmetric if we assume an external field as it leads to different unary potentials. However, we can make the model symmetric simply by assuming a constant external field. Figure 10.1 depicts this situation. This is an example of an over-symmetric approximation of the model, which we will use later in this chapter to do lifted MCMC without biasing the probability estimates.

10.3.2 Relational Model Symmetries

Naturally, there is a close connection between the concept of symmetry and lifted inference. There are deep connections between automorphisms and the statistical notion of exchangeability (Niepert, 2012b, 2013; Bui et al., 2012), which has been used to explain the tractability of exact lifted inference algorithms (Niepert and Van den Broeck, 2014). Moreover, the (fractional) automorphisms of the graphical model representation have been related to lifted inference and exploited for more efficient inference (Niepert, 2012b; Bui et al., 2012; Noessner et al., 2013; Mladenov and Kersting, 2013). For instance, lifted belief propagation identifies and clusters indistinguishable ground atoms and features by keeping track of the messages send and received by each of the corresponding nodes in a factor graph (Singla and Domingos, 2008; Kersting et al., 2009). Bi-simulation type procedures group indistinguishable elements and, therefore, exploit symmetry in the model as well (Sen et al., 2009b). There are a number of sampling algorithms that take advantage of symmetries (Venugopal and Gogate, 2012; Gogate et al., 2012).

The algorithms in this chapter use group theory and, in particular, permutation groups to compactly represent (exact and approximate) symmetries in graphical models (Niepert, 2012b). There are several reasons to consider group theory and permutation groups a natural representation of symmetries in graphical models. First, an irredundant set of generators of a permutation group ensures exponential compression. For instance, for a set of n exchangeable binary random variables, the permutation group acting on the variables is the symmetric group on n which has $n!$ permutations. However, we only need at most $n - 1$ irredundant generators to represent this permutation group. In addition to the compact representation, group theory also provides numerous remarkably efficient algorithms for manipulating and sampling from groups. The product replacement algorithm (Celler et al., 1995), for instance, samples group elements uniformly at random with impressive performance.

Symmetry in statistical relational languages manifests itself at various syntactic levels ranging from the set of constants to the assignment space. There is often symmetry at the level of constants. In the well-known social network model (Singla and Domingos, 2008) without evidence, for example, we have that the constants are indistinguishable meaning that swapping two constants leads to an isomorphic statistical relational model. Now, the permutations on the constant level induce permutations on the level of ground atoms and formulas. From the irredundant generators of the permutation group modeling the symmetries on the constant level we can directly compute the irredundant generators of the permutation group modeling the corresponding symmetries on the ground level. Indeed,

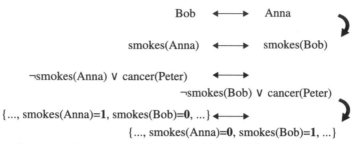

Figure 10.2: Symmetry in the model is observable on different syntactical levels of the relational model. The level of constants, the level of ground atoms (variables), the level of clauses (features) and the level of possible worlds (assignments). Each permutation group acting on the set of constants induces a permutation group acting on the set of ground atoms. The latter induces a permutation group acting on the set of features. This permutation partitions (a) the variables and feature and (b) the assignment space.

it is well-known that isomorphisms between permutation groups always map irredundant generators in one group to irredundant generators in the other. However, symmetry on the ground level does not necessarily lead to symmetry on the constant level. Similarly, while symmetry on the ground level induces symmetry on the space of assignments to the random variables this is not true for the other direction. Figure 10.2 depicts the different syntactical levels on which symmetries can arise.

Niepert (2012b) describes an approach that maps weighted formulas to colored undirected graphs and applies graph automorphism algorithms to compute the symmetries of the log-linear models defined over the weighted formulas (Niepert, 2012b). The resulting permutation groups partition the (exponential) space of variable assignments when acting on it. Since the state space of MCMC approaches is identical to the assignment space of the probabilistic graphical models, we will investigate whether and to what extend the partition induced by the models' symmetries can be leveraged for more efficient MCMC algorithms.

10.4   Lifted MCMC for Symmetric Models

We have seen that symmetries on different syntactical levels of statistical relational formalism ultimately lead to symmetries in the space of joint variable assignments. Now, the space of possible variable assignments is the state space of Monte Carlo Markov chains such as the Gibbs sampler that are often used for approximate probabilistic inference. Since the permutation group modeling the symmetries induces a partition (the so-called orbit partition) on the state space of these Markov chains, we will investigate whether this can be exploited for more efficient MCMC approaches to probabilistic inference. The ba-

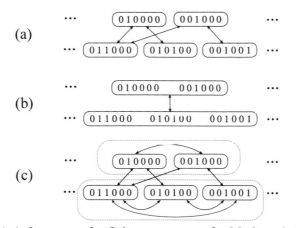

Figure 10.3: (a) A fragment of a finite state space of a Markov chain with non-zero transition probabilities indicated by directed arcs. (b) A lumping of the state space. Instead of moving between individual states, the lumped chain moves between classes of states of the original chain. (c) The benefits of lumping are also achievable by sampling uniformly at random from the implicit equivalence classes (orbits) in each step.

sic idea is that *lifted* MCMC algorithms implicitly or explicitly operate on the partition of the state space instead of the original state space.

10.4.1   Lumping

A *lumping* (also: collapsing, projection) of a Markov chains is a compression of its state space which is possible under certain conditions on the transition probabilities of the original Markov chain (Buchholz, 1994; Derisavi et al., 2003). The following definition formalizes the notion.

Definition 10.2 *Let M be a Markov chain with transition matrix P and state space Ω, and let $C = \{C_1, ..., C_n\}$ be a partition of the state space. If for all $C_i, C_j \in C$ and all $j', j'' \in C_j$*

$$\sum_{i' \in C_i} P(i', j') = \sum_{i' \in C_i} P(i', j'')$$

then M is ordinary lumpable. If, in addition, the stationary distribution has $\pi(j') = \pi(j'')$ for all $j', j'' \in C_j$ and all $C_j \in C$ then M is exactly lumpable.

Let $\hat{\pi}$ be the stationary distribution of the *quotient Markov chain*, that is, the exactly lumped Markov chain whose state space consists of partitions C. Then, the probability $\pi(i)$ of a state $i \in C_i \subseteq \Omega$ of the original chain can be computed as $\pi(i) = \hat{\pi}(i)/|C_i|$.

The benefit of lumping a Markov chain is the potentially much smaller state space and ultimately more rapid mixing. For instance, consider the case of n binary random variables that are exchangeable. Here, the natural choice of a partition of the state space is $\{C_0, C_1, ..., C_n\}$ where each C_i contains the states with Hamming weight i, that is, the states with i non-zeros. Please note that the C_i's are the orbits (equivalence classes) of the orbit partition of the permutation group acting on the set of states (variable assignments). Instead of 2^n states the resulting lumped Markov chain has only $n + 1$ states and mixes more rapidly than the original one. Figure 10.3 depicts (a) a fragment of a finite Markov chain with non-zero probability transitions indicated by arrows and (b) a lumped Markov chain that bundles several states of the original chain into a single one of the lumped chain.

The crucial question is whether the explicit construction of the lumped chain is computationally feasible. After all, if the computation of lumped Markov chains was intractable we would not have gained much. Unfortunately, it turns out that the explicit construction of the lumped state space is indeed intractable. Computing the coarsest lumping quotient of a Markov chain with a bi-simulation procedure is linear in the *number of non-zero probability transitions* of the chain (Derisavi et al., 2003) and, hence, in most cases exponential in the number of random variables. Moreover, other theoretical results show that special cases of the lumping problem are also intractable. The results are negative even for the important special case of partitions resulting from permutation groups acting on the state space of the Markov chains. It is known that, given a permutation group acting on the state space, merely computing the number of equivalence classes of the resulting orbit partition of the state space is a #P-complete problem (Goldberg, 2001).

We hypothesize that the intractability of the explicit construction of the lumped chain's state space is the main reason that the technique of lumping, while well-understood on a theoretical level, has not been seriously considered by communities that apply Markov chain Monte Carlo methods to large-scale applications requiring probabilistic inference. We are not aware of MCMC approaches to probabilistic reasoning that leverage the theory of lumping.

10.4.2   Orbital Markov Chains

Under certain circumstances, the *explicit* computation of the partition of the state space is not necessary to achieve the same computational gains as the lumped chain (Niepert, 2012b). The basic idea is that we only need, for each $\omega \in \Omega$, an efficient way to sample uniformly at random from $[\omega]$ the equivalence class containing ω. The product replacement algorithm (Celler et al., 1995) provides such an efficient method of sampling uniformly from the equivalence classes induced by a permutation group. This novel family of Markov chains is referred to as **orbital Markov chains** (Niepert, 2012b). An orbital Markov chain is always derived from an existing Markov chain so as to leverage the symmetries in the underlying model. In the presence of symmetries orbital Markov chains are able to perform wide-ranging transitions reducing the time until convergence. In the absence of symmetries they are equivalent to the original Markov chains. Orbital Markov chains only require a generating set of a permutation group $\mathfrak{G}$ acting on the chain's state space as additional input. These generators can be computed on a colored graph representation of the distribution at hand, or directly on the relational representation, as described in the previous section.

Let Ω be a finite set, let $\mathcal{M}' = (X_0', X_1', ...)$ be a Markov chain with state space Ω, let π be a stationary distribution of $\mathcal{M}'$, and let $\mathfrak{G}$ be an automorphism group for (Ω, π). The *orbital Markov chain* $\mathcal{M} = (X_0, X_1, ...)$ *for* $\mathcal{M}'$ is a Markov chain which at each integer time $t + 1$ performs the following steps:

1. Let X_{t+1}' be the state of the *original* Markov chain $\mathcal{M}'$ at time $t + 1$;
2. Sample X_{t+1}, the state of the orbital Markov chain $\mathcal{M}$ at time $t+1$, uniformly at random from $X'^{\mathfrak{G}}_{t+1}$, the orbit of X_{t+1}'.

The orbital Markov chain $\mathcal{M}$, therefore, runs at every time step $t \geq 1$ the original chain $\mathcal{M}'$ first and samples the state of $\mathcal{M}$ at time t uniformly at random from the orbit of the state of the original chain $\mathcal{M}'$ at time t. Figure 10.3 (c) depicts a fragment of the orbital Markov chain for the original Markov chain (a). Instead of computing the equivalence of the state space explicitly (b) novel transitions are introduced that make the chain behave *as if it was lumped.*

Given a state X_t and a permutation group $\mathfrak{G}$ orbital Markov chains sample an element from $X_t^{\mathfrak{G}}$, the orbit of X_t, uniformly at random. By the orbit-stabilizer theorem this is equivalent to sampling an element $\mathfrak{g} \in \mathfrak{G}$ uniformly at random and computing $X_t^{\mathfrak{g}}$. Sampling group elements uniformly at random is a well-researched problem (Celler et al., 1995) and computable in polynomial time in the size of the generating sets with product replacement algorithms (Pak, 2000). These algorithms are implemented in several group algebra systems such as GAP (GAP) and exhibit remarkable performance. Once initialized, product replacement algorithms can generate pseudo-random elements by performing, depending on the variant, 1 to 3 group multiplications. We could verify that the overhead of step 2 during the sampling process is indeed negligible.

The following theorem relates properties of the orbital Markov chain to those of the Markov chain it is derived from. A detailed proof can be found in the appendix.

Theorem 10.3 (Niepert (2012b)) *Let Ω be a finite set and let $\mathcal{M}'$ be a Markov chain with state space Ω and transition matrix P'. Moreover, let π be a probability distribution on Ω, let $\mathfrak{G}$ be an automorphism group for (Ω, π), and let $\mathcal{M}$ be the orbital Markov chain for $\mathcal{M}'$. Then,*

(a) *if $\mathcal{M}'$ is aperiodic then $\mathcal{M}$ is also aperiodic;*
(b) *if $\mathcal{M}'$ is irreducible then $\mathcal{M}$ is also irreducible;*
(c) *if π is a reversible distribution for $\mathcal{M}'$ and, for all $\mathfrak{g} \in \mathfrak{G}$ and all $x, y \in \Omega$ we have that $P'(x, y) = P'(x^{\mathfrak{g}}, y^{\mathfrak{g}})$, then π is also a reversible and, hence, a stationary distribution for $\mathcal{M}$.*

The condition in statement (c) requiring for all $\mathfrak{g} \in \mathfrak{G}$ and all $x, y \in \Omega$ that $P'(x, y) = P'(x^{\mathfrak{g}}, y^{\mathfrak{g}})$ expresses that the original Markov chain is compatible with the symmetries captured by the permutation group $\mathfrak{G}$. This weak assumption is met by all of the practical Markov chains we are aware of and, in particular, Metropolis chains and Gibbs sampler.

10.5   Lifted MCMC for Asymmetric Models

This section extends the lifted MCMC framework to construct mixtures of Markov chains where one of the chains operates on the **approximate symmetries** of the probabilistic model. The framework assumes a base Markov chain $\mathcal{M}_B$ such as the Gibbs chain, the MC-SAT chain (Poon and Domingos, 2006), or any other MCMC algorithm. We then construct a mixture of the base chain and an Orbital Metropolis chain which exploits approximate symmetries for its proposal distribution.

10.5.1   Mixing

Two or more Markov chains can be combined by constructing mixtures and compositions of the kernels (Tierney, 1994). Let P_1 and P_2 be the kernels for two Markov chains $\mathcal{M}_1$ and $\mathcal{M}_2$ both with stationary distribution π. Given a positive probability $0 < \alpha < 1$, a *mixture* of the Markov chains is a Markov chain where, in each iteration, kernel P_1 is applied with probability α and kernel P_2 with probability $1 - \alpha$. The resulting Markov chain has π as a stationary distribution. The following result relates properties of the individual chains to properties of their mixture.

Theorem 10.4 (Tierney (1994)) *A mixture of two Markov chains $\mathcal{M}_1$ and $\mathcal{M}_2$ is irreducible and aperiodic if at least one of the chains is irreducible and aperiodic.*

For a more in-depth discussion of combining Markov chains and the application to machine learning, we refer the interested reader to an overview paper (Andrieu et al., 2003).

10.5.2   Metropolis-Hastings Chains

Before we describe the approach in more detail, let us first review Metropolis samplers. The construction of a **Metropolis-Hastings Markov chain** is a popular general procedure for designing reversible Markov chains for MCMC-based estimation of marginal probabilities. Metropolis-Hastings chains are associated with a **proposal distribution** $Q(\cdot|\mathbf{x})$ that is utilized to *propose* a move to the next state given the current state $\mathbf{x}$. The closer the proposal distribution to the distribution π to be estimated, that is, the closer $Q(\mathbf{x} \mid \mathbf{x}_t)$ to $\pi(\mathbf{x})$ for large t, the better the convergence properties of the Metropolis-Hastings chain.

We first describe the Metropolis algorithm, a special case of the Metropolis-Hastings algorithm (Häggström, 2002). Let $\mathbf{X}$ be a finite set of random variables with probability distribution π and let Ω be the set of states of the random variables. The Metropolis chain is governed by a transition graph $G = (\Omega, \mathbf{E})$ whose nodes correspond to states of the random variables. Let $\mathbf{n}(\mathbf{x})$ be the set of neighbors of state $\mathbf{x}$ in G, that is, all states reachable from **x** with a single transition. The Metropolis chain with graph G and distribution π has transition probabilities

$$P(\mathbf{x} \rightarrow \mathbf{y}) = \begin{cases} \frac{1}{|n(\mathbf{x})|} \min\left\{\frac{\pi(\mathbf{y})|n(\mathbf{x})|}{\pi(\mathbf{x})|n(\mathbf{y})|}, 1\right\} & \text{if } x \text{ and } y \text{ are neighbors,} \\ 1 - \sum\limits_{y' \in n(\mathbf{x})} \frac{1}{|n(\mathbf{x})|} \min\left\{\frac{\pi(\mathbf{y}')|n(\mathbf{x})|}{\pi(\mathbf{x})|n(\mathbf{y}')|}, 1\right\} & \text{if } x = y, \\ 0, & \text{otherwise.} \end{cases}$$

Being in state $\mathbf{x}_t$ of the Markov chain $\mathcal{M} = (\mathbf{x}_0, \mathbf{x}_1, ...)$, the Metropolis sampler therefore performs the following steps at time $t + 1$:

1. Select a state **y** from $n(\mathbf{x}_t)$, the neighbors of $\mathbf{x}_t$, uniformly at random;
2. Let $\mathbf{x}_{t+1} = \mathbf{y}$ with probability $\min\left\{\frac{\pi(\mathbf{y})|n(\mathbf{x})|}{\pi(\mathbf{x})|n(\mathbf{y})|}, 1\right\}$;
3. Otherwise, let $\mathbf{x}_{t+1} = \mathbf{x}_t$.

Note that the proposal distribution $Q(\cdot|\mathbf{x})$ is simply the uniform distribution on the set of **x**'s neighbors. It is straight-forward to show that π is a stationary distribution for the Metropolis chain by showing that π is a reversible distribution for it (Häggström, 2002).

Now, the performance of the Metropolis chain hinges on the structure of the graph G. We would like the graph structure to facilitate global moves between high probability modes, as opposed to the local moves typically performed by MCMC chains. To design such a graph structure, we take advantage of approximate symmetries in the model.

10.5.3 Orbital Metropolis Chains

This section describes a class of **orbital Metropolis chains** that move between approximate symmetries of a distribution. The approximate symmetries form an automorphism group $\mathfrak{G}$. We will discuss approaches to obtain such an automorphism group in Section 10.6. Here, we introduce a Markov chain that takes advantage of the approximate symmetries.

Given a distribution π over random variables **X** with state space Ω, and a permutation group $\mathfrak{G}$ acting on Ω, the orbital Metropolis chain $\mathcal{M}_S$ for $\mathfrak{G}$ performs the following steps:

1. Select a state **y** from $\mathbf{x}_t^{\mathfrak{G}}$, the orbit of $\mathbf{x}_t$, uniformly at random;
2. Let $\mathbf{x}_{t+1} = \mathbf{y}$ with probability $\min\left\{\frac{\pi(\mathbf{y})}{\pi(\mathbf{x})}, 1\right\}$;
3. Otherwise, let $\mathbf{x}_{t+1} = \mathbf{x}_t$.

Note that a permutation group acting on Ω partitions the states into disjoint orbits. The orbital Metropolis chain simply moves between states in the same orbit. Hence, two states in the same orbit have the same number of neighbors and, thus, the expressions cancel out in line 2 above. It is straight-forward to show that the chain $\mathcal{M}_S$ is reversible and, hence, that it has π as a stationary distribution. However, the chain is *not* irreducible as it never moves between states that are not symmetric with respect to the permutation group $\mathfrak{G}$. In the binary case, for example, it cannot reach states with a different Hamming weight from the initial state.

10.5.4   Lifted Metropolis-Hastings

To obtain an irreducible Markov chain that exploits approximate symmetries, we construct a mixture of (a) some base chain $\mathcal{M}_B$ with stationary distribution π for which we know that it is irreducible and aperiodic; and (b) an orbital Metropolis chain $\mathcal{M}_S$. We can prove the following theorem.

Theorem 10.5 *Let* **X** *be a set of random variables with distribution π and approximate automorphisms $\mathfrak{G}$. Moreover, let $\mathcal{M}_B$ be an aperiodic and irreducible Markov chain with stationary distribution π, and let $\mathcal{M}_S$ be the orbital Metropolis chain for* **X** *and $\mathfrak{G}$. The mixture of $\mathcal{M}_B$ and $\mathcal{M}_S$ is aperiodic, irreducible, and has π as its unique stationary distribution.*

The mixture of the base chain and the orbital Metropolis chain has several advantages. First, it exploits the approximate symmetries of the model which was shown to be advantageous for marginal probability estimation (Van den Broeck and Darwiche, 2013). Second, the mixture of Markov chains performs wide ranging moves via the orbital Metropolis chain, exploring the state space more efficiently and, therefore, improving the quality of the probability estimates. Figure 10.4 depicts the state space and the transition graph of (a) the Gibbs chain and (b) the mixture of the Gibbs chain and an orbital Metropolis chain. It illustrates that the mixture is able to more freely move about the state space by jumping between orbit states. For instance, moving from state 0110 to 1001 would require 4 steps of the Gibbs chain but is possible in one step with the mixture of chains. The larger the size of the automorphism groups, the more densely connected is the transition graph. Since the moves of the orbital Metropolis chain are between approximately symmetric states of the random variables, it does not suffer from the problem of most proposals being rejected. We will be able to verify this hypothesis empirically.

The general Lifted Metropolis-Hastings framework can be summarized as follows.

1. Obtain an approximate automorphism group $\mathfrak{G}$;
2. Run the following mixture of Markov chains:

    (a) With probability $0 < \alpha < 1$, apply the kernel of the base chain $\mathcal{M}_B$;

    (b) Otherwise, apply the kernel of the orbital Metropolis chain $\mathcal{M}_S$ for $\mathfrak{G}$.

Note that the proposed approach is a generalization of lifted MCMC for symmetric models, as described in the previous section, essentially using it as a subroutine, and that all MH proposals are accepted if $\mathfrak{G}$ is an exact automorphism group of the original model. Moreover, note that the framework allows one to combine multiple orbital Metropolis chains with a base chain.

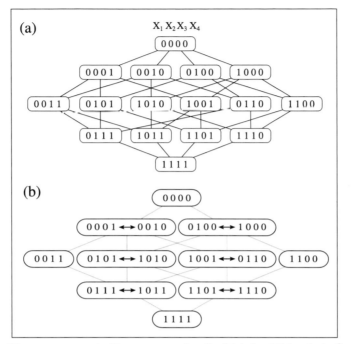

Figure 10.4: The state space (self-arcs are omitted) of (a) the Gibbs chain for four binary random variables and (b) the orbit partition of its state space induced by the permutation group generated by the permutation $(X_1\ X_2)(X_3\ X_4)$. The permutations are approximate symmetries, derived from an over-symmetric approximation of the original model. The Gibbs chain proposes moves to states whose Hamming distance to the current state is at most 1. The orbital Metropolis chain, on the other hand, proposes moves between orbit elements which have a Hamming distance of up to 4. The mixture of the two chains leads to faster convergence while maintaining an unbiased stationary distribution.

10.6   Approximate Symmetries

The Lifted Metropolis-Hastings algorithm assumes that a permutation group $\mathfrak{G}$ is given, representing the approximate symmetries. We now discuss several approaches to the computation of such an automorphism group. While it is not possible to go into technical detail here, we will provide pointers to the relevant literature.

There exist several techniques to compute the *exact symmetries* of a graphical model and construct $\mathfrak{G}$; see (Niepert, 2012b; Bui et al., 2012). The color refinement algorithm is also well-studied in lifted inference (Kersting et al., 2014). It can find (exact) orbits of random variables for a slightly weaker notion of symmetry, called fractional automorphism. These techniques all require some form of exact symmetry to be present in the model.

Detecting *approximate symmetries* is a problem that is largely open. One key idea is that of an *over-symmetric approximations* (OSAs) (Van den Broeck and Darwiche, 2013). Such approximations are derived from the original model by rendering the model more symmetric. After the computation of an over-symmetric model, we can apply existing tools for exact symmetry detection. Indeed, the exact symmetries of an approximate model are approximate symmetries of the exact model. These symmetrization techniques are indispensable to our algorithm.

10.6.1   Relational Symmetrization

Existing symmetrization techniques operate on relational representations, such as Markov logic networks (MLNs). Relational models have numerous symmetries. For example, swapping the web pages A and B in a web page classification model does not change the MLN. This permutation of constants induces a permutations of random variables (e.g., between Page(A, Faculty) and Page(B, Faculty)). Unfortunately, hard and soft evidence breaks symmetries, even in highly symmetric relational models (Van den Broeck and Darwiche, 2013). When the variables Page(A, Faculty) and Page(B, Faculty) get assigned distinct soft evidence, the symmetry between A and B is removed, and lifted inference breaks down.[20] Similarly, when the Link relation is given, its graph is unlikely to be symmetric (Erdős and Rényi, 1963), which in turn breaks the symmetries in the MLN. These observations motivated research on OSAs. Van den Broeck and Darwiche (2013) propose to approximate binary relations, such as Link, by a low-rank Boolean matrix factorization. Venugopal and Gogate (2014a) cluster the constants in the domain of the MLN. Singla et al. (2014) present a message-passing approach to clustering similar constants.

10.6.2   Propositional Symmetrization

A key property of our LMH algorithm is that it operates at the propositional level, regardless of how the graphical model was generated. It also means that the relational symmetrization approaches outlined above are inadequate in the general case. Unfortunately, we are not aware of any work on OSAs of propositional graphical models. However, some existing techniques provide a promising direction. First, basic clustering can group together similar potentials. Second, the low-rank Boolean matrix factorization used for relational approximations can be applied to any graph structure, including graphical models. Third, color passing techniques for exact symmetries operate on propositional models (Kersting et al., 2009, 2014). Combined with early stopping, they can output approximate variable orbits.

---

[20] Solutions to this problem exist if the soft evidence is on a single unary relation (Bui et al., 2012)

10.6.3 From OSAs to Automorphisms

Given an OSA of our model, we need to compute an automorphism group $\mathfrak{G}$ from it. The obvious choice is to compute the exact automorphisms from the OSA. While this works in principle, it may not be optimal. Let us first consider the following two concepts. When a group $\mathfrak{G}$ operates on a set Ω, only a subset of the elements in Ω can actually be mapped to an element other than itself. When Ω is the set of random variables, we call these elements the *moved variables*. When Ω is the set of potentials in a probabilistic graphical model, we call these the *moved potentials*. It is clear that we want $\mathfrak{G}$ to move many random variables, as this will create the largest jumps and improve the mixing behavior. However, each LMH step comes at a cost: in the second step of the algorithm, the probability of the proposed approximately-symmetric state $\pi(\mathbf{y})$ is estimated. This requires the re-evaluation of all potentials that are moved by $\mathfrak{G}$. Thus, the time complexity of an orbital Metropolis step is linear in the number of moved potentials. It will therefore be beneficial to construct *subgroups* of the automorphism group of the OSA and, in particular, ones that move many variables and few potentials.

10.7 Empirical Evaluation

This section empirically evaluates lifted MCMC, both on symmetric models where we use orbital Markov chains, and on asymmetric models with approximate symmetries, where we use then lifted Metropolis-Hastings algorithm.

10.7.1 Symmetric Model Experiments

We conduct experiments with the well-established social network Markov logic network (the smokes-cancer MLN) exactly as specified in (Singla and Domingos, 2008). Here we created two ground MLNs with 50 and 100, respectively, people in the domain, leading to Markov networks with 2600 and 10200 variables, respectively. Building the ground models took only a fraction of a second. We proceeded to apply the symmetry detection algorithm (Niepert, 2012b) taking 24 and 136 ms, respectively, to compute the irredundant generators of the automorphism group of the models. For n people in the domain, there are $n-1$ irredundant generators of the automorphism group and the group has size $n!$ which is exactly the size of the symmetric group on n. Please note that, based on our observation of indistinguishability of objects on different syntactical levels of the model, it is actually not necessary to use symmetry detection algorithms in this case. The irredundant generators of the symmetric group representing the symmetries on the level of constants can be directly used to compute the irredundant generators for the permutation group representing the symmetries on the level of ground atoms and formulas.

We compared the standard Gibbs sampler, Alchemy's MC-SAT algorithm (Poon and Domingos, 2006), and the orbital Gibbs sampler on the models. The overhead of the product replacement algorithm was again negligible and far outweighed by the faster con-

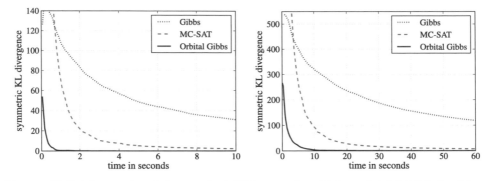

Figure 10.5: The results of the standard Gibbs sampler, Alchemy's MC-SAT algorithm, and the orbital Gibbs sampler for the social network MLN with 50 (top) and 100 (bottom) people in the domain.

vergence of the orbital chain. Figure 10.5 plots the symmetric Kullback-Leibler divergence for the single variable marginals.

Finally, Niepert (2012b) reports additional experiments on using orbital Markov chains to sample independent sets, building on *insert/delete* Markov chains (Luby and Vigoda, 1999; Dyer and Greenhill, 2000).

10.8   Asymmetric Model Experiments

The LMH algorithm is implemented in the GAP algebra system which provides basic algorithms for automorphism groups such as the product replacement algorithm that allows one to sample uniformly from orbits of states (Niepert, 2012b).

For our first experiments, we use the standard WebKB data set, consisting of web pages from four computer science departments (Craven and Slattery, 2001). The data has information about approximately 800 words that appear on 1000 pages, 7 page labels and links between web pages. There are 4 folds, one for each university. We use the standard MLN structure for the WebKB domain, which has MLN formulas of the form shown above, but for all combinations of labels and words, adding up to around 5500 first-order MLN formulas. We learn the MLN parameters using Alchemy.

We consider a collective classification setting, where we are given the link structure and the word content of each web page, and want to predict the page labels. We run Gibbs sampling and the Lifted MCMC algorithm (Niepert, 2012b), and show the average KL divergence between the estimated and true marginals in Figure 10.6. When true marginals are not computable, we used a very long run of a Gibbs sampler for the gold standard marginals. Since every web page contains a unique set of words, the evidence on the word content creates distinct soft evidence on the page labels. Moreover, the link structure is largely asymmetric and, therefore, there are no exploitable exact symmetries and Lifted

MCMC coincides with Gibbs sampling. Next we construct an OSA using a rank-5 approximation of the link structure (Van den Broeck and Darwiche, 2013) and group the potential weights into 6 clusters.

From this OSA we construct a set of automorphisms that is efficient for LMH as follows. First, we compute the exact automorphisms $\mathfrak{G}_1$ of the OSA. Second, we compute the variable orbits of $\mathfrak{G}_1$, grouping together all variables that can be mapped into each other. Then, for every orbit O, we construct a set of automorphisms as follows. We greedily search for a $O' \subseteq O$ such that the symmetric group $\mathfrak{G}_{O'}$ on O' maximizes the ratio between the number of moved variables (i.e., $|O'|$) and the number of moved potentials, while keeping the number of moved potentials bounded by a constant K. This guarantees that $\mathfrak{G}_{O'}$ yields an efficient orbital Metropolis chain. Finally, we remove O' from O and recurse until O is empty. From this set of symmetric groups $\mathfrak{G}_{O'}$, we construct a set of orbital Metropolis chains, each with it own set of moved potentials.

Figure 10.6 shows that the LMH chain, with mixing parameter $\alpha = 4/5$, has a lower KL divergence than Gibbs and Lifted MCMC vs. the number of iterations. Note that there is a slight overhead to LMH because the orbital Metropolis chain is run between base chain steps. Despite this overhead, LMH outperforms the baselines as a function of time. The orbital Metropolis chain accepts approximately 70% of its proposals.

Figure 10.7 illustrates the effect of running Lifted MCMC on OSA, which is the current state-of-the-art approach for asymmetric models. As expected, the drawn sample points produce biased estimates. As the quality of the approximation increases, the bias reduces, but so do the speedups. LMH does not suffer from a bias. Moreover, we observe that its performance is stable across different OSAs (not depicted).

We also ran experiments for two propositional models that are frequently used in real world applications. The first model is a 100x100 ferromagnetic Ising model with constant interaction strength and external field (see Figure 10.1(a) for a 4x4 version). Due to the different potentials induced by the external field, the model has no symmetries. We use the model without external field to compute the approximate symmetries. The automorphism group representing these symmetries is generated by the rotational and reflectional symmetries of the grid model (see Figure 10.1(b)). As in the experiments with the relational models, we used the mixing parameter $\alpha = 4/5$ for the LMH algorithm. Figure 10.8(c) and (d) depicts the plots of the experimental results. The LMH algorithm performs better with respect to the number of iterations and, to a lesser extent, with respect to time.

We also ran experiments on the Chimera model which has recently received some attention as it was used to assess the performance of quantum annealing (Boixo et al., 2013). We used exactly the model as described in Boixo et al. (2013). This model is also asymmetric but can be made symmetric by assuming that all pairwise interactions are identical. The KL divergence vs. number of iterations and vs. time in seconds is plotted in Figure 10.8(a) and (b), respectively. Similar to the results for the Ising model, LMH outperforms Gibbs and

LMCMC both with respect to the number of iterations and wall clock time. In summary, the LMH algorithm outperforms standard sampling algorithms on these propositional models in the absence of any symmetries. We used very simple symmetrization strategies for the experiments. This demonstrates that the LMH framework is powerful and allows one to design state-of-the-art sampling algorithms.

10.9   Conclusions

We have presented a perspective on lifted inference, where instead of directly operating on the space of joint variable assignments, *orbital* Markov chains operate on a symmetry-induced partition of this space. We related lifted MCMC to the notion of lumping of Markov chains. Instead of computing the partition of the state space explicitly which is usually intractable, orbital Markov chains operate on the original state space while having convergence properties identical to the corresponding lumped Markov chain. We want to point out that in the MCMC literature a lifting of a Markov chain (Chen et al., 1999) is *not* the same as what has been coined lifted inference by the statistical relational AI community. Quite the opposite, instead of operating on a more compact state space, lifting in the classical sense introduces additional states. Nevertheless, there might be interesting relationships between lumping, lifting and lifted inference.

We have also presented a Lifted Metropolis-Hastings algorithms capable of mixing two types of Markov chains. The first is a non-lifted base chain, and the second is an orbital Metropolis chain that moves between approximately symmetric states. This allows lifted inference techniques to be applied to asymmetric graphical models.

Numerous extensions to the lifted MCMC framework have been developed in recent years, for example towards exploiting contextual and block-valued symmetries (Anand et al., 2016; Madan et al., 2018), and continuous symmetries (Shariff et al., 2015). Holtzen et al. (2019a) showed how to build an exact lifted inference algorithm that enumerates and counts orbits, and how to sample orbits directly, without the need for Gibbs sampling to move between orbits.

10.10   Acknowledgments

Lifted MCMC for symmetric models first appeared as Niepert (2012b) and Niepert (2012a). Lifted MCMC for asymmetric models first appeared as Van den Broeck and Niepert (2015).

10.11   Appendix: Proof of Theorem 10.3

We first prove (a). Since $\mathcal{M}'$ is aperiodic we have, for each state $x \in \Omega$ and every time step $t \geq 0$, a non-zero probability for the Markov chain $\mathcal{M}'$ to remain in state x at time $t + 1$. At each time $t + 1$, the orbital Markov chain transitions uniformly at random to one of the states in the orbit of the original chain's state at time $t + 1$. Since every state is an element of its own orbit, we have, for every state $x \in \Omega$ and every time step $t \geq 0$, a non-zero probability for the Markov chain $\mathcal{M}$ to remain in state x at time $t + 1$. Hence, $\mathcal{M}$ is aperiodic. The proof of statement (b) is accomplished in an analogous fashion and omitted.

Let $P(x, y)$ and $P'(x, y)$ be the probabilities of $\mathcal{M}$ and $\mathcal{M}'$, respectively, to transition from state x to state y. Since π is a reversible distribution for $\mathcal{M}'$ we have that $\pi(x)P'(x, y) = \pi(y)P'(y, x)$ for all states $x, y \in \Omega$. For every state $x \in \Omega$ let $x^{\mathfrak{G}}$ be the orbit of x. Let $\mathfrak{G}_x := \{\mathfrak{g} \in \mathfrak{G} \mid x^{\mathfrak{g}} = x\}$ be the stabilizer subgroup of x with respect to $\mathfrak{G}$. We have that

$$\sum_{\mathfrak{g} \in \mathfrak{G}} P'(x, y^{\mathfrak{g}}) = \sum_{y' \in y^{\mathfrak{G}}} |\mathfrak{G}_{y'}| P'(x, y')$$

$$= |\mathfrak{G}_y| \sum_{y' \in y^{\mathfrak{G}}} P'(x, y') \tag{10.1}$$

$$= (|\mathfrak{G}|/|y^{\mathfrak{G}}|) \sum_{y' \in y^{\mathfrak{G}}} P'(x, y')$$

where the last two equalities follow from the orbit-stabilizer theorem. We will now prove that $\pi(x)P(x, y) = \pi(y)P(y, x)$ for all states $x, y \in \Omega$. By definition of the orbital Markov chain we have that $\pi(x)P(x, y) = \pi(x)(1/|y^{\mathfrak{G}}|) \sum_{y' \in y^{\mathfrak{G}}} P'(x, y')$ and, by equation (1), we have $\pi(x)(1/|y^{\mathfrak{G}}|) \sum_{y' \in y^{\mathfrak{G}}} P'(x, y')$

$$= \pi(x)(1/|y^{\mathfrak{G}}|)(|y^{\mathfrak{G}}|/|\mathfrak{G}|) \sum_{\mathfrak{g} \in \mathfrak{G}} P'(x, y^{\mathfrak{g}})$$

$$= \pi(x)(1/|\mathfrak{G}|) \sum_{\mathfrak{g} \in \mathfrak{G}} P'(x, y^{\mathfrak{g}})$$

$$= (1/|\mathfrak{G}|) \sum_{\mathfrak{g} \in \mathfrak{G}} \pi(x)P'(x, y^{\mathfrak{g}}).$$

Since P' is reversible and $\pi(x) = \pi(x^{\mathfrak{g}})$ for all $\mathfrak{g} \in \mathfrak{G}$ we have $(1/|\mathfrak{G}|) \sum_{\mathfrak{g} \in \mathfrak{G}} \pi(x)P'(x, y^{\mathfrak{g}}) = (1/|\mathfrak{G}|) \sum_{\mathfrak{g} \in \mathfrak{G}} \pi(y^{\mathfrak{g}})P'(y^{\mathfrak{g}}, x) = \pi(y)(1/|\mathfrak{G}|) \sum_{\mathfrak{g} \in \mathfrak{G}} P'(y^{\mathfrak{g}}, x)$. Now, since $P'(x, y) = P'(x^{\mathfrak{g}}, y^{\mathfrak{g}})$ for all $x, y \in \Omega$ and all $\mathfrak{g} \in \mathfrak{G}$ by assumption, we have that $\pi(y)(1/|\mathfrak{G}|) \sum_{\mathfrak{g} \in \mathfrak{G}} P'(y^{\mathfrak{g}}, x) = \pi(y)(1/|\mathfrak{G}|) \sum_{\mathfrak{g} \in \mathfrak{G}} P'(y, x^{-\mathfrak{g}}) = \pi(y)(1/|\mathfrak{G}|) \sum_{\mathfrak{g} \in \mathfrak{G}} P'(y, x^{\mathfrak{g}})$ and, again by equation (1), we have $\pi(y)(1/|\mathfrak{G}|) \sum_{\mathfrak{g} \in \mathfrak{G}} P'(y, x^{\mathfrak{g}})$

$$= \pi(y)(1/|\mathfrak{G}|)(|\mathfrak{G}|/|x^{\mathfrak{G}}|) \sum_{x' \in x^{\mathfrak{G}}} P'(y, x')$$

$$= \pi(y)(1/|x^{\mathfrak{G}}|) \sum_{x' \in x^{\mathfrak{G}}} P'(y, x') = \pi(y)P(y, x). \quad \square$$

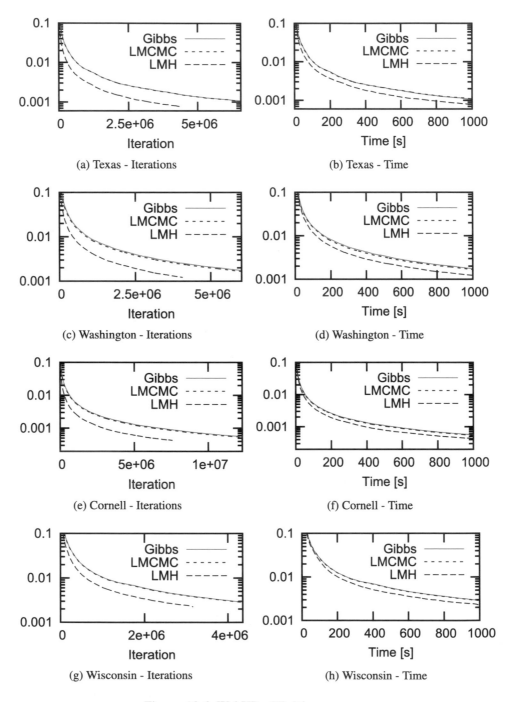

Figure 10.6: WebKB - KL Divergences

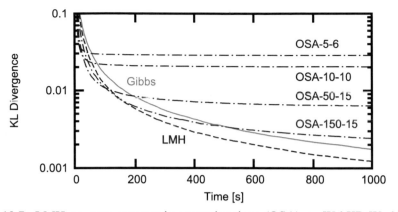

Figure 10.7: LMH vs. over-symmetric approximations (OSA) on WebKB Washington. OSA-*r-c* denotes binary evidence of Boolean rank *r* and *c* clusters of formula weights.

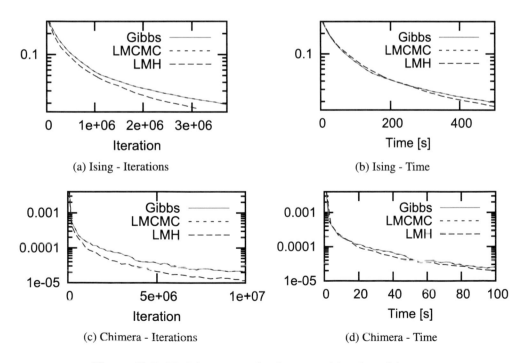

Figure 10.8: KL Divergences for the propositional models.

11 Lifted Message Passing for Probabilistic and Combinatorial Problems

Kristian Kersting, Fabian Hadiji, Babak Ahmadi, Martin Mladenov, and Sriraam Natarajan

Abstract. Inference problems arise in a variety of scientific fields and engineering applications. Probabilistic graphical models provide a scalable framework for developing efficient inference methods, such as message-passing algorithms that exploit the conditional independencies encoded by the given graph. Modern AI applications, however, require inference within large-scale models with hundreds of thousand if not million random variables. Typical ways to handling scaling issue with inference are approximate message passing approaches. Often, however, we encounter inference problems with symmetries and redundancies in the graph structure, e.g. induced by relational models that capture complexity. In this chapter, we review that message-passing approaches can indeed benefit from exploiting symmetries. Specifically, we show that (loopy) belief propagation (BP) can be lifted. That is, a model is compressed by grouping nodes together that send and receive identical messages so that a modified BP running on the lifted graph yields the same marginals as BP on the original one, but often in a fraction of time. We then show that the same idea naturally carries over to solving combinatorial problems such as determining the satisfiability of Boolean formulas. That is, we review lifted survey propagation and its simpler version lifted warning propagation.

11.1 Introduction

Many important inference problems in Machine Learning and Artificial Intelligence can be posed as the computation of certain summarizing statistics (e.g., marginals, modes, means, likelihoods) given a multi-variate probability distribution, where some variables are measured while others must be estimated based on the observed measurements. The practical challenges stem from the fact that, in general, the representation and manipulation of a joint probability distribution scales exponentially with the number of random variables being described. The graphical model formalism provides both a compact representation of large multivariate distributions and a systematic characterization of the associated probabilistic structure to be exploited for computational efficiency. Fundamentally, a graphical model represents a family of probability distributions on the underlying graph: nodes are identified with random variables and edges (or the lack thereof) encode Markov properties among subsets of random variables.

However, having a compact representation of a joint probability distribution is, by itself, not sufficient to solve large-scale inference problems efficiently. The complexity of inference given a graphical model also depends strongly on the underlying graph. For graphs without cycles, or trees, inference can be organized recursively in a manner that scales linearly in the number of nodes. The many variants of this basic idea comprise the class of graph-based **message-passing algorithms** broadly lumped under the term of belief propagation. **Belief Propagation** (BP) algorithms essentially amount to iterating over a certain set of nonlinear fixed-point equations, relating the desired inference solution to so-called messages passed between every node and its immediate neighbors in the graph. Such iterations always converge in a tree-structured graphical model, the final messages into each node representing sufficient statistics of the information at all other nodes.

Many practical applications, however, give rise to graphical models for which exact inference is computationally demanding. **Variational inference** methods for approximate inference, see also the corresponding chapter in this book, start by expressing the intractable solution as a mathematical optimization problem. This view has helped to gain an improved understanding of the popular **loopy belief propagation** (LBP) approach: simply iterate the BP fixed-point equations as if the underlying graph were free of cycles. The efficient algorithm (if it converges) has proven to be considerably successful in a variety of large-scale practical applications. Often, however, we encounter inference problems with symmetries and redundancies in the graph structure. A prominent example are relational models that capture complexity. Exploiting these **symmetries** has traditionally not been considered for scaling inference. In this chapter, we show that (loopy) BP and related approaches such as BP-guided sampling and likelihood maximization can indeed benefit from exploiting symmetries.

In many real-world applications, however, the problem formulation does not neatly fall into the pure probabilistic inference case. The problem may very well have a component that can be well-modeled as a **combinatorial problem**, hence, taking it outside the scope of standard (lifted) belief propagation. Ideally, lifted message-passing should be efficient here, too, exploiting as much symmetries as possible. We therefore revisit lifted message-passing algorithms this time for combinatorial problems. Specifically, we focus on the **Boolean satisfiability** (SAT) problem consisting of a formula F representing a set of constraints over n **Boolean variables**, which must be set so as to satisfy all constraints. It is the most studied problem in Computer Science and AI. All other NP-complete problems can be reduced to it and, hence, "reduction to SAT" is a powerful paradigm for solving Computer Science and AI problems and has applications in several important areas such as automated deduction, verification, and planning, among others. In fact, we review lifted version of **survey propagation** (SP) and **warning propagation** (WP) and apply methods. This provides a complete picture of lifted message-passing algorithms, both for probabilistic inference and for SAT.

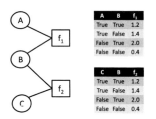

Figure 11.1: An example for a factor graph with associated potentials. Circles denote variables (binary in this case), squares denote factors.

11.2 Loopy Belief Propagation (LBP)

Let $\mathbf{X} = (X_1, X_2, \ldots, X_n)$ be a set of n discrete-valued random variables and let x_i represent the possible realizations of random variable X_i. **Graphical models** compactly represent a joint distribution over $\mathbf{X}$ as a product of **factors** (Pearl, 1988), i.e.,

$$P(\mathbf{X} = \mathbf{x}) = \frac{1}{Z} \prod_k f_k(\mathbf{x}_k) \ . \tag{11.1}$$

Here, each factor f_k is a non-negative function of a subset of the variables $\mathbf{x}_k$, and Z is a normalization constant. As long as $P(\mathbf{X} = \mathbf{x}) > 0$ for all joint configurations $\mathbf{x}$, the distribution can be equivalently represented as a log-linear model:

$$P(\mathbf{X} = \mathbf{x}) = \frac{1}{Z} \exp \left[\sum_i w_i \cdot \varphi_i(\mathbf{x}) \right] , \tag{11.2}$$

where the features $\varphi_i(x)$ are arbitrary functions of (a subset of) the configuration $\mathbf{x}$.

Graphical models can be represented as **factor graphs**. A factor graph, as shown in Fig. 11.1, is a bipartite graph that expresses the factorization structure in Eq. (11.1). It has a variable node (denoted as a circle) for each variable X_i, a factor node (denoted as a square) for each f_k, with an edge connecting variable node i to factor node k if and only if X_i is an argument of f_k. We will consider one factor $f_i(\mathbf{x}) = \exp[w_i \cdot \varphi_i(\mathbf{x})]$ per feature $\varphi_i(\mathbf{x})$, i.e., we will not aggregate factors over the same variables into a single factor.

An important inference task is to compute the conditional probability of variables given the values of some others, the evidence, by summing out the remaining variables. The belief propagation (BP) algorithm is an efficient way to solve this problem that is exact when the factor graph is a tree, but only approximate when the factor graph has cycles. One should note that the problem of computing marginal probability functions is in general hard (#P-complete).

Belief Propagation makes local computations only. It makes use of the graphical structure such that the marginals can be computed much more efficiently. We will now describe the BP algorithm in terms of operations on a factor graph. The computed marginal probability functions will be exact if the factor graph has no cycles, but the BP algorithm is still well-defined when the factor graph does have cycles. Although this loopy belief propagation has no guarantees of convergence or of giving the correct result, in practice it often does, and can be much more efficient than other methods (Murphy et al., 1999).

To define the BP algorithm, we first introduce **messages** between variable nodes and their neighboring factor nodes and vice versa. The message from a variable X to a factor f is

$$\mu_{X \to f}(x) = \prod_{h \in nb(X) \setminus \{f\}} \mu_{h \to X}(x) \qquad (11.3)$$

where $nb(X)$ is the set of factors X appears in. The message from a factor to a variable is

$$\mu_{f \to X}(x) = \sum_{\neg \{X\}} \left(f(\mathbf{x}) \prod_{Y \in nb(f) \setminus \{X\}} \mu_{Y \to f}(y) \right) \qquad (11.4)$$

where $nb(f)$ are the arguments of f, and the sum is over all the values of these except X, denoted as $\neg \{X\}$. The messages are usually initialized to 1.

Now, the **unnormalized belief** of each variable X_i can be computed from the equation

$$b_i(x_i) = \prod_{f \in nb(X_i)} \mu_{f \to X_i}(x_i) \qquad (11.5)$$

Evidence is incorporated by setting $f(\mathbf{x}) = 0$ for states $\mathbf{x}$ that are incompatible with it. Different schedules may be used for message-passing.

Since, loopy belief propagation is efficient, it can directly be used for inference within probabilistic relational models by applying it to the grounded relational model.

11.3   Lifted Message Passing for Probabilistic Inference: Lifted (Loopy) BP

Although already quite efficient, many graphical models produce inference problems with a lot of additional regularities reflected in the graphical structure but not exploited by BP. Probabilistic relational models such as **Markov Logic Networks** (MLNs) are prominent examples (Richardson and Domingos, 2006). As an illustrative example, reconsider the factor graph in Fig. 11.1. The associated potentials are identical. In other words, although the factors involved are different on the surface, they actually share quite a lot of information. Standard BP cannot make use of this information. In contrast, **lifted belief propagation** (lifted BP) – which we will review now – can make use of it and speed up inference by orders of magnitude (Singla and Domingos, 2008; Kersting et al., 2009).

Lifted BP performs two steps: Given a factor graph G, it first computes a **compressed factor graph** $\mathfrak{G}$ and then runs a modified BP on $\mathfrak{G}$. We will now discuss each step in turn using fraktur letters such as $\mathfrak{G}$, $\mathfrak{X}$, and $\mathfrak{f}$ to denote compressed graphs, nodes, and factors.

11.3.1   Step 1 – Compressing the Factor Graph

Essentially, we simulate BP keeping track of which nodes and factors send the same messages, and group nodes and factors together correspondingly.

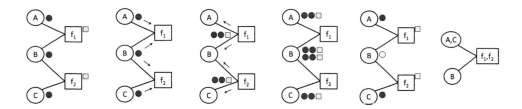

Figure 11.2: From left to right, the steps of CFG compressing the factor graph in Fig. 11.1 assuming no evidence. The shaded/colored small circles and squares denote the groups and signatures produced running CFG. On the right-hand side, the resulting compressed factor graph is shown.

Let G be a given factor graph with Boolean random variable and factor nodes. Initially, all variable nodes fall into three groups (one or two of these may be empty), namely known true, known false, and unknown. For ease of explanation, we will represent the groups by colored/shaded circles, say, magenta/white, green/gray, and red/black. All factor nodes with the same associated potentials also fall into one group represented by colored/shaded squares. For the factor graph in Fig. 11.1 the situation is depicted in Fig. 11.2. As shown on the left-hand side, assuming no evidence, all variable nodes are unknown, i.e., red/black. Now, each variable node sends a message to its neighboring factor nodes saying "I am of **color**/shade red/black". A factor node sorts the incoming colors/shades into a vector according to the order of the variables in its arguments. The last entry of the vector is the factor node's own color/shade, represented as light blue/gray square in Fig. 11.2. This color/shade signature is sent back to the neighboring variables nodes, essentially saying "I have communicated with these nodes". To lower communication cost, identical signatures can be replaced by new colors. The variable nodes stack the incoming signatures together and, hence, form unique signatures of their one-step message history. Variable nodes with the same stacked signatures, i.e., message history can now be grouped together. To indicate this, we assign a new color/shade to each group. In our running example, only variable node B changes its color/shade from red/black to yellow/gray. The factors are grouped in a similar fashion based on the incoming color/shade signatures of neighboring nodes. Finally, we iterate the process. As the effect of the evidence propagates through the factor graph, more groups are created. The process stops when no new colors/shades are created anymore.

The final compressed factor graph $\mathfrak{G}$ is constructed by grouping all nodes with the same color/shade into so-called **clusternodes** and all factors with the same color/shade signatures into so-called **clusterfactors**. In our case, variable nodes A, C and factor nodes f_1, f_2 are grouped together, see the right hand side of Fig. 11.2. Clusternodes (resp. clusterfactors) are sets of nodes (resp. factors) that send and receive the same messages at each step of carrying out BP on G. It is clear that they form a partition of the nodes in G.

Alg. 4 summarizes our approach for computing the compressed factor graph $\mathfrak{G}$ and is an instance of the **color passing** algorithm introduced in the corresponding chapter of the present book. Note that, we have to keep track of the position a variable that appeared in a factor. Since factors in general are not commutative, it matters where the node appeared in the factor. Reconsider our example from Fig. 11.1, where $f_1(A = True, B = False) \neq f_1(A = False, B = True)$. However, in cases where the position of the variables within the factor does not matter, one can gain additional lifting by neglecting the variable positions in the node signatures and by sorting the colors within the factors' signatures. The ordering of the factors in the node signatures, on the other hand, should always be neglected, thus we perform a sort of the node color signatures (Alg. 4, line 15).

The clustering we do here, groups together nodes and factors that are indistinguishable given the belief propagation computations. To better understand the color-passing and the resulting grouping of the nodes, it is useful to think of BP and its operations in terms of its **computation tree** (CT), see e.g. (Ihler et al., 2005). The CT is the unrolling of the (loopy) graph structure where each level i corresponds to the i-th iteration of message passing. Similarly we can view color-passing, i.e., the lifting procedure as a **colored computation tree** (CCT). More precisely, one considers for every node X the computation tree rooted in X but now each node in the tree is colored according to the nodes' initial colors, cf. Fig. 11.3**(a)**. For simplicity edge colors are omitted and we assume that the potentials are the same on all edges. Each CCT encodes the root nodes' local communication patterns that show all the colored paths along which node X communicates in the network. Consequently, color-passing groups nodes with respect to their CCTs: nodes having the same set of rooted paths of colors (node and factor names neglected) are clustered together. For instance, Fig. 11.3**(b)** shows the CCTs for the nodes X_1 to X_4. Because their set of paths are the same, X_1 and X_2 are grouped into one clusternode, X_3 and X_4 into another.[21]

Now we can run BP with minor modifications on the compressed factor graph $\mathfrak{G}$.

11.3.2   Step 2 – BP on the Compressed Factor Graph

Recall that the basic idea is to simulate BP carried out on G on $\mathfrak{G}$. An edge from a clusterfactor $\mathfrak{f}$ to a cluster node $\mathfrak{X}$ in $\mathfrak{G}$ essentially represents multiple edges in G. Let $c(\mathfrak{f}, \mathfrak{X}, p)$ be the number of identical messages that would be sent from the factors in the clusterfactor $\mathfrak{f}$ to each node in the clusternode $\mathfrak{X}$ that appears at position p in $\mathfrak{f}$ if BP was carried out on G. The message from a clustervariable $\mathfrak{X}$ to a clusterfactor $\mathfrak{f}$ at position p is

---

[21] The partitioning of the nodes obtained by color-passing corresponds to the so-called **coarsest equitable partition** of the graph (Mladenov et al., 2012). However, a formal characterization of the symmetries is beyond the scope of the current chapter. We refer to the corresponding chapters in this book.

Algorithm 4 CFG – CompressFactorGraph

Require: A factor Graph G with variable nodes X and factors f, Evidence E
Ensure: Compressed Graph $\mathfrak{G}$ with clustervariable nodes $\mathfrak{X}$ and clusterfactor nodes $\mathfrak{f}$

1: Compute initial clusters of the X_is w.r.t. E
2: **repeat**   ▷ Form color signature for each factor
3:     **for all** factor f_k **do**
4:         $signature_{f_k} = [\ ]$
5:         **for all** node $X_i \in nb(f_k)$ **do**
6:             $signature_{f_k}$.append($X_i.color$)
7:         **end for**
8:         $signature_{f_k}$.append($f_k.color$)
9:     **end for**
10:    Group together all f_ks having the same signature
11:    Assign each such cluster a unique color
12:    Set $f_k.color$ correspondingly for all f_ks   ▷ Form color signature for each variable
13:    **for all** node $X_i \in X, i = 1, \ldots, n$ **do**
14:        $signature_{X_i} = [\ ]$
15:        **for all** factor $f_k \in nb(X_i)$ **do**
16:            $signature_{X_i}$.append($(f_k.color, p(X_i, f_k))$)
17:        **end for**
18:        sort $signature_{X_i}$ according to ordering given by *color*
19:        $signature_{X_i}$.append($X_i.color$)
20:    **end for**
21:    Group together all X_is having the same signature
22:    Assign each such cluster a unique color
23:    Set $X_i.color$ correspondingly for all X_is;
24: **until** grouping does not change

$$\mu_{\mathfrak{x}\to\mathfrak{f},p}(x) = \mu_{\mathfrak{f},p\to\mathfrak{x}}(x)^{c(\mathfrak{f},\mathfrak{X},p)-1} \prod_{\mathfrak{h}\in nb(\mathfrak{X})} \prod_{\substack{q\in P(\mathfrak{h},\mathfrak{X}) \\ (\mathfrak{h},q)\neq(\mathfrak{f},p)}} \mu_{\mathfrak{h},q\to\mathfrak{x}}(x)^{c(\mathfrak{h},\mathfrak{X},q)}, \qquad (11.6)$$

where $nb(\mathfrak{X})$ now denotes the neighbor relation in the compressed factor graph $\mathfrak{G}$ and $P(\mathfrak{h}, \mathfrak{X})$ denotes the positions nodes from $\mathfrak{X}$ appear in $\mathfrak{f}$. The $c(\mathfrak{f}, \mathfrak{X}, p) - 1$ exponent reflects the fact that a clustervariable's message to a clusterfactor excludes the corresponding factor's message to the variable if BP was carried out on G. The message from the factors to neighboring variables essentially remains unchanged. The difference is that we now only send one message per clusternode and position, given by

$$\mu_{\mathfrak{f},p\to\mathfrak{x}}(x) = \sum_{\neg\{\mathfrak{x}\}} \left(\mathfrak{f}(\mathbf{x}) \prod_{\mathfrak{y}\in nb(\mathfrak{f})} \prod_{q\in P(\mathfrak{f},\mathfrak{y})} \mu_{\mathfrak{y}\to\mathfrak{f},q}(y)^{c(\mathfrak{f},\mathfrak{y},q)-\delta_{\mathfrak{xy}}\delta_{pq}} \right), \qquad (11.7)$$

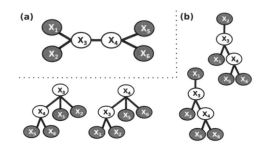

Figure 11.3: **(a):** Original factor graph with colored nodes **(b):** Colored computation trees for nodes X_1 to X_4. As one can see, nodes X_1 and X_2, respectively X_3 and X_4, have the same colored computation tree, thus are grouped together during color-passing.

where $\delta_{\mathfrak{x}\mathfrak{y}}$ and δpq are one iff $\mathfrak{X} = \mathfrak{Y}$ and $p = q$ respectively. The unnormalized belief of $\mathfrak{X}_i$, i.e., of any node X in $\mathfrak{X}_i$ can be computed from the equation

$$b_i(x_i) = \prod_{\mathfrak{f}\in \mathrm{nb}(\mathfrak{X}_i)} \prod_{p\in P(\mathfrak{f},\mathfrak{X}_i)} \mu_{\mathfrak{f},p\to\mathfrak{X}_i}(x_i)^{c(\mathfrak{f},\mathfrak{X}_i,p)} . \tag{11.8}$$

Evidence is incorporated either on the ground level by setting $f(\mathbf{x}) = 0$ or on the lifted level by setting $\mathfrak{f}(\mathbf{x}) = 0$ for states $\mathbf{x}$ that are incompatible with it. [22] Again, different schedules may be used for message-passing. If there is no compression possible in the factor graph, i.e. there are no symmetries to exploit, there will be only a single position for a variable $\mathfrak{X}$ in factor $\mathfrak{f}$ and the counts $c(\mathfrak{f}, \mathfrak{X}, 1)$ will be 1. In this case the equations simplify to Eq.11.3-11.5.

To conclude the section, the following theorem states the correctness of *lifted* BP.

Theorem 11.1 *Given a factor graph G, there exists a unique minimal compressed $\mathfrak{G}$ factor graph, and algorithm* CFG(G) *returns it. Running BP on $\mathfrak{G}$ using Eqs. (11.6) and (11.8) produces the same results as BP applied to G.*

The theorem generalizes the theorem of Singla and Domingos (2008) but can essentially be proven along the same ways. Although very similar in spirit, lifted BP has one important advantage: not only can it be applied to first-order and relational probabilistic models, but also directly to traditional, i.e., propositional models such as Markov networks.

Proof   We prove the uniqueness of $\mathfrak{G}$ by contradiction. Suppose there are two minimal lifted networks $\mathfrak{G}_1$ and $\mathfrak{G}_2$. Then there exists a variable node X that is in clusternode $\mathfrak{X}_1$ in $\mathfrak{G}_1$ and in clusternode $\mathfrak{X}_2$ in $\mathfrak{G}_2$, $\mathfrak{X}_1 \neq \mathfrak{X}_2$; or similarly for some clusterfactor $\mathfrak{f}$. Since all nodes in $\mathfrak{X}_1$, and $\mathfrak{X}_2$ respectively, send and receive the same messages $\mathfrak{X}_1 = \mathfrak{X}_2$.

---

[22] Note that, the variables have been grouped according to evidence and their local structure. Thus all factors within a clusterfactor are indistinguishable and we can set the states of the whole clusterfactor $\mathfrak{f}$ at once.

Following the definition of clusternodes, any pair of nodes $\mathfrak{X}$ and $\mathfrak{Y}$ in $\mathfrak{G}$ send and receive different messages, therefore no further grouping is possible. Hence, $\mathfrak{G}$ is a unique minimal compressed network.

Now we show that algorithm $CFG(G)$ returns this minimal compressed network. The following arguments are made for the variable nodes in the graph, but can analogously be applied to factor nodes. Reconsider the colored computation trees (CCT) which resemble the paths along which each node communicates in the network. Variables nodes are being grouped if they send and receive the same messages. Thus nodes X_1 and X_2 are in the same clusternode iff they have the same colored computation tree. Unfolding the computation tree to depth k gives the exact messages that the root node receives after k BP iterations. $CFG(G)$ finds exactly the similar CCTs. Initially all nodes are colored by the evidence we have, thus for iteration $k = 0$ we group all nodes that are similarly colored at the level k in the CCT. The signatures at iteration $k+1$ consist of the signatures at depth k (the nodes own color in the previous iteration) and the colors of all direct neighbors. That is, at iteration $k + 1$ all nodes that have a similar CCT up to the $(k + 1)$-th level are grouped. $CFG(G)$ is iterated until the grouping does not change. The number of iterations is bounded by the longest path connecting two nodes in the graph. The proof that modified BP applied to $\mathfrak{G}$ gives the same results as BP applied to G also follows from CCTs, Eq.11.6 and 11.7, and the count resembling the number of identical messages sent from the nodes in G.   □

In contrast to the color-passing procedure (Kersting et al., 2009) as presented here, Singla and Domingos (2008) work on the relational representation and lift the Markov logic network in a top-down fashion. While a top-down construction of the lifted network has the advantage of being more efficient for liftable relational models since the model does not need to be grounded, a bottom-up construction has the advantage that we do not rely on a relational model such as Markov logic networks.

11.4 Lifted Message Passing for SAT Problems

The previous section viewed the compression step, i.e., lifting step as a color-passing (CP) approach. This view abstracts from the type of messages sent and, hence, highlights one of the main insights underlying the present chapter: *CP can be used to lift other message-passing approaches.*

In particular, as we will demonstrate now, it can be applied to solving **SAT** problems. Actually, it can be simplified in the SAT context. Specifically, we describe in Sec.11.4.1, the concepts of lifted satisfiability. This will also motivate the requirement of an adapted compression algorithm for factor graphs to fully exploit lifting in SAT. This algorithm is introduced in Sec.11.5 where we also explain how to integrate it into the **lifted SAT** framework and how it can be applied beyond satisfiability. In Sec.11.6, we will show that iterative optimization algorithms based on lifting can often benefit from re-lifting during inference and we present a lifted version of **likelihood maximization** algorithm (Kumar and Zilberstein, 2010).

11.4.1   Lifted Satisfiability

We now turn our attention towards the concepts of lifted satisfiability. Besides the fact that solving satisfiability problems is key to all kinds of AI problems, SAT solving or sampling satisfying assignments also plays a key role in some inference algorithms for relational probabilistic models such as **MC-SAT** (Poon and Domingos, 2006). Using message passing algorithms for satisfiability has been investigated from different points of view. Besides algorithms tailored specifically for SAT solving, approaches based on BP have been used as well. Having observed the success and impact of lifted inference in relational models, we will now show that lifting is also beneficial in SAT solving by developing lifted versions of two message passing algorithms for satisfiability, namely Warning Propagation and Survey Propagation. In addition, we will show that a number of simplifications are possible compared to lifted version of BP when the underlying problem consist of a conjunctive normal form (CNF).

The main drawback of lifted inference approaches such as LBP is the fact that they are not tailored towards combinatorial problems. In many real-world applications, however, the problem formulation does not fall neatly into the pure probabilistic inference case. Problems may very well have a component that can be well-modeled as a combinatorial problem, hence, taking it outside the scope of standard LBP approaches. Besides that, satisfiability of Boolean formulas is one of the classical AI problems and the idea of "reduction to SAT" is a powerful paradigm for solving problems in different areas. Ideally, lifted inference should be efficient here, too, exploiting as much symmetries as possible. Indeed, driven by the same question of reasoning with both probabilistic and deterministic dependencies, but for the non-lifted case, Poon and Domingos (2006) developed MC-SAT that combines ideas from MCMC and satisfiability. It still proceeds by first fully instantiating the first-order theory and then essentially staying at the propositional level. **LazySAT** (Singla and Domingos, 2006b) is a lazy version of **WalkSAT** taking advantage of relational sparsity. Later, (Poon et al., 2008) have shown that this idea of **lazy inference** goes beyond satisfiability and can be combined with other propositional inference algorithms such as BP. However, the link has previously not been explored on the lifted (message passing) level.

In this section, we therefore revisit lifted message passing algorithms this time for combinatorial problems. Specifically, we focus on the Boolean satisfiability problems consisting of a formula F representing a set of constraints over n Boolean variables, which must be set such that all constraints are satisfied. It is one of the most — if not the most – studied problems in computer science and AI. Several other NP-complete problems can be reduced to it and "compilation to SAT" is a powerful paradigm for solving computer science and AI problems. SAT has applications in several important areas such as automated deduction, cryptanalysis, modeling biological networks, hardware and software verification, planning, and machine learning, among others. Hence, pushing the boundary of practical solvability of SAT has far reaching practical consequences.

In fact, we describe the lifted version of an exciting algorithm for solving combinatorial problems, namely **Survey Propagation** (SP). SP was introduced by Mézard et al. (2002) and can easily solve SAT problems with one million variables in a few minutes on a desktop computer. SP has particularly attracted a lot attention because it was able to solve difficult problems. Braunstein and Zecchina (2004) and Maneva et al. (2007) have shown that SP can be viewed as a form of BP, hence, somehow suggesting that lifting SP is possible. As noted by Kroc et al. (2007), SP's remarkable effectiveness — the performance is clearly beyond the reach of mainstream SAT solvers — has created the impression that solving hard combinatorial instances requires SP but BP would have little success. Kroc et al. (2009), however, have shown that "the gap between BP and SP narrows" as the number of variables per clause increases. This is also supported by Montanari et al. (2007) showing good performance of a combination of BP and **Warning Propagation** (WP), a message passing algorithm for SAT that is simpler than SP but also less powerful. Therefore, we also show how to lift WP and, hence, provide a complete picture of lifted message passing algorithms for SAT. Let us again summarize the content of this section:

- The idea of lifted inference is applied to the problem of SAT solving, thus exploiting the structure of combinatorial problems by techniques known from probabilistic inference in relational models.
- We provide a lifted version of WP and derive the lifted update equations in detail.
- In the same spirit as lifting WP, we also provide a lifted version of the more powerful SP algorithm.

These examples significantly advance the understanding of lifted message passing and put an unified, lifted treatment of combinatorial and probabilistic problems within reach.

We proceed as follows. We first provide background on Boolean formulas, show how factor graphs can be used to represent those, and the discuss (lifted) message passing for SAT. Specifically, we show how the Color Passing algorithm from above, which compresses the original factor graph, can be used for SAT problems, and show how a simplified version can be used for lifting WP and SP. We then introduce both Lifted Warning Propagation and Lifted Survey Propagation, i.e., the modified message passing equations used on the lifted factor graph. Again, we argue that they are simpler than the ones for LBP. Before concluding, we present the results of our experimental evaluation where we show that running lifted message passing within the standard decimation approaches for SAT can achieve remarkable efficiency gains.

11.4.2   CNFs and Factor Graphs

Without loss of generality, we assume that a **Boolean formula** is represented in CNF. A **CNF** is a conjunction of disjunctions of Boolean literals. A literal is either a negated or un-negated propositional variable. Specifically, a CNF consists of n variables $\mathbf{X} = \{X_1, \ldots, X_n\}$ with $x_i \in \{\texttt{False}, \texttt{True}\}$ and m clauses $C = \{C_1, \ldots, C_m\}$ constraining the variables. A clause C_a is defined over a subset of the variables $\mathbf{X}_a = \{X_{a,1}, \ldots, X_{a,|C_a|}\}$ and — this will turn out to be convenient for formulating the message passing update formulas — a corresponding list of their signs which is denoted as $s_a = \{s_{a,1}, \ldots, s_{a,|C_a|}\}$ with

$$s_{a,i} = \begin{cases} -1, & \text{if } X_{a,i} \text{ unnegated} \\ 1, & \text{if } X_{a,i} \text{ negated} \end{cases} \tag{11.9}$$

A solution to a CNF is an assignment to all variables in $\mathbf{X}$ such that all clauses in C are satisfied, i.e., evaluate to $\texttt{True}$. As an example, consider the following CNF, consisting of three variables, three clauses, and seven literals: $(X_1 \vee \neg X_2) \wedge (\neg X_1 \vee X_2) \wedge (X_1 \vee X_2 \vee X_3)$. A possible solution to that formula is $X_1 = \texttt{True}, X_2 = \texttt{False}, X_3 = \texttt{True}$.

Every CNF can be represented as a factor graph. The factor graph representing the example CNF from above is shown in Fig. 11.4(left). For CNFs, there is a one-to-one mapping between the variables in the CNF and the variable nodes in the factor graph, as well as the clauses and factor nodes. Therefore, we will not distinguish between logical and probabilistic variables explicitly. We have a factor f_a for each clause $C_a \in C$. We use a dashed line between a variable X_i and a factor f_a whenever the variable appears negated in a clause, i.e., $s_{i,a} = 1$, otherwise a full line. We want to stress the fact that variables can only appear negated or unnegated in a factor by introducing the dashed lines for negated variables. The factor functions $f_a(\mathbf{x}_a)$ are defined in such a way that they return a value of 1 if the corresponding clause evaluates to $\texttt{True}$ for $\mathbf{x}_a$ and 0 otherwise. With this definition, the normalization constant Z amounts to the number of solutions of the CNF and each satisfying assignment of $\mathbf{X}$ has the same probability, i.e., $1/Z$.

So, before deriving lifted message updates for WP and SP in the next sections, let us first illustrate lifting in the case of CNFs. As we will see, CP and its outcome, the lifted factor graph, are simpler for the CNF case compared to the LBP case. As we have seen above, we can use the CP algorithm to lift a ground factor graph. Of course, this is also possible for a factor graph representing a CNF. Again, for the lifted satisfiability approach, CP on factor graph representing a CNF is the first step. For such a factor graph, the domain of each variable has size $d = 2$ and the initial coloring of the factors just distinguishes clauses with different number of negated and unnegated variables. i.e., two clauses fall into the same group if both have the same number of positive and negative literals because the logic operator $\vee$ is commutative. But more importantly, as we will show next, the lifting can actually be simplified in the case of CNFs.

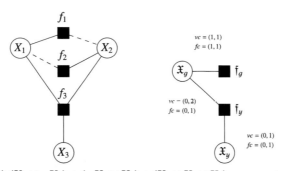

Figure 11.4: (left) $(X_1 \vee \neg X_2) \wedge (\neg X_1 \vee X_2) \wedge (X_1 \vee X_2 \vee X_3)$ represented as a factor graph. Circles denote variable nodes and squares denote clauses. A dashed line indicates that the variable appears negated in a clause; otherwise we use a full line. (right) The resulting lifted CNF after running the Color Passing algorithm as illustrated earlier.

First, we can employ a more efficient color signature coding scheme. The initial grouping of the factors solely depends on the zero state of their corresponding clause. Additionally, the factor nodes do not have to sort the incoming colors according to the positions of the variables, instead only the sign of a variable matters, i.e., only two positions exist. One position for the variables appearing unnegated in the clause and a second position for the negated variables. We have seen in the case of the BP lifting that we have to distinguish arbitrarily many positions.

Second, we can employ simplified counts. Because there are only two positions, we store two values: counts for the negated position and for the unnegated position. We do not have to manage a complex structure with counts for different positions as in the case of LBP anymore. Reconsider Fig. 11.4(right). Here, the count associated with the edge between $\mathfrak{X}_1$ and $\mathfrak{f}_1$ is $(1, 1)$ because $\mathfrak{X}_1$ represents ground variables that appeared positive and negative in ground factors represented by $\mathfrak{f}_1$. Second, there is the **variable count** vc. It corresponds to the number of ground variables that send messages to a factor at each position. In the ground network, the factor f_3 is connected to three positive variables, but two of them are represented by a single clusternode in the lifted network. Hence, we have $vc = (0, 2)$ for the edge between $\mathfrak{X}_1$ and $\mathfrak{f}_2$ because $\mathfrak{f}_2$ is now the representative of f_3. So, for lifted CNFs, all counts in the factor graph are of dimension two only.

Having the **lifted CNF** at hand, we can now run LBP on this lifted graph. However, there are other ground message passing algorithms available that are more appropriate for solving CNFs. We will show that these algorithms can be lifted too. First, we will describe the approach based on LBP, before we will then derive the lifted versions of WP and SP which are designed to operate on factor graphs for CNFs directly.

11.4.3   Belief Propagation for Satisfiability

Typically, iteratively solving a SAT problem requires a heuristic, which is applied to each unclamped variable in each iteration. One possible heuristic can be the single node marginal probabilities $p(X_i)$. If we chose the potentials of the factors as described above, running sum-product BP on a factor graph representing a CNF has a nice and intuitive explanation. The beliefs of a state give the approximate marginal probability of a variable being in that state in a solution to the underlying CNF. Additionally, the partition function estimates the number of solutions for a given CNF. Of course, these values are only guaranteed to be correct for tree-structured CNFs. Nevertheless, BP can still be used to solve satisfiability problems. In particular, the approximate BP marginals have been used successfully in decimation procedures to find a satisfying solution (Montanari et al., 2007). Such decimation procedures make iteratively use of the BP marginals. Besides this, BP can also be used for approximative model counting (Kroc et al., 2011). We will now describe the concept of BP-guided decimation in detail.

11.4.4   Decimation-based SAT Solving

Having solely BP at hand is usually not sufficient to solve complex CNFs. Nevertheless, we can run BP, or similar message passing algorithms as we will see in the next sections, on this problem and use the fixed points of the algorithm to obtain insights about satisfying configurations. However, in most cases the information about the fixed points is not sufficient to obtain an entire satisfying configuration. Still, the output of the algorithm, i.e., the beliefs, will point us in the right direction to find a solution in many cases. The general paradigm of applying this idea iteratively is known as decimation and will now be explained in detail. Many complex inference tasks, such as finding a satisfying assignment to a Boolean formula, or computing joint marginals of random variables, can be cast into a sequence of simpler ones. This is achieved by selecting and **clamping** variables one at a time and running inference after each selection. SAT problems can be solved using a **decimation** procedure based on BP (Montanari et al., 2007). For SAT problems, decimation is the process of iteratively assigning a truth value to k variables (often a single variable only) in **X**. This results in a factor graph which has these variables clamped to a specific truth value. We can then simplify the factor graph accordingly, obtaining a smaller factor graph representing a simplified formula with $n - k$ variables. Now, we repeatedly decimate the formula in this manner, until all variables have been clamped. In the end, either a satisfying assignment has been found or a contradiction has been detected.

This procedure heavily depends on the strategies for choosing variables to fix and the corresponding states. In a **BP guided decimation**, BP is run on the (simplified) factor graph and a variable is picked based on the approximated marginals. Usually, the variables with the highest probability for one state is picked and clamped accordingly. This variable is often referred to as the **most magnetized variable**. This popular approach can be seen as a general idea of turning the complex inference task into a sequence of simpler ones by selecting and clamping variables one at a time and running inference again after each selection.

The concept of decimation can be lifted as well. In the following sections, we will describe how an efficient **lifted decimation** procedure can be realized. In first place, we will be using lifted message passing to speed up each iteration. However, this requires lifting from scratch in each iteration, possibly canceling some of the benefits of lifting. Therefore, we will also introduce a more efficient lifting algorithm in the second step that makes use of the current lifting when producing the lifted graph for the following iteration. Additionally, BP is not the only message passing algorithm that can be used to obtain insights about the state of a variable in a satisfying configuration. We will see how Survey Propagation is capable of computing indicators for each variable upon which one can pick a variable to clamp as well. We will propose a lifted variant of this algorithm which also requires the introduction of counts. As described above, fixing variables often allows a simplification of the formula. This simplification can be casted into a message passing algorithm as well and amounts to running Warning Propagation on the factor graph. Having Warning Propagation at hand, we implement the decimation completely by means of a factor graph and message passing. BP, or Survey Propagation, can be combined with Warning Propagation to simplify the CNF after having variables clamped. We will show that Warning Propagation is liftable as well and can be run on the lifted factor graph, yielding a decimation with all operations on the lifted level.

11.4.5 Lifted Warning Propagation

Clauses consisting of a single literal only, i.e., factors with an edge to exactly one variable X only, are called **unit clauses**. A unit clause fixes the truth value of X. The process of fixing all variables appearing in unit clauses and simplifying the CNF correspondingly is called **Unit Propagation** (Zhang and Stickel, 1996). Casted into the message passing framework (Braunstein et al., 2005; Montanari et al., 2007), it is known under the name of **warning propagation**.

Intuitively, a message $\mu_{a \to i} = 1$ sent from a factor to a variable says: "Warning! The variable i must take on the value satisfying the clause represented by the factor a." As shown by Braunstein et al. (2005), this can be expressed mathematically as follows:

$$\mu_{a \to i} = \prod_{j \in nb(a) \setminus \{i\}} \theta(\mu_{j \to a} s_{aj})$$

where $nb(a)$ is the set of variables that a constrains and $\theta(x) = 0$ if $x \le 0$ and $\theta(x) = 1$ if $x > 0$. s_{aj} is the sign of the edges and defined in Eq. 11.9. The message from a variable to a factor is:

$$\mu_{i \to a} = \left(\sum_{b \in nb_+(i) \setminus \{a\}} \mu_{b \to i} \right) - \left(\sum_{b \in nb_-(i) \setminus \{a\}} \mu_{b \to i} \right)$$

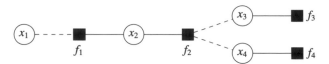

Figure 11.5: CNF containing unit clauses depicted as factor graph.

where $nb_+(i)$, respectively $nb_-(i)$, is the set of factors in which i appears unnegated, respectively negated. The exchange of message is repeated until the messages converge or a maximum number of iterations has been reached. If the message passing converges, a set of warnings can be read off this fixed point.

To be more precise, let us assume that $\mu^*_{a \to i}$ are the fixed point warnings. We can now calculate for every variable the **local field** H_i and the **contradiction number** c_i, which are defined as follows:

$$H_i = -\sum_{b \in nb(i)} s_{b,i} \cdot \mu^*_{b \to i},$$

and

$$c_i = \begin{cases} 1, & \text{if } \left(\sum_{b \in nb_+(i)} \mu^*_{b \to i}\right)\left(\sum_{b \in nb_-(i)} \mu^*_{b \to i}\right) > 0 \\ 0, & \text{otherwise} \end{cases}$$

The local field indicates the preferred state of a variable. If $H_i > 0$, X_i should be set to True, for $H_i < 0$, X_i prefers False, otherwise no preference is expressed by the warnings. However, it can happen that a variable receives conflicting warnings which is indicated by the contradiction number. The contradiction number is non-zero if and only if a variable receives warnings for both possible states.

The initialization of the messages can either be done at random and each message is set to $\mu_{a \to i} \in \{0, 1\}$ with equal probability. Alternatively, we set all messages to zero which shows to be beneficial in the case of lifting and exploiting symmetries. Similar to BP, we have to specify a schedule which is used to update the messages. We will here use a parallel schedule and hence, a fixed order of the variables.

Example 11.1 To illustrate the effectiveness of WP, we consider a small example. Given the CNF $(\neg X_1 \vee X_2) \wedge (X_2 \vee \neg X_3 \vee \neg X_4) \wedge (X_3) \wedge (X_4)$ (see Fig. 11.5 for the corresponding factor graph). One can easily verify that this CNF constraints variables X_2, X_3, and X_4 to be positive to satisfy the problem and X_1 can take on any value. Assume that all messages are initially being set to 0. Let us take a look at the WP messages sent from factors f_3 and f_4 to its neighbors. Since these factors have only a single neighbor, the product is over the empty set, and hence the factors can only be satisfied by the neighbors taking on the satisfying state. This results in a warning $\mu_{a \to i} = 1$ being send from f_3 and f_4 to its neighbors. This warning now effects the message being send from f_2 to X_2. Now, the product over the neighbors of f_2 excluding X_2 is equal to one, indicating a warning being sent from f_2 to X_2. Factor f_2 informs X_2 that is has to take on the value satisfying f_2. On the other hand, the message $\mu_{f_1 \to X_1}$ remains zero, since f_1 is automatically satisfied by X_2 and does not further constrain X_1. The algorithm converges to a fixed point after two iterations. Calculating the local fields, we obtain $H_1 = 0$, $H_2 = -(-1 \cdot 1) = 1$, $H_3 = -((1 \cdot 0) + (-1 \cdot 1)) = 1$, and similarly $H_4 = 1$.

If we run sum-product BP on this graph, the marginals and partition function are exact because of the tree structure. We can read off the solution from the beliefs in a similar way. The marginals of X_2, X_3, and X_4 are all equal to $(0.0, 1.0)$, indicating that these variables have to be `True` in a solution. Additionally, the marginal of X_1 is equal to $(0.5, 0.5)$ which shows in this problem that X_1 can take on both values with equal probability. Lastly, the partition function amounts to $Z = 2$ because there are only two solution.

WP is also guaranteed to converge if the underlying factor graph is a tree. Additionally, if at least one of the contradictory numbers c_i is equal to 1 in the tree case, the represented problem is not satisfiable.

Example 11.1 also exemplifies how symmetries in an CNF lead to identical messages. We can see that the messages being sent from X_2 to f_1 and from X_3 to f_1 are identical. The same holds for the messages the other way around. We can now apply the Color Passing algorithm to any CNF and use modified WP messages on the lifted graph. The **lifted warning propagation** (LWP) equations take the following form[23]:

$$\mu_{a \to i,p} = \prod_{j \in nb(a)} \prod_{p' \in pos(a,j)} \theta(\mu_{j \to a,p'} s_{p'})^{vc(a,j,p') - \delta_{ij}\delta_{pp'}}$$

with $pos(a,j) \subseteq \{0,1\}$ referring to a negated or unnegated position of a variable X_j in a clause f_a. Furthermore, we have $s_{p'} = 1$ if $p' = 0$ and $s_{p'} = -1$ if $p' = 1$. This is a position dependent version of the $s_{a,j}$ from above. Note that the set of positions $pos(a,j)$ contains only those positions with a variable count (vc) greater than zero. Here, the variable count corresponds to the number of ground variables that send messages to a factor at each position. Because there are only two positions — a variable may occur at any position in negated and unnegated form — we store two values: counts for the negated position and for the unnegated position. The Kronecker deltas take care of the "\{i}" in the product's range of the ground formula. They reduce counts by one when a message is sent to the node itself at the same position, i.e., they prevent double counting of messages.

To exemplify the idea of positions and variable counts, reconsider the lifted factor graph in Fig.11.4(right), "pos" contains a non-zero value for both positions for the variable count of edge $\mathfrak{X}_g - \mathfrak{f}_g$. This is due to the fact that the factors f_1 and f_2 contain a negated and an unnegated variable in the ground factor graph. On the other hand, for $\mathfrak{X}_g - \mathfrak{f}_y$ the variable counts are only positive for the unnegated position because f_3 contains only unnegated variables in the ground network. In the ground network, the clause f_3 was connected to

---

[23] We define $0^0 = 1$ in cases where $\theta(x) = 0$ and $vc(a,j,p') - \delta_{ij,pp'} = 0$.

three positive variables, but two of them are now represented by a single clusternode ($\mathfrak{X}_g$) in the lifted network. Hence, we have variable count vc = $(0, 2)$ for the edge between $\mathfrak{X}_g$ and $\mathfrak{f}_y$. Remember that ground factor f_3 is now represented by the clusterfactor $\mathfrak{f}_y$. Intuitively, this already indicates that the lifted formula simulates WP on the ground network. It even becomes more clear when we run it on the ground network shown in Fig.11.4(left). In this case, there is exactly one position with a variable count greater than zero for each neighbor j of factor a. In other words, pos(a, j) is a set containing only a single position for which the count is always 1. In turn, the second product can be dropped, none of the $\delta_{..}$ can be 1, and both formulas coincide. Similarly, one can intuitively check that the following formula in the other direction is correct for the lifted case:

$$\mu_{i \to a, p} = \left(\sum_{\substack{b \in \text{nb}(i), \\ \text{fc}(b,i,1)>0}} \mu_{b \to i} \Big(\text{fc}(b, i, 1) - \delta_{ab}\delta_{p1} \Big) \right) - \left(\sum_{\substack{b \in \text{nb}(i), \\ \text{fc}(b,i,0)>0}} \mu_{b \to i} \Big(\text{fc}(b, i, 0) - \delta_{ab}\delta_{p0} \Big) \right)$$

Again the δs take care of the "\{a\}" in the original ground formulation and fc is a factor count. The factor count represents the number of equal ground factors that send messages to the variable at a given position. The "fc$(b, j, 1/0) > 0$" simulates the ground sets "nb$_{+/-}(j)$".

The role of the factor count can again be understood best when looking at Fig.11.4. Here, the factor count associated with the edge between $\mathfrak{X}_1$ and $\mathfrak{f}_1$ is $(1, 1)$ because $\mathfrak{X}_1$ represents the ground variables corresponding to X_1 and X_2. Both ground variables were involved once negated and once unnegated in the two factors represented by $\mathfrak{f}_1$. This leads to a positive factor count for both positions. Using this correspondence, the proof that applying the lifted formulas to the lifted graph gives the same results as WP applied to the ground network follows essentially from the one for the LBP, see above and also (Singla and Domingos, 2008).

If we are interested in solving CNFs, we make use of WP in every iteration. By fixing one or several variables, other variables in the formula may now be implied directly by unit clauses. WP allows us to find these variables easily and at lower computational costs then running a more expensive inference algorithm such as BP. Of course, we could just run unit propagation on the simplified formula. However, with help of the message passing formulation, we were able to develop a lifted variant that directly operates on the lifted factor graph. This enables us now to do the simplification on the lifted level, getting us closer to a completely lifted decimation.

11.4.6   Lifted Survey Propagation

After we have seen the WP algorithm and its lifted version, we will now shift our focus to **survey propagation** (Mézard et al., 2002; Braunstein et al., 2005) and its lifted equations. SP has been very successful in solving large and difficult **random CNFs**. The difficulty of an k-CNF is often measured by the ratio of the number of clauses and the number of variables. A CNF with only few constraints and many variables is said to be under-constraint and it is likely that the problem has many solution where one of them can easily be found. Contrary, a CNF with a very large number of clauses is over-constraint and more likely to be unsatisfiable. It is conjectured that this ratio splits the space of all **k-SAT** problems in mainly satisfiable and unsatisfiable instances. This threshold or crossover point is often referred to as a_c. The ratio a_c plays an important role in k-SAT satisfiability and its theoretical analysis. Usually, SAT solvers have the longest running time in the region around a_c. There have been various experimental studies for determining a_c. For instance, Crawford et al. (1996) estimated a_c to be around 4.258 for random 3-SAT problems.

However, we are here interested in applying SP to non-random instances in order to speed up the message passing. SP has a nice interpretation in terms of warnings being sent in WP. SP assume that WP has multiple fixed points and surveys represent the probability of a warning being send. More precisely, the intuition of a message $\mu_{a\to i}$ sent from a factor to a variable is that of a "survey" which represents the probability that a warning, as in WP, is sent from a to i. In comparison to WP, where the value of a warning was binary only, i.e., $\mu_{a\to i} \in \{0, 1\}$, the corresponding message in SP is real $\mu_{a\to i} \in [0, 1]$. The ground SP message being sent from a factor to a variable look as follows (Braunstein et al., 2005):

$$\mu_{a\to i} = \prod\nolimits_{j\in nb(a)\backslash\{i\}} \left[\frac{\mu_{j\to a}^{u}}{\mu_{j\to a}^{u} + \mu_{j\to a}^{s} + \mu_{j\to a}^{*}} \right]$$

This message is a ratio of three different types of messages being send from variables to factors. Besides the messages for the unsatisfied (μ^u) and the satisfied (μ^s) case, there exists also a message for the arbitrary/undecided case (μ^*). This is often referred to as the "don't care"-state. The three messages from variables to factors are defined the following way:

$$\mu_{i\to a}^{u} = \left[1 - \prod\nolimits_{b\in nb_a^u(i)} (1 - \mu_{b\to i}) \right] \prod\nolimits_{b\in nb_a^s(i)} (1 - \mu_{b\to i})$$

$$\mu_{i\to a}^{s} = \left[1 - \prod\nolimits_{b\in nb_a^s(i)} (1 - \mu_{b\to i}) \right] \prod\nolimits_{b\in nb_a^u(i)} (1 - \mu_{b\to i})$$

$$\mu_{i\to a}^{*} = \prod\nolimits_{b\in nb(i)\backslash\{a\}} (1 - \mu_{b\to i})$$

where $nb_a^u(i) = nb_+(i)$ and $nb_a^s(i) = nb_-(i) \backslash \{a\}$ if $s_{a,i} = 1$, while $nb_a^u(i) = nb_-(i)$ and $nb_a^s(i) = nb_+(i) \backslash \{a\}$ if $s_{a,i} = -1$. Similar to BP and WP, the initialization of messages can be done at random or with an initial fixed value. The different update schedules for BP which were describe earlier can also be used for SP. Braunstein et al. (2005), for example, suggested sequential random updates. If the message passing converges, we want to read off assignments from the surveys in the same spirit as it can be done from the warnings in

WP. For SP, we calculated positive and negative biases of the variables:

$$W_i^+ = \frac{b_i^+}{b_i^+ + b_i^- b_i^*},$$

$$W_i^- = \frac{b_i^-}{b_i^+ + b_i^- b_i^*},$$

where b_i^+, b_i^-, and b_i^* are defined as follows:

$$b_i^+ = \left[1 - \prod_{a \in nb_+(i)} (1 - \mu_{a \to i}^*)\right] \prod_{a \in nb_-(i)} (1 - \mu_{a \to i}^*)$$

$$b_i^- = \left[1 - \prod_{a \in nb_-(i)} (1 - \mu_{a \to i}^*)\right] \prod_{a \in nb_+(i)} (1 - \mu_{a \to i}^*)$$

$$b_i^* = \prod_{a \in nb(i)} (1 - \mu_{a \to i}^*)$$

Each bias expresses the tendency of a variable to be either in the negated or unnegated state. Looking at the absolute difference of both biases, one can find the most biased variable which has the strongest tendency to be in a particular state.

It is difficult to give a simple example of SP akin to Fig.11.1 for WP. On trees, it has been shown that SP reduces to WP (Braunstein et al., 2005), and is hence exact. For cyclic graphs with $a_c < 3.9$, it is has been reported that SP converges to a set of trivial messages, i.e., "don't care" states (Braunstein et al., 2005). Therefore, meaningful examples quickly grow large and we now continue with the equations for **lifted survey propagation** (LSP) and explain how the ground equations from above are rewritten to run on the lifted level. Essentially, the SP equations can be lifted in a similar way as for WP. That is, we apply rewriting rules such as "\{i}" $\mapsto$ "$\delta_{ij}\delta_{pp'}$" to the ground SP equations. This yields the following **lifted survey propagation** (LSP) equations. A message sent from factor a to a variable i at position p is defined as follows:

$$\mu_{a \to i,p} = \prod_{\substack{j \in nb(a), \\ p' \in pos(a,j)}} \left[\frac{\mu_{j \to a,p'}^u}{\mu_{j \to a,p'}^u + \mu_{j \to a,p'}^s + \mu_{j \to a,p'}^*}\right]^{vc(a,j,p') - \delta_{ij}\delta_{pp'}}$$

Messages sent from variable j to factor a at position p' becomes:

$$\mu_{i \to a,p}^u = \left[1 - \prod_{\substack{b \in nb(i), \\ p' \in pos(b,i), p' \neq p}} (1 - \mu_{b \to i})^{fc(b,i,p') - \delta_{ab}}\right] \prod_{\substack{b \in nb(i), \\ p' \in pos(b,i), p' = p}} (1 - \mu_{b \to i})^{fc(b,i,p') - \delta_{ab}}$$

$$\mu_{i \to a,p}^s = \left[1 - \prod_{\substack{b \in nb(i), \\ p' \in pos(b,i), p' = p}} (1 - \mu_{b \to i})^{fc(b,i,p') - \delta_{ab}}\right] \prod_{\substack{b \in nb(i), \\ p' \in pos(b,i), p' \neq p}} (1 - \mu_{b \to i})^{fc(b,i,p') - \delta_{ab}}$$

$$\mu_{i \to a,p}^* = \prod_{\substack{b \in nb(i), \\ p' \in pos(b,i)}} (1 - \mu_{b \to i})^{fc(b,i,p') - \delta_{ab}\delta_{pp'}}$$

Indeed, we have to be a little bit more careful. For messages from factors to variables, we use exactly the same rules. For the messages from variables to factors, however, we use slightly different rules. The set "$b \in nb_a^{u/s}(i)$" maps to "$b \in nb(i), p' \in pos(b,i), p' \neq$

Algorithm 5 Lifted Decimation

Require: A factor graph G, a list of query vars $\mathbf{Y}$, e.g., the variables in G
Ensure: An assignment $\mathbf{a}$ to the variables in $\mathbf{Y}$

1: $\mathbf{a} \leftarrow \emptyset$
2: $\mathfrak{G} \leftarrow \text{LIFT}(G)$
3: **while** $\mathbf{Y} \neq \emptyset$ **do**
4:     $\mathbf{b} \leftarrow \text{RUNLIFTEDINFERENCE}(\mathfrak{G})$ /* **run LBP or SSP** */
5:     $\mathbf{y}_t \leftarrow \text{PICKVARSTOCLAMP}(\mathbf{b})$ /* **based on beliefs or surveys** */
6:     $\mathbf{a} = \mathbf{a} \bigcup \mathbf{y}_t$
7:     $\mathbf{Y} = \mathbf{Y} \setminus \mathbf{Y}_t$
8:     $G \leftarrow \text{CLAMP}(G, \mathbf{y}_t)$
9:     $\mathfrak{G} \leftarrow \text{LIFT}(G)$
10:     $\mathfrak{G} \leftarrow \text{SIMPLIFY}(\mathfrak{G})$ /* **based on LWP** */
11:     **if** CONTAINSCONTRADICTION$(\mathfrak{G})$ **then**
12:         **return** None
13: **end while**
14: **return a**

---

p", respectively to "$\ldots, p' = p$", for the message sent at position p. This ensures that we consider exactly those neighbors of i which tend to make the clause a unsatisfied, respectively satisfied. In the same way as in WP, we can now apply these lifted equations to the lifted factor graph. In the following, we will investigate different SAT solving scenarios using message passing algorithms.

11.4.7 Lifted Decimation

We have described the idea of BP-based decimation already in Sec.11.4.4 and we also indicated that SP could be used as an alternative inference algorithm to pick variables to clamp. The first way of lifting the decimation is lifting the factor graph before running the inference algorithm. Then we run the corresponding lifted variant of the inference algorithm on the compressed graph. After picking one or more variables to clamp based on the beliefs or surveys, we clamp the variables in the ground factor graph, lift the modified graph again and run LWP on the lifted graph to simplify the graph subsequently. This process is iterated until either a solution has been found, or alternatively a contradiction has been detected during the simplification. The pseudo code for the lifted decimation is depicted in Alg.5. Especially with LBP, LSP, and LWP at hand, we can execute every step now on the lifted level. We will analyze the performance gains by lifting in the next section.

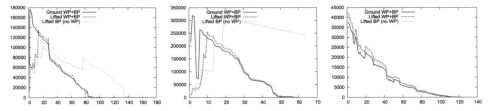

Figure 11.6: Results for Lifted Warning Propagation: Total number of messages sent (y-axis) in each iteration (x-axis) of decimation for (lifted) WP+BP and lifted BP on several CNF benchmarks. (left, `2bitmax_6`) Lifted WP+BP saves 13% of the messages ground WP+BP sent. (middle, `ls8-norm`) Lifted WP+BP saves 16% of the messages ground WP+BP sent. (right, `wff.3.150.525`) A small lifting overhead occurs on this random CNF.

11.4.8   Experimental Evaluation

Our intention in this section is to empirically investigate the correctness of the lifted message updates for WP and SP, and compare their performance to the corresponding ground versions. We want to investigate whether message passing for SAT can benefit from lifting. Therefore, we define the following questions for this section:

(Q1)   Do ground and lifted decimation return the same results?

(Q2)   Does lifting speed up decimation procedure for SAT solving?

To this aim, we implemented LWP and LSP in C++ using the LIBDAI library (Mooij, 2010) for the underlying data structures. Based on the C++ code, we implemented the (lifted) decimation in Python which uses the C++ message passing as a subroutine. More precisely, we implemented the decimation approaches as described in Sec.11.4.4 and Sec.11.4.7. As a proof of concept, we compared the performances of lifted message passing approaches with the corresponding ground versions on three CNFs from a standard benchmark (Gomes et al., 2007). We use the decimation to measure the effectiveness and efficiency of the lifted algorithms.

To assess performance, we report run times. In order to abstract from the underlying physical hardware and programing language, run time is measured by the number of messages being sent. For the typical message sizes, e.g., for binary random variables with low degree, computing color messages is essentially as expensive as computing the actual messages of the inference algorithm. Therefore, we view the run time as the sum of both types of messages, i.e., color and (modified) propagation messages, treating individual message updates as atomic unit time operations. BP messages were damped by a factor of 0.4, and the convergence threshold was set to 10^{-3}. We used parallel update schedules for (L)WP, (L)BP, and (L)SP. While the (L)BP messages were initialized uniformly, the (L)WP and (L)SP messages were initialized with zero.

We evaluated (lifted) WP+BP decimation on a circuit synthesis problem (2bitmax_6). The formula has 192 variables and 766 clauses. The resulting factor graph has 192 variable nodes, 766 factor nodes, and 1,800 edges. The statistics of the different variants are shown in Fig. 11.6. As one can see, lifting yields significant improvement in efficiency especially in the first iterations: LIFTED WP+BP reduces the messages sent by 13% compared to GROUND WP+BP when always fixing the most magnetized variable, i.e., the variable with the largest difference between negated and unnegated marginals. Additionally, the ground and the lifted version found the same solution to the problem.

Then, we investigated a Latin square construction problem of size eight (ls8-norm). The formula has 301 variables and 1,601 clauses, resulting in a factor graph with 3,409 edges. The statistics of running lifted message passing are shown in Fig. 11.6. Again, the lifting yields improvement in efficiency: LIFTED WP+BP saved 16% of the messages sent by the ground version. Both variants were also able to find the same solution.

Finally, we also applied the lifted message passing algorithms to a random 3-CNF, namely wff.3.150.525. The CNF contains 150 variables, 525 clauses and 1,575 edges. We did not expect lifting on this problem instance because random structure usually do not provide any significant symmetries. As expected, no lifting was possible as shown in Fig. 11.6.

We also ran ground and lifted SP on these three problem instances. Due to the properties of the CNFs, SP always converged to a paramagnetic state, i.e., the trivial solution, in the first iteration. In such cases the problem is usually passed to a simpler SAT solver such as WALKSAT. Nevertheless LSP saved roughly 45% of the messages the ground version sent on the 2bitmax_6 problem. On the latin square, LSP sent only 44% of messages ground SP sent. As expected, the lifting produced again a small overhead on the random 3-CNF for LSP.

To summarize, questions **(Q1)** and **(Q2)** can both be answered in favor of the introduced algorithms. The experimental results clearly show that a unified lifted message passing framework, including BP, WP, and SP, is indeed useful for SAT. Not only are significant efficiency gains obtainable by lifting, but lifted WP-guided BP clearly outperforms unguided (lifted) BP. On top of that, the assignments returned by the lifted decimation were identical to the results of the ground decimation.

11.4.9   Summary: Lifted Satisfiability

In this section we presented the first lifted message passing algorithms for determining the satisfiability of Boolean formulas. Triggered by the success of Lifted Belief Propagation approaches, our algorithms construct a lifted CNF, represented as a lifted factor graph. The clusternodes and clusterfactors correspond to sets of variables and clauses that are sending and receiving the same messages. On these lifted data structures, we can then apply Lifted Warning Propagation, respectively Lifted Survey Propagation, yielding (approximate) information on satisfying assignments upon converge. The experimental results validate the correctness of the resulting lifted approaches as they coincide with the ground results. Significant efficiency gains are obtainable when employing lifted instead of standard WP, respectively SP. In particular, lifted decimation approaches based on LBP and LWP, or alternatively LSP and LWP, are faster than "just" using LBP to determine the satisfiability of Boolean formulas.

Of course, with decades of research in SAT solving, lifted SAT is not the first approach exploiting symmetries in CNFs. Symmetries have been studied in propositional and first-order logic for a long time and have already been related to graph isomorphism decades ago, e.g. (Crawford, 1992) and further references in there. In SAT solving, symmetries are often considered to be undesirable because a search-based solver should avoid checking two symmetric branches. To do so, in symmetry breaking, e.g. (Aloul et al., 2006), additional predicates are inserted that break symmetries but preserve the satisfiability of the CNF. Instead, the framework we presented in this section clusters symmetries to reduce run time. In addition, it nicely integrates itself into the lifted inference paradigm and benefits from existing implementation and ongoing research. Relating the idea of symmetry breaking back to probabilistic inference, one should further investigate an operation the other way around, i.e., adding evidence to the factor graph which makes the problem more symmetric without heavily changing the underlying probability distribution.

The SP experiments are by far not exhaustive yet and it is still an open question, how LSP performs on further difficult but symmetric problems. For additional experiments, we require more suitable large-scale CNFs that contain appropriate symmetries. In particular, running lifted decimation based on SP+WP on structured problems, while enforcing that non-trivial solutions are returned by SP, can make this approach significantly more applicable.

Several extensions have been proposed to SP in recent years, extending SP to the maximum satisfiability (Max-SAT) or weighted Max-SAT case. For example, the *SP-y* algorithm (Battaglia et al., 2004) generalizes SP to handle Max-SAT problems. The *Relaxed Survey Propagation* (RSP) algorithm introduced in (Chieu and Lee, 2009) allows to run a modified SP algorithm on weighted Max-SAT instances. Using extensions of SP that can handle weighted SAT problems, such as RSP, are also interesting for lifting because they could be used in inference on relational models such as MLNs. In particular MLNs with deterministic and soft clauses asks for a tight integration of lifted SAT and lifted probabilistic inference. In many real-world applications, the problem formulation does not fall neatly into one of the classes. The problem may have one component that can be well-modeled as a pure SAT problem, while it contains another part better modeled as a probabilistic inference problem. We suggest to partition a problem into corresponding subnetworks, run the corresponding type of lifted message passing algorithm on each subnetwork, and to combine the information from the different subnetworks. For the non-lifted case, this is akin to Duchi et al. (2007)'s COMPOSE approach that has been proven to be successful for problems involving matching problems as subproblems.

Looking at the lifted decimation framework more closely, we can already identify a possible drawback of the naive lifting. As we have mentioned above, the approaches presented here apply the lifting as pre-processing step in every iteration. As we will see in Sec.11.5, the lifting can be integrated more tightly into the decimation which yields a more elaborated lifted decimation approach, however, also requiring a more complex lifting scheme. Our current decimation approach does not take backtracking strategies into account. Many SAT solvers though make extensive use of backtracking which allows reverting earlier decisions and exploring other paths to the solutions. Naturally, one could integrate backtracking in our approach as well. But more interestingly, one would like to investigate how well backtracking can benefit from lifting, i.e., connecting it again closer to symmetry breaking. Nevertheless, this section focused on the feasibility of lifted satisfiability. Naturally, many concepts and improvements from the past years or even decades could be included in the decimation as well, to further improve the results and reduce run time additionally.

As a last note, it is interesting to observe that WP can be seen as a simplified version of max-product BP. More precisely, WP is an instance of the *min-sum* algorithm with only binary variables and a particular initialization of the messages. We obtain the min-sum algorithm by running the message passing on the energy function based formulation of the MRF. If we assume that $\mathcal{E}_c(\mathbf{x}_c) = 0$ if the constraint c is satisfied and 1 otherwise, it can be shown that the messages in the energy minimization form are equivalent to the WP updates if the messages are initialized with values from $\{0, 1\}$. Further details can be found in (Mézard and Montanari, 2009, Chapter 14.3). Bearing this in mind, the existence of LWP can be derived from LBP, however, we showed in this section that we can also lift the simpler WP equations directly. Similarly, it has also been shown in (Braunstein and Zecchina, 2004) that the SP equations can be derived from the sum-product BP equations.

11.5 Lifted Sequential Clamping for SAT Problems and Sampling

As we have seen in the previous section and as it has been reported in the literature, BP and other lifted message passing algorithms can be extremely fast at computing approximate marginal probability distributions over single variable nodes. These marginals were used in the decimation procedure to solve SAT problems. They allow us to determine the magnetization of a variable or alternatively to sample a variable according to its marginal distribution. This computation works similarly well for marginals over neighboring nodes in the underlying graphical model. However, if we want to solve an entire CNF, sample a complete assignment to all variables, or sample an arbitrary subset of the variables, (lifted) message passing approaches leave space for improvement. For instance, in the BP+WP-guided decimation as described above, or in sampling configurations from a joint configuration (Montanari et al., 2007; Mézard and Montanari, 2009), one is essentially computing these probabilities in the same network but with changing evidence in a sequential procedure with re-lifting in every iteration.

While we have already discussed the importance of decimation approaches in satisfiability, obtaining samples from a joint distribution is equally important to AI and another example of a task fitting into the decimation framework. For example, this kind of sampling is often used in parameter learning. In the decimation framework, we do not set the variable to its most magnetized state but instead sample a state according to its marginal distribution. We then clamp the variable in the network to the sampled state and continue as before.

In such cases where we want to apply lifted message passing to the same network with sequentially changing evidence, we want to avoid the naive solution that recomputes the lifted network in each step from scratch. This can possibly cancel the benefits of lifted inference and we also ignore information from the previous iterations.

As we will argue in the present section, we can do considerably better. Especially in cases where only additional evidence is added in each iteration, we can exploit the knowledge from the previous lifting to obtain the current lifted network more efficiently. To this end, we present **shortest-path-sequence lifting** for Lifted Belief Propagation (SPS-LBP). This is a scalable lifted inference approach for approximating solutions of complex inference tasks that is based on shortest paths sequences between variables in the factor graph. It does not lift in every iteration from scratch but instead use the already known liftings from previous iterations. That is, SPS-LBP avoids the repetitive lifting for each subtask by efficiently computing the corresponding lifted network directly from the one already known for the unconditioned case. This is different to the approach of Nath and Domingos (2010), which essentially memorizes the intermediate results of previous liftings and, for new evidence, makes a warm start from the first valid intermediate lifting.

We will now explain Shortest-Path-Sequence Lifting (SPS-lifting) in detail and provide a running example along with it. We will also give an in-depth analysis of it and sketch the soundness of SPS-lifting. As our experimental results will highlight, SPS-lifting is capable of saving further messages sent in the decimation compared to the naive approach.

11.5.1   Lifted Sequential Inference

When we turn a complex inference task into a sequence of simpler tasks, we are repeatedly answering slightly modified queries on the same graph. Because LBP and LWP generally lack the opportunity of adaptively changing the lifted graph and using the updated lifted graph for efficient inference, they are doomed to lift the original model in each of the k iterations again from scratch. Each Color Passing run scales $O(h \cdot |E|)$ where $|E|$ is the number of edges in the factor graph and h is the number of iterations required. Hence, we can spend up to $O(k \cdot h \cdot |E|)$ time just on lifting if we clamp k variables in a sequential fashion. The previous section already showed how BP-guided decimation fixed one variable after

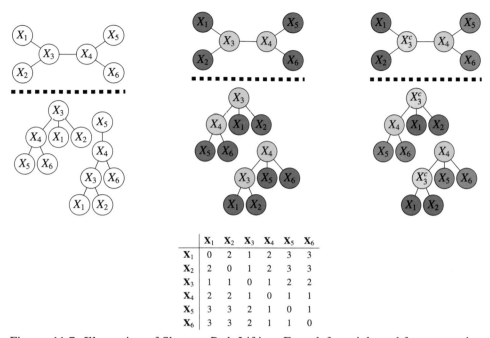

	$\mathbf{X_1}$	$\mathbf{X_2}$	$\mathbf{X_3}$	$\mathbf{X_4}$	$\mathbf{X_5}$	$\mathbf{X_6}$
$\mathbf{X_1}$	0	2	1	2	3	3
$\mathbf{X_2}$	2	0	1	2	3	3
$\mathbf{X_3}$	1	1	0	1	2	2
$\mathbf{X_4}$	2	2	1	0	1	1
$\mathbf{X_5}$	3	3	2	1	0	1
$\mathbf{X_6}$	3	3	2	1	1	0

Figure 11.7: Illustration of Shortest-Path-Lifting: From left to right and from top to bottom: (a) Original factor graph. (b) Prior lifted network, i.e., lifted factor graph without clamped variables. (c) Lifted factor graph when X_3 is clamped, indicated by the "c" superscript. Factor graphs are shown (top) with corresponding colored computation trees (bottom). For the sake of simplicity, we assume identical factors but omit them in the graphical representation. The shades in (b) and (c) encode the clusternodes. Finally, the table (d) shows that shortest path distances of the nodes. The *i*-th row will be denoted as dist$_i$ in the text.

another. Depending on the nature of the CNF, or more generally on the structure of the factor graph, this can require up to $k = n$ iterations.

Let us now consider **BP-guided sampling**, which can also be casted into the framework of decimation. When we want to sample from the joint distribution over k variables, this can be reduced to a sequence of one-variable samples conditioned on a subset of the other variables (Mézard and Montanari, 2009). Thus, to get a sample for $\mathbf{X} = (X_1, \ldots, X_k)$, we first compute $p(X_1)$, then $p(X_2|X_1), \ldots, p(X_k|X_1, \ldots X_{k-1})$.

Example 11.2 Assume we want to sample from the joint distribution $p(X_1, \ldots, X_6)$, given the network in Fig.11.7(a, top). Further assume that we begin our sequential process by first computing $p(X_3)$ from the prior lifted network, i.e., the lifted network when no evidence has been set, cf. Fig. 11.7(b,top). After sampling a state x_3, we want to compute $p(X_{\backslash 3}|x_3)$ as shown in Fig. 11.7(c, top).

To do so, it is useful to describe BP and its operations in terms of its **computation tree** (CT) as already used earlier. To recap, the CT is the unrolling of the (loopy) graph structure where each level i corresponds to the i-th iteration of message passing. Similarly we can view Color Passing as a **colored computation tree** (CCT). More precisely, one considers for every node X the CT rooted in X but now each node in the tree is colored according to the nodes' initial colors, cf. Fig. 11.7(a-c, bottom). Each CCT encodes the root nodes' local communication patterns that show all the colored paths along which node X communicates in the network. Consequently, CP groups nodes with respect to their CCTs: nodes having the same set of rooted paths of colors (node and factor names neglected) are clustered together.

Example 11.3 For instance, Fig. 11.7(a, bottom) shows the CTs rooted in X_3 and X_5. Because their set of paths are different, X_3 and X_5 are clustered into different clusternodes as indicated by different colors in Fig. 11.7(b, top). For this prior lifted network, the light green nodes exhibit the same communication pattern in the network which can be seen in identical CCTs in Fig. 11.7(b, bottom), and were consequently grouped together. Now, when we clamp the node X_3 to a value x_3, we change the communication pattern of every node having a path to X_3. Specifically, we change X_3's, and only X_3's, color in all CCTs where X_3 is involved, as indicated by the "c" in Fig. 11.7(c). This affects nodes X_1 and X_2 differently than X_4, respectively X_5 and X_6, for two reasons:

1. they have different communication patterns as they belong to different clusternodes in the prior network
2. more importantly, they have different paths connecting them to X_3 in their CCTs

The **shortest path** is the shortest sequence of factor colors connecting two nodes. Since we are not interested in the paths but whether the paths are identical or not, these sets might as well be represented as colors. Note that in Fig. 11.7 we assume identical factors for simplicity. Thus in this case path colors reduce to distances. In the general case, however, we compare the paths, i.e., the sequence of factor colors.

Example 11.4 The prior lifted network can be encoded as the vector $l = (0, 0, 1, 1, 0, 0)$ of node colors. Thus, to get the lifted network for $p(X_{\backslash 3}|x_3)$, as shown in Fig. 11.7(c, top), we only have to consider the vector dist_3 of shortest paths distances to X_3 (see Fig. 11.7(d)) and refine the initial clusternodes correspondingly. This is done by

1. the **element-wise concatenation** of two vectors: $l \oplus \text{dist}_3$
2. viewing each resulting number as a new color

For our example, we obtain:

$$(0, 0, 1, 1, 0, 0) \oplus (1, 1, 0, 1, 2, 2) =_{(1)} (01, 01, 10, 11, 02, 02) =_{(2)} (3, 3, 4, 5, 6, 6) \, ,$$

which corresponds to the lifted network for $p(X_{\backslash 3}|x_3)$ as shown in Fig. 11.7(c). Thus, having the shortest path matrix, we can directly update the prior lifted network in linear time without taking the detour through running CP on the ground network. Now, we run inference, sample a state $X_4 = x_4$ afterwards, and compute the lifted network for $p(X_{\backslash\{3,4\}}|x_4, x_3)$ to draw a sample for $p(X_1|x_4, x_3)$. Essentially, we proceed as before: compute $l\oplus(\text{dist}_3 \oplus \text{dist}_4)$.

However, the resulting network might be suboptimal in cases when more than one variables is clamped and variables from the same initial cluster are sampled identically, i.e., take on the same value in the sample.

Example 11.5 The concatenation in the previous example assumed $x_3 \neq x_4$ and, hence, X_3 and X_4 cannot be in the same clusternode. For $x_4 = x_3$, they could be placed in the same clusternode because they were in the same clusternode in the prior network. If X_3 and X_4 are clamped, this can be checked by $\text{dist}_3 \odot \text{dist}_4$, the element-wise sort of two vectors. In our case, this yields $l \oplus (\text{dist}_3 \odot \text{dist}_4) = l \oplus l = l$: the prior lifted network.

In general, we compute

$$l \oplus \left(\bigoplus_{\mathfrak{x}} (\bigoplus_{s} \text{dist}_{\mathfrak{x},s})\right),$$

where

$$\text{dist}_{\mathfrak{x},s} = \left(\bigodot\right)_{i\in\mathfrak{x}:x_i=s} \text{dist}_i ,$$

with clusternodes $\mathfrak{x}$ and the truth values s. For an arbitrary network, however, the shortest paths might be identical although the nodes have to be split, i.e., they differ in a longer path, or in other words, the shortest paths of other nodes to the evidence node are different. Consequently, we apply the shortest paths lifting iteratively. Let CN_E denote the clusternodes given the set E as evidence. By applying the shortest paths procedure, we compute $CN_{\{X_1\}}$ from $CN_\emptyset$. This step might cause initial clusternodes to be split into newly formed clusternodes. To incorporate these changes in the network structure the shortest paths lifting procedure has to be iteratively applied. Thus in the next step we compute $CN_{\{X_1\}\cup\Delta_{X_1}}$ from $CN_{\{X_1\}}$, where Δ_{X_1} denotes the changed clusternodes of the previous step. This procedure is iteratively applied until no new clusternodes are created. We exemplify this issue by introducing another example where a single run of shortest paths lifting fails, however, an iterative application returns the correct lifting.

Example 11.6 The initial lifting of the graph depicted in Fig. 11.8(left) can be encoded as $l = (0, 1, 2, 2, 1, 0)$. If we now clamp variable X_3 as shown in Fig. 11.8(middle), we can use the distances in Fig. 11.8(right) to concatenate l and dist_3:

$$(0, 1, 2, 2, 1, 0) \oplus (1, 1, 0, 1, 2, 1) = (01, 11, 20, 21, 12, 01) = (0, 1, 2, 3, 4, 0) .$$

Yet, this lifting is not correct, as we have to distinguish X_1 and X_6. We also observe that the clusternodes of X_2, X_4, and X_5 have changed. Therefore, we now have to iteratively apply

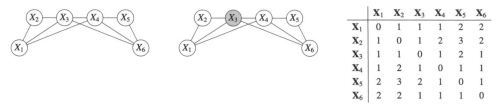

	$\mathbf{X}_1$	$\mathbf{X}_2$	$\mathbf{X}_3$	$\mathbf{X}_4$	$\mathbf{X}_5$	$\mathbf{X}_6$
$\mathbf{X}_1$	0	1	1	1	2	2
$\mathbf{X}_2$	1	0	1	2	3	2
$\mathbf{X}_3$	1	1	0	1	2	1
$\mathbf{X}_4$	1	2	1	0	1	1
$\mathbf{X}_5$	2	3	2	1	0	1
$\mathbf{X}_6$	2	2	1	1	1	0

Figure 11.8: Example highlighting the need for iterative shortest-path-lifting. From left to right: Toy example of a graph where a single run of Shortest-Paths-lifting fails to return the correct lifting.

the shortest paths lifting based on the nodes that changed in the previous iteration:

$$(0, 1, 2, 3, 4, 0)$$
$$\oplus(1, 0, 1, 2, 3, 2)$$
$$\oplus(1, 2, 1, 0, 1, 1)$$
$$\oplus(2, 3, 2, 1, 0, 1)$$
$$=(0112, 1023, 2112, 3201, 4310, 0211) = (0, 1, 2, 3, 4, 5) \; .$$

Since we are now at the ground level anyhow, we can stop iterating.

The description above together with Example.11.6 essentially sketch the proof of the following Theorem. Originally, the theorem in a slightly different way, together with its proof, were presented by Ahmadi et al. (2010).

Theorem 11.2 *If the **shortest path colors** among all nodes and the prior lifted network are given, computing the lifted network for $p(\mathbf{X}|X_k, \ldots, X_1)$, $k > 0$, takes $O((k + h) \cdot n)$, where n is the number of nodes and h is the number of required iterative applications of the concatenation. Furthermore, running LBP produces the same results as running BP on the original model.*

Proof   Assuming a graph $G = (V, E)$. We have seen above that the concatenation is a linear operation in the number of nodes. When k nodes have been set, we have to concatenate k distance vectors. However, this concatenation can result in supernodes being changed. Consequently, this requests the concatenation of additional nodes. This iterative application can result in h additional concatenations.

When we set new evidence for a node $X \in V$ then for all nodes within the network the color of node X in the *CCTs* is changed. If two nodes $Y_1, Y_2 \in V$ were initially clustered together and belonged to the same clusternode, i.e., $\mathfrak{X}(Y_1) = \mathfrak{X}(Y_2)$), they have to be split if the *CCTs* differ. Now we have to consider two cases:

1. If the difference in the *CCTs* is on the shortest path connecting X with Y_1 and Y_2, respectively, then shortest path lifting directly provides the new clustering.

2. If the coloring along the shortest paths is identical, the nodes' *CCTs* might change in a longer path. Since $\mathfrak{X}(Y_1) = \mathfrak{X}(Y_2)$ there exists a mapping between the paths of the respective *CCTs*. In particular $\exists Z_1, Z_2$, s.t. $\mathfrak{X}(Z_1) = \mathfrak{X}(Z_2)$ from a different clusternode, i.e., $\mathfrak{X}(Z_i) \neq \mathfrak{X}(Y_i)$, and

$$Y_1, \ldots, \underbrace{Z_1, \ldots, X}_{\Delta_1} \in CCT(Y_1), Y_1, \ldots, \underbrace{Z_2, \ldots, X}_{\Delta_2} \in CCT(Y_2) ,$$

and $\Delta_1 \in CCT(Z_1) \neq \Delta_2 \in CCT(Z_2)$ are the respective shortest paths for Z_1 and Z_2. Thus, by iteratively applying shortest-path lifting as explained above, the evidence propagates through and we obtain the new clustering.   □

In fact, SPS lifting can be quite fast and we will now show in our experimental evaluation how the lifting time can be decreased for tasks with evidence arriving sequentially, e.g., lifted satisfiability or lifted sampling of joint configurations. If, however, lifting does not pay off, computing the pairwise distances may produce an overhead.

11.5.2   Experimental Evaluation

After we have described the idea of SPS-lifting, we will now show how it can be integrated into an improved lifted decimation framework and we want to investigate the following questions: To see whether this can improve lifted decimation we investigated the following questions: **(Q3)** Can we further support the results from the previous section and show that lifted decimation solves satisfiability problems more efficiently? **(Q4)** Does lifting improve sequential inference approaches beyond lifted satisfiability? **(Q5)** Is SPS-lifting even more beneficial than naive lifting based on CP?

Therefore, we run experiments on two AI tasks in which sequential clamping is essential. Namely, BP-guided decimation for satisfiability problems and sampling configurations from MLNs. With SPS-lifting, both tasks essentially follow the decimation strategy presented earlier in this chapter. However, in the case of sampling, we do not check for contradictions and the variable is not clamped based on the magnetization. Instead, we sample the value of the variable based on its marginal belief. The previous section already contained initial experiments for lifted satisfiability, however, the lifted decimation was implemented in a naive way. This initial investigation will now be extended to additional problem instances and supported by a secondary set of experiments based on a different inference task.

11.5.2.1   Lifted Satisfiability   As indicated above, our lifted satisfiability based on decimation fits well into the sequential setting. However, we now use a more elaborated variant compared to the one depicted earlier. Now, however, it distinguishes most importantly from avoiding the entire re-lifting from scratch in each iteration. We first run LBP on the lifted

CNF Name	Iters	Ground	Naive	SPS	Walksat
ls8-normalized	26	3.17	1.12	0.95	540
ls9-normalized	13	5.47	1.65	1.47	1,139
ls10-normalized	14	10.27	1.84	1.59	1,994
ls11-normalized	26	38.82	11.51	10.64	4,500
ls12-normalized	35	60.83	13.15	11.57	10,351
ls13-normalized	21	55.39	9.99	8.21	30,061
ls14-normalized	22	83.30	10.22	8.30	104,326
2bitmax_6	55	2.35	1.25	1.05	379
5_100_sd_schur	53	111.19	75.98	64.91	1,573,208
wff.3.100.150	54	0.19	0.26	0.22	17
wff.4.100.500	78	1.73	2.04	1.89	33
wff.3.150.525	126	6.36	6.76	6.56	284

Figure 11.9: Experimental Results for Lifted Decimation: Total messages sent (in millions) in SAT experiments and number of average flips needed by Walksat (last column).

graph and use its marginals to fix the next variable. Based on this clamping, the lifting is updated using SPS-lifting. We then run LWP on the modified lifted factor graph to clamp directly implied variables as before. Again, LWP is also used to detect possible contradictions. When LWP finds a contradiction, the algorithm stops and does not return a satisfying configuration. Otherwise, we continue by running LBP again.

We continue to compare the performance of lifted message passing approaches with the corresponding ground versions on the previously used CNF benchmark from (Gomes et al., 2007). We use the decimation as described to measure the effectiveness of the algorithms. To assess performance, we again report the number of messages sent. As before, for the typical message sizes, e.g., for binary random variables with low degree, computing color messages is essentially as expensive as computing the actual messages. Therefore, we report both color and (modified) BP messages, treating individual message updates as atomic unit time operations. We use the parallel message protocol for (L)BP and (L)WP where messages are passed from each variable to all corresponding factors and back at each step. The convergence threshold was set to 10^{-8} for (L)BP and all messages were initialized uniformly. In the case of (L)WP, all messages were initialized with zero, i.e., no warning being sent initially. As mentioned above, it is usually necessary to iteratively apply the SPS-lifting to obtain the correct adapted lifted graph. The number of required iterations, however, can be high if long paths occur in the network. Therefore, we use the SPS-lifting only once but then continued with standard CP to determine the new lifting.

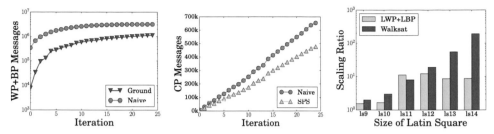

Figure 11.10: Experimental results for lifted decimation on Latin squares. (left) Comparison of ground decimation with naive lifted decimation on `ls8-normalized`. (middle) Comparison of naive lifting with SPS-lifting on `ls8-normalized`. (right) Comparison of the growth in computational costs on increasing problem sizes measured relative to the smallest problem.

This can still save several passes of CP because the SPS-lifting provides us with a good head start.

We evaluated (LIFTED) WP+BP decimation on different CNFs, ranging from problems with about 450 up to 78,510 edges. The CNFs contain structured problems as well as random instances. The statistics of the runs are shown in Tab. 11.9. As one can see, naive lifting already yields significant improvement, further underlining the experiments from the previous section. When applying the SPS-lifting, we can do even better by saving additional messages in the compression phases. The savings in messages are visible in running times as well. Looking at Fig. 11.10(left), we compare ground decimation with its lifted counterpart. In the lifted case, we only send 33% of the total ground messages. When using the SPS-lifting, we can save up to an additional 10% of messages sent in the compression phase, cf. Fig. 11.10(middle). In the decimation, we always clamp the most magnetized variable which is the variable having the largest difference between the probability of the `True` and `False` state. We also applied the lifted message passing algorithms to random CNFs (last three rows in Tab. 11.9). As expected, no lifting was possible because random instances usually do not contain symmetries. In our experiments, we were able to find satisfying solutions for all problems which validates the effectiveness of the (lifted) decimation.

Although we are not aiming at presenting a state-of-the-art SAT solver, we solved all problems using Walksat (Selman et al., 1995) as well. Hence, we report results measured in variable flips in Tab. 11.9 too. Although Walksat requires fewer flips than we send messages, one can see that our lifted decimation strategy still scales well. In Fig. 11.10(right) we have compared the computational effort on increasing problem sizes for Walksat and our lifted decimation. The results indicate that our approach can handle large problem instances without employing complex heuristics and code optimization but exploiting symmetries in the problems.

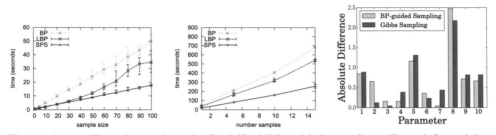

Figure 11.11: Experimental results for Lifted BP guided sampling. From left to right: (left) (L)BP-guided sampling for varying sample size (middle) (L)BP-guided sampling for varying number of samples. (right) Absolute difference of the learned parameters

In combination with the results from the previous section, both questions **(Q3)** and **(Q5)** can clearly been answered in favor of our algorithms for the task of lifted SAT solving. The next experiments will show that this also holds in the case of lifted sampling.

11.5.2.2 Lifted Sampling   We investigated BP, LBP and SPS-LBP for sequentially sampling a joint configuration over a set of variables, i.e., for a sequence of one-variable samples conditioned on a subset. Thus, to get a sample for $\mathbf{X} = X_1, \ldots, X_n$, we first compute $p(X_1)$, then $p(X_2|X_1), \ldots, p(X_n|X_1, \ldots X_{n-1})$. In contrast to the decimation for SAT solving, the procedure of picking a variable to clamp is solely based on the index of the variables. Additionally, we now sample a state from the computed BP marginals instead of clamping a variable to the most magnetized state. Fig. 11.11 summarizes the results for the "Smokers-and-Friends" dynamic MLN with ten people over ten time steps.

In our first experiment, we randomly chose 1, 5, 10, 20, 30, ..., 100 "cancer" nodes over all time steps, and sampled from the joint distribution. As one can see in Fig. 11.11(left), LBP already provides significant improvement compared to BP, however, as the sample size increases, the speed-up is lower. The more different evidences we have in the network, the less lifting is possible. SPS-LBP comes with an additional reduction in runtime as we do not need to perform the lifting in each step from scratch.

In our second experiment we fixed the sample size to 100, i.e., we sampled from the joint distribution of all `cancer(x, t)` predicates for all persons x in the domain and all time steps `t`. We drew 1, 5, 10 and 15 samples and the timings are averaged over five runs. Fig. 11.11(right) shows that LBP is only slightly advantageous compared to BP, as the sample size is 100, especially in the later iterations we have lots of evidence and long chains to propagate the evidence. Repeatedly running CP almost cancels the benefits. SPS-LBP on the other hand, shows significant speed-ups.

To evaluate the quality of the samples, we drew 100 samples of the joint distribution of all variables using the BP-guided approach and Gibbs sampling respectively. We learned the parameters of the model maximizing the conditional marginal log-likelihood using scaled conjugate gradient. Fig.11.11(right) shows the absolute difference of the learned weights from the model the datacases were drawn. To summarize the error into a single number, we calculated the RMSE. The RMSE for the BP parameters was 0.31 and for Gibbs parameters 0.3. Thus, parameter learning with BP-guided samples performs as good as with samples drawn by Gibbs sampling.

We have seen that lifting, and in particular SPS-lifting, speeds up the sampling of joint configurations and hence, confirms a positive answer to question **(Q4)**. At the same time, BP-guided sampling achieves comparable quality to an alternative approach base on Gibbs sampling.

11.5.3   Summary Lifted Sequential Clamping

In this section, we presented an adaption to the original Color Passing algorithm for lifting. To avoid the lifting from scratch in each iteration of the naive realization, we employed an efficient sequential clamping approach. The SPS-lifting also gives a novel characterization of the main information required for lifting in terms of shortest paths in a given network. The experimental results were conducted on two tasks for lifted inference, namely Boolean satisfiability and sampling joint configurations from MLNs. These experiments validate the correctness of the proposed lifted decimation framework and demonstrate that instantiations can actually be faster than "just" using standard lifting.

Since lifting SAT solvers themselves and the exploitation of efficient sequential lifting for them is a major advance, one should further look into the integration of SPS-lifting for Survey Propagation because SP has proven to be very powerful for solving certain classes for CNFs. Initial experiments were conducted in the previous section already, but for structured CNFs, we are often facing the problem of trivial surveys returned by SP. One should investigate if the choice of a variable to clamp can directly influence the results of SP and its convergence.

As mentioned above, in the lifted sampling experiments, the variable to clamp was solely picked based on its index in the problem formulation. Other algorithms use the idea of conditioning to construct approximate inference algorithm based on BP, e.g., CBP-BBP presented in (Eaton and Ghahramani, 2009), and use more elaborated methods to select a variable to clamp. Both, lifting the CBP-BBP-algorithm, as well incorporating their ideas on picking a variable to clamp into our decimation framework are interesting research venues and still open questions.

Lastly, if we are clamping the factor graph over the iterations, we will most likely end up with a factor graph with almost all variables being set and little opportunities for lifting. Hence, one should also look for heuristics that stop compressing when almost all variables are clamped and lifting is too costly.

While this section showed how lifting can be efficiently exploited over several runs of inference with changing evidence, we will now turn our attention to an atomic inference run. For iterative MAP inference algorithms, one should also consider adapting the lifting over the iterations of message passing to possibly speed up run time.

11.6   Lifted Message Passing for MAP via Likelihood Maximization

The previous sections focused on marginal inference and SAT. To conclude the chapter, we will turn our attention towards **MAP inference**. It was already mentioned that MAP inference is an important, yet challenging, problem and it was discussed how an approximate MAP assignment can be obtained by max-product BP. In this section, we will go into further detail how an approximate MAP assignment can be obtained by other means than BP. Specifically, we will consider the known **likelihood maximization** (LM) approach (Kumar and Zilberstein, 2010) ,which was empirically shown to have a convergence rate often significantly higher than approaches such as MPLP (Globerson and Jaakkola, 2007). LM increases a lower bound on the MAP value while being fairly simple to implement. Intuitively, Kumar and Zilberstein approximate the original distribution by a simpler one and show that this approximation often yields satisfying results. This approximation imposes the constraint on the model distribution that any probability over subsets of variables is equal to the product of their single node marginals. Kumar and Zilberstein derive EM (Dempster et al., 1977) equations to optimize the objective and show that their algorithm can be implemented using a message passing paradigm. LM quickly finds high quality solutions and yet the message updates are remarkably simple. Specifically, as our first contribution, we show that LM is liftable in a bottom-up fashion akin to the lifting of BP and the message passing algorithms for satisfiability presented in the previous section.

Specifically, we introduce lifted LM equations that are used on top of the lifted problem formulation. Here, we also obtain the lifted model by applying a variation of the CP algorithm to the ground model. This contribution already significantly extends the family of known lifted MAP approaches. However, one common issue in lifting is evidence which often destroys symmetries in the model. The more evidence is given, the more propositional the problem becomes, ultimately canceling the benefits of lifted inference. It is easy to generate evidence cases that actually cancel the benefits of lifted inference completely. Hence, the main contribution of this section is to show, how to employ additional structure for optimization provided by LM, to further speed up lifted inference. In particular, we will introduce the concept of **pseudo evidence**.

For LM, it has been recognized that **pseudo marginals** may converge quickly when forcing updates to be as greedy as possible (Toussaint et al., 2008). We interpret this greedy update rule as ultimately inducing pseudo evidence: if a pseudo marginal is close to one state, clamp the corresponding variable accordingly. As we will show empirically, this can already considerably speed up inference in the propositional case and reduce the number of messages being sent. However, the model could potentially be simplified even further over the inference iterations, but the naive LLM approach cannot make use of this new evidence to speed up lifted inference in later iterations. Consequently, we propose an efficient LLM version for updating the structure of the lifted network with pseudo evidence over the iterations.

To do so, one is tempted to just lift the network again when new pseudo evidence is "observed". However, we can do significantly better. Existing online lifting and clamping approaches simply compute a more shattered model, should the new evidence require to do so. But note that the current lifted model is always valid for the remaining iterations if we always set entire clusters as pseudo evidence. Therefore we can simply lift the current lifted model and obtain models monotonically decreasing in size. Indeed, this is akin to Bui et al. (2012) who show how soft evidence affects lifting. Bui et al. exploit symmetries before soft evidence is obtained. So, in contrast to existing lifted inference approaches, we add evidence over the iterations and use this additional evidence for lifting. Indeed, this **bootstrapped**[24] LM using pseudo evidence is akin to lifted **relax, compensate and then recover** (RCR), see (Van den Broeck et al., 2012) and Chapter 12 in this book, because messages to clamped variables do not have to be calculated anymore, hence these edges could be deleted. However, whereas RCR first relaxes and then compensates, we just "reduce and re-lift" over the iterations. In any case, our experimental results on MLNs, Ising models, image segmentation and entity resolution show that BLLM can yield considerable efficiency gains compared to standard LM and naive LLM.

We proceed as follows. We start off by reviewing the LM approach from (Kumar and Zilberstein, 2010). We then develop its naive lifted variant. Afterwards, we show how to obtain pseudo evidence and how it can be used to efficiently re-lift the already lifted model. Before discussing the new approach and putting it into the context of the previous sections, we will present our experimental evaluation.

11.6.1   Lifted Likelihood Maximization (LM)

Inspired by the equivalence between MAP inference and solving a linear program, Kumar and Zilberstein place a constraint on the probability distribution in the definition of the marginal polytope $\mathcal{M}(G)$ (we refer to the chapter on lifted variational inference for more details). More precisely, they constrain the original distribution in such a way that the following holds

$$p(\mathbf{x}) = \prod\nolimits_{i=1}^{n} p_i(x_i)$$

and afterwards maximize over the resulting set of mean parameters. This set of parameters defines an inner bound of the marginal polytope. Kumar and Zilberstein show that this maximization problem can be reformulated as likelihood maximization in a finite-mixture

---

[24] Here, we refer to bootstrapping as the capability of self-improvement during the inference stage. In learning settings, bootstrapping also refers to iteratively improved learning where the learner adds previously learned concepts to the training set, e.g. (Morik and Kietz, 1989)

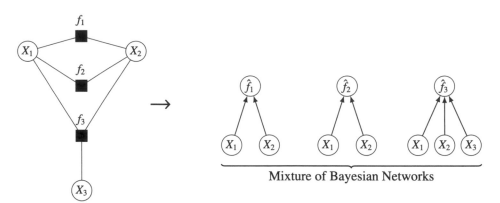

Figure 11.12: Kumar and Zilberstein's LM approach relies on the fact that a factor graph is decomposed into a mixture of Bayesian networks. Here, the factor graph composed of three variables is split into three Bayesian networks, each containing a reward variable $\hat{f}_a$ for the corresponding factor.

of simple **Bayesian networks** (BNs). The hidden variables in the mixture are the X_i in the original Markov random field (MRF). The potentials f_a are incorporated via "binary reward variables" $\hat{f}_a$ and the conditional probability distribution of $\hat{f}_a$ is proportional to f_a, i.e., for every potential f_a in the MRF, there exists a BN with a reward variable and its parents being x_a. Fig.11.12 shows how an example factor graph is transformed into a mixture of BNs. The optimization of this new objective cannot be done using linear programming techniques anymore because the restrictions on the probability distribution are now non-linear. Therefore, Kumar and Zilberstein introduce an EM approach that monotonically increases the lower bound of the MAP. While in the original work only the equations for pairwise MRFs were derived, we will here give the EM message passing equations in an adapted form for arbitrary factor graphs. A message sent from a factor a to one of its neighboring variables i is defined as:

$$\mu_{a \to i}(x_i) = \sum_{\neg\{i\}} f_a(\mathbf{x}_a) \prod_{j \in \mathrm{nb}(a) \backslash \{i\}} p_j(x_j),$$

as before, $\neg\{i\}$ denotes all possible instantiations of the variables in the domain of factor a with variable X_i fixed to x_i and $\mathrm{nb}(a)$ denotes the variables connected to factor i. Compared to the BP message from a factor to a neighboring variable, the LM message looks very similar. However, we do not have a message in the other way around involved but instead solely the current beliefs $p_j(x_j)$ of the marginal probabilities of the neighboring variables. This greatly simplifies the message passing. The M-Step updates these beliefs in every iteration as follows:

$$p_i'(x_i) = \frac{1}{Z_i} p_i(x_i) \sum_{a \in \mathrm{nb}(i)} \mu_{a \to i}(x_i) ,$$

where Z_i is a normalization constant for variable X_i. Again as in BP, or the other message passing algorithms we have seen so far, the process of sending messages is iterated until convergence. In the case of LM, the stopping criterion is defined over the difference of the beliefs in two subsequent iterations. For what follows, it is important to note that a modification of this M-Step is common in order to speed up convergence: the so called *soft greedy M-Step,* originally proposed in (Toussaint et al., 2008), is used to weight the current expectation stronger. Its usefulness has been shown empirically in the literature.

We now derive the formulas for the lifted variant of LM. As we have explained above, lifting can be viewed as a color passing procedure that finds symmetries in a given factor graph by sending color messages. For BP, the idea of the CP algorithm was to find variables that have the same marginal distributions under the message passing algorithm following a parallel update schedule. Similarly, this held for warnings in WP and surveys in SP. For the lifted LM this means that one identifies variables that have the same optimizing distribution in the EM algorithm.

If we want to run the LM algorithm on the lifted model, we have to adapt the messages sent from factors to variables and the M-step, to reflect merged groups of variables and factors, i.e., nodes with the same color after running CP. Again, this is achieved by adding counts to the formulas. We save messages sent from factors to nodes by observing that factors connected to k variables of the same color, only need to calculate a single message instead of k messages because all k messages are identical. With a similar argument, we multiply the incoming messages by k in the M-step if a variable is connected to k factors of the same color. In a lifted representation, the variables would only appear in one single Bayesian network instead of k different nets. Hence, the lifted message equation of the M-Step is:

$$p_i'(\mathfrak{x}_i) = \frac{1}{Z_i} p_i(\mathfrak{x}_i) \sum_{a \in \mathrm{nb}(i)} \mu_{a \to i}(\mathfrak{x}_i) \cdot \mathrm{vc}(a, i), \qquad (11.10)$$

where $\mathfrak{x}_i$ is a clustervariable in the lifted network. As opposed to the ground case, the lifted equation sums over the lifted network ($a \in \mathrm{nb}(i)$) and introduces the count $\mathrm{vc}(a, i)$. This count is the number of times, ground variable X_i is connected to factors with the color of f_a in the ground network. This M-Step has only to be done once for every distinct variable color. This provides an intuitive interpretation of what the equation do in the case of **lifted likelihood maximization** (LLM): LLM can be thought of sorting the Bayesian Networks of the mixture (cf. 11.12) into buckets and each bucket corresponds to a clusterfactor. Initially, every Bayesian Network is assigned to a bucket according to the conditional probability table of $\hat{f}_a$. Then variables compute their color signatures as before. The Bayesian Networks are now put into different buckets depending on the colors of their child nodes.

11.6.2   Bootstrapped Likelihood Maximization (LM)

While lifting is an exact method to speed up message-passing inference, we will now introduce an additional, approximate modification to the LM approach that reduces run times even further. Adjusting the current estimation of a variable's distribution by a small amount should only have a small effect on its direct neighbors and the effect on other variables decays with the distance. For example, such an effect of errors in BP-messages has been analyzed in (Ihler et al., 2005). On the other hand, if we clamp a variable in the graph, we do not have to care about messages passed to that variable anymore and additionally, potentials over this variable can be simplified. On top of that, the influence on the other variables does not necessarily have to be negative, instead these variables can converge faster to their correct maximizing state. We will now describe, how we make use of this observation.

When LM iteratively computes the new beliefs p'_i, the uncertainty about the MAP state decreases over the iterations. Once the probability for one state is above a threshold π, one can assume that it will not change its state in the final MAP assignment anymore. Since LM essentially implements a gradient-based solver for the underlying mathematical MAP program (Kumar and Zilberstein, 2010), being close to one state makes it very unlikely ever turning to a different one. In such cases, we fix the distribution in such a way that all states have zero probability except for the most likely state which is set to one. More formally, if $x_i^* = \text{argmax}_{x_i} p'_i(x_i)$ and $p'_i(x_i^*) > \pi$, we set

$$p'_i(x_i) = \begin{cases} 1, & \text{if } x_i = x_i^*, \\ 0, & \text{otherwise} \end{cases} \tag{11.11}$$

We call such states pseudo evidence (PE) because they do not belong to knowledge that is available beforehand but instead becomes available during the inference. Therefore BLM will refer to the LM approach that adds evidence over the iterations, i.e., **bootstrapped LM**. More importantly, PE has major implications on future message updates.

- It simplifies the MRF because it cancels out states from the potentials.
- It allows skipping messages to these variables because for clamped variables the M-step is obsolete.

Recall that the idea of PE is inspired by the soft greedy rule for the M-Step in LM and also reduces the number of iterations required to converge. On top of that, we can now combine PE and lifting. We can re-lift the network during inference based on this new evidence. Thereby we obtain a more compact lifted network. This is intuitively the case because PE often introduces additional independencies in the graphical model which can be exploited via lifting. However, as we will see below (Fig.11.13), standard CP is not always capable of exploiting all of these newly introduced independencies. Hence, fully exploiting PE for lifting requires an adapted form of CP. Before we explain this bootstrapped lifted LM, we want to give some intuition into the approximative nature of the maximization when PE is present. Essentially there are two errors that can be introduced by PE:

(**A.1**) In rare cases a variable will change its state again, even though its belief was already above the threshold π. Therefore, we also propose a variant of BLM where we do not directly clamp a variable on observing a belief above π for the first time but instead we only mark it for clamping. We clamp the variable if we see that its state remains the same for the following T iterations and proceed as before. To estimate the effect of this lag, we simply count the number of state changes during the iterations for different T. However, as we will show in our experiments below, these state changes occur rarely.

(**A.2**) Clamping a variable can be seen as introducing an error to the messages. However, as shown in (Ihler et al., 2005) for BP, the influence of the error on other variables in the network decays with the distance to the clamped node. Additionally, this influence does not have to be negative. In fact, it can also lead to faster convergence to the correct state. Optimally, one would clamp a variable only if the effect on neighboring variables is negligible or even positive. Determining the exact influence of PE is difficult but we estimate it by comparing the final MAP solution obtained by BLM to the result of standard LM. One can think of more elaborated ways to measure the effect of clamping. For example, comparing the messages exchanged between direct neighbors with and without PE, one can calculate the error made by setting the variable. Inspired by the idea of backtracking, one could even calculate the change in beliefs of the neighboring nodes and omit the clamping if the direction of the gradient changes. However, this requires even more computational effort and dispossesses LM one of its key features, namely its simplicity.

In the experimental section, we will empirically show that issues A1 and A2 do not have critical influence on the quality. Instead, BLM results are of high quality and we hence, do not use more enhanced ways for clamping which requires higher computational demand.

11.6.3   Bootstrapped Lifted Likelihood Maximization (LM)

We now combine PE and lifting. After setting PE, we may now get more opportunities for lifting since variables and factors may fall into the same cluster though being in different ones in earlier iterations. A naive lifting approach takes the ground network, clamps it according to the PE, uses CP or any online variant to lift it, and continues message passing on this new lifted network. Although this approach sounds promising, we will now give an example to illustrate the downside of this naive approach.

Example 11.7 Rows 1-3 in Fig.11.13 depict the naive approach of combining PE and lifting based on CP. The figure assumes a binary, pairwise MRF with identical pairwise potentials, i.e., $f_{ij}(x_i, x_j)$ is the same for all pairs $(i, j) \in E$. The line color of the nodes denotes different unary potentials $f_i(x_i)$. In our example, all unary potentials are identical (red) except for the one associated with the variable in the upper left corner (green). However, for simplicity let us further assume that these priors are not converse but only different in

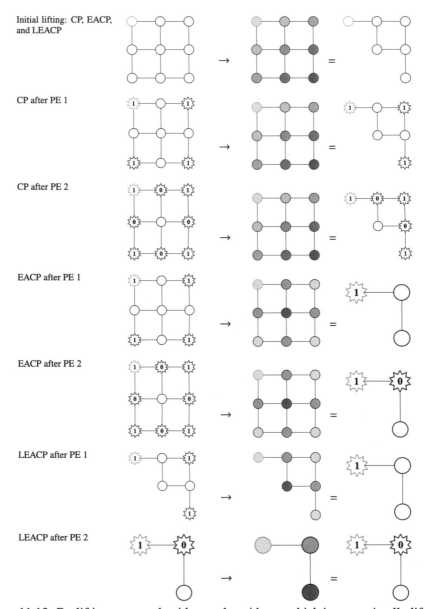

Figure 11.13: Re-lifting a network with pseudo evidence which is *not* optimally lifted via standard Color Passing once pseudo evidence is present.

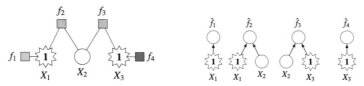

Figure 11.14: Factor graph and corresponding Bayesian netwoork. X_1 and X_3 are clamped. f_1 and f_4 are different (denoted by different colorings of the factor nodes) while f_2 and f_3 are identical (same color of factor nodes).

their parameter strength. For example, green could encode something like $(0.2, 0.8)$ while red encodes $(0.25, 0.75)$. The first row in Fig.11.13 shows the ground network without any PE. After running CP, we get the graph shown in the center column (on the right hand side of the →) with nodes colored by six different gray shades according to CP. This graph can be compressed to the lifted network next to the equality sign (right column). The lifted network contains only six nodes as opposed to nine ground nodes. Now assume that at some point we obtain identical PE for all four corners (depicted as stars instead of circles for the variable nodes). Running CP on the ground network with the new evidence in mind, we obtain the very same lifting as in the previous iteration, namely, a network containing six variables (Fig. 11.13, second row). As we will explain now, this behavior of CP is suboptimal. Since the upper left corner is set, its unary factor does not have any influence on any message being sent in future iterations. This also holds for the other three corners. This means that their unary factors can actually be neglected and all four corner nodes should receive the same color in the final coloring of the graph. We will below show how to modify CP in order to recognize the new symmetries induced by PE. However, let us first finish our running example. In the third row of Fig. 11.13 we assume that all remaining unclamped variables are now being fixed except for the center variable. Again, CP is not able to exploit additional symmetries and the compressed model is simply remains the initial lifting.

   To sum up, the lifting obtained from **color passing** (CP) is not always optimal if evidence is present. To overcome this issue, we will now propose an adapted CP algorithm and we will show how to modify CP to achieve a higher compression due to **edge-deletion** induced by PE. **Edge-deletion adapted CP** (EACP) is similar to the original CP but capable of returning a lifted network that respects PE. We begin by illustrating the underlying idea by an example.

Example 11.8 Assume the factor graph in Fig. 11.14 has identical and symmetric pairwise potentials f_2 and f_3. The unary factors of X_1 and X_3 are different, denoted by the different colors of the factor nodes f_1 and f_2. We also assume that X_1 and X_3 have just been set to 1 and we stick to the notation that variables with PE are denoted as stars. Fig. 11.14 also

---

Algorithm 6 The EACP algorithm also accounts for existing evidence during the color passing and thus finds more compact liftings.

Require: A factor graph fg
Ensure: Variable and factor colors
1: initialize_colors
2: propagate evidence to factors
3: **while** not converged **do**
4:     **for all** factor f_k **do**
5:         **if** free_vars$f_k > 1$ **then**
6:             build_signaturef_k
7:     **end for**
8:     factor_colors = new_newcolors
9:     **for all** variable $X_i \in \mathbf{X}$ **do**
10:         **if** not_setX_i **then**
11:             build_signatureX_i
12:     **end for**
13:     variable_colors = new_newcolors
14: **end while**

---

shows the BNs associated with the factor graph. The message sent from f_2 to X_2 ($\mu_{f_2 \to X_2}$) is calculated as follows:

$$\mu_{f_2 \to X_2}(x_2) = \sum_{x_1} f_2(x_1, x_2) \cdot p(x_1) = f_2(0, x_2) \cdot \underbrace{p(0)}_{=0} + f_2(1, x_2) \cdot \underbrace{p(1)}_{=1} = f_2(1, x_2)$$

One can see that this message is affected by the evidence on X_1. The message reduces to the value of $f_2(1, x_2)$ because the probability of X_1 being in state 0 becomes zero due to the PE, i.e., $p(x_1 = 0) = 0$. Similarly, the message sent from f_3 to X_2 can be simplified to $\mu_{f_3 \to X_2} = f_3(1, x_2)$. In a network without PE, f_2 and f_3 send different messages to X_2 because these messages are dependent on the current beliefs of X_1 and X_3 which are different due to their unary factors. Since the messages are now identical, we can put f_2 and f_3 into the same cluster, i.e., the corresponding BNs can be put into the same bucket. The key insight is that standard CP assigns too many colors, i.e., factors that actually send identical messages are colored differently. CP initializes X_1 and X_3 identically, but f_1 and and f_4 have different colors which are propagated to X_1 and X_3 in the initial color exchange.

Essentially, evidence creates independencies in the underlying factor graph and hence, blocks influence. If there is no flow between two nodes due to evidence, their colors should not be affected by each other. To overcome this problem, EACP differs from CP in several aspects and produces a coloring of the ground factor graph that respects the PE. The pseudo code for EACP is given in Alg.6. In line 2, the evidence of the variables is propagated to

---

Algorithm 7 BLLM

Require: A factor graph *fg*
Ensure: MAP assignment
 1: cfg ← compressfg
 2: **while** not converged **do**
 3:      calc_deltas
 4:      calc_marginals
 5:      clamp_variables
 6:      **if** new bootstrapped evidence **then**
 7:          leacp
 8:          average_marginals
 9: **end whilereturn** calc_map

---

the neighboring factors. This includes pseudo and observational evidence. This step allows us to distinguish identical factors that are connected to variables with different evidence. In turn, it can distinguish f_2 and f_3 in the factor graph of Fig.11.14 if X_1 and X_3 had been set to different states. The other two main modifications compared to CP are as follows. In line 5, factors only create new color signatures if they are connected to more than one unclamped variable. Otherwise they keep their current color. Secondly, in line 9, evidence variables never receive new colors. We now show that EACP indeed returns a correct lifting.

Theorem 11.3 *EACP is sound, i.e., it returns a valid lifting of the ground network.*

Proof   [sketch] All evidence variables that have the same range and are set to the same state keep their colors because their distribution is fixed. They all have the same influence and do not have to be distinguished. Nodes that have not been set behave as in standard Color Passing. Without loss of generality, let us assume pairwise factors but the argument generalizes to the arbitrary case. If one variable of a factor is clamped, messages to the other variable become independent of any of the marginal distributions; they solely depend on the potential of the factor. Hence, two factors with identical potentials receive the same colors initially and do not have to adapt their colors; the messages they send are identical which satisfies the requirement for the same color. By propagating the evidence of the variables before the actual Color Passing, we ensure that factors can be distinguished that are connected to PE variables in different states. □

The main idea is now to replace CP with EACP. As our experiments will demonstrate, considerably higher compression rates are obtainable because we exploit the PE. This is also illustrated in rows four and five in Fig.11.13. Here, EACP obtains better compression than CP after PE is observed. After clamping the corner variables, EACP returns a model containing three variables instead of six. However, we can still do considerably better and become more efficient because EACP operates on the ground network. As we will show now, it can also operate on the current lifted network.

LEACP works similar to EACP but it requires some modifications of the color signatures. We have to distinguish variables that are connected to a different number of factors of the same color. This is exactly the information contained in the counts $vc(a, i)$. LEACP now simulates the ground color signature for a variable X_i by appending $vc(a, i)$ times the color c of a factor f_a, i.e.,

$$\underbrace{\langle c, c, \ldots, c \rangle}_{|\langle \cdot \rangle| = vc(a, i)} .$$

This idea is only correct for lifted networks that have non-fractional counts. This makes it non-trivial to employ other approximate lifting approaches such as informed lifting (Kersting et al., 2010) and early stopping (Singla et al., 2010). To see that LEACP returns a valid lifting, first note that setting PE will never require a cluster in the lifted network to be split. This is simply because all variables in a cluster have the same expectation of their marginal distribution so that we will always set an entire cluster to evidence. In turn, the variables in that cluster will remain to be indistinguishable in all later iterations and we will never have to split them. This is clearly a distinction from the decimation-based framework where usually just a subset of the variables, in the extreme case only a single variable, were clamped. This essentially proves the soundness of LEACP.

Theorem 11.4 *LEACP is sound, i.e., it returns a valid re-lifting of the current lifting containing additional evidence.*

Proof   Every partition is valid for all following iterations. A partition remains valid after setting evidence because we always set an entire cluster, i.e., all ground variables in a cluster are set to the same state. This means that we never require more colors than we have clusternodes in the current lifting. Only merging two clusternodes is a possible action which reduces the number of required colors. We can merge two clusters $\mathfrak{X}_i$ and $\mathfrak{X}_j$ iff all corresponding ground variables behave the same in the ground network. Similar to the BP message passing in the lifted network, we now have to simulate the color signatures in the case of Color Passing. Here, the counts come into play which then produces the same color signature as in the ground case. Hence, via Color Passing on the lifted graph and adapting the color signatures by the counts, we obtain the same color signatures as if we were running CP on the ground graph. $\square$

Putting all pieces together, we obtain **bootstrapped lifted likelihood maximization** (BLLM) as summarized in Alg. 7. BLLM exploits all enhancements we have introduced above. It uses standard CP to obtain an initial lifting (line 1) for a factor graph without PE. It then runs on the lifted level until PE is encountered (lines 2-8). In cases, where PE is set, it uses LEACP to re-lift the model on the lifted level (line 7).

To complete the discussion of BLLM, we have to explain one final issue. Indeed, LEACP returns a valid lifting. However, when two clusternodes are merged, they may have different beliefs about the current estimation of the marginals. Hence, we need to calculate a common marginal in the re-lifted network to continue with the iterative process. Reconsider Fig. 11.13. When the PE for all four corner variables is encountered, their neighboring variables have different marginals but BLLM puts them together into one cluster. We propose to to simply take the average of the marginals of the involved variables to make use of the calculations of previous iterations (line 8 in Alg.7). Intuitively, we do not initialize the variables again since we already know their tendency and they should get same beliefs in the end. Although this is not exact, it does not sacrifice the estimation quality as we will validate in our experimental results below and it also justifies the re-lifting. On top of that, it is also likely to reduce the variance of the estimates by this averaging. BLLM saves messages compared to the ground inference by avoiding the calculation of indistinguishable messages. Additionally, LEACP is more efficient than EACP and CP because it solely operates on the lifted network to obtain the lifting for the following iterations. That is we never have to return to the ground again after the initial lifting. This is also illustrated in rows six and seven in Fig. 11.13.

Finally, we want to comment on the time complexity of EACP and LEACP. The idea of bootstrapped lifting only works if the repeated call to lifting is efficient. Since (L)EACP involves only linear-time modifications of CP, each of their iterations also runs in linear time.

11.6.4   Experimental Evaluation

Here, we investigate the question whether our bootstrapped and lifted LM variants can be considerably faster than the baselines without decreasing performance. We will answer the following main questions: **(Q6)** Does LM benefit from PE? **(Q7)** Can PE combined with lifting speed up LM even further? **(Q8)** How does the quality of B(L)LM's results compare to standard LM?

To do so, we implemented the following set of algorithms: LM, the propositional approach by Kumar and Zilberstein; LLM, the naively lifted LM using only standard Color Passing; BLM, Bootstrapped LM using PE but no lifting; BLLM, the bootstrapped lifted LM. We then evaluated them on lifting benign (friendly) and malignant (unfriendly) models. In particular, we consider synthetic Ising grids and relational models, as well as real-world models for simple image segmentation and relational entity resolution.

Before we present the quantitative and qualitative results, we give some further technical notes on the implementation and the parameters used. Since clamping a single variable is unlikely to achieve significant gains in lifting, we re-lift after we have obtained PE for a batch of $b\%$ of the original number of variables. In our experiments we set b to 10. We stop the algorithm after 2,000 iterations if the algorithm does not converge earlier. The other parameters are set to $\pi = 0.9$ (PE-threshold) and $T = 0$ (lag-parameter).

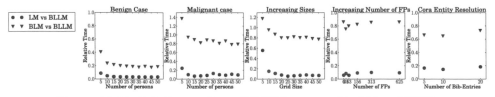

Figure 11.15: Relative run time of BLLM compared to LM and BLM.

11.6.4.1   Smokers-and-Friends-MLN   For our first experiment, we generated MLNs of varying domain sizes, i.e., for an increasing number of persons (see e.g. (Richardson and Domingos, 2006) for more details on the S&F-MLN). More precisely, this means that we increased the domain size of the person predicate in the MLN from 5 up to 50. Resulting in factor graphs with up to 2,600 variables, 5,510 factors, and 10,150 edges. We call this setting the *benign* S&F-MLN because it does not contain any evidence and hence is well suited for lifting. The first plot in Fig.11.15 shows that BLLM only requires a fraction of the time compared to LM (red circles). Additionally, the blue triangles highlight that lifting reduces the running required by BLM significantly. For a clearer picture, we have omitted the running times of LLM. But for this problem, LLM outperforms BLM while BLLM is the fastest method among all. This is not surprising, as this scenario is very well suited for lifting.

We now add 25% random evidence to the MLNs by randomly choosing predicates and their state. We call this setting the *malignant* S&F-MLN because the random evidence prevents CP from lifting the network initially. The results are depicted in the second plot in Fig.11.15. Again, BLLM only requires a fraction of LM's running time. But compared to the benign case, lifting BLM does not help as much as before. Nevertheless, lifting is still beneficial for the larger problems and BLLM is capable of exploiting symmetries in models for which standard CP resorts to the ground network. We see that lifting is an overhead in the small problem instances but as the problems become larger, BLLM becomes faster than BLM, exploiting symmetries obtained from pseudo evidence. For these malignant MLN problems, LLM basically reduces to the ground LM case as the evidence destroys initial symmetries. We now also want to compare the quality of the obtained MAP configurations. To this end, we computed the log-likelihood score of the obtained assignments. As shown in Tab.11.16, all solutions obtained by BLLM are qualitatively as good as the results produced by LM.

11.6.4.2   Ising Models   The next set of experiments were conducted on Ising grids. Ising grids are pairwise MRFs with $x_i \in \{-1, +1\}$ and

$$p(\mathbf{x}) = \frac{1}{Z} \prod_{i \in V} f_i(x_i) \prod_{(i,j) \in E} f_{ij}(x_i, x_j)$$

	S&F-MLNs Benign Case									
	5	10	15	20	25	30	35	40	45	50
(L)LM	103.5	362.0	775.5	1344.0	2067.5	2946.0	3979.5	5168.0	6511.5	8010.0
B(L)LM	**103.5**	**362.0**	**775.5**	**1344.0**	**2067.5**	**2946.0**	**3979.5**	**5168.0**	**6511.5**	**8010.0**
	S&F-MLNs Malignant Case									
	5	10	15	20	25	30	35	40	45	50
(L)LM	85.2	333.0	687.3	1234.4	1854.9	2682.3	3655.6	4744.0	5895.6	7251.8
B(L)LM	**85.2**	**333.0**	**687.3**	**1234.4**	**1854.9**	**2682.3**	**3655.6**	**4744.0**	**5895.6**	**7251.8**
	Ising-Grids of Varying Sizes									
	5	10	15	20	25	30	35	40	45	50
(L)LM	73.1	309.1	499.1	892.0	1991.6	2447.8	3095.0	3227.0	5391.7	5480.4
B(L)LM	**72.8**	**294.4**	**481.4**	**859.4**	**1913.0**	**2352.4**	**2997.2**	**3118.9**	**5201.2**	**5248.8**
	Ising-Grids of Varying Field Parameters									
	6	31	63	156	313	625				
(L)LM	1693.9	1808.8	1991.6	1566.9	1093.0	1748.1				
B(L)LM	**1663.0**	**1739.8**	**1913.0**	**1520.4**	**1061.5**	**1681.8**				

Figure 11.16: The log-scores (the higher, the better) show that the B(L)LM versions of the algorithms achieve high quality results. Bold log-scores are in 95% of the standard LM score.

The unary potentials are $f(x_i) = \exp(x_i w_i)$ where w_i is called the *field parameter* and is drawn from $[-1, 1]$. Pairwise potentials are defined as $f(x_i, x_j) = \exp(x_i x_j w_{ij})$ with the *interaction parameter* w_{ij}. We call the grid *attractive* if all $w_{ij} > 0$. We consider different Ising models in the classical sense which have a single interaction parameter w_{ij} only. Motivated by applications in image processing, we use attractive interaction parameters and a limited set of field parameters instead off drawing all w_i randomly.

The first problem set consists of grids of increasing size. We generated grids from 5×5 up to 50×50 with a fixed ratio of different field parameters equal to 10%. We generated ten grids for every size and averaged the results over these runs. The results are shown in the third plot in Fig.11.15. BLLM is by far the fastest method and lifting of BLM achieves around 20% speed up on the larger instances.

The second experiment keeps a fixed grid size of 25×25 and varies the number of field parameters. For every grid, we generated a set of field parameters beforehand. The size of this set is chosen relative to the number of variables in the grid. The fourth plot in Fig.11.15 again shows that BLLM is the fastest method and requires less than 10% of LM's running time. Similar to the malignant S&F-MLN, lifting can here speed up BLM only little. This can be explained by the fact that the effects of unary factors vanish once a variable is set and pairwise potentials with only a single free variable can be clustered for all distinct values of the clamped variables. On the other hand, standard lifting does not help in this scenario and requires the same time as the ground variant plus an small overhead due to the

Figure 11.17: Input and output for the image segmentation task. In summary, BLLM outperforms all other variants in terms of run time while achieving accurate results.

lifting at the beginning. Tab.11.16 contains average log-scores for a qualitative comparison of both experiments on Ising grids. The results of BLLM are close to LM's solutions and always above 95% of the LM score.

   To obtain another qualitative measure, we compared the log-scores of LM and BLM with max-product BP on different grids. One observes that the scores of LM and BLM are often close to the ones obtained by max-product. For example, on a grid with 156 field parameters, we have LM and BLM both with log-scores of 391 while max-product achieved a score of 393. However, when the strength of the interaction parameters increases, we observe that the quality of LM solutions decrease compared to max-product, e.g., we obtain a log-score of 2,544 for LM and 3,162 for max-product BP (LM is not above 95% of the BP-score anymore). A similar observation has already been made in the original work. Kumar and Zilberstein state that the LM approach can tend to poor performance in cases where the gap between the minimum and maximum reward is large.

   In general, these random grids are not well suited for exact lifted inference as there is no structural symmetry. However, BLLM is still able to perform better than BLM. This clearly shows the benefits of exploiting PE and shows that lifting can become beneficial even in settings that are adversarial towards lifting in general.

11.6.4.3   Image Segmentation   Many tasks in computer vision can be encoded as binary pairwise grids, e.g., simple forms of foreground/background classification. To show that the BLLM algorithm actually results in high quality solutions on real world problems, we ran our algorithm on an instance of an image segmentation task. We have used Ising models in the previous experiment already in a set of synthetic experiments. Now, we will use such binary discrete MRFs to label pixels as foreground or background in an image. Having variable $x_i = -1$ assigns the corresponding pixel to the background and correspondingly, for $x_i = 1$ the pixel belongs to the foreground. As described in (Mooij, 2008), the unary factors $f_i(\cdot)$ are based on the difference of two video camera images and defined as follows:

$$f_i(x_i) = \exp(x_i w_i) = \exp\left(x_i w \tanh \frac{h_i - c}{r}\right)$$

Here, h_i is the absolute difference in hue of pixel i in the two images. The other parameters, i.e., w, c, and r can be fined tuned based on a training set. This function penalizes a large

Cora Entity Resolution		
5	10	20
(L)LM	0.98	-103.6	-689.2
B(L)LM	**0.98**	-116.5	**-702.8**

Figure 11.18: The log-scores on the Cora dataset (the higher, the better) also show that the B(L)LM versions of the algorithms achieve high quality results on a real-world problem dataset. Bold log-scores are in 95% of the standard LM score.

difference in the input and the reference image. The pairwise factors $f_{ij}(\cdot)$ encode the intuition that neighboring pixels are more likely to belong to the same part of the image:

$$f_{ij}(x_i, x_j) = \exp(x_i x_j w_{ij})$$

with $w_{ij} > 0$. These pairwise factors are defined as in attractive Ising models and act as a smoothing function. We used the scripts with their default parameters contained in libDAI (Mooij, 2010) to generate a factor graph for this task and we also used their example images. The inputs were the first two images depicted in Fig.11.17. The script calculates the difference between both images to define the unary factors as described above. The potentials of the pairwise factors are defined such that neighboring pixels are more likely to belong to the same segment. The third and fourth image in Fig.11.17 show the results after running inference (LM on the left, BLLM on the right). The images show that the result obtained by BLLM is almost as good as the one obtained by LM. One has to look very carefully to see differences which visually validates the PE based approach. In fact, the score achieved by BLLM is only slightly lower compared to LM.

11.6.4.4   Cora   We conduct our final experiments on the Cora entity resolution dataset. We obtained the datasets from the ALCHEMY repository and also used ALCHEMY for weight learning. We did not evaluate on the entire dataset since subsets are already sufficient to highlight our results. Instead we randomly sampled k bib-entries and added all predicates describing their properties to the evidence. We ran experiments for $k - 5$, 10, and 20. The largest problem instance contains 1,110 variables, 32,965 factors, and 71,209 edges. The results in Fig.11.15, 5th plot, show that BLLM clearly outperforms LM in terms of runtime, while the scores are very similar (see Tab.11.18). Here, lifting achieves good performance increases over BLM again. This also holds for the naive LLM which is again omitted in the plot for clearness. One should not get confused about the negative values for the datasets with 10 and 20 bibliographic entries. This is due to the fact that the Cora-MLN has positive and negative weights. If not sufficiently many clauses with positive weights are satisfied, the sum of the log-weights becomes negative.

We also analyzed the behavior of BLM on the Cora problem with respect to the nature of the approximation as motivated in A1 and A2 above (see Sec.11.6.2). We have observed

Figure 11.19: If we assume that unary and pairwise factors are identical, this fully connected graph is compressed to a single node if Color Passing is applied.

that state changes, as described for A1, during the iterations almost never occur. One has to set the lag-parameter $T > 50$ to observe any state changes at all. Similarly, as one can already expect from the log-scores, the approximate MAP solutions are very similar. The differences seem to result from variables changing their states only but not due to wrong influence on neighboring variables that results form setting PE.

The experimental results clearly show that LM is boosted by lifting and PE can indeed considerably reduces run times without sacrificing accuracy. Moreover, BLLM scales well and is most beneficial on relational models. Nevertheless, BLLM can also improve state-of-the-art on propositional models that so far were challenging for lifted inference approaches. To summarize, questions (Q6-8) can all answered in favor of BLLM.

11.6.5   Summary of Lifted Likelihood Maximization for MAP Inference

We introduced the idea of pseudo evidence (PE) for MAP inference. We showed that it can considerably speed up MAP inference and provides novel structure for optimization. Specifically, we bootstrapped lifting using PE and empirically demonstrated that this "reduce and re-lift" concept can decrease run time even further and achieve lifting in situations where standard lifting fails completely.

The approach exploits additional symmetries by adding evidence during the optimization. Essentially, we try to fix broken symmetries by PE. The fact that evidence breaks symmetries is well known and other authors have dealt with that issue as well. For example, Van den Broeck and Darwiche approximate evidence by low-rank boolean matrix factorization to speed up lifted inference. It is shown how a large class of binary relations can be encoded with the help of unary relations only. More precisely, the unary relations are represented by boolean vectors and their product represents the original binary relation. This vector-product can be extended to the matrix-product case as well. Based upon this insight, approximate boolean matrix factorization is used to find a decomposition of the binary evidence relation. Instead of a decomposition, our approach iteratively adds evidence and could be combined with a decomposition approach to further compress the evidence, including the pseudo evidence.

It is important to note that LLM and BLLM currently rely on the parallel protocol for message updates. We have seen this requirement for the other lifted algorithms as well and we cannot run asynchronous updates because the Color Passing schedule relies on parallel updates. This restriction prevents us from solving problems such as the one depicted in Fig.11.19 correctly. Think of a fully connected pairwise MRF with all unary and pairwise potentials being identical. We further assume distractive pairwise potentials with $w_{ij} < 0$. Any of the CP algorithms will reduce the graph to one node. If we now run LM on this graph, we will end up with marginals $(0.5, 0.5)$. An optimal assignment would consists of any combination with the states not being identical, e.g., $(-1, 1, 1)$. We cannot read off any optimal assignment from our calculated marginals, although the marginals are computed correctly in terms of the underlying model. Besides using a decimation approach based on conditioning, one heuristic would be to add small noise to every unary potential. In this case, however, we could not lift the model anymore. On the other hand, the parallel update protocol results in satisfying solutions in many applications and also allows easy parallelization.

Another limitation induced by the lifting is the fact that we do not have the complete freedom to randomly initialize the EM algorithm. Similar to the message passing algorithms for satisfiability solving, e.g., SP in Sec.11.4.6, LM can benefit from random initialization. However, in lifted LM all variables have identical message updates that fall into the same cluster, i.e., we would initialize all these variables identically. We additionally rely on the fact that we always set all variables in one cluster as PE. If clusters were only partially clamped, we are back in the decimation framework and we have to use methods such as the sequential clamping described in Sec.11.5.

We used only exact CP in our experiments and Lifted EACP also provided an exact extension for BLLM. Combining BLLM with informed lifting approaches described earlier is an interesting avenue for future work. At the beginning, these algorithms will be more efficient than BLLM but once evidence has been set, only BLLM can exploit the additional symmetries. It would also be interesting, to use approximate lifting techniques, such as early stopping. However, as mentioned above, this is not obvious to implement due to fractional counts that can arise. Fractional counts prevent us from using the LEACP algorithm for lifting because signatures cannot be copied in the current form.

After PE has been set, it is also possible to further analyze the factors. Some configurations are now impossible and hence factor can be reduced. This might also introduce new symmetries on the factor level which can be exploited in the upcoming iterations. Similarly, setting PE can introduces conditional independencies between variables. In the extreme case, this can result in variables for which the entire Markov blanket is set. It is not necessary anymore to update these variables in future iterations and one can therefore immediately clamp these messages as well based on the Markov blanket. Additionally, one can also imagine to calculate the conditional probabilities exactly for other small sets of variables that are independent from the rest of the network and for which this calculation is feasible. This is a common pattern in many approximate inference algorithms to increase the quality of the results. While these two adjustments focus more on the implementation of each iteration, there are other, more general, additional research questions. For example, one should also consider connecting PE to the use of extreme probabilities used for lifting (Choi and Amir, 2012). In general, exploring PE for other inference and learning approaches is highly interesting.

11.7   Conclusions

This chapter has reviewed lifted message passing approaches, from belief propagation over survey propagation to likelihood maximization. The idea of moving beyond probabilistic queries and combing them with combinatorial queries fits well view that Statistical Relational AI covers the whole AI spectrum. Here, the information at hand often changes and naive approaches that re-lift every time from scratch often completely cancel the benefits of lifting. Resulting from these observations, we have presented alternative lifting approaches for the online settings. The lifted message passing algorithms presented in this chapter ran some of form of the CP algorithm, hence, the implementation of CP and the number of iterations influence the run time of those algorithms strongly. Actually, there is close connection between CP and computing the so-called **coarsest equitable partition** (CEP), see e.g. (Kersting et al., 2015) but also the corresponding chapter in this book: partition obtained by CP amounts to the CEP. CEPs are frequently computed in algorithms for graph isomorphism testing and therefore have been studied for a long time. This research has resulted in highly efficient implementations for CEP computations. All lifted message passing algorithms in this chapter can benefit from more efficient lifting algorithm. Nevertheless, the fact that LBP, LWP, LSP, and LLM all require an adaption of the "ground" algorithm represents one common drawback of these lifted message passing algorithms. The introduction of different counts and explicit positions make the algorithms more complex and to some extend cumbersome to implement. More elegant lifted inference approaches only preprocess the data and then run an unmodified algorithm on the lifted model. For example, Mladenov et al. (2014) show how to avoid the adaption of the message passing algorithm for lifted MAP inference based on linear program relaxations in pairwise MRFs. Their approach follows the idea of a reparametrized model which is used instead of the original MRF and is covered in another chapter of the present volume.

The approaches we have used in this chapter all relied on exact lifting. First, initial factors were only clustered if there potentials were identical based on a syntactical comparison. Second, variables and factors were only clustered if their color signatures exactly matched. To show the complete picture of lifted inference, one should actually employ approximate lifting algorithms. We can argue that such algorithms will work better than standard lifting in several cases as it has been shown previously in the literature. For example, grids with identical attractive interaction parameters and field parameters sampled from a small range. While exact CP will not recognize that the marginals are possibly very similar, informed lifting would still achieve lifting. This also holds for LEACP in the case of BLLM. Still, the underlying ideas are rather orthogonal, but it is an interesting avenue to combine both.

All of the experiments in this section focused on binary MRFs which included CNFs, grounded MLNs, and Ising grids for image segmentation, among others. However, non-binary problems, such as Potts models, are of great interest as well. We will see in the next chapter how lifting is applied to a labeling tasks with finite discrete random variables that have a large but finite domain. Motivated by such applications, generalizing BLLM to MRFs with non-binary variables is an interesting avenue and the restriction to binary variables is currently rather of technical nature. In general, BLLM is not limited to the binary case and CP works with finite ranges anyhow. Following up this research question, it is natural to ask for lifting approaches in infinite domains or hybrid models as well. Besides the work on lifted Gaussian BP or the lifted variational inference for hybrid models in (Choi and Amir, 2012).

12 Lifted Generalized Belief Propagation: Relax, Compensate and Recover

Guy Van den Broeck, Arthur Choi, and Adnan Darwiche

Abstract. This chapter proposes an approach to lifted approximate inference for statistical relational models, that is based on performing exact lifted inference in a simplified first-order model, which is found by relaxing first-order constraints, and then compensating for the relaxation. These simplified models can be incrementally improved by carefully recovering constraints that have been relaxed, also at the first-order level. This leads to a spectrum of approximations, with lifted belief propagation on one end, and exact lifted inference on the other. We discuss how relaxation, compensation, and recovery can be performed, all at the first-order level, and show empirically that our approach substantially improves on the approximations of both propositional solvers and lifted belief propagation.

12.1 Introduction

This chapter proposes an approach to *approximate* lifted inference that is based on performing *exact* lifted inference in a simplified first-order model. Namely, we simplify the structure of a first-order model until it is amenable to exact lifted inference, by relaxing first-order equivalence constraints in the model. Relaxing equivalence constraints ignores (many) dependencies between random variables, so we compensate for this relaxation by restoring a weaker notion of equivalence, in a lifted way. We then incrementally improve this approximation by recovering first-order equivalence constraints back into the model.

In fact, our proposal corresponds to an approach to approximate inference, called **relax, compensate and then recover** (RCR) for (ground) probabilistic graphical models (Choi and Darwiche, 2011).[25] For such models, the RCR framework gives rise to a spectrum of approximations, with iterative belief propagation on one end (when we use the coarsest possible model), and exact inference on the other (when we use the original model). We show how relaxations, compensations and recovery can all be performed at the first-order

---

[25] A solver based on the RCR framework won first place in two categories evaluated at the UAI'10 approximate inference challenge, in the most demanding sub-category of 20-second response time (Elidan and Globerson, 2010).

level, giving rise to a *spectrum of first-order approximations*, with lifted first-order belief propagation on one end, and exact lifted inference on the other.

The propositional RCR framework operates on probabilistic graphical models. In order to motivate our Lifted RCR algorithm, we first adapt RCR to work with propositional, or ground Markov logic networks in Section 12.2. A Lifted RCR algorithm that operates on first-order MLNs is presented in Section 12.3. An essential component in the Lifted RCR framework is an algorithm that partitions equiprobable equivalences. Section 12.4 discusses this problem in more detail. Section 12.5 discusses related work on approximate lifted inference. It identifies propositional inference algorithms whose approximations are in the spectrum of Lifted RCR, as well as existing lifted inference algorithms that fit into this framework. In Section 12.6, we evaluate our approach on benchmarks from the lifted inference literature. Experiments indicate that recovering a small number of first-order equivalences can improve on the approximations of lifted belief propagation by several orders of magnitude. We show that, compared to Ground RCR, Lifted RCR can recover many more equivalences in the same amount of time, leading to better approximations.

12.2 RCR for Ground MLNs

Although our main objective is to present a lifted version of the RCR framework, we start by adapting RCR to ground Markov logic networks (MLNs). This is meant to both motivate the specifics of Lifted RCR and to provide a basis for its semantics (i.e., the correctness of Lifted RCR will be against the results of Ground RCR).

In addition to the standard syntax and semantics of MLNs (Domingos and Lowd, 2009), we will here further assume that all logical variables in the MLN are free. This can always be achieved by transforming existential quantifiers into disjunctions and universal quantifiers into conjunctions. As a consequence of this assumption, all instances of MLN formulas are ground. The grounding of an entire MLN Δ can simply be obtained by replacing each formula in Δ with all its instances (using the same weight). The *ground distribution* of MLN Δ is then the probability distribution induced by the grounding of Δ.

Example 12.1 As a running example, consider the following MLN due to Singla and Domingos (2008).

$$1.5 \quad \text{Smokes}(x) \wedge \text{Friends}(x, y) \Rightarrow \text{Smokes}(y) \qquad (12.1)$$

$$1.3 \quad \text{Smokes}(x) \Rightarrow \text{Cancer}(x) \qquad (12.2)$$

For the domain {A, B}, this first-order MLN represents the ground MLN

$$1.5 \quad \text{Smokes}(A) \wedge \text{Friends}(A, A) \Rightarrow \text{Smokes}(A)$$

$$1.5 \quad \text{Smokes}(A) \wedge \text{Friends}(A, B) \Rightarrow \text{Smokes}(B)$$

$$1.5 \quad \text{Smokes}(B) \wedge \text{Friends}(B, A) \Rightarrow \text{Smokes}(A)$$

$$1.5 \quad \mathsf{Smokes(B)} \wedge \mathsf{Friends(B, B)} \Rightarrow \mathsf{Smokes(B)}$$
$$1.3 \quad \mathsf{Smokes(A)} \Rightarrow \mathsf{Cancer(A)} \tag{12.3}$$
$$1.3 \quad \mathsf{Smokes(B)} \Rightarrow \mathsf{Cancer(B)}$$

The RCR algorithm can be understood in terms of three steps: Relaxation (R), Compensation (C), and Recovery (R). Next, we examine each of these steps and how they apply to ground MLNs.

12.2.1 Ground Relaxation

Relaxation is the process of ignoring interactions between the formulas of a ground MLN. An interaction takes place when the same ground atom a_g appears in more than one ground formula in an MLN. We can ignore this interaction via a two step process. First, we rename one occurrence of a_g into, say, a_g', through a process that we call *cloning*. We then assert an equivalence constraint between the original ground atom a_g and its clone, $a_g \Leftrightarrow a_g'$. At this point, we can ignore the interaction by simply dropping the equivalence constraint, through a process that we call *relaxation*. Bringing back the equivalence is known as *recovery* and will be discussed in more detail later.

Example 12.2 The Smokes(A) atom in Formula 12.3 leads to an interaction between this formula and some of the other five formulas in the ground MLN. To ignore this interaction, we first rename this atom occurrence into $\mathsf{Smokes_2(A)}$ leading to the modified formula

$$1.3 \quad \mathsf{Smokes_2(A)} \Rightarrow \mathsf{Cancer(A)} \tag{12.4}$$

which replaces Formula 12.3 in the MLN. The corresponding equivalence is

$$\mathsf{Smokes_2(A)} \Leftrightarrow \mathsf{Smokes(A)} \tag{12.5}$$

Dropping this equivalence constraint amounts to removing the interaction between Formula 12.4 and the rest of the MLN.

12.2.2 Ground Compensation

When relaxing an equivalence constraint $a_g \Leftrightarrow a_g'$, the connection between the ground atoms a_g and a_g' is lost. One can *compensate* for this loss by adding two weighted atoms

$$w \quad a_g \qquad \text{and} \qquad w' \quad a_g'.$$

If the weights w and w' are chosen carefully, one can reestablish a weaker connection between the ground atoms. For example, one can choose these weights to ensure that the ground atoms have the same probability, establishing a weaker notion of equivalence.

We will now suggest a specific compensation scheme based on a key result from Choi and Darwiche (2006). Suppose that we relax a single equivalence constraint, $a_g \Leftrightarrow a_g'$, which splits the MLN into two disconnected components, one containing atom a_g and another containing atom a'_g. Suppose further that we choose the compensations w and w' such that

$$\Pr(a_g) = \Pr(a'_g) = \frac{e^{w+w'}}{1 + e^{w+w'}}. \tag{12.6}$$

We now have a number of guarantees. First, the resulting MLN will yield exact results when computing the probability of any ground atom. Second, the compensations w and w' can be identified by finding a fixed point for the following equations:

$$w_{i+1} = \log\left(\Pr{}_i(a'_g)\right) - \log\left(\Pr{}_i(\neg a'_g)\right) - w'_i$$
$$w'_{i+1} = \log\left(\Pr{}_i(a_g)\right) - \log\left(\Pr{}_i(\neg a_g)\right) - w_i. \tag{12.7}$$

Following Choi and Darwiche (2006), we will seek compensations using these update equations even when the relaxed equivalence constraint does not disconnect the MLN, and even when relaxing multiple equivalence constraints. In this more general case, a fixed-point to the above equations will still guarantee the weak equivalence given in Equation 12.6. However, when computing the probabilities of ground atoms, we will only get approximations instead of exact results.[26]

Searching for compensations using Equations 12.7 will lead to the generation of a sequence of MLNs that differ only on the weights of atoms added during the compensation process. The first MLN in this sequence is obtained by using zero weights for all compensating atoms, leading to an initial ground distribution $\Pr_0$. Each application of Equations 12.7 will then lead to a new MLN (with new compensations) and, hence, a new ground distribution, $\Pr_{i+1}$. Upon convergence, the resulting MLN and its ground distribution will then be used for answering queries. This is typically done using an exact inference algorithm as one usually relaxes enough equivalence constraints to make the ground MLN amenable to exact inference. Note that applying Equations 12.7 also requires exact inference, as one must compute the probabilities $\Pr_i(a_g)$ and $\Pr_i(a'_g)$.

12.2.3   Ground Recovery

Now that we can relax equivalences and compensate for their loss, the remaining question is which equivalences to relax. In general, deciding which equivalences to relax is hard, because it requires inference in the original model, which is intractable. Instead, Choi and Darwiche (2006) take the approach of relaxing every equivalence constraint and then incrementally recovering them as time and exact inference allow.

It follows from their results that when (i) relaxing all equivalence constraints, (ii) using the above compensation scheme and (iii) doing exact inference in the approximate model,

---

[26] In this more general case, there is no longer a guarantee that Equations 12.7 will converge to a fixed point. Convergence can be improved by using damping in Equations 12.7. Damping sets the new parameters to a weighted sum of the parameters from the previous iteration, and the updates as defined in Equations 12.7.

the approximate marginals found correspond to the approximations found by iterative **belief propagation** (IBP) (Pearl, 1988; Yedidia et al., 2003). The connection to IBP is even stronger: the compensating weights computed in each iteration of Equations 12.7 exactly correspond to the messages passed by IBP.

Several heuristics have been proposed to decide which equivalences to recover, by performing inference in the relaxed model. We will work with the *residual recovery* heuristic (Choi and Darwiche, 2011). It is based on the practical observation that when IBP converges easily, the quality of its approximation is high. The heuristic tries to recover those edges whose convergence was most difficult in previous compensation steps. It recovers the equivalence that least satisfies Equation 12.6 after the compensation algorithm has converged. This is measured by taking the three-way symmetric KL divergence between the three terms of Equation 12.6.

12.3   Lifted RCR

We now introduce a lifted version of the relax, compensate and recover framework, which is meant to operate directly on first-order MLNs without having to ground them. **Lifted RCR** is based on first-order relaxation, compensation and recovery.

12.3.1   First-order Relaxation

We begin with a first-order notion of relaxation where the goal is to ignore interactions between ground MLN formulas, yet without having to fully ground the MLN. This requires a first-order version of atom cloning and first-order equivalences.

Definition 12.1 (first-order cloning)   *Cloning an atom occurrence in an MLN formula amounts to renaming the atom by concatenating its predicate with (i) an identifier of the formula, (ii) an identifier of the occurrence of the atom within the formula, and (iii) the logical variables appearing in the atom's formula.*

Example 12.3   The first-order cloning of the atom occurrence $\mathsf{Smokes}(y)$ in Formula 12.1 gives

$$1.5 \quad \mathsf{Smokes}(x) \wedge \mathsf{Friends}(x, y) \Rightarrow \mathsf{Smokes}_{1b\langle x,y \rangle}(y) \tag{12.8}$$

Here, 1 is an identifier of the formula, b is an identifier of the atom occurrence in the formula, and $\langle x, y \rangle$ are the logical variables appearing in the formula.

As in the ground case, each first-order cloning is associated with a corresponding equivalence between the original atom and its clone, except that the equivalence is first-order in this case.

Example 12.4   The first-order cloning of atom occurrence

$$\mathsf{Smokes}(y) \qquad \text{into} \qquad \mathsf{Smokes}_{1b\langle x,y \rangle}(y)$$

in the example above leads to introducing the following first-order equivalence:

$$\text{Smokes}(y) \Leftrightarrow \text{Smokes}_{1b\langle x,y\rangle}(y) \tag{12.9}$$

Let us now consider the groundings of Formulas 12.8 and 12.9, assuming a domain $\{A, B\}$:

1.5 $\text{Smokes}(A) \wedge \text{Friends}(A, A) \Rightarrow \text{Smokes}_{1b\langle A,A\rangle}(A)$

1.5 $\text{Smokes}(A) \wedge \text{Friends}(A, B) \Rightarrow \text{Smokes}_{1b\langle A,B\rangle}(B)$

1.5 $\text{Smokes}(B) \wedge \text{Friends}(B, A) \Rightarrow \text{Smokes}_{1b\langle B,A\rangle}(A)$

1.5 $\text{Smokes}(B) \wedge \text{Friends}(B, B) \Rightarrow \text{Smokes}_{1b\langle B,q\rangle}(B)$

$\qquad\;\;\text{Smokes}(A) \Leftrightarrow \text{Smokes}_{1b\langle A,A\rangle}(A)$

$\qquad\;\;\text{Smokes}(B) \Leftrightarrow \text{Smokes}_{1b\langle A,B\rangle}(B)$

$\qquad\;\;\text{Smokes}(A) \Leftrightarrow \text{Smokes}_{1b\langle B,A\rangle}(A)$

$\qquad\;\;\text{Smokes}(B) \Leftrightarrow \text{Smokes}_{1b\langle B,B\rangle}(B)$

We have a few observations on the proposed cloning and relaxation techniques. First, the four groundings of Formula 12.8 contain distinct groundings of the clone $\text{Smokes}_{1b\langle x,y\rangle}(y)$. Second, if we relax the equivalence in Formula 12.9, the ground instances of Formula 12.8 will no longer interact through the clone $\text{Smokes}_{1b\langle x,y\rangle}(y)$. Third, if we did not append the logical variables $\langle x, y\rangle$ during the cloning process, the previous statement would no longer be correct. In particular, without appending logical variables, the four groundings of Formula 12.8 would have contained only the two distinct clone groundings, $\text{Smokes}_{1b}(A)$ and $\text{Smokes}_{1b}(B)$. This would lead to continued interactions between the four groundings of Formula 12.8.

Removing all interactions among groundings of the same formula is necessary for the following reasoning. Consider an MLN with the single formula,

$$w \quad \text{Friends}(x, y) \wedge \text{Friends}(y, z) \Rightarrow \text{Friends}(x, z)$$

If we clone all atom occurrences, yet without appending logical variables, and then relax all equivalences, we would obtain the MLN $(w, \text{Friends}_a(x, y) \wedge \text{Friends}_b(y, z) \Rightarrow \text{Friends}_c(x, z))$, which still has many interacting random variables at the ground level. In particular, there is currently no exact lifted inference algorithm that can handle this MLN without grounding it first (Van den Broeck, 2011). Furthermore, because this ground model has high treewidth for non-trivial domain sizes, it is also not amenable for exact inference by propositional algorithms. However, by cloning atoms in Lifted RCR, we are able to perform approximate lifted inference in this model.

The proposed cloning technique leads to MLNs in which one quantifies over predicate names (as in second-order logic). This can be avoided, but it leads to less transparent semantics. In particular, we can avoid quantifying over predicate names by using ground predicate names with increased arity. For example, $\text{Smokes}_{1b(x,y)}(y)$ could have been written as $\text{Smokes}_{1b}(x,y)$ where we pushed $\langle x, y \rangle$ into the predicate arguments. The disadvantage of this, however, is that the semantics of the individual arguments is lost as the arguments become overloaded.

We now have the following key theorem.

Theorem 12.1   *Let Δ^r be the MLN resulting from cloning all atom occurrences in MLN Δ and then relaxing all introduced equivalences. Let Δ^g be the grounding of Δ^r. The formulas of Δ^g are then fully disconnected (i.e., they share no atoms).*

Proof   *First, concatenating each predicate occurrence with an identifier for its formula causes all formulas to become disconnected. Second, cloned atoms contain all logical variables in the formula. For two groundings g_1, g_2 of a formula, there is a difference in the assignment to at least one logical variable. Hence, there is a difference in the arguments of any pair of atoms a_1 from g_1 and a_2 from g_2. Therefore, a_1 and a_2 are distinct random variables and the groundings g_1 and g_2 are disconnected.*

With this theorem, the proposed first-order cloning and relaxation technique allows one to fully disconnect the grounding of an MLN by simply relaxing first-order equivalences in the first-order MLN.

12.3.2   First-order Compensation

In principle, one can just clone atom occurrences, relax some equivalence constraints, and then use the resulting MLN as an approximation of the original MLN. By relaxing enough equivalences, the approximate MLN can be made arbitrarily easy for exact inference. Our goal in this section, however, is to improve the quality of approximations by compensating for the relaxed equivalences, yet without making the relaxed MLN any harder for exact inference. This will be done through a notion of first-order compensation.

12.3.2.1   Equiprobable Equivalences   The proposed technique is similar to the one for ground MLNs, that is, using *weighted atoms* whose weights will allow for compensation. The key, however, is to use first-order weighted atoms instead of ground ones. For this, we need to define the notion of **equiprobable equivalences**, based on the notion of equiprobable sets of random variables

Definition 12.2 (Equiprobable Set)   *A set of random variables V is called equiprobable w.r.t. distribution $\Pr$ iff for all $v_1, v_2 \in V : \Pr(v_1) = \Pr(v_2)$.*

Definition 12.3 (Equiprobable Equivalence)   *Let Δ be an MLN from which a first-order equivalence $a \Leftrightarrow a'$ was relaxed. Let $a_1 \Leftrightarrow a'_1, \ldots, a_n \Leftrightarrow a'_n$ be all groundings of $a \Leftrightarrow a'$. The equivalence $a \Leftrightarrow a'$ is equiprobable iff the sets $\{a_1, \ldots, a_n\}$ and $\{a'_1, \ldots, a'_n\}$ are both equiprobable w.r.t the ground distribution of MLN Δ.*

The basic idea of first-order compensation is that when relaxing an equiprobable equivalence $a \Leftrightarrow a'$, under certain conditions, one can compensate for its loss using only two weighted first-order atoms of the form:

$$w \quad a \qquad \text{and} \qquad w' \quad a'$$

This follows because if we were to fully ground the equivalence into $a_1 \Leftrightarrow a'_1, \ldots, a_n \Leftrightarrow a'_n$ and then apply ground compensation, the relevant ground atoms will attain the same weights. That is, by the end of ground compensation, the weighted ground atoms,

$$w_i \quad a_i \qquad \text{and} \qquad w'_i \quad a'_i$$

will have $w_i = w_j$ and $w'_i = w'_j$ for all i and j.

12.3.2.2 Partitioning Equivalences   To realize first-order compensation, one must address two issues. First, a relaxed first-order equivalence may not be equiprobable to start with. Second, even when the equivalence is equiprobable, it may cease to be equiprobable as we adjust the weights during the compensation process. Recall that equiprobability is defined with respect to the ground distribution of an MLN. Yet, this distribution changes during the compensation process, which iteratively changes the weights of compensating atoms and, hence, also iteratively changes the ground distribution.

Example 12.5 Consider the following relaxed equivalences: $R(x) \Leftrightarrow S(x)$ and $S(x) \Leftrightarrow T(x)$. Suppose the domain is $\{A, B\}$ and the current ground distribution, $\Pr_i$, is such that $\Pr_i(R(A)) = \Pr_i(R(B))$, $\Pr_i(S(A)) = \Pr_i(S(B))$, and $\Pr_i(T(A)) \neq \Pr_i(T(B))$. In this case, the equivalence $R(x) \Leftrightarrow S(x)$ is equiprobable, but $S(x) \Leftrightarrow T(x)$ is not equiprobable.

If an equivalence constraint is not equiprobable, one can always partition it into a set of equiprobable equivalences — in the worst case, the partition will include all groundings of the equivalence. In the above example, one can partition the equivalence $S(x) \Leftrightarrow T(x)$ into the equivalences $S(A) \Leftrightarrow T(A)$ and $S(B) \Leftrightarrow T(B)$, which are trivially equiprobable.

Given this partitioning, the compensation algorithm will add distinct weights for the compensating atoms $S(A)$ and $S(B)$. Therefore, the set $\{S(A), S(B)\}$ will no longer be equiprobable in the next ground distribution, $\Pr_{i+1}$. As a result, the equivalence $R(x) \Leftrightarrow S(x)$ will no longer be equiprobable w.r.t. the ground distribution $\Pr_{i+1}$, even though it was equiprobable with respect to the previous ground distribution $\Pr_i$.

12.3.2.3 Strongly Equiprobable Equivalences   To attain the highest degree of lifting during compensation, one needs to dynamically partition equivalences after each iteration of the compensation algorithm, to ensure equiprobability. We defer the discussion on dynamic partitioning to Section 12.4.3, focusing here on a strong version of equiprobability that allows one to circumvent the need for dynamic partitioning.

The mentioned technique is employed by our current implementation of Lifted RCR, which starts with equivalences that are *strongly equiprobable*. An equivalence is strongly equiprobable if it is equiprobable w.r.t all ground distributions induced by the compensation algorithm (i.e., ground distributions that result from only modifying the weights of compensating atoms).

Example 12.6 Consider again Formula 12.1 where we cloned the atom occurrence Smokes(y) and relaxed its equivalence, leading to the MLN:

$$1.5 \; \mathsf{Smokes}(x) \wedge \mathsf{Friends}(x, y) \Rightarrow \mathsf{Smokes}_{1b\langle x,y\rangle}(y)$$

and relaxed equivalence

$$\mathsf{Smokes}(y) \Leftrightarrow \mathsf{Smokes}_{1b\langle x,y\rangle}(y) \tag{12.10}$$

Suppose we partition this equivalence as follows:[27]

$$x = y: \; \mathsf{Smokes}(y) \Leftrightarrow \mathsf{Smokes}_{1b\langle x,y\rangle}(y)$$
$$x \neq y: \; \mathsf{Smokes}(y) \Leftrightarrow \mathsf{Smokes}_{1b\langle x,y\rangle}(y)$$

These equivalences are not only equiprobable w.r.t. the relaxed MLN, but also strongly equiprobable. That is, suppose we add to the relaxed model the compensating atoms

$$w_1 \quad \mathsf{Smokes}(x)$$
$$w_1' \quad x = y: \; \mathsf{Smokes}_{1b\langle x,y\rangle}(y)$$
$$w_2 \quad \mathsf{Smokes}(x)$$
$$w_2' \quad x \neq y: \; \mathsf{Smokes}_{1b\langle x,y\rangle}(y)$$

The two equivalences will be equiprobable w.r.t. any ground distribution that results from adjusting the weights of these compensating atoms.

We will present an equivalence partitioning algorithm in Section 12.4 that guarantees strong equiprobability of the partitioned equivalences. This algorithm is employed by our current implementation of Lifted RCR and will be used when reporting experimental results later.

---

[27] We are using an extension of MLNs that allows constraints, such as $x \neq y$, to be associated with logical variables. Our implementation is in terms of parfactor graphs (Poole, 2003), which do allow for the representation of such constraints. In extended MLNs, we will write $cs : \phi$ to mean that cs is a constraint set that applies to formula ϕ.

12.3.3   Count-Normalization

We are one step away from presenting our first-order compensation scheme. What is still missing is a discussion of **count-normalized** equivalences, which are also required by our compensation scheme.

Example 12.7 Consider Equivalence 12.10, which has four groundings

$$\text{Smokes}(A) \Leftrightarrow \text{Smokes}_{1b\langle A,A\rangle}(A)$$
$$\text{Smokes}(B) \Leftrightarrow \text{Smokes}_{1b\langle A,B\rangle}(B)$$
$$\text{Smokes}(A) \Leftrightarrow \text{Smokes}_{1b\langle B,A\rangle}(A)$$
$$\text{Smokes}(B) \Leftrightarrow \text{Smokes}_{1b\langle B,B\rangle}(B)$$

for the domain $\{A, B\}$. There are two distinct groundings of the original atom $\text{Smokes}(y)$ in this case and each of them appears in two groundings. When each grounding of the original atom appears in exactly the same number of ground equivalences, we say that the first-order equivalence is count-normalized.

Now consider a constrained version of Equivalence 12.10

$$x \neq B \vee y \neq B : \ \text{Smokes}(y) \Leftrightarrow \text{Smokes}_{1b\langle x,y\rangle}(y)$$

which has the following groundings

$$\text{Smokes}(A) \Leftrightarrow \text{Smokes}_{1b\langle A,A\rangle}(A)$$
$$\text{Smokes}(B) \Leftrightarrow \text{Smokes}_{1b\langle A,B\rangle}(B)$$
$$\text{Smokes}(A) \Leftrightarrow \text{Smokes}_{1b\langle B,A\rangle}(A)$$

This constrained equivalence is not count-normalized since the atom $\text{Smokes}(A)$ appears in two ground equivalences while the atom $\text{Smokes}(B)$ appears in only one.

Generally, we have the following definition.

Definition 12.4 *Let $(cs : \ a \Leftrightarrow a')$ be a first-order equivalence. Let θ be an instantiation of the variables $\mathbf{x}$ in the original atom a for which the constraint set cs is satisfiable. The equivalence is* count-normalized *iff for each instantiation θ, $cs\,\theta$ has the same number of solutions for the remaining variables $\mathbf{y}$. More formally, the condition is as follows.*

$$\forall \theta_1, \theta_2 \in \text{solutions}(cs, \mathbf{x}) : |\,\text{solutions}(cs\,\theta_1, \mathbf{y})| = |\,\text{solutions}(cs\,\theta_2, \mathbf{y})|$$

Moreover, the number of groundings for $cs : \ a$ is called the *original count* and the number of groundings for $cs : \ a'$ is called the *clone count*.

Count-normalization can only be violated by constrained equivalences. Moreover, for a certain class of constraints, count-normalization is always preserved. The algorithm we shall present in Section 12.4 for partitioning equivalences takes advantage of this observation. In particular, the algorithm generates constrained equivalences whose constraint structure guarantees count-normalization.

12.3.4 The Compensation Scheme

We now have the following theorem.

Theorem 12.2 *Let Λ_i be an MLN with relaxed equivalences $(cs : a \Leftrightarrow a')$ and, hence, corresponding compensating atoms:*

$$w_i \quad cs : a \qquad and \qquad w'_i \quad cs : a'$$

Suppose that the equivalences are count-normalized and strongly equiprobable. Let $(a_g \Leftrightarrow a'_g)$ be one grounding of equivalence $(cs : a \Leftrightarrow a')$, let n be its original count and n' be its clone count. Consider now the MLN Δ_{i+1} obtained using the following updates:

$$w_{i+1} = \frac{n'}{n} \left(\log \left(\Pr{}_i(a'_g) \right) - \log \left(\Pr{}_i(\neg a'_g) \right) - w'_i \right)$$
$$w'_{i+1} = \log \left(\Pr{}_i(a_g) \right) - \log \left(\Pr{}_i(\neg a_g) \right) - w_i \tag{12.11}$$

The ground distribution of MLN Δ_{i+1} equals the one obtained by applying Ground RCR to MLN Δ_i.

Proof   *It follows from Equations 12.7 that Ground RCR assigns the same compensating weights to the groundings of $a \Leftrightarrow a'$, because it is equiprobable. Each grounding of a' occurs in a single grounding of $a \Leftrightarrow a'$ and gets the same weight from Lifted or Ground RCR. From Definition 12.4 for count-normalization, each grounding of a occurs in n'/n groundings of $a \Leftrightarrow a'$. Ground RCR would add n'/n compensating weighted atoms which all get the same weight. Lifted RCR aggregates these ground weighted atoms in a single first-order weighted atom, and aggregates the weights for each grounding into a single weight by multiplying it with n'/n. Finally, because of* strong *equiprobability, the equivalences that were equiprobable in Δ_0 will also be equiprobable in Δ_i and the computed weights correspond to the weights of Ground RCR in each iteration i.*

Note that first-order compensation requires exact inference on the MLN Δ_i, which is needed for computing $\Pr_i(a_g)$ and $\Pr_i(a'_g)$. Moreover, these computations will need to be repeated until one obtains a fixed point of the update equations given by Theorem 12.2. The key, however, is that one does not need to change the set of compensating atoms during the compensation process, given the strong equiprobability of relaxed equivalences.

12.3.5 First-order Recovery

Recovering a first-order equivalence $(cs : a \Leftrightarrow a')$ amounts to removing its compensating atoms

$$w_i \quad cs : a \qquad and \qquad w'_i \quad cs : a'$$

and then adding the equivalence back to the MLN.

Adapting the ground recovery heuristic suggested earlier, one recovers the first-order equivalence that maximizes the symmetric pairwise KL-divergence[28]

$$n' \cdot \text{KLD}\left(\Pr(a_g), \Pr(a'_g), \frac{e^{w_i + w'_i}}{1 + e^{w_i + w'_i}}\right),$$

where n' is the clone count of the equivalence. Note here that n' is also the number of equivalence groundings since, by definition, the clone atom contains all logical variables that appear in the equivalence.

Please note that recovering first-order equivalences may destroy the equiprobability of equivalences that continue to be relaxed. Hence, one generally needs to re-partition these relaxed equivalences.

12.4   Partitioning Equivalences

We will now discuss a method for partitioning first-order equivalences, which guarantees both strong equiprobability and count-normalization. It builds on the preemptive shattering algorithm of Poole et al. (2011) to partition atoms into equiprobable sets. First, we review this procedure, which is a conceptually simpler version of the influential *shattering* algorithm proposed by Poole (2003) and de Salvo Braz et al. (2005) in the context of exact lifted inference. Second, we present the partitioning method that is used by our current implementation of Lifted RCR. It uses the atom partitions to partition first-order equivalences. Third, we discuss the problem of dynamic partitioning of equivalences.

12.4.1   Partitioning Atoms by Preemptive Shattering

Preemptive shattering looks at the constants K, domains $\mathcal{D}$ and free variables V that appear in the model. When using MLN models, the set of domains $\mathcal{D}$ and free variables V is always empty.[29] Hence, for the special case of MLN models, the preemptive shattering algorithm adds (in)equality constraints such that each logical variable in the MLN is either equal to exactly one constant in K, or different from all of them. It furthermore adds (in)equality constraints between logical variables that appear in the same literal, so that there is an inequality constraint between each pair of variables.

Example 12.8 Consider the formula $\text{Smokes}(x) \Leftrightarrow \text{Smokes}_{(x,y)}(x)$ and assume we have evidence $\text{Smokes}(A)$ and therefore $K = \{A\}$. Preemptive shattering of the input atom

---

[28] This is the sum of the KL-divergences between all pairs of arguments, in both directions.

[29] All variables are quantified when transforming the MLN to a WFOMC representation, on which preemptive shattering is defined.

Smokes(x) returns back

$$x = \mathsf{A} : \; \mathsf{Smokes}(x)$$
$$x \neq \mathsf{A} : \; \mathsf{Smokes}(x)$$

Preemptive shattering of the input atom $\mathsf{Smokes}_{\langle x,y \rangle}(x)$ returns back

$$x = \mathsf{A} \wedge y = \mathsf{A} : \; \mathsf{Smokes}_{\langle x,y \rangle}(x)$$
$$x = \mathsf{A} \wedge y \neq \mathsf{A} : \; \mathsf{Smokes}_{\langle x,y \rangle}(x)$$
$$x \neq \mathsf{A} \wedge y = \mathsf{A} : \; \mathsf{Smokes}_{\langle x,y \rangle}(x)$$
$$x \neq \mathsf{A} \wedge y \neq \mathsf{A} \wedge x = y : \; \mathsf{Smokes}_{\langle x,y \rangle}(x)$$
$$x \neq \mathsf{A} \wedge y \neq \mathsf{A} \wedge x \neq y : \; \mathsf{Smokes}_{\langle x,y \rangle}(x)$$

We will next show how this shattering procedure forms the basis of a method for partitioning equivalence constraints, with the aim of ensuring both strong equiprobability and count-normalization.

12.4.2   Partitioning Equivalences by Preemptive Shattering

Consider an MLN which results from cloning some atom occurrences and then adding corresponding equivalence constraints. Let K be all the constants appearing explicitly in the MLN.

To partition a first-order equivalence $a \Leftrightarrow a'$, our method will first apply preemptive shattering to the original atom a and clone atom a', yielding a partition for each. Suppose that $\{(cs_1 : a_1), \ldots, (cs_n : a_n)\}$ is the partition returned for original atom a. Suppose further that $\{(cs'_1 : a'_1), \ldots, (cs'_m : a'_m)\}$ is the partition returned for clone atom a'. By definition of cloning, all variables that appear in original atom a must also appear in clone atom a'. This implies the following property. For every (original) constraint cs_i, there is a corresponding set of (clone) constraints cs'_j that specialize, or partition cs_i. Each pair of constraints cs_i and cs'_j will then generate a member of the equivalence partition: $(cs_i \wedge cs'_j : \; a_i \Leftrightarrow a'_j)$. Note that $cs_i \wedge cs'_j \equiv cs'_j$ since cs'_j implies cs_i.

Example 12.9 Suppose we are partitioning the equivalence

$$\mathsf{Smokes}(x) \Leftrightarrow \mathsf{Smokes}_{\langle x,y \rangle}(x).$$

Example 12.8 illustrated the preemptive shattering of the atoms $\mathsf{Smokes}(x)$ and $\mathsf{Smokes}_{\langle x,y \rangle}(x)$. These give rise to the following equivalence partition:

$$x = \mathsf{A} \wedge y = \mathsf{A} : \; \mathsf{Smokes}(x) \Leftrightarrow \mathsf{Smokes}_{\langle x,y \rangle}(x)$$
$$x = \mathsf{A} \wedge y \neq \mathsf{A} : \; \mathsf{Smokes}(x) \Leftrightarrow \mathsf{Smokes}_{\langle x,y \rangle}(x)$$
$$x \neq \mathsf{A} \wedge y = \mathsf{A} : \; \mathsf{Smokes}(x) \Leftrightarrow \mathsf{Smokes}_{\langle x,y \rangle}(x)$$
$$x \neq \mathsf{A} \wedge y \neq \mathsf{A} \wedge x = y : \; \mathsf{Smokes}(x) \Leftrightarrow \mathsf{Smokes}_{\langle x,y \rangle}(x)$$

$$x \neq \mathsf{A} \wedge y \neq \mathsf{A} \wedge x \neq y : \; \mathsf{Smokes}(x) \Leftrightarrow \mathsf{Smokes}_{\langle x,y \rangle}(x)$$

We now have the following results.

Lemma 12.1 *Partitioning by preemptive shattering returns equiprobable equivalences.*

Proof   *[Proof outline] Let $cs \wedge cs' : a \Leftrightarrow a'$ be an element of the partition of an equivalence constraint found by preemptive shattering. The constrained atoms ($cs : a$) and ($cs' : a'$) themselves were found by preemptive shattering of the MLN Δ. It follows from Van den Broeck (2013, Prop. 5.4) that the groundings of ($cs : a$) and ($cs' : a'$) are equiprobable.*

Lemma 12.2 *Partitioning by preemptive shattering returns count-normalized equivalences.*

Proof   *[Proof outline] Let $cs : a \Leftrightarrow a'$ be an element of the partition of an equivalence constraint found by preemptive shattering. Let $\mathbf{x}$ be the set of logical variable arguments of a and $\mathbf{x} \cup \mathbf{y}$ be the logical variable arguments of its clone a'. For a each grounding of a, that is, each substitution of the variables $\mathbf{x}$ by constants that satisfy cs, the clone a' has the same number of groundings of the variables $\mathbf{y}$ that satisfy cs. This follows from the fact that each logical variable that is not bound to a constant has the same set of inequality constraints associated with it.*

Lemma 12.3 *Partitioning by preemptive shattering returns strongly equiprobable equivalences.*

Proof   *[Proof outline] The preemptive shattering procedure neither depends on the weight parameters of the MLN model Δ, nor on the precise formulas in Δ. It only depends on the constants K that appear in it. The partition returned by preemptive shattering does not introduce any constants that were not in the input Δ. Therefore, the MLNs Δ_i that are constructed in each iteration of the compensation algorithm (with compensating weighted atoms) do not introduce additional constants and each call to preemptive shattering returns identical partitions in each iteration of the compensation algorithm.*

Theorem 12.3 *Partitioning by preemptive shattering returns count-normalized, strongly equiprobable equivalences.*

Proof   *Follows from Lemmas 12.1, 12.2 and 12.3.*

12.4.3   Dynamic Equivalence Partitioning

To attain the highest degree of lifting, one may need to dynamically partition equivalences after each iteration of the compensation scheme. Moreover, one would need to find the smallest possible partition for each considered equivalence, while guaranteeing both equiprobability and count-normalization.[30]

Dynamic partitioning leads to a higher degree of lifting as it removes the need for strong equiprobability, which usually leads to larger partitions of equivalences. We do not employ dynamic partitioning in our current implementation of Lifted RCR, leaving this to future work. We point out, however, that dynamic partitioning requires a slight adjustment to the compensation scheme given by Theorem 12.2.

Suppose that the first-order equivalence $cs : a \Leftrightarrow a'$ was equiprobable in MLN Δ_i, but ceases to be equiprobable in MLN Δ_{i+1} due to the adjustment of weights for compensating atoms. A dynamic partitioning scheme will then partition this equivalence into a set of equiprobable and count-normalized equivalences. One implication of this partitioning is that the two compensating atoms associated with the equivalence $cs : a \Leftrightarrow a'$ in MLN Δ_i will now have to be partitioned as well. That is, the compensating atoms in MLN Δ_i

$$w_i : \quad cs : a \qquad \text{and} \qquad w_i' : \quad cs : a'$$

will need to be replaced in MLN Δ_{i+1} by two compensating atoms for each equivalence in the computed partition. Moreover, the initial weights of these new compensating atoms will be precisely w_i (for original atoms) and w_i' for cloned atoms.

12.5   Related and Future Work

In this section, we discuss related propositional and lifted approximate inference algorithms and identify future work on the problem of equivalence shattering.

12.5.1   Relation to Propositional Algorithms

The RCR framework has previously been used to characterize iterative belief propagation and some of its generalizations. In the case that the simplified model is fully disconnected, the approximate marginals of RCR correspond to the approximate marginals provided by *iterative belief propagation* (Pearl, 1988; Choi and Darwiche, 2006). The approximation to the partition function further corresponds to the **Bethe free energy** approximation (Yedidia et al., 2003; Choi and Darwiche, 2008). When equivalence constraints have been recovered, RCR corresponds to a class of **generalized belief propagation** (GBP) approximations (Yedidia et al., 2003), in particular **iterative joingraph propagation** with the corresponding joingraph free energies (Aji and McEliece, 2001; Dechter et al., 2002). RCR further inspired a system that was successfully employed in a recent approximate inference competition (Elidan and Globerson, 2010; Choi and Darwiche, 2011). **Mini-bucket approximation** can be viewed as an instance of RCR where no compensations are used (Kask and Dechter, 2001; Dechter and Rish, 2003; Choi et al., 2007). In these cases, RCR is

---

[30] There is a unique smallest partition satisfying these properties.

also capable of providing upper bounds on the partition function. Since every approximate MLN found by Lifted RCR corresponds to a solution found by Ground RCR on the ground MLN, the above mentioned results carry over to the lifted setting.

12.5.2   Relation to Lifted Algorithms

The motivation for calling our approach *lifted* is threefold. First, in the compensation phase, we are compensating for many ground equivalences at the same time. Computing compensating weights for all of these requires inferring only a *single* pair of marginal probabilities. Second, computing marginal probabilities is done by an *exact lifted* inference algorithm. Third, we relax and recover first-order equivalence constraints, which correspond to *sets* of ground equivalences.

The work on lifted approximate inference has mainly focused on lifting the IBP algorithm. The correspondence between IBP and RCR carries over to their lifted counterparts: Lifted RCR compensations on a fully relaxed model correspond to *lifted belief propagation* (lifted BP) (Jaimovich et al., 2007; Singla and Domingos, 2008). Starting from a first-order model, Singla and Domingos (2008) proposed **lifted network construction** (LNC), which partitions the random variables and factors of a factor graph into so-called *supernodes* and *superfeatures*. The ground atoms represented by these supernodes send and receive the same messages when running IBP. Kersting et al. (2009) proposed a *color-passing* (CP) algorithm that achieves similar results as LNC, only starting from a ground model, where the first-order structure is not apparent.

Bisimulation-based lifted inference (Sen et al., 2009a) uses a mini-bucket approximation on a model that was compressed by detecting symmetries. Because of the correspondence between Ground RCR and mini-buckets mentioned above, also this approach can be seen as an instance of Lifted RCR with the compensation phase removed.

12.5.3   Opportunities for Equivalence Partitioning

For models that contain few explicit constants (K is small), preemptive shattering will find partitions of equivalence constraints that are close to minimal. When K is large, however, it will create large partitions, defeating the purpose of lifted inference.

The work on lifted BP provides us with alternative partitioning algorithms that works correctly on a fully relaxed model. These algorithms construct lifted networks whose nodes send and receive the same messages when running IBP. This means that they partition the atoms into equiprobable sets (w.r.t. the approximate distribution) and that LNC can be used for equivalence partitioning in Lifted RCR, when the model is fully relaxed.

However, preemptive shattering is the only partitioning algorithm to our knowledge that works for any level of relaxation. One way to make the partitions found by preemptive shattering smaller (with fewer sets) is by exploiting the actual structure of the model Δ itself. Knowing that there are certain independences in Δ might remove the need to ground atoms for all constants in K. This requires reasoning on the model, as is done by the vanilla shattering algorithm (Poole, 2003; de Salvo Braz et al., 2005). We believe our work can motivate future work on finding more efficient general partitioning algorithms.

Much of the work on lifted BP has looked at more intelligently constructing lifted networks. Given the strong connection between LNC and partitioning of equiprobable equivalences, these high-level ideas can directly be applied to Lifted RCR, as we discuss next.

The problem of dynamic equivalence partitioning is related to **anytime lifted BP** (de Salvo Braz et al., 2009; Freedman et al., 2012), which gradually shatters its partitions of random variables and factors with each iteration of lifted BP. In addition, it makes these partitions specific to a query and propagates bounds on the marginals, in order to stop message passing early, when the desired approximation quality has been reached. This would correspond to a version of RCR where compensating weights are not numbers but intervals that are tightened with each iteration of the compensation algorithm.

Several ways of constructing an *approximate lifted network* were proposed (Singla et al., 2010; Kersting et al., 2010). This corresponds to partitioning into approximately equiprobable equivalences in the context of Lifted RCR. Approximating the partition can be done by ignoring long-range dependencies in the model or by looking at the actual parameters of the model, instead of only its logical syntax. Another line of work looks at efficient lifted network construction for multiple queries with *changing evidence* (Nath and Domingos, 2010; Ahmadi et al., 2010; Hadiji et al., 2011; Ahmadi et al., 2011). These techniques also allow for the efficient computation of joint marginals.

12.6 Experiments

In this section, we evaluate the Lifted RCR algorithm on common benchmarks from the lifted inference literature. The experiments were set up to answer the following questions:

(Q1) To what extent does recovering first-order equivalences improve the approximations found by Lifted RCR?

(Q2) Can IBP be improved considerably through the recovery of a small number of equivalences?

(Q3) Is there a significant advantage to using Lifted RCR over Ground RCR?

(Q4) What is the run time behavior of the compensation algorithm?

12.6.1 Implementation

We implemented Lifted RCR in the WFOMC tool.[31] To compute exact marginal probabilities in Equations 12.11, we use exact lifted inference by first-order knowledge compilation (Van den Broeck et al., 2011). In combination with preemptive shattering, we compile a first-order circuit once for every equiprobable set of random variables and re-evaluate it

---

[31] https://dtai.cs.kuleuven.be/software/wfomc

in each iteration of the compensation algorithm. This is possible because the structure of the compensated MLNs does not change between iterations, only their parameters change. Re-evaluating an already compiled first-order circuit can be done very efficiently. In the compensation phase, we use damping with a weight of 50%.

In practice, the Ground RCR algorithm does not start off with a fully relaxed model. Instead, it starts with some equivalences intact, such that the relaxed model forms a spanning tree, and exact inference is still efficient. As long as the relaxed model is a tree, the set of approximate single marginals that can be found with RCR correspond to the set of marginals that the loopy BP algorithm can converge to in the original model. Relaxing equivalences beyond a spanning tree neither makes inference more tractable, nor does it change the approximations made by RCR.

For these reasons, we do not clone all atoms occurrences and atom groundings in the relaxation step of Lifted RCR. Instead, we clone all but one atom per MLN formula. This guarantees that the relaxed model is still a tree (but not necessarily spanning). To select the atom that is not cloned, we choose one with a high number of logical variable arguments, to have as few equivalences relaxed as possible overall. As a consequence of this approach to relaxation, weighted unit clauses are never relaxed.

12.6.2   Results

To answer **(Q1-2)** we ran Lifted RCR on MLNs from the exact lifted inference literature, where computing exact marginals is tractable. This allows us to evaluate the approximation quality of Lifted RCR for different degrees of relaxation. We used the models *p-r* and *sick-death* (de Salvo Braz et al., 2005), *workshop attributes* (Milch et al., 2008), *smokers* (Singla and Domingos, 2008), *smokers and drinkers* (Van den Broeck et al., 2011) and *symmetric smokers* (Van den Broeck, 2011). Each of these models represents a new advance in exact lifted inference. They are incrementally more complex and challenging for lifted inference.

The results are shown in Figure 12.1, where we ran Lifted RCR on the above models for two sets of domain sizes: a small and a large set, where the number of random variables is on the order of 100 and 10,000 respectively. We plot the symmetric KL divergence between the exact marginals and the approximations found by Lifted RCR, as a percentage of the KL divergence of the fully relaxed approximation (the IBP approximation). The horizontal axis shows the level of relaxation in terms of the percentage of recovered ground equivalences. The 0% recovery point corresponds to the approximations found by (lifted) IBP. The 100% recovery point corresponds to exact inference.[32]

We see that each recovered first-order equivalence tends to improve the approximation quality significantly, often by more than one order of magnitude, answering **(Q1)**. In the case of *smokers* with a large domain size, recovering a single equivalence more than the

---

[32] All experiments ran up to the 100% point, which is not shown in the plot because it has a KL divergence of 0.

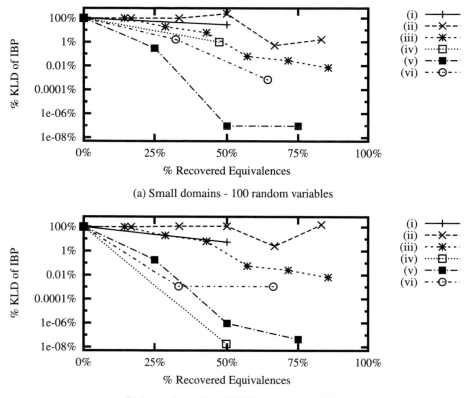

(a) Small domains - 100 random variables

(b) Large Domains - 10,000 random variables

Figure 12.1: Normalized approximation error of Lifted RCR for different levels of approximation on the models (i) *p-r*, (ii) *sick-death*, (iii) *workshop attributes*, (iv) *smokers*, (v) *smokers and drinkers* and (vi) *symmetric smokers*.

IBP approximation reduced the KL divergence by 10 orders of magnitude, giving a positive answer to **(Q2)**. The *sick-death* model is the only negative case for **(Q2)**, where recovering equivalences does not lead to approximations that are better than IBP.[33]

To answer **(Q3)**, first note that as argued in Section 12.5, the 0% recovery point of Lifted RCR using LNC or CP to partition equivalences corresponds to lifted BP. For this case, the work of Singla and Domingos (2008) and Kersting et al. (2009) has extensively shown that Lifted BP/RCR methods can significantly outperform Ground BP/RCR. Similarly, computational gains for the 100% recovery point were shown in the exact lifted inference

---

[33] Interestingly, it is also the only example where some compensations failed to converge without using damping.

literature. For intermediate levels of relaxation, we ran Ground RCR on the above models with large domain sizes. On these, Ground RCR could never recover more than 5% of the relaxed equivalences before exact inference in the relaxed model becomes intractable. This answers **(Q3)** positively.

To answer **(Q4)**, we report on the quality of the approximations and the convergence of the compensation algorithm, both as a function of run time and the number of iterations. The results for the four most challenging benchmarks (small and large *smokers and drinkers* and *symmetric smokers*) are shown in Figure 12.2.

The figures on the left show convergence as a function of the number of iterations of the compensation algorithm.

· The solid line shows the KL divergence between the approximate marginals (according to the relaxed MLN and compensating weights of that iteration) and the exact marginals (computed with exact lifted inference).
· The dashed line shows the three-way pairwise symmetric KL divergence between the three terms of Equation 12.6. Our compensation algorithm aims to satisfy this equation, lowering the reported KLD.

We see that the compensation KLD increases dramatically between certain iterations. This happens when we recover a relaxed first-order equivalence. After a recovery step, the compensating weights still stem from the previous compensation iteration. These weights no longer satisfy Equation 12.6. The compensation KLD is increased and the compensation algorithm starts searching for a better set of compensating weights. In the iterations where recovery happens, we also see a change in the quality of the approximation. It can improve, because of the added equivalence, but it can also decrease, because the compensating weights no longer satisfy Equation 12.6. In either case, we see that running the compensation algorithm after recovering an equivalence improves both the compensation objective and the quality of the approximations. The KLD of the approximation right before each recovery corresponds to the errors reported in Figure 12.1.

The figures on the right show the same quality of approximation (solid line to the left), this time as a function of time. In between reported data points, there are large gaps. These again correspond to recovery steps, where the algorithm needs to compile a set of new circuits to perform exact inference in the new approximate, relaxed model. Evaluating these circuits in the compensation phase (to compute all marginals) is typically fast, leading to a quick succession of iterations. For one experiment, in Figure 12.2d, the compilation is not the bottleneck. This is because evaluating the compiled circuits for its large domain size becomes expensive when most equivalences have been recovered.

As a final observation, the iterations on the left side become increasingly time-consuming. For example, in Figure 12.2a, the first 80 iterations out of 160 take less than 10 seconds out of 100 in Figure 12.2b. Similarly, in Figure 12.2c, the first 150 iterations out of 180 take 300 seconds out of 1800 in Figure 12.2d. To complement our earlier claim that it suf-

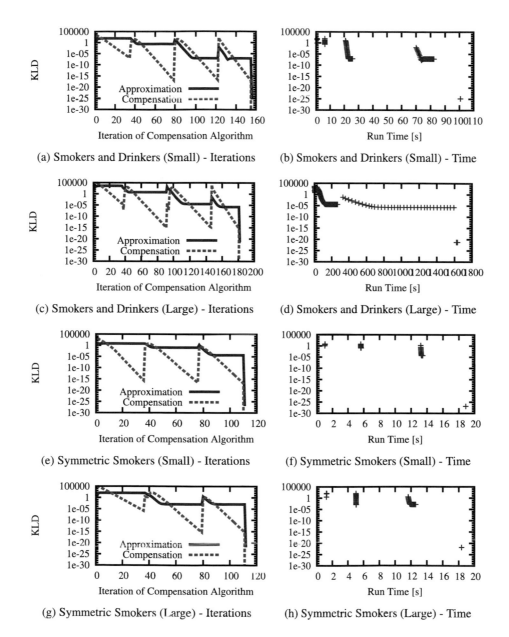

(a) Smokers and Drinkers (Small) - Iterations

(b) Smokers and Drinkers (Small) - Time

(c) Smokers and Drinkers (Large) - Iterations

(d) Smokers and Drinkers (Large) - Time

(e) Symmetric Smokers (Small) - Iterations

(f) Symmetric Smokers (Small) - Time

(g) Symmetric Smokers (Large) - Iterations

(h) Symmetric Smokers (Large) - Time

Figure 12.2: Convergence Rate of Lifted RCR

fices to recover a small number of equivalences in order to significantly improve upon IBP approximations, this also shows that recovering these first few equivalences can be done very efficiently.

12.7   Conclusions

This chapter presented Lifted RCR, a lifted approximate inference algorithm that performs exact inference in a simplified model. We showed how to obtain a simplified model by relaxing first-order equivalences, compensating for their loss, and recovering them as long as exact inference remains tractable. The algorithm can traverse an entire spectrum of approximations, from lifted iterative belief propagation to exact lifted inference. Inside this spectrum is a family of lifted joingraph propagation (and GBP) approximations. We empirically showed that recovering first-order equivalences in a relaxed model can substantially improve the quality of an approximation. We also remark that Lifted RCR relies on an exact lifted inference engine as a black box, and that with Lifted RCR, any future advances in exact lifted inference have immediate impact in lifted approximate inference.

13 Liftability Theory of Variational Inference

Martin Mladenov, Hung Bui, and Kristian Kersting

Abstract. In this chapter, we lay down a theory of lifted inference for approximate variational inference methods. The main building block is group theory and in particular the concept of "lifting partitions". These are partitions of the variables and constraints of the corresponding variational optimization problem that allow one to reduce the model in size by replacing all equivalent variables/constraint by class representatives such that every solution of the reduced problem corresponds to a solution of the original one. This can lead to efficiency gains by solving the reduced model instead and then expanding the solution back to the original one. As we will demonstrate, a large class of existing variational inference algorithms can directly be made aware of symmetries without modifications. Actually, the coarseness of lifting partitions (respectively the degree of reduction of the variational problem) depends not only on the model being solved, but also on the approximation strategy. Any symmetry (automorphism) present in the original model induces a symmetry of the variational problem, however, in many cases the solution of the variational problem may contain additional symmetries not present in the "true" solution, e.g., if it enforces constraints only among neighboring vertices of the graphical model.

13.1 Introduction

By making use of **symmetries**, lifted inference can significantly reduce the cost of inference in highly connected and structured (symmetric) models, where standard inference methods such as variable elimination would fail due to the high induced width of the (probabilistic) graphical model. Yet, even in cases where exact lifted inference guarantees theoretical tractability, the resulting algorithms may not naturally be used and scaled in practice, due to factors such as limited parallelizability and the inability to make use of existing, efficient code and/or hardware. This motivates the design of this **lifted approximate inference** algorithms that yield practical, although inexact alternatives.

Recent years have witnessed **variational inference** (Wainwright and Jordan, 2008; Heskes, 2006; Meshi et al., 2009) as a prominent framework for designing approximate inference algorithms in general. In variational inference, one aims to approximate the distribution induced by a graphical model by a simpler one, i.e. , one whose marginals are easier to read off; if a good fit is found, the marginals of the approximating distribution tend to be close to the marginals of the original one. To this aim, the variational design principle typically involves two major components:

1. the selection of an approximation strategy (that is, a family of distributions from which to approximate the input distribution and a measure of "closeness"), and
2. the construction of an efficient optimizer to find the closest approximation.

For many popular choices of approximations, the optimizers take the form of a "**message passing**" algorithm. They propagate information only among adjacent nodes of the graphical model in order to compute the overall result and, hence, typically scale well and can easily be parallelized. Nevertheless, they also suffer from a combinatorial explosion when applied to highly symmetric models as often induced by probabilistic relational models. Understanding how to apply ideas from lifted inference within the variational framework is therefore crucial for designing approximative inference algorithms for highly symmetric models, and the goal of the present chapter.

More precisely, the chapter lays down a theory of lifted inference for approximate variational inference methods. The main building block is **group theory** and in particular the concept of "**lifting partitions**". These are partitions of the variables and constraints of the corresponding variational optimization problem that allow one to reduce the model in size by replacing all equivalent variables/constraint by class representatives such that every solution of the reduced problem corresponds to a solution of the original one. This can lead to efficiency gains by solving the **reduced model** instead and then expanding the solution back to the original one. As it turns out, the coarseness of lifting partitions (respectively the degree of reduction of the variational problem) depends not only on the model being solved, but also on the approximation strategy. Any symmetry (**automorphism**) present in the original model induces a symmetry of the variational problem, however, in many cases the solution of the variational problem may contain additional symmetries not present in the "true" solution, e.g., if it enforces constraints only among neighboring vertices of the graphical model.

The chapter is structured as follows. We will start off by reviewing the basics of group theory as the mathematical language of symmetries, the variational principle for inference in exponential families, and by summarizing the core tool for lifting the variational principle—lifting partitions. Afterwards, we will instantiate lifting partitions for exact and approximate variational inference in the exponential family resulting in automorphisms respectively fractional automorphisms of the members of the exponential family. Before concluding, we will show how to reparametrize the reduced models in order to eliminate the necessity of using a lifted solver.

13.2   Lifted Variational Inference

Let us start by reviewing the basics of group theory and variational inference required as well as by introducing the core tool for lifting variational inference—lifting partitions.

13.2.1 Symmetry: Groups and Partitions

Since our model reduction framework is based on symmetries, we start off by reviewing basic concepts about group theory and partitions.

Partitions. A **partition** $\Delta = \{\Delta_1, \ldots, \Delta_k\}$ of a set V is a collection of nonempty, disjoint subsets of V whose union is V. Each element Δ_i is called a **cell**. $|\Delta|$ is thus the number of cells or the **size** of the partition. A partition Δ defines an **equivalence relation** $\overset{\Delta}{\sim}$ on V by letting $v \overset{\Delta}{\sim} v'$ if and only iff v and v' are in the same cell. We refer to the cell containing v as the *class* of v, denoted by $\Delta[v]$. A partition Δ' is **finer** than Δ if every cell of Δ' is a subset of some cell of Δ. Finally, given a function f whose domain is V, we say that Δ *respects* f if for each $v, v' \in V$, the equivalence $v \overset{\Delta}{\sim} v'$ implies $f(v) = f(v')$.

Groups. A **group** $(G, \circ)$ consists of a ground set G together with a binary operation $\circ : G \times G \to G$ having the following properties:

(a) $\circ$ is *associative*, i.e., $a \circ (b \circ c) = (a \circ b) \circ c$ for any $a, b, c \in G$;

(b) G contains an *identity* element $\mathbf{1}$, i.e., $a \circ \mathbf{1} = \mathbf{1} \circ a = a$ for any $a \in G$;

(c) for each $a \in G$, there exists an *inverse* element $a^{-1} \in G$ having the property that $a \circ a^{-1} = a^{-1} \circ a = \mathbf{1}$.

A group containing only an identity element is called **trivial**. Now, groups $(G, \circ)$ and $(G', \diamond)$ are called **isomorphic** whenever G can be mapped to G' via a bijection $i : G \to G'$ such that i maps the identity in G to the identity in G', $i(\mathbf{1}) = \mathbf{1}'$ and i commutes with the group operation, i.e., $i(a \circ b) = i(a) \diamond i(b) = a' \diamond b'$. In such cases, the groups are essentially identical up to a renaming of their elements. Any subset of G that contains its identity and is also closed under the group operation is a **subgroup** of G. We use the notation $G' \leq G$ to denote that G' is isomorphic to a subgroup of G. Note that this covers the situation where G' is a subgroup of G by way of the identity bijection.

A **permutation** of a set V is a bijection from V to itself. Two permutations can be composed together via the usual composition of two mappings. Any set of permutations (on V) that contains the identity permutation and is closed under composition and taking inverse thus forms a group. The set of all permutations of V is called the *symmetric group* $\mathbb{S}(V)$. The **symmetric group** $\mathbb{S}_n$ is the set of all permutations of $\{1, 2, \ldots, n\}$. Let $\pi \in \mathbb{S}_n$ be a permutation. If x is a vector in $\mathbb{R}^n$, x permuted by π, denoted by x^π, is $[x_{\pi(1)}, \ldots, x_{\pi(n)}]$; for a set $A \subseteq \mathbb{R}^n$, A permuted by π, denoted by A^π is defined as $\{x^\pi | x \in A\}$.

A **subgroup** G of $\mathbb{S}(V)$ induces the following equivalence relation on V: $v \sim v'$ if and only if there exists $g \in G$ such that $g(v) = v'$ (the fact that $\sim$ is an equivalence relation follows from the definition of a group). G therefore induces a partition on V, called the **orbit partition**, denoted by $\mathrm{Orb}_G(V)$. The **orbit** of an element $v \in V$ is its cell (or class) in the orbit partition: $\mathrm{Orb}_G[v] = \{v' \in V | v' \sim v\}$.

A group G can induce an *orbit partition* on any set U as long as members of G can be viewed as (not necessarily distinct) permutations of U. In this case, there is a group

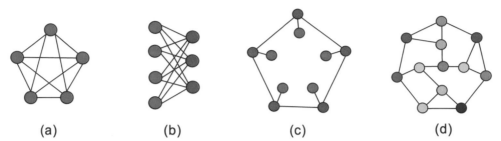

(a)   (b)   (c)   (d)

Figure 13.1: Graphs and their automorphism groups: (a) $\mathrm{Aut}(K_5) = \mathbb{S}_5$; (b) $\mathrm{Aut}(K_{4\times3}) = \mathbb{S}_4 \times \mathbb{S}_3$; (c) this graph can be rotated or flipped, yielding the automorphism dihedral group $\mathbb{D}_5$; and (d) this is known as the Frucht graph, a regular but asymmetric graph. Different colors denote different node orbits.

homomorphism from G to a subgroup of $\mathbb{S}(U)$, and the group G is said to *act* on the set U. A subgroup G' of G also acts on U and induces a finer orbit partition. Given a set element $u \in U$ and a group element $g \in G$, if $g(u) = u$ then g is said to stabilize u. If for all group elements $g \in G$, $g(u) = u$, then the group G is said to stabilize u.

Actions. The concept of group **actions** is powerful since it allows the same group G to act (and in turn induce orbit partitions) on many different sets. For example, $\mathbb{S}_n$ acts on the set of n-dimensional vectors $\mathbb{R}^n$ via the action $\pi(x) = x^\pi$. $\mathbb{S}_n$ also acts on the set of n-vertex graphs in the following way. Every permutation $\pi \in \mathbb{S}_n$ transforms a graph $\Gamma = (V, E)$ to its isomorphic variant $\Gamma' = (V', E')$ (i.e., $\{i,j\} \in E$ if and only if $\{\pi(i), \pi(j)\} \in E'$). Hence, it can be viewed as a bijection (permutation) on the set of n-vertex graphs. If $\pi(\Gamma) = \Gamma$, then π stabilizes Γ and is called an **automorphism of the graph**. The set of all automorphisms forms a group, named the **automorphism group** of Γ and denoted by $\mathrm{Aut}(\Gamma)$ (the concept is illustrated on Figure 13.1). It is clear that for finite graphs, $\mathrm{Aut}(\Gamma)$ is a subgroup of $\mathbb{S}_n$. The cardinality of $\mathrm{Aut}(\Gamma)$ indicates the level of symmetry in Γ. If $\mathrm{Aut}(\Gamma)$ is the trivial group then Γ is asymmetric; if $\mathrm{Aut}(\Gamma) = \mathbb{S}_n$ then Γ either is fully connected or has no edges. This concept of graph automorphism directly generalizes to graphs with additional structures such as directions, colors, etc.

If we now ask what elements of Γ are indistinguishable up to symmetry, the automorphism group $\mathrm{Aut}(\Gamma)$ can give us the precise answer. For example, if a node v' can be obtained from a node v via some permutation π in $\mathrm{Aut}(\Gamma)$, then these two nodes are indistinguishable and must have the same the graph properties (e.g., degree, averaged distance to other nodes, etc.). $\mathrm{Aut}(\Gamma)$ thus partitions the set of nodes V into the node-orbits $\mathrm{Orb}_{\mathrm{Aut}(\Gamma)}(V)$, where each node orbit is a set of vertices equivalent to one another up to some node relabeling. Furthermore, $\mathrm{Aut}(\Gamma)$ also acts on the set of graph edges E of Γ by letting $\pi(\{u, v\}) = \{\pi(u), \pi(v)\}$, and this action partitions E into a set of edge-orbits $\mathrm{Orb}_{\mathrm{Aut}(\Gamma)}(E)$.

Similarly, we can also obtain the set of arc-orbits $\text{Orb}_{\text{Aut}(\Gamma)}(\vec{E})$, where an arc is a directed edge indicated by $\vec{E}$.

13.2.2 Exponential Families and the Variational Principle

Probabilistic **graphical models** provide a unifying framework for capturing complex dependencies among random variables. The majority of current applications of them as well as variational inference methods involve distributions in the *exponential family* (Wainwright and Jordan, 2008).

More formally, let $\{x_i | i \in \mathcal{V}\}$ be a set of n random varaibles, with $\mathcal{V} = \{1, \ldots, n\}$. Assume each x_i has domain X. We consider probability mass functions over X^n specified in the form of the **exponential family**, i.e.,

$$\mathcal{F}(x|\theta) = h(x) \exp\left(\langle \Phi(x), \theta \rangle - A(\theta)\right),$$

where h is the base density, $\Phi(x) = (\Phi_j(x))_{j \in \mathcal{I}}$ is a feature functions with $\mathcal{I} = \{1, \ldots, m\}$ being an m-dimensional feature vector, $\theta \in \mathbb{R}^m$ are the natural parameters, and $A(\theta)$ is the log-partition function, defined as

$$A(\theta) = \sum_{x \in X^n} \langle \Phi(x), \theta \rangle.$$

To any exponential family, we associate the sets $\Theta = \{\theta | A(\theta) < \infty\}$, the set of **natural parameters**, and $\mathcal{M} = \{\mu \in \mathbb{R}^m | \exists p, \mu = E_p \Phi(x)\}$, the set of realizable mean parameters. The function $A(\theta) : \Theta \to \mathbb{R}$ is convex and its conjugate dual $A^* : \mathcal{M} \to \mathbb{R}$ is defined as

$$A^*(\mu) = \sup_{\theta \in \Theta} \langle \mu, \theta \rangle - A(\theta).$$

Conversely, the **log-partition function** can be expressed as

$$A(\theta) = \sup_{\mu \in \mathcal{M}} \langle \mu, \theta \rangle - A^*(\mu). \tag{13.1}$$

Moreover, $\mathbf{m} : \Theta \to \mathcal{M}$ is the mean parameter mapping that maps $\theta \mapsto \mathbf{m}(\theta) = \mathbb{E}_\theta \Phi(x)$. Note that $\mathbf{m}(\Theta) = \text{ri}\mathcal{M}$ is the relative interior of M.

If the feature function Φ_i depends only on a subset of the variables in $\mathcal{V}$, we will write Φ_i more compactly in factorized form as $\Phi_i(x) = f_i(x_{l_1}, \ldots, x_{l_k})$, where the indices l_j are distinct, $i_1 < i_2, \ldots < i_k$, and f_i cannot be reduced further, i.e., it must depend on all of its arguments. To keep track of variable indices of the arguments of f_i, we let $scope(f_i)$ denote its set of arguments, $\eta_i(j) = i_j$, the j-th argument and $|\eta_i|$ its arity. Factored forms of features can be encoded as a hypergraph $\Gamma(\mathcal{F})$ of $\mathcal{F}$ (called the graph structure or graphical model of $\mathcal{F}$) with nodes $\mathcal{V}$, and hyperedges (clusters) $\{C | \exists i : scope(f_i) = C\}$. For models with pairwise features, Γ is a standard graph.

For finite random variables (i.e., the domain X is finite), it is sometimes advantageous to work with the **overcomplete family** $\mathcal{F}^o$. The set of overcomplete features $\mathcal{I}^o$ are indicator functions on the nodes and edges of the graphical model Γ of $\mathcal{F}$: $\Phi^o_{u:t}(x) = \mathbb{I}\{x_u = t\}$, $t \in X$ for each node $v \in V(\Gamma)$; and $\Phi^o_{u:t,v:t'}(x) = \mathbb{I}\{x_u = t, x_v = t'\}$, $t, t' \in X$ for each edge $\{u, v\} \in E(\Gamma)$. The set of overcomplete realizable mean parameters M^o is also called the **marginal polytope** since the overcomplete mean parameters correspond to node and edge marginal probabilities. Given parameters θ, the transformation of $\mathcal{F}(x|\theta)$ to its overcomplete representation is done by letting θ^o be the corresponding parameters in the overcomplete family: $\theta^o_{u:t} = \sum_{i:scope(f_i)=\{u\}} f_i(t)\theta_i$ and (assuming $u < v$) $\theta^o_{u:t,v:t'} = \sum_{i:scope(f_i)=\{u,v\}} f_i(t, t')\theta_i$. One can verify that $\mathcal{F}^o(x|\theta^o) = \mathcal{F}(x|\theta)$ for all values of $x \in X^n$.

A fundamental result, cf. (Wainwright and Jordan, 2008), states that the supremum in (13.1) is attained at $\mu = E_\theta[\Phi(x)]$, that is, μ is the vector of expected feature values under the distribution $\mathcal{F}(x|\theta)$. This fact suggests the following approach to **marginal inference**: using the overcomplete representation, we solve

$$\mu^* = \underset{\mu \in M}{\operatorname{argmax}} \; \langle \mu, \theta \rangle - A^*(\mu) \tag{13.2}$$

using the tools from convex optimization, obtaining the marginals vector μ^*. To see how this corresponds to marginal inference, consider the overcomplete representation $\mathcal{F}^o$. The values $(\mu^*_{u:t})_{u \in V, t \in X}$ are precisely the single node marginals of $\mathcal{F}$. Indeed, one can arrive at a similar formulation for MAP inference, as computing the configuration $x \in X^n$ that maximizes $\log(\mathcal{F}(x|\theta))$ is equivalent to

$$\mu^* = \underset{\mu \in M^o}{\operatorname{argmax}} \; \langle \mu, \theta \rangle + \text{const.} \tag{13.3}$$

This approach, often refered to as the **variational principle**, is usually not tractable in practice: A^* may not be representable in explicit form and deciding membership in M or M^o may involve checking exponentially many inequalities. Nevertheless, the variational principle is useful in two major ways. First, tractability issues can be addressed by designing suitable relaxations of (13.1). This will be of interest in later sections. Second, it allows the use of convex analysis to derive non-trivial insights about the marginals of $\mathcal{F}$. In fact the relationship between convex sets and symmetry is at the heart of lifted variational inference. Let us flesh this point out in more detail.

13.2.3   Lifting the Variational Principle

In a nutshell, the main idea in lifted variational inference is to find a partition of the variables and/or constraints used to describe the inference problems in (13.2) and (13.3) such that replacing all variables in the same cell by a single variable preserves at least one solution. The hope is that we can find a coarse enough partition, as to reduce the inference problems to a "tractable" size. We call these partitions *lifting partitions*.

Definition 13.1 (Lifting partition)   *Consider a convex problem $C \equiv \inf_{x \in S} J(x)$, where $S \subseteq \mathbb{R}^m$ is a convex set and J is a convex function. A partition Δ of $\{1, \ldots, m\}$ is a **lifting**

partition *for C if it holds for at least one of its solutions that* $i \overset{\Delta}{\sim} i'$ *implies* $x_i = x_{i'}$. *In other words, a partition is a lifting partition for a convex program if it respects at least one if its solutions.*

The usefulness of lifting partitions comes from their ability to narrow down the feasible region of convex programs in a structured fashion. Let $\mathbb{R}^m_\Delta$ denote the subspace

$$\left\{ r \in \mathbb{R}^m \,\middle|\, r_j = r_{j'} \text{ if } j \overset{\Delta}{\sim} j' \right\}.$$

For any set $S \subset \mathbb{R}^m$, let $S_\Delta = S \cap \mathbb{R}^m_\Delta$. If we now restrict the feasible set S to S_Δ, which is a subset of S, we know that at least one solution must be contained in that restriction. Thus, instead of solving the original problem, we can solve $\inf_{x \in S_\Delta} J(x)$, the **lifted convex program**. Since all equivalent entries of x are tied, we reduce the degrees of freedom from m to $|\Delta|$ by replacing each tied entry with a single variable representing the entire cell. If $|\Delta|$ is much smaller than m, considerable speed-ups can be achieved.

An important question that remains to be answered is how to compute lifting partitions. This is the major challenge of lifted variational inference. On one hand, the lifting partitions must be coarse enough so that a meaningful reduction is achieved; on the other, the algorithms to compute these partitions must be considerably faster end-to-end than answering the inference query by traditional means. The techniques developed in the remainder of this chapter focus on exploiting symmetry for the computing lifting partitions. This is a sensible idea since symmetry detection can be done efficiently in practice, while it often yields coarse partitions for the model at hand. Moreover, symmetry and convexity enjoy a rather productive relationship as illustrated in the following argument.

Lemma 13.1 *Let G act on $I = \{1, \ldots, m\}$, so that every $g \in G$ corresponds to some permutation on $\{1, \ldots, m\}$. If $S^g = S$ and $J(x^g) = J(x)$ for every $g \in G$ (i.e., G stabilizes both S and J) then the induced orbit partition $\text{Orb}_G(I)$ is a lifting partition for $\inf_{x \in S} J(x)$.*

Proof   Suppose x_0 is some feasible solution to the convex program. Let us consider the centroid of the orbit of x_0 under G, i.e., $\bar{x}_0 = \frac{1}{|G|} \sum_{g \in G} x_0^g$. The permuting action of G stabilizes $\bar{x}_0$ since for any $g' \in G$, $\bar{x}_0^{g'} = \left(\frac{1}{|G|} \sum_{g \in G} x_0^g \right)^{g'} = \sum_{g \in G} x_0^{g \circ g'} = \sum_{g \in G} x_0^g$, as $G \circ g = G$. Now, we will show that $\bar{x}_0$ both: (a) is a feasible solution and (b) decreases the value of J, $J(\bar{x}_0) \leq J(x)$. Both facts follow from the convexity of S and J. Recall that $J(x_0^g) = J(x_0)$ for every $g \in G$. The convexity of J dictates that J evaluated at the average of all x_0^g must be less than the average of their values, i.e. $J(\bar{x}_0) = J\left(\frac{1}{|G|} \sum_{g \in G} x_0^g \right) \leq \frac{1}{|G|} \sum_{g \in G} J(x_0^g) = J(x_0)$. Similarly, since all x_0^g belong to S, so does their average by the convexity of S. Now, applying the above argument to an optimal solution x_*, we conclude that $\bar{x}_*$ must also be an optimal solution. Moreover it is stabilized by G, so $\text{Orb}_G(I)$ is a lifting partition. $\square$

To summarize, the **orbits** of groups stabilizing (13.2) and/or (13.3) are an excellent source for lifting partitions. As we will see in the next section, such partitions are connected intimately with the structure of the exponential family $\mathcal{F}$ itself.

13.3   Exact Lifted Variational Inference: Automorphisms of $\mathcal{F}$

Now we have everything at hand to instantiate lifting partitions for exact variational inference in the exponential family $\mathcal{F}$. This will establish the notion of **automorphisms of members of the exponential family**.

13.3.1   Automorphisms of Exponential Family

We define the symmetry of an exponential family $\mathcal{F}$ as the group of transformations that preserve $\mathcal{F}$ (hence preserve h and Φ). The transformation used will be a pair of permutations (π, γ), where π permutes the set of variables and γ permutes the feature vector.

Definition 13.2 (Automorphism of exponential family) *An **automorphism of the exponential family** $\mathcal{F}$ is a pair of permutations (π, γ) where $\pi \in \mathbb{S}_n$, $\gamma \in \mathbb{S}_m$, such that for all vectors x: $h(x^\pi) = h(x)$ and $\Phi(x^\pi) = \Phi^\gamma(x)$ (or equivalently, $\Phi^{\gamma^{-1}}(x^\pi) = \Phi(x)$).*

Showing that the set of all automorphisms of $\mathcal{F}$, denoted by $\mathrm{Aut}(\mathcal{F})$, forms a subgroup of $\mathbb{S}_n \times \mathbb{S}_m$ is straightforward. This group acts on $\mathcal{I}$ by the permuting action of γ, and on $\mathcal{V}$ by the permuting action of π. In the following, h is always a symmetric function (e.g., $h \equiv 1$) therefore, the condition $h(x^\pi) = h(x)$ automatically holds.

Example 13.1 Let $\mathcal{V} = \{1, \ldots, 4\}$ and $\Phi = \{f_1, \ldots, f_5\}$, where $f_1(x_1, x_2) = x_1(1 - x_2)$, $f_2(x_1, x_3) = x_1(1 - x_3), f_3(x_2, x_3) = x_2 x_3, f_4(x_2, x_4) = x_4(1 - x_2), f_5(x_3, x_4) = x_4(1 - x_3)$. Then $\pi = (1 \leftrightarrow 4)(2 \leftrightarrow 3), \gamma = (1 \leftrightarrow 5)(2 \leftrightarrow 4)$ form an automorphism of $\mathcal{F}$, since

$$\Phi^{\gamma^{-1}}(x^\pi) =$$
$$= (\phi_5(x_4, \ldots, x_1), \phi_4(x_4, \ldots, x_1), \ldots, \phi_1(x_4, \ldots, x_1))$$
$$= (f_5(x_2, x_1), f_4(x_3, x_1), f_3(x_3, x_2), f_2(x_4, x_2), f_1(x_4, x_3))$$
$$= (x_1(1 - x_2), x_1(1 - x_3), x_3 x_2, x_4(1 - x_2), x_4(1 - x_3))$$
$$= \Phi(x_1, \ldots, x_4).$$

Proposition 13.1 (π, γ) *is an automorphism of $\mathcal{F}$ if and only if the following conditions are true for all $i \in \mathcal{I}$: (1) $|\eta_i| = |\eta_{\gamma(i)}|$; (2) π is a bijective mapping from* $\mathrm{scope}(f_i)$ *to* $\mathrm{scope}(f_{\gamma(i)})$; (3) *let $\alpha = \eta_\gamma^{-1}(i) \circ \pi \circ \eta_i$ then $\alpha \in \mathbb{S}_{|\eta_i|}$ and $f_i(t^\alpha) = f_{\gamma(i)}(t)$ for all $t \in X^{|\eta_i|}$.*

Proof   (Part 1) We first prove that if $(\pi, \gamma) \in \mathrm{Aut}(\mathcal{F})$, then the conditions in the proposition hold. Pick $i \in \mathcal{I}$ and let $\gamma(i) = j$. Since $\Phi^\gamma(x) = \Phi(x^\pi), f_j(x) = f_i(x^\pi)$. Expressing the feature f_i and f_j in their factorized forms, we have $f_j(x_{j_1}, \ldots, x_{j_{|\eta_j|}}) = f_i(x_{\pi(i_1)}, \ldots, x_{\pi(i_{|\eta_i|})})$. Since each f_j cannot be reduced further, it must depend on all the distinct arguments in $\{j_1, \ldots, j_{|\eta_j|}\}$. This implies that the set of arguments on the LHS $\{\pi(i_1), \ldots, \pi(i_{|\eta_i|})\} \supseteq \{j_1, \ldots, j_{|\eta_j|}\}$. Thus $|\eta_i| \geq |\eta_j|$. Applying the same argument with the automorphism (π^{-1}, γ^{-1}), and noting that $\gamma^{-1}(j) = i$, we obtain $|\eta_j| \geq |\eta_i|$. Thus $|\eta_i| = |\eta_j| = K$. Furthermore, $\{\pi(i_1), \ldots, \pi(i_K)\} = \{j_1, \ldots, j_K\}$. This implies that π is a bijection from $\mathrm{scope}(f_i) = \{i_1, \ldots, i_K\}$ to $\mathrm{scope}(f_j) = \{j_1, \ldots, j_K\}$. For the third condition, from $f_j(x_{j_1}, \ldots, x_{j_{|\eta_j|}}) = f_i(x_{\pi(i_1)}, \ldots, x_{\pi(i_{|\eta_i|})})$, we let $t_k = x_{j_k}$ so that $t_{\eta_j^{-1}(k)} = x_k$ (since $j_k = \eta_j(k)$) to arrive at $f_j(t_1, \ldots, t_K) = f_i(t_{\eta_j^{-1} \circ \pi \circ \eta_i(1)}, \ldots, t_{\eta_j^{-1} \circ \pi \circ \eta_i(K)})$, or in short form $f_j(t) = f_i(t^\alpha)$. Here, α is a bijection since all the mappings η_j, η_i and π are bijections.

(Part 2) Let (π, γ) be a pair of permutations such that the three conditions are satisfied, we will show that they form an automorphism of $\mathcal{F}$. Picking $i \in I$ and let $j = \gamma(i)$ and $K = |\eta_i| = |\eta_j|$. From $f_j(t) = f_i(t^\alpha)$, we have $f_j(x_{j_1}, \ldots, x_{j_K}) = f_i(x_{j_{\alpha(1)}}, \ldots, x_{j_{\alpha(K)}})$. Note that $j_{\alpha(k)} = \eta_j \circ \alpha(k) = \pi(i_k)$. Thus $f_j(x_{j_1}, \ldots, x_{j_K}) = f_i(x_{\pi(i_1)}, \ldots, x_{\pi(i_K)})$, so $f_j(x) = f_i(x^\pi)$ and hence $\Phi^\gamma(x) = \Phi(x^\pi)$.   □

Consider automorphisms of the type $(\mathbf{1}, \gamma)$: γ must permute between the features having the same scope: $\mathrm{scope}(f_i) = \mathrm{scope}(f_{\gamma(i)})$. Thus if the features do not have redundant scopes (i.e., $\mathrm{scope}(f_i) \neq \mathrm{scope}(f_j)$ when $i \neq j$) then γ must be $\mathbf{1}$. More generally when features do not have redundant scopes, π uniquely determines γ. Next, consider automorphisms of the type $(\pi, \mathbf{1})$: π must permute among variables in a way that preserve all the features f_i. Thus if all features are asymmetric functions then π must be $\mathbf{1}$; more generally, γ uniquely determines π. As a consequence, if the features do not have redundant scopes and are asymmetric functions then there exists a one-to-one correspondence between π and γ that form an automorphism in $\mathrm{Aut}(\mathcal{F})$.

An automorphism as defined above preserves a number of key characteristics of the exponential family F (such as its natural parameter space, its mean parameter space, and its log-partition function), as shown in the following theorem.

Theorem 13.1   *If $(\pi, \gamma) \in \mathrm{Aut}(\mathcal{F})$, then*

1. $\pi \in \mathrm{Aut}(G[\mathcal{F}])$, *i.e., π is an automorphism of the graphical model graph $G[F]$.*
2. $\Theta^\gamma = \Theta$ *and $A(\theta^\gamma) = A(\theta)$ for all $\theta \in \Theta$.*
3. $\mathcal{F}(x^\pi | \theta^\gamma) = \mathcal{F}(x | \theta)$ *for all $x \in X^n$, $\theta \in \Theta$.*
4. $\mathbf{m}^\gamma(\theta) = \mathbf{m}(\theta^\gamma)$ *for all $\theta \in \Theta$.*
5. $M^\gamma = M$ *and $A^*(\mu^\gamma) = A^*(\mu)$ for all $\mu \in M$.*

Proof   Part (1) To prove that π is an automorphism of Γ, the hypergraph representing the structure of the exponential family graphical model, we need to show that $c \subseteq \mathcal{V}$ is a hyperedge (cluster) of Γ iff $\pi(c)$ is a hyperedge.

If c is a hyperedge, $\exists i \in I$ such that $c = \mathrm{scope}(f_i)$. Let $j = \gamma(i)$, by Proposition 13.1, $\pi(c) = \mathrm{scope}(f_j)$, so $\pi(c)$ is also a hyperedge.

If $\pi(c)$ is a hyperedge, apply the same reasoning using the automorphism (π^{-1}, γ^{-1}), we obtain $\pi^{-1}(\pi(c)) = c$ is also a hyperedge.

Part (2)-(5). We first state some identities that will be used repeatedly throughout the proof. Let $x, y \in \mathbb{R}^n$. The first identity states that permuting two vectors do not change their inner products

$$\langle x, y \rangle = \langle x^\pi, y^\pi \rangle.$$

As a result if $(\pi, \gamma) \in \mathrm{Aut}[\mathcal{F}]$,

$$\langle \Phi(x^\pi), \theta^\gamma \rangle = \langle \Phi^{\gamma^{-1}}(x^\pi), \theta \rangle = \langle \Phi(x), \theta \rangle.$$

The next identity allows us to permute the integrating variable in a Lebesgue integration

$$\int_S f(x) d\lambda = \int_{S^{\pi^{-1}}} f(x^\pi) d\lambda \, ,$$

where λ is a **counting measure**, or a **Lebesgue measure** over $\mathbb{R}^n$. The case of counting measure can be verified directly by establishing a bijection between summands of the two summations, and the case of Lebesgue measure is direct result of the property of linearly transformed Lebesgue integrals (Theorem 24.32, page 616).

Part (2). By definition of the log-partition function,

$$A(\theta^\gamma) =$$

$$= \int_{X^n} h(x) \exp\langle \Phi(x), \theta^\gamma \rangle d\lambda$$

$$= \int_{X^n} h(x^\pi) \exp\langle \Phi(x^\pi), \theta^\gamma \rangle d\lambda (by 11.3)$$

$$= \int_{X^n} h(x) \exp\langle \Phi(x), \theta \rangle) d\lambda (by 11.2)$$

$$= A(\theta).$$

As a result, $\Theta^\gamma = \{\theta^\gamma | A(\theta) < \infty\} = \{\theta^\gamma | A(\theta^\gamma < \infty\} = \Theta$.

Part (3). $\mathcal{F}(x^\pi | \theta^\gamma) = h(x^\pi) \exp\langle \Phi(x^\pi), \theta^\gamma \rangle = h(x) \exp\langle \Phi(x), \theta \rangle = \mathcal{F}(x|\theta)$.

Part (4). Expanding $\mathbf{m}^\gamma(\theta)$ gives

$$\mathbf{m}^\gamma(\theta) = \mathbf{E}_\theta \Phi^\gamma(x) = \int_{X^n} \Phi^\gamma(x) \mathcal{F}(x|\theta) d\lambda = \int_{X^n} \Phi^\gamma(x^{\pi^{-1}}) \mathcal{F}(x^{\pi^{-1}}|\theta) d\lambda,$$

where the last equality follows from (11.3). Since (π^{-1}, γ^{-1}) is also an automorphism, $\Phi(x^{\pi^{-1}}) = \Phi^{\gamma^{-1}}(x)$, thus $\Phi^\gamma(x^{\pi^{-1}}) = \Phi(x)$. Further, by part (3), $\mathcal{F}(x^{\pi^{-1}}|\theta) = \mathcal{F}(x|\theta^\gamma)$. Thus $\mathbf{m}^\gamma(\theta) = \int_{X^n} \Phi(x)\mathcal{F}(x|\theta^\gamma) = \mathbf{m}(\theta^\gamma)$.

Part (5). Let $\mu \in M$, so $\mu = \int_{X^n} p(x)\Phi(x)d\lambda$ for some probability density p. Expanding μ^γ gives

$$\mu^\gamma = \int_{X^n} p(x)\Phi^\gamma(x)d\lambda = \int_{X^n} p(x)\Phi(x^\pi)d\lambda = \int_{X^n} p(x^{\pi^{-1}})\Phi(x)d\lambda$$

Let $p'(x) = p(x^{\pi^{-1}})$ and observe that $\int p'(x)d\lambda = \int p(x)d\lambda = 1$, so p' is also a probability density. Thus $\mu^\gamma \in M$, hence $M^\gamma \subseteq M$. Applying similar reasoning to the automorphism $(\pi - 1, \gamma - 1)$, we have $\mu^{\gamma^{-1}} \in M$. Thus, every $\mu \in M$ can be expressed as μ'^γ for some $\mu' \in M$, but this means $M \subseteq M^\gamma$. Thus, $M = M^\gamma$.

For $\mu \in \text{ri}M$, there exists $\theta \in \Theta$ such that $\mu = \mathbf{m}(\theta)$. The negative entropy function becomes $A_*(\mu) = \langle \mu, \theta \rangle - A(\theta)$. From part (4), $\mu^\gamma = \mathbf{m}(\theta^\gamma)$, thus $A_*(\mu^\gamma) = \langle \mu^\gamma, \theta^\gamma range -$

$A(\theta^\gamma) = \langle \mu, \theta \rangle - A(\theta) = A_*(\mu)$. For $\mu \in \text{border}\mathcal{M} \backslash \text{ri}\mathcal{M}$, $A_*(\mu^\gamma) = A_*(\mu)$ holds by a continuity argument. □

13.3.2   Parameter Tying and Lifting Partitions

Having automorphisms of $\mathcal{F}$ at hand, we can now start to develop lifted *variational inference algorithms*. Intuitively, —keywordlifted variational inference is all about tying indistinguishable variables together. More formally, let us assume a partition Δ of I such that the partition equivalence $j \overset{\Delta}{\sim} j'$ implies that the corresponding parameters are equal, $\theta_j = \theta_{j'}$. We call such a partition **parameter-tying**. Our goal now is to study how this *parameter-tying*—coupled with the symmetry of the family $\mathcal{F}$—can lead to more efficient variational inference. Restricting the natural parameters to Θ_Δ is equivalent to parameter tying, and hence, equivalent to working with a different exponential family where $|\Delta|$ aggregates corresponding features $\left(\sum_{j \in \Delta_i} f_j \right)$. While this family has fewer parameters, we note that it is not obvious how it would help inference; moreover, in working directly with the aggregated features, the structure of the original family is actually lost.

To gain insights as to the effect of parameter tying on the efficiency of inference, we turn towards the question of how to characterize the image of Θ_Δ under the mean mapping $\mathbf{m}$. At first, note that in general $\mathbf{m}(\Theta_\Delta) \neq \mathcal{M}_\Delta$: taking Δ to be the singleton partition $\{I\}$ will enforce all natural parameters to be the same, but clearly this does not guarantee that all mean parameters are the same. However, one can hope that perhaps some mean parameters are forced to be the same due to the symmetry of the graphical model. More precisely, we ask the following question: is there a partition ϕ of I such that for all $\theta \in \Theta_\Delta$ the mean parameters are guaranteed to lie inside $\mathcal{M}_\phi$, and therefore the domain of the variational problem in (13.1) can be restricted accordingly to $\mathcal{M}_\phi$? We will now show the existence of such partitions for convex optimization problems in general.

From Theorem 13.1, we know that $\text{Aut}(\mathcal{F})$ stabilizes $\mathcal{M}$ and A^*; however, this group does not take the parameters θ into account. Given a partition Δ, a permutation γ on I is consistent with Δ if and only if γ permutes only among elements of the same cell of Δ. Such permutations are of special interest since for every $\theta \in \Theta_\Delta$, $\theta^\gamma = \theta$. If G is a group acting on I, we denote as G_Δ the set of group elements whose actions are consistent with Δ, that is $G_\Delta = \left\{ g \in G \,\middle|\, \forall u \in I, g(u) \overset{\Delta}{\sim} u \right\}$. It is straightforward to verify that G_Δ is a subgroup of G. With this notation, $\text{Aut}_\Delta(\mathcal{F})$ is the subgroup of $\text{Aut}(\mathcal{F})$ whose members' action is consistent with Δ. The group $\text{Aut}_\Delta(\mathcal{F})$ thus stabilizes not only the family $\mathcal{F}$, but also every parameter $\theta \in \Theta_\Delta$. It is straightforward to verify $\text{Aut}_\Delta(\mathcal{F})$ stabilizes both the constraint set and the objective functions of (13.3) and (13.2). Therefore, by Lemma 13.1, its induced orbit yields a lifting partition:

Corollary 13.1.1 *Let $\phi = \phi(\Delta) = \text{Orb}_{\text{Aut}(\mathcal{F})}(\mathcal{I})$. Then for all $\theta \in \Theta_\Delta$, ϕ is a lifting partition for the variational problems* (13.3) *and* (13.2), *that is*

$$\sup_{\mu \in \mathcal{M}} \langle \theta, \mu \rangle - A^*(\mu) = \sup_{\mu \in \mathcal{M}_\phi} \langle \theta, \mu \rangle - A^*(\mu) . \qquad (13.4)$$

In (13.4), we call the left-hand side the **ground** formulation of the variational problem, and the right-hand side the **lifted** formulation. Let $l = |\phi|$ be the number of cells of ϕ. Then, the lifted feasible set $\mathcal{M}_\phi$ effectively lies inside an l-dimensional subspace where $l \leq m$. This formalizes the core idea of the principle of lifted variational inference: to perform optimization over the lower dimensional (and hopefully easier to optimize over) feasible set $\mathcal{M}_\phi$ instead of $\mathcal{M}$. Indeed, the above result also holds for any subgroup G of $\text{Aut}_\Delta(\mathcal{F})$ since $\phi_G = \text{Orb}_G(\mathcal{I})$ is finer than ϕ. Thus, it is obvious that ϕ_G is also a lifting partition. However, the smaller the group G is, the finer is the lifting partition ϕ_G, and the less symmetry can be exploited. In the extreme, G can be the trivial group. In this case, ϕ_G is the discrete partition on $\mathcal{I}$ putting each element in its own cell, and $\mathcal{M}_{\phi_G} = \mathcal{M}$, which corresponds to no lifting.

13.3.3   Computing Automorphisms of $\mathcal{F}$

We have just introduced automorphisms of the exponential family and established that they induce lifting partitions. Now we show how to **compute the lifting partitions**. More precisely, we discuss the computation of the lifting group $\text{Aut}_\Delta(\mathcal{F})$ and its orbit partitions. We would like to stress that, in practice, computing and working with a subgroup of the lifting group is sufficient.

Our first approach is to construct a suitable graph whose automorphism group is guaranteed to be a subgroup of $\text{Aut}_\Delta(\mathcal{F})$, and thus any tool and algorithm for computing graph automorphism can be applied. The constructed graph resembles a factor graph representation of $\mathcal{F}$. We will make use of colors of factor nodes to mark feature functions that are both identical and in the same cell of Δ, and colors of edges to encode symmetry of the feature functions themselves.

Definition 13.3 (Colored factor graph) *The **colored factor graph** induced by $\mathcal{F}$ and Δ, denoted by $\Gamma_\Delta[\mathcal{F}]$ is a **bipartite graph** with nodes $\{x_1, \ldots, x_n\} \cup \{f_1, \ldots, f_m\}$ and edges $\{\{x_{\eta_i(k)}, f_i\} | i \in \mathcal{I}, k = 1, \ldots, |\eta_i|\}$. Variable nodes are assigned the same color which is different from the colors of factor nodes. Factor nodes f_i and f_j have the same color iff $f_i = f_j$ and $i \overset{\Delta}{\sim} j$. If the function f_i is symmetric, then all edges adjacent to f_i have the same color; otherwise, they are colored according to the argument number of f_i, i.e., $\{x_{\eta_i(k)}, f_i\}$ is assigned the k-th color.*

Figure 13.2 shows the construction of the colored factor graph for the exponential family in Example 13.1, where we have assumed that all the parameters are the same.

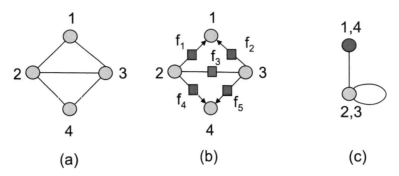

Figure 13.2: Graph construction for computing the lifting group and its orbits: (a) original graphical model of example 13.1; (b) constructed **colored factor graphs**, assuming all parameters are the same (arrows represent first arguments of the asymmetric factors); and (c) lifted graphical model with nodes representing node orbits and edges representing edge orbits of the original graphical model.

Theorem 13.2 *The **automorphism group** $\mathrm{Aut}(\Gamma_\Delta[\mathcal{F}])$ of $\Gamma_\Delta[\mathcal{F}]$ is a subgroup of $\mathrm{Aut}_\Delta(\mathcal{F})$, i.e., $\mathrm{Aut}(\Gamma_\Delta[\mathcal{F}]) \leq \mathrm{Aut}_\Delta(\mathcal{F})$.*

The theorem says that finding the automorphism group $\mathrm{Aut}(\Gamma_\Delta[\mathcal{F}])$, the colored factor graph induced by $\mathcal{F}$ and Δ, yields a procedure to compute a subgroup $\mathrm{Aut}_\Delta(\mathcal{F})$, our target of interest. Nauty, for example, directly implements operations of computing the automorphism group of a graph and extracting the induced node orbits and edge orbits.

Given an automorphism of a member of the exponential family (and its induces lifting partition), it is natural to ask how it affects the **marginal polytope**. We will show now that its dimensionality can be reduced and that its size can be characterized in terms of the number of orbits found.

13.3.4   The Lifted Marginal Polytope

Let us consider $\mathcal{M}_\phi$ in the case of (finite) discrete random variables. Note that $\mathcal{M}$ is the convex hull *conv* $\{\Phi(x) \,|\, x \in X^n\}$, which is a polytope in $\mathbb{R}^m$, and $\mathrm{Aut}(\mathcal{F})$ acts on the set of configurations X^n by the permuting action of π, $x \mapsto x^\pi$.

Theorem 13.3 *Let $O = \mathrm{Orb}_{\mathrm{Aut}_\Delta(\mathcal{F})}(X^n)$ be the set of X-configuration orbits. For each orbit $C \in O$, let $\bar{\Phi}(C) = \frac{1}{|C|} \sum_{x \in C} \Phi(x)$ be the feature-centroid of all the configurations $x \in C$. Then $\mathcal{M}_{\phi(\Delta)} = conv\{\bar{\Phi}(C) \,|\, C \in O\}$.*

As a consequence of the theorem, the **lifted polytope** $\mathcal{M}_\phi$ can have at most $|O|$ extreme points. The number of configuration orbits $|O|$ can be much smaller than the total number of configurations $|X|^n$ when the model is highly symmetric. For example, for a fully connected graphical model with identical pairwise and unary potentials and $X = \{0, 1\}$, every permutation $\pi \in \mathbb{S}^n$ is an automorphism; thus, every configuration with the same number of 1's belongs to the same orbit, hence $|O| = n + 1$. In general, however, $|O|$ often is still exponential in n. We discuss approximations of $\mathcal{M}_\phi$ later on.

A representation of the lifted polytope $\mathcal{M}_\phi$ by a set of constraints in $\mathbb{R}^{|\phi|}$ can be directly obtained from the constraints of the polytope $\mathcal{M}$. For each cell ϕ_j ($j = 1, \ldots, |\phi|$) of ϕ, let $\bar{\mu}_j$ be the common value of the variables μ_i, $i \in \phi_j$. Recalling the notation $\phi[i]$, which maps i to its class in ϕ (i.e., $\phi[i] = j$ if $i \in \phi_j$), we obtain $\mathcal{M}_\phi$ substitute each μ_i by a class representative variable $\bar{\mu}_{\phi[i]}$ in the constraints of $\mathcal{M}$, and thus obtain a set of constraints in $\bar{\mu}$ (in vector form, we substitute μ by $D\bar{\mu}$ where D is an $n \times |\phi|$ matrix having $D_{ij} = 1$ if $i \in \phi_j$ and 0 otherwise). In doing this, some constraints will become identical and thus redundant. In general, the number of non-redundant constraints can still be exponential.

With this at hand, we can now state analogous results in lifting the overcomplete variational problems in (13.2) and (13.3) when X is finite. To simplify notation, we will consider only the case where features are unary or pairwise. As before, the group $\mathrm{Aut}_\Delta(\mathcal{F})$ will be used to induce a lifting partition. However, we need to define the action of this group on the set of overcomplete features $\mathcal{I}^o$.

Recall that if $(\pi, \gamma) \in \mathrm{Aut}(\mathcal{F})$ then π is an **automorphism of the graphical model** graph Γ. Since overcomplete features naturally correspond to nodes and edges of Γ (as defined in Section 13.2.2), π has a natural action on $\mathcal{I}^o$ in terms of $v : t \mapsto \pi(v) : t$ and $\{u : t, v : t'\} \mapsto \{\pi(u) : t, \pi(v) : t'\}$. Let us define $\phi^o = \mathrm{Orb}_{\mathrm{Aut}_\Delta(\mathcal{F})}(\mathcal{I}^o)$ to be the induced orbits of $\mathrm{Aut}_\Delta(\mathcal{F})$ on the set of overcomplete features.

Corollary 13.3.1 *For all $\theta \in \Theta_\Delta$, ϕ^o is a lifting partition for the variational problems in* (13.2) *and* (13.3).

Thus, the optimization domain can be restricted to $\mathcal{M}^o_{\phi^o}$, which we call the **lifted marginal polytope**. The cells of ϕ^o are intimately connected to the node, edge and arc orbits of the graph G induced by $\mathrm{Aut}_\Delta(\mathcal{F})$. We now list all the cells of ϕ^o in the case where $X = \{0, 1\}$: each node orbit $\bar{v}$ corresponds to 2 cells, $\{v : t | v \in \bar{v}\}, t \in \{0, 1\}$; each edge orbit $\bar{e}$ corresponds to 2 cells $\{\{u : t, v : t\} | \{u, v\} \in \bar{e}\}, t \in 0, 1$; and each arc orbit $\bar{a}$ corresponds to the cell $\{\{u : 0, v : 1\} | (u, v) \in \bar{a}\}$. The orbit mapping $\phi_o[\cdot]$ maps each element of $\mathcal{I}^o$ to its orbit as follows: $\phi^o[v : t] = \bar{v} : t$, $\phi[\{u : t, v : t\}] = \overline{\{u, v\}} : t$, $\phi[\{u : 0, v : 1\}] = \overline{(u, v)} : 01$.

The total number of cells of ϕ^o is $O(|\bar{V}| + |\bar{E}|)$ where $|\bar{V}|$ and $|\bar{E}|$ are the number of node and edge orbits of Γ (each edge orbit corresponds to at most 2 arc orbits). Thus, in working with $\mathcal{M}^o_{\phi^o}$, the big-O order of the number of variables is reduced from the number of nodes and edges in Γ to the number of node and edge orbits.

Let us summarize what we have achieved so far. we have introduced exact lifted variational inference for the exponential family in general. As we have argued, symmetries—automorphisms—of an exponential family distribution may reduce the dimension of its marginal polytope and in turn speed up variational inference. Now, we will move beyond exact inference and consider liftability of approximate variational inference problems. It turns out that relaxing the inference problem also allows one to relax the form of symmetries. That is, lifting approximate inference my result in even more compressed models.

13.4   Beyond Orbits of $\mathcal{F}$:   Approximate Lifted Variational Inference

An exponential family $\mathcal{F}$ is capable of describing quite complex dependencies among random variables. Given these complex dependencies, it is not surprising that probabilistic inference is often computationally intractable. Variational inference is no exception. To achieve **tractability**, we hit two two major roadblocks. First, the description of $\mathcal{M}$ (resp. the marginal polytope $\mathcal{M}^o$) may be exponentially sized, making feasibility queries inefficient. Second, the conjugate of the log-partition function A^* may not be representable in closed form, or its representation may also be of prohibitive size. In both cases, lifting variational inference may help to tackle the roadblock, and in many cases the achieved reductions indeed make up for the effort spend to compute the automorphisms. However, to make exact variational inference really tractable, we would need an exponential reduction in problem size. Currently this can be guaranteed only in very specific situations.

Given these roadblocks to tractability, it is natural to ask the question whether we can still enjoy the benefits of **dimensionality reduction**, even when the reduction is not exponential. In the following, we will show that this is indeed the case. Akin to successful approximation schemes based on relaxations of the original inference optimization problem—**dual decompositions** and **linear program** (LP) **relaxations** are two prominent examples for this—we show that lifting can also be relaxed in a similar fashion: we relax some of the constraints in an attempt to turn both the inference and the lifting problem jointly into easier ones to solve. In other words, we develop combinations of lifting and inference algorithms, which are tractable to begin with, i.e. approximations.

To analyze the liftability of **approximate variational inference** problems, we will introduce a new class of partitions, called equitable partition. These partitions strictly generalize **orbit partitions** of exponential families in the sense that every orbit partition is equitable, yet completely asymmetric models could still admit an equitable partition. Remarkably, equitable partitions of exponential families relate to local relaxations of variational inference problems in exactly the same way as automorphism (orbits) relate to the variational inference problem. That is, equitable partitions identify a non-empty lower-dimensional subset of the local polytope of the exponential family, such that if the parameter is fixed by the partition, a solution to the problem is guaranteed to exist in that slice. That is, they act as lifting partitions.

We will now proceed with formalizing all these notions and developing the corresponding theory.

13.4.1   Approximate Variational Inference

Approximations to the **MAP** respectively **marginal variational inference** problem must overcome both of the aforementioned issues, hence they generally involve a tractable approximation of the marginal polytope $\mathcal{M}^o$, and, in the case of marginal inference, a tractable approximation to A^*.

To approximate the **marginal polytope**, the standard strategy is to construct some compactly representable polytope containing $\mathcal{M}^o$. The most popular choice of these outer bounds is the so-called **local polytope**. It is obtained by relaxing an natural **binary linear program** encoding of **MAP inference**:

$$
\mathcal{M}^L(\mathcal{F}) = \left\{ \tau \geq 0 \;\middle|\; \begin{array}{l} \sum_{t \in X} \tau_{u:t} = 1 \text{ for all } u \in \mathcal{V} \\ \sum_{t' \in X} \tau_{u:t,v:t'} = \tau_{u:t} \text{ for all } \{u,v\} = \mathrm{scope}(f_i), i \in \mathcal{I}, t \in X \end{array} \right\} . \quad (13.5)
$$

The variables τ in $\mathcal{M}^L$, which we refer to as **pseudo marginals**, behave like marginals locally: if we look at a variable x_u, $\tau_{u:t}$ is a probability distribution; moreover, pairwise pseudo marginals $\tau_{u:t,v:t'}$, corresponding to joint configurations of the scopes of features, are consistent with singleton marginals, as marginalizing out one of the variables produces the pseudo marginal of the remaining variable. It is clear that any vector of true marginals has these properties (hence $\mathcal{M}^o \subseteq \mathcal{M}^L$), however, depending on the structure of $\mathcal{F}$, there might be feasible τ's that do not correspond to the marginals of any distribution. With the exception of some graph families such as trees, the inclusion is strict. Nevertheless, this outer bound often gives a good approximation to the true marginals. If needed, the approximation can be tightened using additional constraints.

The second ingredient is an approximation to A^*. Formally, $-A^*(\mu)$ is the **entropy** of the **maximum entropy distribution** with marginals μ. A prominent class of approximations to this quantity come in the form of a weighted sum of the marginal entropies, i.e.

$$
\widehat{A_c^*}(\mathcal{F})(\tau) = \widehat{A_c^*}(\tau) = \sum_{u \in \mathcal{V}} c_u H(\tau_u) - \sum_{i \in \mathcal{I}} c_i I(\tau_i) , \quad (13.6)
$$

where $H(\tau_u) = \sum_{t \in X} \tau_{u:t} \log_2(\tau_{u:t})$ and $I(\tau_i) = \sum_{t,t' \in X} \tau_{u:t,v:t'} \log_2(\tau_{u:t,v:t'})$ for $\{u,v\} = \mathrm{scope}(f_i)$.

Since speed is crucial for approximate inference, one typically prefers approximations where the optimization is efficient. A natural class of efficiently optimizeable approximations results from $\widehat{A_c^*}$ being concave. In particular, we are interested in values of c that make $\widehat{A_c^*}$ concave. One sufficient condition for $\widehat{A_c^*}$ being concave is that nonnegative auxiliary numbers c_{ui} exist for all $u, i, x_u \in \mathrm{scope}(f_i)$), such that

$$
C(\mathcal{F}) = \left\{ c_u, c_i \;\middle|\; \begin{array}{l} \exists c_{ui} \geq 0 : \\ c_u - \sum_{u:x_u \in \mathrm{scope}(f_i)} c_{ui} \geq 0, \\ c_i + \sum_{i:x_u \in \mathrm{scope}(f_i)} c_{ui} \geq 0 \end{array} \right\} . \quad (13.7)
$$

With this at hand, the approximate variational inference problem can be written as follows:

$$\tau^* = \underset{\tau \in \text{OUTER}}{\text{argmax}} \underbrace{\left[\langle \tau, \theta \rangle + T \cdot \widehat{A}_c^*(\tau) \right]}_{=:F_c(\tau)}, \tag{13.8}$$

where OUTER is $\mathcal{M}^L$ (or later on one of its tightened versions). The term $F_c(\tau) := \langle \tau, \theta \rangle + T \cdot \widehat{A}_c^*(\tau)$ will be referred to as the **energy** and will be assumed to be concave by means of $c \in C(\mathcal{F})$. Note that θ refers to the parameters according to the overcomplete representation, otherwise the expression $\langle \tau, \theta \rangle$ would be ill-defined.

This formulation unifies approximate marginal inference, when $T = 1$, and MAP inference if $T = 0$. Since for $T = 0$ the objective becomes linear (subject to linear constraints), the MAP problem is typically referred to as MAP-LP. Note also that for $c_u = 1 - |\{i : x_u \in \text{scope}(f_i)\}|$ and $c_i = 1$, F becomes the **Bethe energy**, F_{Bethe}. In this case, solving the set of saddle-point equations of (13.8) by means of fixed-point iteration yields the popular **loopy belief propagation** algorithm. The Bethe Energy often gives surprisingly good approximations to the true marginals, however, it is difficult to optimize over.

In any case, we will focus here on approximating the marginal polytope. For this, it is natural to ask whether we can also approximate the notion of **automorphism** for lifting the marginal polytope. In the following we will show that this is indeed the cases. Instead of using orbit partitions we can use equitable partitions as lifting partitions. They are a generalization of orbit partitions and correspond to **fractional automorphisms** of the **exponential family**. That is, some of the constraints describing automorphisms are relaxed.

13.4.2   Equitable Partitions of the Exponential Family

Let start off by defining equitable partitions. Let U and V be a finite sets and $A : U \times V \to \mathbb{R}$ be a real-valued function on pairs of elements of U and V. Moreover, let Δ^U and Δ^V be partitions of U and V.

Definition 13.4 (Equitable partition) *A pair $\Delta = (\Delta^U, \Delta^V)$ is an **equitable partition** of A if:*

* *for each class C of Δ^U and for each pair of elements $u, u' \in C$,*

$$\sum_{v \in D} A(u, v) = \sum_{v \in D} A(u', v)$$

   for every class D of Δ^V ;
* *for each class D of Δ^V and for each pair of elements $v, v' \in D$,*

$$\sum_{u \in C} A(u, v) = \sum_{u \in C} A(u, v')$$

   for every class C of Δ^U .

There are several special cases of this general definition of equitable partitions. Since U and V are finite, they could be the indices of a real-valued matrix. For example, if A is the adjacency matrix of a bipartite graph Γ, the definition collapses to the following.

Definition 13.5 (Equitable partition of a bipartite graph) *A partition (Δ^U, Δ^V) is equitable for a bipartite graph Γ if for each pair of vertices w, w' in the same class C and any other class D, $|\text{nb}(w) \cap D| = |\text{nb}(w') \cap D|$.*

If we term the quantity $|\text{nb}(w) \cap D|$ as $\deg(w, D)$, then the above definition tells us that $\deg(w, D)$ must be constant over the elements w of a class. Therefore, we can introduce the term class degree as $\deg(C, D) := \deg(C, D)$, where w is any element of the class C. Note that in general $\deg(D, C)$ is not equal to $\deg(C, D)$. However, what is true is that $|D| \deg(D, C) = |C| \deg(C, D)$, as both sides of the equation express the same quantity—the total number of edges between elements of C and D.

We will now instantiate the definition for equitable partition of an exponential family.

Definition 13.6 (Equitable partition of exponential family) *Given an exponential family $\mathcal{F} = ((x_v))_{v \in \mathcal{V}}, (f_i)_{i \in \mathcal{I}})$, an equitable partition of $\mathcal{F}$ is a partition $\Delta = (\Delta^{\mathcal{V}}, \Delta^{\mathcal{I}})$ of $\mathcal{V}$ and $\mathcal{I}$ that obeys the following:*

a) *for each pair $u \overset{\Delta}{\sim} v$ in $\mathcal{V}$ there exists a permutation $\gamma : \{i : x_u \in \text{scope}(f_i)\} \to \{j : x_v \in \text{scope}(f_j)\}$ such that $i \overset{\Delta}{\sim} \gamma(i)$.*

b) *for each feature pair $i \overset{\Delta}{\sim} j$ in $\mathcal{I}$, and for each pair, $u \in \text{scope}(f_i) \overset{\Delta}{\sim} v \in \text{scope}(f_j)$ there exists a permutation $\pi : \{1, \ldots, |\eta^i|\} \to \{1, \ldots, |\eta^j|\}$ such that: if $u = \eta^i(m)$ and $v = \eta^j(n)$, then $\pi(m) = n$; $f_i(t_1, \ldots, t_k) = f_j(t_{\pi(1)}, \ldots, t_{\pi(k)})$ for all $t_1, \ldots, t_k \in X^k$, and $\eta^i(m) \overset{\Delta}{\sim} \eta^j(\pi(m))$ for all $m \in \{1, \ldots, n\}$.*

This definition might seem cumbersome at first sight, especially part b). However, the high level idea is rather intuitive, seen as a relaxation of the characterization in Proposition 13.1. Recall that π, γ is a symmetry of $\mathcal{F}$ if $\Phi(x^\pi) = \Phi^\gamma(x)$ for all $x \in X^{|\mathcal{V}|}$. The orbit partition then groups feature f_i with feature f_j if $i = \gamma(j)$ for some symmetry γ and likewise, x_u is grouped with x_v if $u = \pi(v)$ for some symmetry π. Roughly speaking, equitable partitions emulate orbit partitions in a local sense. That is, if we fix some $u \overset{\Delta}{\sim} v$, then by a), their neighborhood will appear symmetric by virtue of some local permutation γ. If we were to fix $i \overset{\Delta}{\sim} j$, then each pair of variables in their scope will appear to have been grouped together because of some symmetry π. This symmetry π, however, acts only locally on their scopes. It is easy to verify that orbit partitions according to any subgroup of the automorphism group is equitable. Yet, the converse is not true. There are asymmetric exponential families which have nontrivial equitable partitions.

To see how the Definition 13.6 relates to Definition 13.5, observe that since γ and π are class-preserving bijections, the following holds.

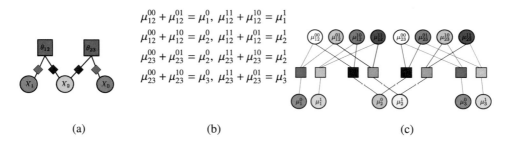

$$\mu_{12}^{00} + \mu_{12}^{01} = \mu_1^0, \ \mu_{12}^{11} + \mu_{12}^{10} = \mu_1^1$$
$$\mu_{12}^{00} + \mu_{12}^{10} = \mu_2^0, \ \mu_{12}^{11} + \mu_{12}^{01} = \mu_2^1$$
$$\mu_{23}^{00} + \mu_{23}^{01} = \mu_2^0, \ \mu_{23}^{11} + \mu_{23}^{10} = \mu_2^1$$
$$\mu_{23}^{00} + \mu_{23}^{10} = \mu_3^0, \ \mu_{23}^{11} + \mu_{23}^{01} = \mu_3^1$$

(a)	(b)	(c)

Figure 13.3: Model symmetry of $\mathcal{F}$ propagates to its local polytope. (a) $\mathcal{F}$ – colored by the classes of the CEP. (b) The local polytope of $\mathcal{F}$ (nonnegativity and normalization constraints as well as position nodes have been omitted for clarity). (c) The factor graph of the local polytope of $\mathcal{F}$ colored by its CEP. The colors indicate that the classes can be deduced from the CEP of $\mathcal{F}$. Note that the darker and lighter version of each color are not grouped together. (Best viewed in color)

Proposition 13.2 *If $\Delta = (\Delta^V, \Delta^I)$ is an equitable partition of $\mathcal{F}$, then for any $u \overset{\Delta}{\sim} v$, and for any feature class $\phi \in \Delta^I$, $\deg(u, \phi) = \deg(v, \phi)$. For any $i \overset{\Delta}{\sim} j$ and variable class $v \in \Delta^V$, $\deg(i, v) = \deg(j, v)$.*

The final fact, which we state here and that is discussed in the chapter on color refinement and its applications, will be the central tool in the theory we are about to develop. It provides the analysis of the symmetries of the local polytope.

Theorem 13.4 *Let U, V be finite sets, $A : U \times V \to \mathbb{R}$, $b \in \mathbb{R}^{|U|}$, and L be the polyhedron $L = \{x \in \mathbb{R}^{|V|} \mid \forall u \in U : \sum_{v \in V} A(u, v)x_v \geq b_u\}$. If A admits an equitable partition $\Delta = (\Delta^U, \Delta^V)$ stabilizing b ($u \overset{\Delta^U}{\sim} u' \Rightarrow b_u = b_{u'}$), then Δ is a lifting partition of L, i.e. there exists an extreme point in L, $\bar{x}$, stabilized by Δ ($v \overset{\Delta^V}{\sim} v' \Rightarrow \bar{x}_v = \bar{x}_{v'}$).*

13.4.3 Equitable Partitions of Concave Energies

With this, we can now prove the following theorem, following similar argument as for Corollary 13.1.1 in the case of lifted exact variational inference.

Theorem 13.5 (Equitable Partions of Concave Energies Are Lifting Partitions) *Let $\mathcal{F}$ be an exponential family and $\Delta = (\Delta^V, \Delta^I)$ an equitable partition of $\mathcal{F}$ (in the sense of Def. 13.6). If Δ is a tying partition for the parameter $\theta \in \Theta$ and for the **counting** vector $\mathbf{c} \in C$, then there exists an according lifting partition for the approximate variational problem in (13.8).*

Since the proof of Theorem 13.5 goes along similar lines as the proof of Corollary 13.1.1, let us briefly recap the main argument: from Lemma 13.1, we know that if a convex program admits a group of automorphisms, then it also admits a solution that is stabilized by that partition (i.e. variables in the same orbit have the same values). Then, Theorem 13.1, together with the assumption that Δ is parameter-tying allows us to argue that the automorphism group of $\mathcal{F}$ also stabilizes the (convex) variational inference problem.

Unfortunately, we no longer have group-theoretic tools at hand, especially the concept of a group action. To get around, will use the averaging operator of a partition:

Definition 13.7 (Averaging operator) *Let* $\Delta = (\Delta_1, \dots, \Delta_k)$ *be a partition of* $\{1, \dots, n\}$. *The **averaging operator** of* Δ, $X_\Delta = \mathbb{R}^n \to \mathbb{R}^n$ *is defined as*

$$(X_\Delta y)_i = \frac{1}{|\Delta[y]|} \sum_{j \stackrel{\Delta}{\sim} i} y_j \; .$$

Observing now that the averaging operator is idempotent, i.e. $X_\Delta X_\Delta y = X_\Delta y$, the averaging operator allows one to generalize Lemma 13.1 as follows.

Lemma 13.2 *Let* $\inf_{y \in S} J(y)$ *be a convex optimization problem, where* $S \subseteq \mathbb{R}^m$. *Let* Δ *be a partition of* $\{1, \dots, m\}$. *If* $J(y) \geq J(X_\Delta y)$ *and* $y \in S$ *implies* $X_\Delta y \in S$ *for all* $y \in \mathbb{R}^m$, *then* Δ *is a lifting partition for that problem.*

Proof Let y_0 be any optimal solution. By hypothesis $X_\Delta y_0$ is also optimal and feasible. Since X_Δ is idempotent, $X_\Delta y_0$ is stabilized by Δ, hence Δ is a lifting partition.□ Similarly to the case of Corollary. 13.1.1, to prove Theorem 13.5 we will need to argue that if $\mathcal{F}$ admits an equitable partition, we can derive a partition for the variables of the **concave energy** from it, such that its averaging operator satisfies the hypotheses of Lemma 13.2.

We will now start to address the situation with the constraints. I.e. we will show that an equitable partition of $\mathcal{F}$ induces a partition on the variables of $\mathcal{M}^L$ whose averaging operator preserves feasibility. The approach will be to translate the equitable partition Δ of $\mathcal{F}$ (in the sense of Def. 13.6) to an equitable partition (in the sense of Def. 13.4) of $\mathcal{M}^L$. Then we apply Theorem 13.4 to establish the result.

Looking at Figure 13.3, it is clear that $\mathcal{M}^L$ has more variables and constraints than $\mathcal{F}$ has variables and features, hence directly using the equitable partition of $\mathcal{F}$ as an equitable partition of $\mathcal{M}^L$ directly cannot possibly work. However, we also see that on a high level, the structure is indeed very similar.

Lemma 13.3 *Let* $\mathcal{F} = ((x_v)_{v \in \mathcal{V}}, (f_i)_{i \in I})$ *be an exponential family. Every equitable partition* $(\Delta^I, \Delta^{\mathcal{V}})$ *of* $\mathcal{F}$ *induces an equitable partition* ∇ *on the constraints matrix of* $\mathcal{M}^L$.

Proof We first observe that the description of $\mathcal{M}^L$ consists of the following elements:

- one pseudo marginal LP variable for each $v \in \mathcal{V}$ and $t \in X$, e.g. $\tau_{v:t}$;
- one pseudo marginal LP variable for each joint configuration of variables in the scope of some f_i, e.g. $\tau_{\eta_i(1):t_1, \dots, \eta_i(k):t_k}$ (k is the arity of f_i);

- one marginalization constraint for each pair of v, i such that $x_v \in \text{scope}(f_i)$, denoted as $c_{v,t,i}$.

In addition to these, the description of $\mathcal{M}^L$ contains non-negativity and normalization constraints are omitted from the analysis as their structure is trivial. Given an equitable partition $\Delta = (\Delta^\mathcal{V}, \Delta^\mathcal{I})$ of $\mathcal{F}$, we construct the partition ∇ of $\mathcal{M}^L$ as follows:

a) $\tau_{v:t} \overset{\nabla}{\sim} \tau_{v':t'}$ if and only if $t = t'$ and $v \overset{\Delta}{\sim} v'$;

b) $\tau_{\eta_i(1):t_1,\dots\eta_i(k):t_k} \overset{\nabla}{\sim} \tau_{\eta_j(1):t'_1,\dots\eta'_j(k):t'_k}$ if and only if $i \overset{\Delta}{\sim} j$ (including the case of $j = i$) and there exists a bijection $\pi : \{1, \dots, k\} \to \{1, \dots, k\}$ (where $k = |\eta^i| = |\eta^j|$) such that: $t'_1, \dots, t'_k = t_{\pi(1)}, \dots, t_{\pi(k)}, f_j(t'_1, \dots, t'_k) = f_i(t_{\pi(1)}, \dots, t_{\pi(k)})$ and $\eta^i(q) \overset{\Delta}{\sim} \eta^j(\pi(q))$ for all $q \in \{1, \dots, k\}$;

c) $c_{v,t,i} \overset{\nabla}{\sim} c_{v',t',i'}$ if and only if $v \overset{\Delta}{\sim} v'$, $t = t'$ and $i \overset{\Delta}{\sim} i'$.

Let us now argue that ∇ is equitable. Since the coefficients in the constraints matrix of $\mathcal{M}^L$ are in $\{-1, 0, 1\}$, where only one entry per row is equal to -1 (the coefficient of $\tau_{v:t}$ in $c_{v,t,i}$, see e.g. Fig. 13.3b), matrix equitability (Def. 13.4), which we would need for Theorem 13.4, collapses to graph equitability (Def. 13.5), so we only need to establish that the same number of variables of each class participate in equivalent constraints, and any equivalent variables participate in the same number of constraints for each class. We verify this now, which established the result.

Each single-node pseudo marginal participates in one constraint for each of its features, i.e. $\tau_{v:t}$ is adjacent only to constraints of type $c_{v,t,i}$. Hence by construction (a), for any $\tau_{v:t} \overset{\nabla}{\sim} \tau_{v':t}$ the classes of constraints they participate in correspond to classes of features, e.g. $\deg(\tau_{v:t}, C) = \deg(v, \phi)$ and $\deg(\tau_{v':t}, C) = \deg(v', \phi)$, where ϕ is the class of the feature of the constraint. By Proposition 13.2, the two quantities are equal. Moreover, each constraint $c_{v,t,i}$ involves exactly one single-node marginal $\tau_{v:t}$. Since $c_{v,t,i} \overset{\nabla}{\sim} c_{v',t,i'}$ implies $t = t'$ and $v \overset{\Delta}{\sim} v'$, $\deg(c_{v,t,i}, S) = \deg(c_{v',t',i'}), S) = 1$, if S is the class of $\tau_{v:t}$ and $\tau_{v':t}$ in ∇ and 0 otherwise. Each joint pseudo marginal $\tau_{\eta_i(1):t_1,\dots,\eta_i(k):t_k}$ participates in the constraints $c_{\eta_i(1),t_1,i}, \dots, c_{\eta_i(k),t_k,i}$. Let $\tau_{\eta_i(1):t_1,\dots\eta_i(k):t_k} \overset{\nabla}{\sim} \tau_{\eta_j(1):t'_1,\dots\eta'_j(k):t_k}$. By construction (b), $i \overset{\Delta}{\sim} j$ (also including $j = i$) and there exists a bijection π with $t'_1, \dots, t'_k - t_{\pi(1)}, \dots, t_{\pi(k)}$. Therefore, for all $q \in 1, \dots, k$ we have: $t'_q = t_{\pi(q)}, \eta^i(q) \overset{\Delta}{\sim} \eta^j(\pi(q))$. By (c), this implies $c_{\eta_i(q),t_q,i} \overset{\nabla}{\sim} c_{\eta_j(\pi(q)),t_{\pi(q)},j}$. In other words, π maps the constraints of $\tau_{\eta_i(1):t_1,\dots\eta_i(k):t_k}$ to those of $\tau_{\eta_j(1):t'_1,\dots\eta'_j(k):t_k}$ by class, hence $\deg(\tau_{\eta_i(1):t_1,\dots\eta_i(k):t_k}, C) = \deg(\tau_{\eta_j(1):t'_1,\dots\eta'_j(k):t_k}, C)$ for any constraint class C. Finally, by (c), $c_{v,t,i} \overset{\nabla}{\sim} c_{v',t',i'}$ implies $v \overset{\Delta}{\sim} v'$, $t = t'$ and $i \sim i'$. The joint pseudo marginals that participate in $c_{v,t,i}$ are all $\tau_{\eta_i(1):t_1,\dots,\eta_i(k):t_k}$ such that $\eta^i(q) = v$ and $t_q = t$ for some q. Since $i \sim i'$, there exists $\pi : \{1, \dots, |\eta^i|\} \to \{1, \dots, |\eta^j|\}$ (Def. 13.6) such that: if $u = \eta^i(m)$ and $v = \eta^j(n)$, then $\pi(m) = n$; for all $t_1, \dots, t_k \in X^k$, and $\eta^i(m) \overset{\Delta}{\sim} \eta^j(\pi(m))$ for all $m \in \{1, \dots, n\}$. By (b), this implies $\tau_{\eta_i(1):t_1,\dots,\eta_i(k):t_k} \overset{\nabla}{\sim} \tau_{\pi(\eta_i(1)):t_{\pi(1)},\dots,\pi(\eta_i(k)):t_{\pi(k)}}$. That is, π acts as a class-preserving bijection between the neighbors of $c_{v,t,i}$ and $c_{v',t',i'}$. $\square$ The following is a consequence of Theorem 13.4:

Corollary 13.5.1 *If τ is feasible in $\mathcal{M}^L$, then so is $X_\nabla \tau$.*

Let us now turn to the objective function. Ignoring the semantics of τ, $\widehat{A_c^*}$ is just a linear combination of $x \log x$ terms with c as coefficients, i.e. $\widehat{A_c^*}(x) = \sum_k c_k x_k \log x_k$. Observe that any permutation mapping x_i to x_j such that $c_i = c_j$ leaves $\widehat{A_c^*}$ invariant. In other words, the automorphism group of $\widehat{A_c^*}$ is the group $\mathrm{Aut}(\widehat{A_c^*}) = \{\pi | c^\pi = c\}$. This has as consequence the following Lemma.

Lemma 13.4 *Let Δ be an equitable partition of $\mathcal{F}$ such that: a) Δ respects the parameter θ; b) Δ respects c. Then, for the partition ∇ induced on the pseudo marginal vector τ as described in Lemma 13.3, $F_c(X_\nabla \tau) \geq F_c(\tau)$.*

Proof We start by observing that X_Δ as defined in Def 13.7 corresponds to a symmetric doubly stochastic matrix as $(X_\Delta)_{ij} = \frac{1}{|\Delta[i]|}$ if $i \overset{\Delta}{\sim} j$ and 0 otherwise. The **Birkhoff-von Neumann Theorem** (Marshall et al., 2011) allows one to decompose the doubly stochastic X as a convex combination of permutation matrices $\sum_k \lambda_k \Pi_k$. Since Δ respects both c and θ, any of these permutations matrices will exchange only variables that have the same θ and c. This follows from the fact that a) λ's form a convex combination, b) if $c_i \neq c_j$ or $\theta_i \neq \theta_j$, $X_{ij} = 0$, hence $(\Pi_k)_{ij} = 0$ for all k. Thus, all Π_k are automorphisms of F_c. With this in mind, the concavity of F_c yields the result: $F_c(\bar{\tau}) = F(\sum_k \lambda_k \Pi_k \tau) \geq \sum_k \lambda_k F_c(\Pi_k \tau) = F_c(\tau)$. $\square$

Putting everything together, we finally prove Theorem 13.5: we have satisfied all the hypotheses of Lemma 13.2, hence an equitable partition of $\mathcal{F}$ induces a lifting partition of the **concave energy**.

13.4.4   Lifted Counting Numbers

While equitable partitions of concave energies are lifting partitions for the variational problem of interest, it is not clear whether they actually exist. In other words, the question arises: if we want to perform inference, can we find counting numbers that permit lifting at all? As it turns out, not only do such counting numbers c exist, but also most heuristics known from the literature for finding them will yield c that respect equitable partitions of $\mathcal{F}$. Let us first show the existence of **liftable counting numbers**.

Theorem 13.6 (Existence of Lifted Counting Numbers) *Let $\Delta = (\Delta^I, \Delta^V)$ be an equitable partition of $\mathcal{F}$. If C is not empty, there exists a c-vector that respects (Δ^I, Δ^V).*

Proof We only sketch the proof as it is analogous to the proof of Lemma 13.3. Observe that the set of counting numbers is defined by a linear program, as indicated by (13.7). Thus, we can again rely on the fact that equitable partitions of linear programs are lift-

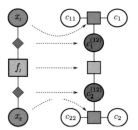

Figure 13.4: (a) Illustration of Lemma 5. (b) Commutative diagram underlying the "lifted inference by reparametrization" paradigm. (Best viewed in color)

ing partitions, as long as we can translate (Δ^I, Δ^V) into an equitable partition of C. The translation is as follows:

- c_v is grouped with $c_{v'}$ if and only if $v \stackrel{\Delta}{\sim} v'$ and c_i is grouped with $c_{i'}$ if and only if $i \stackrel{\Delta}{\sim} i'$;
- c_{vi} is grouped with $c_{v'i'}$ if and only if $v \stackrel{\Delta}{\sim} v'$ and $i \stackrel{\Delta}{\sim} i'$;
- the constraints are grouped according to the classes of the nodes that generate them.

The resulting partition is equitable on C. Moreover, it is also compatible with the partition ∇ induced on $\mathcal{M}^L$. □

Thus, if $\mathcal{F}$ admits an equitable partition Δ, then there will be at least one energy that admits an equitable partition based on Δ. This leads to the question: how do we compute the appropriate counting numbers? Naturally, we can take any vector of counting numbers, any equitable partition of $\mathcal{F}$ and simply average c over the classes by applying the appropriate averaging operator. However, as C is a polyhedron, there are infinitely many vectors of counting numbers and the usefulness of the resulting energies in terms of the approximate inference problem will vary. Luckily, several existing heuristics for picking counting numbers naturally yield counting numbers that allow nice equitable partitions of the energy. The heuristics that we discuss are the following two:

(a) (Hazan and Shashua, 2008) following the principle of insufficient reason one tries to make c's as uniform as possible by minimizing either $\sum_{i \in I}(c_i - 1)^2$ or $\sum_{i \in I} c_i \log c_i$ over C;

(b) (Meshi et al., 2009) finding the least-squares projection of the c's of the Bethe energy onto C, i.e. $c^* = \text{argmin}_{c \in C} \sum_{i \in I}(c_i - 1)^2 + \sum_{v \in \mathcal{V}}(c_v - |\text{nb}(x_v)| + 1)^2$.

Both of them lead to liftable energies.

Proposition 13.3 (Finding lifted counting numbers) *Let Δ be an equitable partition of $\mathcal{F}$ and ∇ be the partition of C obtained as in Thm. 13.6. Then ∇ is a lifting partition of both (a) and (b).*

Proof The proof is identical to the proof of Thm. 13.5. Due to Theorem 13.6, given any $c \in C$, we obtain a new $c' \in C$ by applying the averaging operator X_∇. Now we need to show that this averaging does not increase the respective objective values. We argue in the same way as in Lemma. 13.4. For both cases in (**a**), the automorphism group consists of the product of two symmetric groups - we can exchange any two c_u, c_v and any two c_i, c_j independently. Thus averaging over ∇ is equivalent to averaging over automorphisms of the objectives, which by convexity does not increase the value. For the case of (**b**), the automorphism group is smaller. We are allowed to exchange any two c_i, c_j independently, but we can exchange c_u, c_v only if $|\text{nb}(x_u)| = |\text{nb}(x_v)|$. However, this property is indeed implied if u and v are grouped in any equitable partition of $\mathcal{F}$. Thus, the averaging matrix X_∇ as in Lemma. 13.4 will have a non-zero value only if $|\text{nb}(x_u)| = |\text{nb}(x_v)|$, and the resulting Birkhoff-von-Neumann decomposition will consist of automorphisms of the objective. □

In essence, this proposition tells us that one can find optimal numbers w.r.t. the heuristics (**a**) and (**b**) that turn any equitable partition of $\mathcal{F}$ into an equitable partition of the energy.

Unfortunately, this does not hold for all heuristics. Some heuristics impose restrictions on what partitions can be EPs of their respective energies. To see this, consider the (**c**) **tree-reweighted belief propagation** (in pairwise models) heuristic. We obtain $c \in C$ by setting c_i to be the number of spanning trees passing through f_i divided by the number of all spanning trees in $\mathcal{F}$. If we translate the result of Bui et al. (2014) in the language of the present chapter, it states that the equitable partition of $\mathcal{F}$ coarser than its orbit partition may generally not be turned into an equitable partition of the **tree-reweighted energy**. However, for the orbit partition, they give an efficient algorithm to produce the appropriate c. Or consider the heuristic (**d**) due to Meshi et al. (2009) : instead of approximating the numbers of F_{Bethe}, we project F_{Bethe} itself onto the set of concave energies $F_{c \in C}$. More precisely, we take $c^* = \text{argmin}_{c \in C} \int_{\tau \in \mathcal{M}^L} (F_{\text{Bethe}}(\tau) - F_c(\tau))^2 \, d\tau$. As a matter of fact, it is even an open question whether F_{Bethe} admits stationary points that respect[34] any Δ. It should be noted that if we want a coarser partition than what (**c**) or (**d**) permit, we could still take the counting numbers and average them over a coarser equitable partition of $\mathcal{F}$. However, the question of whether this operation preserves the quality of approximation remains open.

13.4.5   Computing Equitable Partitions of Exponential Families

We conclude this section with a brief discussion of the computational aspect of equitable partitions. The set of equitable partitions forms a lattice with respect to the refinement operation. Since the orbit partition (of a colored, oriented graphical structure) is also equitable, computing an equitable partition with prescribed properties is assumed to not be doable in polynomial time. Equitable partitions form a lattice under the refinement operation, implying that there exists a unique coarsest equitable partition. Surprisingly, the

---

[34] Works on lifted loopy belief propagation, see e.g. (Jaimovich et al., 2007; Singla and Domingos, 2008; Kersting et al., 2009) and also the corresponding chapter in this book, do indeed show that BP admits such solutions. However, the question of whether this is due to F_{Bethe} or an artifact of the way BP optimizes it is unclear.

coarsest equitable partition of a structure is computable in quasi-linear time via an intuitive algorithm called **color refinement**, see the corresponding chapter in the book.

For exponential families, it is straightforward to show that the coarsest equitable partition of the colored factor graph (Definition 13.3) is also an equitable (possibly not the coarsest) partition of $\mathcal{F}$. In contrast to the corresponding orbit partition, the coarsest equitable partition is both tractably computable and offers more compression.

13.5 Unfolding Variational Inference Algorithms via Reparametrization

So far, we have established that any convex energy that admits an equitable partition also admits a solution, which is homogenous over the classes of the partition. Thus, having found such a partition, we can safely restrict the solution space accordingly and obtain a smaller optimization problem, which can be solved with an off-the-shelf convex solver.

The main drawback of doing so is that by resorting to an off-the-shelf convex solver, we give up on a wide variety of very efficient algorithms developed for convex variational inference. Efficient solvers, such as MPLP (Globerson and Jaakkola, 2007; Sontag et al., 2008b,a) for MAP and TRW-BP for marginals (Wainwright et al., 2005), which have been reported to often minimize a convex energy significantly faster than generic convex solvers, are derived by applying a given optimization strategy—block coordinate descent, fixed-point iteration, etc.—to the algebraic form of the variational problem. Due to the structure of the equations, a single descent step results in only a local update to the solution vector, which can be interpreted as passing a message between variables resp. features.

Fortunately, as we will show now, the theory of equitable partitions provides us with a way around these issues. The main insight is that a given exponential family induces actually a whole family of exponential familis of different sizes sharing essentially the same MAP-LP solution. From these, one can select the smallest one where MAP beliefs can be computed using off-the-shelf MP approaches. These beliefs then are also valid—after a simple transformation—for the original problem. Moreover, this incurs no major overhead: the selected exponential family is at most twice as large than the fully lifted feature graph. In this way we eliminate the need for using off-the-shelf convex solver only or to modified existing MP algorithms.

Thus, the take-home message is two-fold:

1. By making use of lifted linear programming, we show that lifted inference over convex energies can be formulated as ground inference on a reparametrized exponential family.

2. There is an efficient algorithm that, given a ground exponential family, finds the smallest reparametrized exponential family and show that its size is not more than twice the size of the fully lifted model.

This has several implications for lifted inference. For instance, using MPLP (Globerson and Jaakkola, 2007) results in the first convergent MP approach for MAP-LP relaxations, and using other MP approaches such as TRW-BP (Wainwright et al., 2005) actually spans a whole family of lifted MP approaches. This suggests a novel view on lifted probabilistic inference: *lifted inference can be vifewed as standard inference in a reparametrized model.*

We start our discussion by examining the structure of the lifted approximate variational inference problem.

13.5.1   The Lifted Approximate Variational Problem

We start by describing the structure of the lifted approximate variational problem of (13.8), i.e., the compressed optimization problem obtained by unifying all partition-equivalen variables (replacing each of them with a single representative variable) and removing all redundant constraints (it can be shown that after variable unification, all constraints in the same class become identical, but we will not do this here). Specifically, given an equitable partition Δ of $\mathcal{F}$, we consider the structure of

$$\tau^* = \underset{\tau \in \mathcal{M}_\Delta^L}{\operatorname{argmax}} F_\Delta^c(\tau) \,, \tag{13.9}$$

where $\mathcal{M}_\Delta^L$ and F_Δ^c are the restrictions $\mathcal{M}^L$ and F^c to the space $\mathbb{R}_\nabla$ (where ∇ corresponds to Δ as in Lemma 13.3.). For the purposes of exposition, we will assume that the variables of $\mathcal{F}$ take values in $\{0, 1\}$ and the features of $\mathcal{F}$ have scope size at most 2, i.e., we are looking at Ising models of arbitrary topology. Both of these assumptions can be removed easily. In fact, the domain of the variables will not matter at all later on, whereas increasing the arity of features will incur some cost as will become clear in the following.

Suppose we are given an equitable partition $\Delta = (\Delta^\mathcal{V}, \Delta^I)$ of $\mathcal{F}$. If we follow the translation rules of Lemma 13.3, the lifted local polytope $\mathcal{M}_\Delta^L$ becomes:

$$\mathcal{M}_\Delta^L(\mathcal{F}) = \left\{ \tau \geq 0 \left| \begin{array}{l} \tau_{v:0} + \tau_{v:1} = 1 \text{ for all } v \in \Delta^\mathcal{V} \text{for all } \omega \in \Delta^I : \\ \tau_{v:0,\omega:0} + \tau_{v:0,\mathit{ff}:1} = \tau_{v:0} \\ \tau_{v:1,\omega:1} + \tau_{v:1,\omega:0} = \tau_{v:1} \\ \tau_{v:0,\omega:0} + \tau_{v:1,\omega:0} = \tau_{\omega:0} \\ \tau_{v:1,\omega:1} + \tau_{v:0,\omega:1} = \tau_{\omega:1} \end{array} \right. \right\}, \tag{13.10}$$

where ϕ is a feature class, and v and ω are the classes of the variables in its scope. At first glance, the above looks as if it might be the local polytope of a smaller (lifted) exponential family. This would indeed be the case, save for one very important detail. Namely, ω and v might actually be the same element, i.e., there could be a feature whose two variables have been grouped together by the equitable partition. In this case, the derived partition ∇ will group $\tau_{u:0,v:1}$ with $\tau_{u:1,v:0}$ for every $u, v \in \omega$, if there is a feature f with $f(u, v) = f(v, u)$ (item (b) in Lemma 13.3). This leaves us with the peculiar set of constraints:

$$\tau_{v:0,v:0} + \tau_{v:*,v:*} = \tau_{v:0} \text{ and } \tau_{v:1,v:1} + \tau_{v:*,v:*} = \tau_{v:1} \,,$$

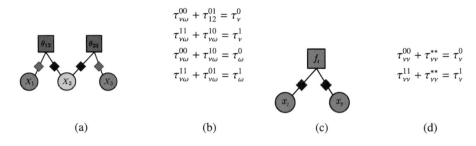

$$\tau_{v\omega}^{00} + \tau_{12}^{01} = \tau_{v}^{0}$$
$$\tau_{v\omega}^{11} + \tau_{v\omega}^{10} = \tau_{v}^{1}$$
$$\tau_{v\omega}^{00} + \tau_{v\omega}^{10} = \tau_{\omega}^{0}$$
$$\tau_{v\omega}^{11} + \tau_{v\omega}^{01} = \tau_{\omega}^{1}$$

$$\tau_{vv}^{00} + \tau_{vv}^{**} = \tau_{v}^{0}$$
$$\tau_{vv}^{11} + \tau_{vv}^{**} = \tau_{v}^{1}$$

(a)   (b)   (c)   (d)

Figure 13.5: Degeneracy of the lifted local polytope: (a) a factor graph of an exponential family colored by the classes of an equitable partition, which does not unify neighboring variables (b); the corresponding lifted local polytope corresponds to the local polytope of a smaller model exponential family. (c) The factor graph of an exponential family colored by an equitable partition that does merge neighboring variables; (d) the corresponding lifted local polytope does not correspond to the local polytope of any exponential family.

where $\tau_{v:*,v:*} = \tau_{v:1,v:0} = \tau_{v:0,v:1}$. This situation is illustrated on Figure 13.5.

It is clear that such a constraint could never be generated according to the description of $\mathcal{M}^{L}$ in (13.5). In other words, there is no exponential family whose local polytope is $\mathcal{M}_{\Delta}^{L}$. Consequently the set of all local polytopes of exponential families is not closed under the lifting operation. Similarly, the lifted energy after compression becomes:

$$F = \sum_{t \in X, v \in \Delta} \theta_{v:t} \tau_{v:t} + \sum_{t,t' \in T} \sum_{v,\omega = \text{scope}(\phi), \phi \in \Delta} \theta_{v:t,\omega:t'} \tau_{v:t,\omega:t'}$$
$$+ \sum_{t \in X, v \in \Delta} c_{v} \tau_{v:t} \log_{2}(\tau_{v:t}) - \sum_{t,t' \in X, v,\omega \in E(\Gamma)} c_{\phi} \tau_{v:t,\omega:t'} \log_{2}(\tau_{v:t,\omega:t'}) ,$$

where

$$\theta_{v} = |v| \theta_{v:t}, \quad \theta_{\phi} = |\phi| \theta_{u,v:t,t'}, \quad c_{v} = \sum_{v:t \in v'} c_{v}, \text{ and } c_{\phi} = \sum_{u,v \in \phi} c_{u,v} . \tag{13.11}$$

Given that parameters θ are tied in every class, we can reformulate the above directly in terms of the classes of Δ. If we denote as Δ^{V} the set of all classes consisting to single-node pseudomarginals and Δ^{I} as those consisting to pairwise pseudo-marginals, then, F can be expressed as

$$F = \sum_{v \in \Delta^{V}} \theta_{v} \tau_{v} + \sum_{\phi \in \Delta^{I}} \theta_{\phi} \tau_{\phi} + \sum_{v \in \Delta^{V}} c_{v} \tau_{v} \log_{2}(\tau_{v}) - \sum_{\phi \in \Delta^{I}} c_{\phi} \tau_{\phi} \log_{2}(\tau_{\phi}) . \tag{13.12}$$

If our goal were to apply "ground" algorithms to lifted models, an obvious solution would be to just prevent variables in the scope of a feature from being grouped when computing an equitable partition. This is easily implementable with tools such as Saucy, as they

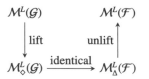

Figure 13.6: Commutative diagram established by Alg. 8 underlying our reparametrization approach to lifted inference.

allow for providing such constraints in the form of an initial partition that will be respected (the computed equitable parition will be a refinement of the initial partition). While this does indeed solve the problem, guaranteeing that the lifted energy will be the energy of some exponential family, there is a rather severe drawback. The approach prevents the compression of cliques. Given that the families generated by relational models tend to have many large cliques, this has a large impact on lifting, as the main potential for compression is lost. To overcome this, we will now present an alternative.

13.5.2 Energy Equivalence of Exponential Families

To recap, the problem we are facing is as follows: On the one hand, we would like to have lifted structures, which correspond to energies of exponential families, so that we can run convex energy-based message-passing algorithms without modifications. On the other hand, if we forbid unification of neighboring variables as to guarantee the former, we end up losing significant compression. So, is there a way to both run unmodified message-passing and keep a reasonable amount of compression?

Fortunately, the answer is yes. The key idea is to identify as set of models of different sizes and with different energies, which, however, after compression via appropriate equitable partitions give the exact same lifted energy. If two probability distributions have the same lifted energy, this automatically entails they share at least one optimal pseudo-marginal vector in common. To see this, consider two distributions $p \in \mathcal{F}$ and $g \in \mathcal{G}$ having the same lifted energy. We solve the approximate variational problem for p using any solver. Any solution of the ground energy of p can be averaged and compressed to a solution of the lifted energy of p (as described by Lemma 13.4 and Corollary 13.5.1). This is in turn identical to the lifted energy of q, hence, also its solution. This solution can now be unlifted to solve the ground energy of q. This is sketched in Fig. 13.6 and outlines a method for lifted inference (Algorithm 8) which does not require a modified lifted solver: lifting can be viewed as reparametrization.

It now remains to flesh out this method in full detail. We start with a definition of lifted equivalence among exponential families, then precisely characterize the classes of lifted equivalence models. Finally, we use this characterization to construct an algorithm for finding the smallest of such equivalent models.

Algorithm 8 Solving exponential family $\mathcal{F}$ using an equivalent exponential family $\mathcal{G}$

Require: Exponential family $\mathcal{F}$ and partition-equivalent exponential family $\mathcal{G}$ with respect to equitable partitions Δ and $\diamond$

Ensure: Vector of pseudo-marginals solving a convex energy.
  1: **Solve** the energy of exponential family $\mathcal{G}$.
  2: **Lift** the solution to the lifted energy of $\mathcal{G}$ w.r.t. $\diamond$; Due to partition equivalence, this is also a solution to the lifted energy of $\mathcal{F}$ w.r.t. Δ.
  3: **Recover** the solution of $\mathcal{F}$ by unlifting with respect to Δ.

Definition 13.8 (Partition Equivalence) *Let* $\mathcal{F} = ((x_v)_{v \in V}, (f_i)_{i \in I})$ *and* $\mathcal{G} = \left((y_w)_{w \in W}, (g_j)_{j \in J}\right)$ *(the number of variables and features may be different, i.e.* $|V| \neq |W|$ *resp.* $|I| \neq |J|$*) be exponential families.* $\mathcal{F}$ *and* $\mathcal{G}$ *are **partition equivalent** if and only if there exists an equitable partition* Δ *of* $\mathcal{F}$ *and an equitable partition* $\diamond$ *of* $\mathcal{G}$ *such that* $M_\Delta^L = M_\diamond^L$.

In order to realize the lifted inference schema suggested in Figure 13.6, we need to establish two key points: a) that distributions (with tied parameters) from parition equivalent families share solutions; b) a method for generating such equivalent families. There is one bit of insight into partition equivalence that will be indispensible in both endeavours.

Proposition 13.4 *Let there be two distributions* $\mathcal{F}$ *and* $\mathcal{G}$ *in the exponential family, and let* Δ *and* $\diamond$ *be corresponding equitable partitions.* $\mathcal{F}$ *and* $\mathcal{G}$ *are equivalent with respect to* Δ *and* $\diamond$ *if and only if there exists a bijection* $\sigma : \Delta \to \diamond$ *between the classes of* Δ *and* $\diamond$ *such that for each feature and variable class pair* $\phi, v \in \Delta$, $\deg(\phi, v) = \deg(\sigma(\phi), \sigma(v))$.

This enables us to approach our goals (a) and (b). We start with (a) and defer (b) to the next section.

Theorem 13.7 *If* $\mathcal{F}$ *and* $\mathcal{G}$ *are partition equivalent with respect to equitable partitions* Δ *and* $\diamond$*, then for each parameter vector* $\theta \in \Theta_\Delta(\mathcal{F})$*, and each counting vector* $c \in C_\Delta(\mathcal{F})$*, there exists a parameter* $\widehat{\theta} \in \Theta_\diamond(\mathcal{G})$ *and counting vector* $\widehat{c} \in C_\diamond(\mathcal{G})$ *such that* $F_\Delta^c(\theta) = F_\diamond^{\widehat{c}}(\widehat{\theta})$.

Proof   The proof constructs one possible $\widehat{\theta}, \widehat{c}$ pair given a pair θ, c. We first construct $\widehat{\theta}$.

The linear part of F_Δ^c is the one that depends on the parameter, and is equal to $\sum_{\phi \in \Delta} |\phi| \theta_\phi \tau_\phi$, where θ_ϕ is the parameter of the members of class ϕ (as the partition is parameter-tying). Since σ is a bijection between the classes of Δ and $\diamond$ (Proposition 13.4), the linear part of $F_\diamond^{\widehat{c}}$ can be expressed as $\sum_{\phi \in \Delta} |\sigma(\phi)| \widehat{\theta}_{\sigma(\phi)} \tau_{\sigma(\phi)}$. Therefore, if for each feature j of $\mathcal{G}$ in class $\sigma(\phi)$ we set $\widehat{\theta}_j = \widehat{\theta}_{\sigma(\phi)} := \frac{|\phi|}{|\sigma(\phi)|} \theta_\phi$, setting the linear parts of both lifted energies equal.

The counting vector $\widehat{c}$ is constructed similarly. That is, for each feature j of $\mathcal{G}$ in class $\sigma(\phi)$, we set $\widehat{c}_j := \widehat{c}_{\sigma(\phi)} = \frac{|\phi|}{|\sigma(\phi)|} c_\phi$. For each variable x_u of $\mathcal{G}$, we set $\widehat{c}_u := \widehat{c}_{\sigma(v)} = \frac{|v|}{|\sigma(v)|} c_v$. Finally, $\widehat{c}_{uj}$ is set as $\widehat{c}_{uj} := \widehat{c}_{\sigma(v)\sigma(\phi)} = \frac{|\phi|}{|\sigma(\phi)|} c_{\phi v}$. It now remains to be argued that $\widehat{c}$ is a valid element of $C(\mathcal{G})$.

Let ϕ, ν be classes of Δ. In any equitable partition it holds that $|\phi| \deg(\phi, \nu) = |\nu| \deg(\nu, \phi)$ (the quantity on both sides is the total number of edges between ϕ and ν written in two different ways), resp. $|\sigma(\phi)| \deg(\sigma(\phi), \sigma(\nu)) = |\sigma(\nu)| \deg(\sigma(\nu), \sigma(\phi))$. Taking the quotient of both yields

$$\frac{|\nu| \deg(\nu, \phi)}{|\sigma(\nu)| \deg(\sigma(\nu), \sigma(\phi))} = \frac{|\phi| \deg(\phi, \nu)}{|\sigma(\phi)| \deg(\sigma(\phi), \sigma(\nu))} = \frac{|\phi|}{|\sigma(\phi)|}, \qquad (13.13)$$

where the second equality is due to Proposition 13.4.

We have assumed c respects Δ. This allows us to rewrite the conditions of (13.7) as

$$c_\phi \geq - \sum_{\nu \in \Delta} \deg(\phi, \nu) c_{\nu\phi} \quad \text{and,} \ c_\nu \geq \sum_{\phi \in \Delta} \deg(\nu, \phi) c_{\nu\phi} \ .$$

The above describes the lifted linear program of $C(\mathcal{F})$ after unifying all equivalent variables. Now, observe that

$$c_{\sigma(\phi)} = \frac{|\phi|}{|\sigma(\phi)|} c_\phi \geq - \frac{|\phi|}{|\sigma(\phi)|} \sum_{\nu \in \Delta} \deg(\phi, \nu) c_{\nu\phi} = - \sum_{\nu \in \Delta} \frac{|\phi|}{|\sigma(\phi)|} \deg(\phi, \nu) c_{\nu\phi}$$

$$= - \sum_{\sigma(\nu) \in \diamond} \deg(\sigma(\phi), \sigma(\nu)) \frac{|\phi|}{|\sigma(\phi)|} c_{\nu\phi} = - \sum_{\sigma(\nu) \in \diamond} \deg(\sigma(\phi), \sigma(\nu)) c_{\sigma(\nu)\sigma(\phi)} \ .$$

Note, we were allowed to switch the index in the last line due to (13.13). Similarly,

$$c_{\sigma(\nu)} = \frac{|\nu|}{|\sigma(\nu)|} c_\nu \geq \frac{|\nu|}{|\sigma(\nu)|} \sum_{\phi \in \Delta} \deg(\nu, \phi) c_{\nu\phi} = \sum_{\phi \in \Delta} \frac{|\nu|}{|\sigma(\nu)|} \deg(\nu, \phi) c_{\nu\phi}$$

$$= \sum_{\phi \in \Delta} \deg(\sigma(\nu), \sigma(\phi)) \frac{|\phi|}{|\sigma(\phi)|} c_{\nu\phi} \qquad \text{(due to (13.13))}$$

$$= \sum_{\sigma(\phi) \in \diamond} \deg(\sigma(\nu), \sigma(\phi)) c_{\sigma(\nu)\sigma(\phi)} \ .$$

Thus, the claim that $\widehat{c}$ is in $C(\mathcal{G})$ is established. We can finally verify that both lifted dual partition function approximations are equal:

$$\widehat{A_{\widehat{c},\diamond}} = \sum_{\sigma(\nu) \in \diamond} |\sigma(\nu)| c_{\sigma(\nu)} H(\tau_{\sigma(\nu)}) - \sum_{\sigma(\phi) \in \diamond} |\sigma(\phi)| c_{\sigma(\phi)} I(\tau_{\sigma(\phi)})$$

$$= \sum_{\sigma(\nu) \in \diamond} |\sigma(\nu)| \frac{|\nu|}{|\sigma(\nu)|} c_\nu H_\nu + \sum_{\sigma(\phi) \in \diamond} |\sigma(\phi)| \frac{|\phi|}{|\sigma(\phi)|} c_\phi H_\phi$$

$$= \sum_{\nu \in \Delta} |\nu| c_\nu H_\nu + \sum_{\phi' \in \Delta} |\phi| c_\phi H_\phi = \widehat{A_{c,\Delta}} \ .$$

To recap, we have now established the following. If $\mathcal{F}$ and $\mathcal{G}$ are partition equivalent with respect to partitions Δ and $\diamond$, then for any distribution in $\mathcal{F}$ with parameters and counting numbers respecting Δ, there is a counterpart in $\mathcal{G}$ such that the lifted energies of both are identical. Moreover, Theorem 13.7 gives us an explicit (and efficient) method of constructing this counterpart. Hence, we have accomplished part (a) of the program. Now,

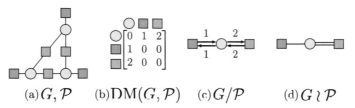

(a) $G, \mathcal{P}$ (b) $\mathrm{DM}(G, \mathcal{P})$ (c) $G/\mathcal{P}$ (d) $G \wr \mathcal{P}$

Figure 13.7: Lifted Structures: Example of a feature graph and its partitions and quotients. (a) A feature graph G (variables in circles, features in squares) and its coarsest EP $\mathcal{P}$ represented by the colors. (b) The degree DM matrix of G according to v. (c) The corresponding quotient graph $G/\mathcal{P}$. (d) The feature quotient $G \wr \mathcal{P}$. (Best viewed in color)

we turn towards part (b), that is, given a family $\mathcal{F}$ and partition Δ, we address the problem of computing $\mathcal{G}$ and $\diamond$.

13.5.3   Finding Partition-Equivalent Exponential Families

Given an exponential family $\mathcal{F}$ and an equitable partition Δ, Algorithm 9 finds the smallest partition equivalent family $\mathcal{G}$ in linear time. This section is dedicated to the explanation and analysis of Algorithm 9, namely proving its soundness and giving a bound on the size $\mathcal{G}$ (in terms of variables and features) of $2|\mathcal{F} \wr \Delta|$.

Before we proceed with the algorithm, let us introduce a visual tool for understanding partition equivalence, namely, the so-called degree matrix DM of the exponential family $\mathcal{F}$ relative to its equitable partition Δ. Recall that in an equitable partition, given classes P, Q, the quantity $\deg(p, Q)$ is fixed for all $p \in P$ and vice-versa. Thus, for an equitable partition Δ, we can visualize all class degrees as a square matrix of size $|\Delta| \times |\Delta|$. We specialize this to exponential families as follows:

Definition 13.9 (Degree matrix)   *Let $\mathcal{F}$ be an exponential family and equitable partition Δ. The **degree matrix** DM with respect ot Δ is an $|\Delta| \times |\Delta|$ matrix of natural numbers, such that for any two classes $\delta, \gamma \in \Delta$, $DM_{\delta, \gamma} = \deg(\delta, \gamma)$.*

For bipartite structures such as exponential families, the degree matrix assumes the block form $\mathrm{DM}(\mathcal{F}, \Delta) = \begin{pmatrix} 0 & \mathrm{DV} \\ \mathrm{DF} & 0 \end{pmatrix}$, where DV represents the relationship of thevariable classes to the feature classes and DF vice-versa. As a shorthand, we use the notation $\mathrm{DM} = (\mathrm{DF}, \mathrm{DV})$. Graphically (see Figure 13.7(c)), a degree matrix can be visualized as a quotient graph $\mathcal{F}/\Delta$, which is a directed multi-graph. In $\mathcal{F}/\Delta$ there is a node for every class of v. Given two nodes u, v we have $|P(u) \cap P_j|, u \in P_i$ many edges going from u to v. $\mathrm{DM}(G, \Delta)$ is essentially the weighted adjacency matrix of $G/\mathcal{P}$. Of particular interest to this development is the part of DM which captures the degrees from feature classes to variable classes. We will denote DF as $\mathcal{F} \wr \Delta$ and term it the *feature quotient*. To see the importance of $\mathcal{F} \wr \Delta$, we note the following direct corollary of Proposition 13.4.

Proposition 13.5 *Exponential families $\mathcal{F}$ and $\mathcal{G}$ are partition equivalent with respect to partitions Δ and $\Diamond$ if and only if $\mathcal{F} \wr \Delta = \mathcal{G} \wr \Diamond$.*

Recall that the lifted local polytope of $\mathcal{F}$ is fully defined by the feature f $\mathcal{F} \wr \Delta$. Hence, a necessary and sufficient condition for partition-equivalence the families exhibit the same feature quotients for some equitable partitions. Thus, the problem of finding a partition-equivalent family boils down to finding $\mathcal{G}$ such that $\mathcal{F} \wr \Delta = \mathcal{G} \wr \Diamond$ for some partition $\Diamond$. Moreover, to maximize the compression, we want Δ to be the coarsest EP of $\mathcal{F}$ (resulting in the smallest possible $\mathcal{F} \wr \Delta$) and that $\mathcal{G}$ is the smallest possible partition-equivalent exponential family. Let us now discuss how to find $\mathcal{G}$. As a running example, we will use the feature graph in Figure 13.8(a).

Suppose $\mathcal{F} \wr \Delta$ is given, e.g. computed using color-refinement. For our running example, $\mathcal{F} \wr \Delta$ is shown in Figure 13.8(b). Let us divide the superfeatures and supervariables of $\mathcal{F} \wr \Delta$ into classes based on their connectivity. A superfeature connected to a supervariable via a double edge is called a **(2)-superfeature**. In Figure 13.8(b), these are the cyan and orange superfeatures. Correspondingly, we call a variable connected to a superfeature via a double edge a **(2)-supervariable** (red and yellow in Figue 13.8(b)). Next, a superfeature connected to at least one (2)-supervariable via a *single* edge is called a **(1, 2)-superfeature** (violet and pink in Figure 13.8(b)). Finally, all other superfeatures and supervariables are **(1)-superfeatures** and **(1)-supervariables** respectively (e.g. the green supervariable).

We then compute $\mathcal{G}$ in the following way as also illustrated in Figure 13.8(c)-(e). We start with an empty graph. Then, **Step (A)** as illustrated in Figure 13.8(c) consists of adding for every (2)-superfeature in $\mathcal{F} \wr \Delta$ exactly one representative feature to $\mathcal{G}$. Furthermore, for every (2)-supervariable, we add two representatives in $\mathcal{G}$ and connect them to the corresponding (2)-superfeature representatives whenever the supernodes they represent are connected in $\mathcal{F} \wr \Delta$. In **Step (B)**, see Figure 13.8(d), for every (1, 2)-superfeature, we instantiate two representatives. Moreover, for every (2)-supervariable (all of them are already represented in $\mathcal{G}$), we match the two (1, 2)-superfeature representatives to the two (2)-supervariable representatives whenever the represented supernodes are connected in $\mathcal{F} \wr \Delta$. Finally, **Step (C)** as shown in Figure 13.8(e) introduces one representative for every other supernode and connects it to other representatives based on $\mathcal{F} \wr \Delta$. If it happens that the represented supernode is connected to a (2)-supervariable or (1, 2)-superfeature in $\mathcal{F} \wr \Delta$, we connect the representative to both representatives of the corresponding neighbor.

This is summarized in Algorithm 9 and provably computes a minimal structure of an partition-equivalent exponetial family.

Theorem 13.8 **(Soundness)** *$\mathcal{G}$ as computed above is partition-equivalent to $\mathcal{F}$.*

Proof   Following Def. 13.8 we must show that given $\mathcal{F}$ and its EP Δ, there is a partition $\Diamond$ of $\mathcal{G}$ such that the lifted local polytopes are equal. We take the partition $\Diamond$ to be the one induced by Algorithm 9. $\Diamond$ is equitable on $\mathcal{G}$ by construction: we can go through

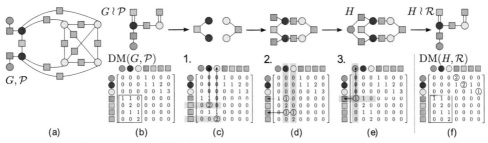

(a)   (b)   (c)   (d)   (e)   (f)

Figure 13.8: Illustration of "lifted inference by reparametrization" as summarized in Algorithm 9 and used as input to Algorithm 8. (a) input: a graph G and its coarsest equitable partition v; (b) the feature quotient $\mathcal{F} \wr \Delta$ and the corresponding degree matrix; (c) **Step (A)**: representing all degree 2 features and the corresponding degree 2 variables. (d) **Step (B)**: instantiating degree 1 features adjacent to degree 2 variables; (e) **Step (C)**: instantiating the remaining classes. After this step we terminate and output the result H; (f) soundness: $H \wr \mathcal{R}$ is identical to $G \wr \Delta$. Observe that the upper-right corner of $DM(G, \Delta)$ is different from that of $DM(H, \mathcal{R})$. (Best viewed in color.)

Algorithm 9 to verify that every two nodes in G representing the same supernode of $\mathcal{F} \wr \Delta$ are connected to the same number of representatives of every other supernode of $\mathcal{F} \wr \Delta$. Now, to show that $\mathcal{M}^L_\Delta(\mathcal{F})$ has the same constraints as $\mathcal{M}^L_\diamond(\mathcal{G})$, we need $\mathcal{F} \wr \Delta = \mathcal{G} \wr \diamond$. To see that this holds, observe that Algorithm 9 connects p to q in G if only if v is connected to ϕ in $\mathcal{F} \wr \Delta$: if ϕ is a (2)-superfeature, v is a (2)-supervariable – q will be connected to p in **Step (A)**. If v is a (2)-supervariable and ϕ is (1, 2)-superfeature, p and q will be connected in **Step (B)**. If ϕ is (1, 2)- of a (1)-superfeature and v is a (1)-supervariable, p and q will be connected in **Step (C)**. There are no other possible combinations. Hence, as v' consists of all representatives of v and ϕ' consists of all representatives of ϕ, v' and ϕ' are connected in $\mathcal{G} \wr \diamond$ iff v is connected to ϕ. Moreover, representatives of (2)-superfeatures are the only ones connected to two representatives of the same supervariable in $\mathcal{G}$, hence ϕ' is connected to v' via a double edge in $\mathcal{G} \wr \diamond$ if and only if ϕ is connected to v via a double cdge in $\mathcal{T} \wr \Delta$.

   We have just shown that Algorithm 9 and parameters computed as described in the proof of Theorem 13.7 together produce an partition-equivalent exponential family. We will now show that this exponential family is the smallest partition-equivalent exponential family to the original.

Theorem 13.9 (Minimality) *Let $\mathcal{F}$ be an exponential family and $\mathcal{G}$ be computed as above. Then there is no other partition-equivalent exponential family with less features or less vertices than $\mathcal{G}$. Moreover, $|V(\mathcal{G})| \le 2|V(\mathcal{F} \wr \Delta)|$ and $\mathcal{G}$ and $|E(\mathcal{G})| \le 2|E(\mathcal{F} \wr \Delta)|$, i.e., the size of I' is at most twice the size of the fully lifted model.*

Algorithm 9 Computing the smallest partition-equivalent exponential family.

Require: Fully lifted feature graph $\mathcal{F} \wr \Delta$ of G

Ensure: $\mathcal{G}$ such that $\exists \Diamond : \mathcal{G} \wr \Diamond = \mathcal{F} \wr \Delta$.

  1: Initialize $\mathcal{G} \leftarrow \emptyset$, i.e., the empty graph

  2: /* **Step (A): Treat double edges** */

  3: **for all** (2)-superfeature ϕ in $\mathcal{F} \wr \Delta$ with neighboring (2)-supervariable ν **do**

  4:     Add a feature q representing ϕ to $\mathcal{G}$

  5:     Add two variables p, p' representing ν in $\mathcal{G}$ and connect them to the feature q

  6: **end for**

  7: /* **Step (B): To preserve degrees, treat now single edges in $\mathcal{G}$ incident to double edges** */

  8: **for all** every (1,2)-superfeature ϕ in $\mathcal{F} \wr \Delta$ **do**

  9:     Add two features q, q' representing ϕ to $\mathcal{G}$

 10:     Connect q to p and q' to p' where p, p' are the representatives of a (2)-supervariable ν in $\mathcal{F} \wr \Delta$ that is neighboring ϕ

 11: **end for**

 12: /* **Step (C): Add remaining nodes and edges to $\mathcal{G}$** */

 13: **for all** supervariables variables ν and superfeature ϕ in $\mathcal{F} \wr \Delta$ not represented in $\mathcal{G}$ so far **do**

 14:     Add a single variable p resp. feature q to $\mathcal{G}$

 15:     Connect p to all representatives of superfeature ϕ neighboring to ν in $\mathcal{F} \wr \Delta$

 16: **end for**

Proof   Let $\mathcal{I}$ be any graph with the same feature quotient as $\mathcal{F}$. Then, let ϕ be a (2)-superfeature in $\mathcal{F} \wr \Delta$ adjacent to some (2)-supervariable ν. Due to equivalence, ϕ' is a (2)-superfeature in $H \wr \Diamond$ as well and ν' is a (2)-supervariable. Hence, the class $\nu' \in \Diamond$ must have at least two ground variables from $\mathcal{I}$. Next, let ϕ be a (1,2)-feature in $\mathcal{F} \wr \Delta$ adjacent to a (2)-supervariable. Analogously, ϕ' is a (1,2)-feature in $\mathcal{I} \wr \Diamond$ and ν' is a (2)-supervariable. As we have established ν' must have at least two ground elements in $\mathcal{I}$. Since ν' is connected to ϕ' via a single edge, the same holds on the ground level: any $p \in P'$ is connected to $q \in \phi'$ via a single edge. This means that there are at least as many $q \in \phi'$ as there are $p \in P'$, that is, at least two. All other supernodes must have at least one representative. These conditions are necessary for any partition-equivalent $\mathcal{I}$.

Now, let $\mathcal{G}$ be computed from Algorithm 9 and $\Diamond$ be the corresponding partition. To see why $\mathcal{G}$ is minimal, observe that $\mathcal{G}$ has *exactly* two representatives of any (2)-supervariable in $\mathcal{F} \wr \Delta$ (step 1) and *exactly* two representatives of any (1, 2)-superfeature (step 2). All other supernodes have *exactly* one representative (steps 1 and 3). Therefore, $\mathcal{G}$ meets the conditions with equality and is thus minimal. Finally, since we represent any supernode of

$\mathcal{F} \wr \Delta$ by at most 2 nodes in $\mathcal{G}$, $\mathcal{G}$ can have at most twice as many features and variables as $\mathcal{F} \wr \Delta$.

Since Algorithm 9 makes only one pass over the lifted feature graph, the overall time to compute the partition-equivalent exponential family (which is then input to Algorithm 8) is dominated by color-refinement, which is quasi-linear in the size of G.

13.6   Symmetries of Markov Logic Networks (MLNs)

Before concluding the chapter, let us touch upon the symmetries of Markov Logic Networks. **Markov Logic Networks (MLNs)** (Richardson and Domingos, 2006) are a first-order probabilistic model that defines an exponential family on random structures (i.e., random graphs, hypergraphs, or more generally random Herbrand models of the first-order language). In this case, a subgroup of the lifting group can be obtained via the symmetry of the unobserved constants in the domain without the need to consider the **ground** graphical model.

More precisely, an MLN is prescribed by a list of weighted formulas $\mathcal{F}_{\mathrm{MLN}} = \{F_1, \ldots, F_K\}$ (consisting of a set of predicates, logical variables, constants, and a weight vector w) and a logical domain $\mathcal{D} = \{a_1, \ldots, a_{|\mathcal{D}|}\}$. Let $\mathcal{D}_0$ be the set of objects appearing as constants in these formulas, then $\mathcal{D}_* = \mathcal{D} \setminus \mathcal{D}_0$ is the set of objects in $\mathcal{D}$ that do not appear in these formulas. Let Gr be the set of all ground predicates $p(a_1, \ldots, a_l)$. Given a substitution s, $F_i[s]$ denotes the result of applying the substitution s to F_i and is a grounding of F_i if it does not contain any free logical variables. The set of all groundings of F_i is $Gr(F_i)$, and let $Gr(F) = Gr(F_1) \cup \ldots \cup Gr(F_K)$. Furthermore, let ω be a truth assignment to all the ground predicates in Gr and w_i be the weight of the formula F_i. The MLN corresponds to an exponential family $\mathcal{F}_{\mathrm{MLN}}$ where Gr is the variable index set and each grounding $F_i[s] \in Gr(F_i)$ is a feature function $\phi_{F_i[s]}(\omega) = \mathbb{I}(\omega \models F_i[s])$ with the associated parameter $\theta_{F_i[s]} = w_i$. Since all the ground features of the formula F_i have the same parameter w_i, the MLN also induces the parameter-tying partition $\Delta_{\mathrm{MLN}} = \{\{\phi_{F_1[s]}(\omega)\}, \ldots, \phi_{F_K[s]}(\omega)\}$. Let a renaming permutation r be a permutation over $\mathcal{D}$ that fixes every object in $\mathcal{D}_0$ (i.e., r only permutes objects in $\mathcal{D}_*$). Thus, the set of all such renaming permutations is a group $\mathbb{G}^{\mathrm{re}}$ isomorphic to the symmetric group $\mathbb{S}(\mathcal{D}_*)$. Consider the following action of $\mathbb{G}^{\mathrm{re}}$ on $Gr: \pi_r : p(a_1, \ldots, a_l) \mapsto p(r(a_1), \ldots, r(a_l))$, and the action on GrF, $\gamma_r : F_i[s] \mapsto F_i[r(s)]$ where $r(s = (x_1/a_1, \ldots, x_k/a_k)) = (x_1/r(a_1), \ldots, x_k/r(a_k))$. Intuitively, π_r and γ_r rename the constants in each ground predicate $p(a_1, \ldots, a_l)$ and ground formula $F_i[s]$ according to the renaming permutation r. With this in mind, the following is a consequence of Lemma 1 in (Bui et al., 2012).

Theorem 13.10 *For every **renaming permutation** r, $(\pi_r, \gamma_r) \in \mathrm{Aut}(\mathcal{F}_{\mathrm{MLN}})$. Furthermore, the **renaming group** $\mathbb{G}^{\mathrm{re}}$ is isomorphic to a subgroup of the MLN's lifting group: $\mathbb{G}^{\mathrm{re}} \leq \mathrm{Aut}_{\Delta_{\mathrm{MLN}}}(\mathcal{F}_{\mathrm{MLN}})$.*

That means that **orbit partitions** induced by $\mathbb{G}^{re}$ on the set of predicate groundings can be derived directly from the first-order representation of an MLN without considering its ground graphical model. The size of this orbit partition depends only on the number of observed constants $|\mathcal{D}_o|$, and does not depend on actual domain size $|\mathcal{D}|$. For example, if $q(\cdot, \cdot)$ is a binary predicate and there is one observed constant a, then we obtain the following partition of the groundings of q: $\{q(a, a)\}, \{q(x, x)|x \neq a\}, \{q(a, x)|x \neq a\}, \{q(x, a)|x \neq a\}, \{q(x, y)|x \neq y, x \neq a, y \neq a\}$. Similar partitions on the set of factors and variable clusters can also be obtained with complexity polynomial in $|\mathcal{D}_o|$ and independent of $|\mathcal{D}|$.

13.7   Conclusions

We have laid down a theory of lifted inference for approximate variational inference methods. The main building block is group theory and in particular the concept of "lifting partitions". These are partitions of the variables and constraints of the corresponding variational optimization problem that allow one to reduce the model in size by replacing all equivalent variables/constraint by class representatives such that every solution of the reduced problem corresponds to a solution of the original one. This can lead to efficiency gains by solving the reduced model instead and then expanding the solution back to the original one. Actually, a large class of existing variational inference algorithms can directly be made aware of symmetries without modifications. As it turns out, the coarseness of lifting partitions (respectively the degree of reduction of the variational problem) depends not only on the model being solved, but also on the approximation strategy. Any symmetry (automorphism) present in the original model induces a symmetry of the variational problem, however, in many cases the solution of the variational problem may contain additional symmetries not present in the "true" solution, e.g., if it enforces constraints only among neighboring vertices of the graphical model.

14 Lifted Inference for Hybrid Relational Models

Jaesik Choi, Kristian Kersting, and Yuqiao Chen

Abstract. It is important to answer questions on large-scale probabilistic graphical models in an efficient manner. This chapter presents efficient lifted inference algorithms for large-scale probabilistic graphical models with hybrid (discrete and continuous) variables and interesting insights which transform probabilistic first-order languages into compact representations which allow efficient ways of inference. It demonstrates the effectiveness in the inference problem of relational linear dynamical systems, Kalman filtering, and in the PageRank problem of personalize recommendations.

14.1 Introduction

Hybrid Relational Models (HRMs) represent relationships among sets of random variables with continuous and discrete domains in a concise manner. The intuition of HRMs is that each set of random variables has the same numbers and types of relationships as other sets. For example, prices of houses in the same residential district may change together. Two random variables, representing the prices of a house A and a neighboring house B, may have the same relationship with another random variables, the mortgage rate and the existence of a school. Thus, one may also model the relationship with the same factor, e.g., having the same Gaussian noise.

In this chapter, probabilistic first-order language describes relationships among sets of discrete and continuous random variables for the HRMs. The probabilistic language handles uncertainty using probability theory and exploits structure using first-order logic. The language provides an expressive formalism that represents the joint probability distribution of a large number of random variables. The language first defines a first-order logic sentence over the universe of random variables. Any set of random variables satisfied by the first-order logic sentence has the same factor over the set of random variables. In this way, the relational models can compactly represent the joint probability distribution without redundancies.

The language allows the utilization of the first-order structure for efficient inference. It is well known that first-order logic allows for efficient reasoning procedures by enumerating first-order logic sentences without referring to all propositional, or individual, elements.

Lifted inference algorithms can calculate the conditional and marginal probabilities for HRMs by uplifting the model structures and referring only to first-order relationships, not all propositional variables.

Many real-world systems in finance, environmental engineering, and robotics include continuous domains. One cannot avoid dealing with continuous random variables when answering questions about the systems. Unfortunately, most principles devised for discrete RMs, e.g. (Poole, 2003; de Salvo Braz et al., 2005; Milch and Russell, 2006; Richardson and Domingos, 2006), are not applicable to such complex continuous systems and require discretizing continuous domains. Furthermore, discretization and usage of discrete lifted inference algorithms is highly imprecise. Therefore, the first fundamental challenge addressed in this chapter is building probabilistic representation languages and efficient inference algorithms for RMs with continuous variables.

Another key challenge is handling individual attributes of random variables in relational models. For example, an RM can represent a housing market model in a country. The models should be able to handle the price of each house in the country. However, many lifted inference algorithms force random variables to have the exact same attributes so they are in the same group. Thus, whenever new attributes are given or observed in an individual random variable, the attributes force the lifted inference algorithms to deteriorate the first-order structures into fine-grained propositional structures. Given such propositional structures, lifted inference algorithms refer to all ground random variables, and do as badly as propositional inference algorithms.

This chapter introduces efficient lifted inference algorithm for large-scale probabilistic graphical models with continuous variables (Choi et al., 2010) and interesting insights which transform probabilistic first-order language into compact approximations which allow efficient ways of inference (Choi and Amir, 2012; Ahmadi et al., 2011; Choi et al., 2015). It demonstrates the effectiveness in the inference problem of relational linear dynamical systems, Kalman filtering, and in the PageRank problem of personalize recommendations. exact and approximate inference algorithms for HRMs (Ahmadi et al., 2011; Choi et al., 2011b).

14.2   Relational Continuous Models

Many real world systems are described by continuous variables and relations among them. Such systems include measurements in environmental-sensors networks (Hill et al., 2009), localizations in robotics (Limketkai et al., 2005), and economic forecastings in finance (Niemira and Saaty, 2004). Once a relational model among variables is given, inference algorithms can solve value prediction problems and classification problems.

At a ground level, inference with a large number of continuous variables is non-trivial. Typically, inference is the task of calculating a marginal over variables of interest. Suppose that a market index has a relationship with n variables, revenues of n banks. When marginalizing the market index out, the marginal is a function of n variables (revenues of banks), thus marginalizing remaining variables out becomes harder. When n grows, the computation becomes expensive. For example, when relations among variables follow Gaussian distributions, the computational complexity of the inference problem is $O(|U|^3)$ (U is a set of ground variables). Thus, the computation with such models is limited to moderate-size models, preventing its use in the many large, real-world applications.

To address these issues, Probabilistic Relational Models (PRMs) (Ng and Subrahmanian, 1992; Koller and Pfeffer, 1997; Pfeffer et al., 1999; Friedman et al., 1999; Poole, 2003; de Salvo Braz et al., 2005; Milch et al., 2005; Richardson and Domingos, 2006; Milch and Russell, 2006; Getoor and Taskar, 2007) describe probability distributions at a relational level with the purpose of capturing larger models. PRMs combine probability theory for handling uncertainty and relational models for representing system structures. Thus, they facilitate construction and learning of probabilistic models for large systems. Previously, (Poole, 2003; de Salvo Braz et al., 2005; Milch et al., 2008; Singla and Domingos, 2008) showed that such models enable more efficient inference than possible with propositional graphical models, when inference occurs directly at the relational level.

Exact lifted inference algorithms (Poole, 2003; de Salvo Braz et al., 2006; Milch et al., 2008) and those developed in the efforts above are suitable for discrete domains, thus can in theory be applied to continuous domains through discretization. However, the precision of discretizations deteriorates exponentially in the number of dimensions in the model, and the number of dimensions in relational models is the number of ground random variables. Thus, discretization and usage of discrete lifted inference algorithms is highly imprecise.

Here, we present exact lifted inference algorithms for **Relational Continuous Models (RCMs)**, a relational probabilistic language for continuous domains. The main insight is that, for some classes of potential functions (or potentials), marginalizing out a ground random variable in a RCM can yield a RCM representation that does not force other random variables to become propositional (Section 14.2.1.1). Further, relational pairwise models, i.e. the product of relational potentials of arity 2, remain relational pairwise models after eliminating out ground random variables in those models. Thus, it leads to the compact representations and the efficient computations. We explain Gaussian potentials which satisfy the conditions for relational pairwise models (Section 14.2.2).

We also adapt principles of *Inversion Elimination*, a method devised by (Poole, 2003), to continuous models. *Inversion Elimination*'s step essentially takes advantage of an ability to exchange sums and products. The lifted exchange of sums and products translates directly to continuous domains.

Given a RCM, an inference algorithm marginalizes continuous variables by analytically integrating out random variables except query variables. It does so by finding a variable, and eliminating it by *Inversion Elimination*. If such elimination is not possible, *Relational*

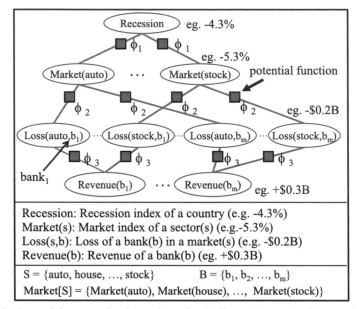

Figure 14.1: A model among banks and market indices. *Recession* is a random variable. *Market[S]*, *Gain[S,B]* and *Revenue[B]* are relational atoms. The variable and atoms have continuous domain $[-\infty, \infty]$. For example, *Market(stock)* is -5.3%, and *Loss(stock,B_m)* is $-\$0.2B$.

Atom Elimination eliminates each pairwise form in a linear time. If the marginal is not in a pairwise form, it converts the marginal into a pairwise form.

This chapter presents a relational model for continuous variables, Relational Continuous Models (RCMs). Relations among attributes of objects are represented by *Parfactor* models.[35] Each *parfactor* (L, C, A_R, ϕ) is composed of a set of logical variables (L)[36], constraints on L (C), a list attributes of objects (A_R), and a potential on A_R (ϕ). Here, each attribute is a random variable with a continuous domain.

We define a *Relational Atom* to refer the set of ground attributes compactly. For example, *Revenue[B]* is a *relational atom* which refers to revenues of banks (e.g. $B = \{$'Pacific Bank', 'Central Bank', $\cdots \}$). To make the *parfactor* compact, a list of relational atoms is used for A_R. To refer to an individual random variable, we use a substitution θ. For example, if a substitution $(B = $ 'Pacific Bank'$)$ is applied to a relational atom, then the relational atom

---

[35] Part of its representation and terms are based on (Poole, 2003; de Salvo Braz et al., 2005; Milch and Russell, 2006).

[36] Instead of objects, we use the general term, logical variables.

Revenue[*B*] becomes a ground variable *Revenue*('Pacific Bank').[37] Formally, applying a substitution θ to a parfactor $g = (L, C, A_R, \phi)$ yields a new parfactor $g\theta = (L', C\theta, A_R\theta, \phi)$, where L' is obtained by renaming the variables in L according to θ. If θ is a ground substitution, $g\theta$ is a factor. Θ_g is a set of all substitution for a parfactor g. The set of *groundings* of a parfactor g is represented as $gr(g) = \{g\theta : \theta \in \Theta_{gr(L:C)}\}$. We use $RV(X)$ to enumerate the random variables in the relational atom X. Formally, $RV(\alpha) = \{\alpha[\theta] : \theta \in gr(L)\}$. $LV(g)$ refers the set of logical variables (L) in g.

The joint probability over random variables is defined by *factors* in a *parfactor*. A *factor* f is composed of A_g and ϕ. A_g is a list of ground random variables (i.e. $(X_1(\theta), \cdots, X_N(\theta))$). ϕ is a *potential* on A_g: a function from $range(A_g) = \{range(X_1(\theta)) \times \cdots \times range(X_N(\theta))\}$ to non-negative real numbers. The factor f defines a weighting function on a valuation $(v = (v_1, \cdots, v_m))$: $w_f(v) = \phi(v_1, \cdots, v_m)$). The weighting function for a *parfactor* F is the product of weighting function of all factors, $w_F(v) = \prod_{f \in F} w_f(v)$. When G is a set of *parfactors*, the density is the product of all factors in G:

$$w_G(v) = \prod_{g \in G} \prod_{f \in gr(G)} w_f(v). \tag{14.1}$$

For example, consider the model in Figure 14.1. S and B in L are two logical variables which represent markets and banks respectively. For example, S can be substituted by a specific market sector (e.g. S = 'stock'). A parfactor $f_1 = (\{Market[S], Gain[S, B]\}, \phi_2)$ is defined over two relational atoms, *Market*[*S*] and *Gain*[*S, B*]. *Market*(*s*) (one variable in *Market*[*S*]) represents the quarterly market change (e.g. *Market*(*auto*)=−3.1%). *Gain*(*s, b*) represents the gain of bank b in the market s. Given two values, a potential $\phi_1(Market(s), Gain(s, b))$ provides a numerical value. Given all valuations of random variables, the product of potentials is the probability density.

14.2.1   Inference Algorithms for RCMs

RCMs represent large real-world systems in a compact way. One inference task with such models is to find the conditional probability of query variables given observations of some variables.

14.2.1.1   Inference with Gaussian Potentials   This section presents an efficient variable elimination algorithms for relational Gaussian models. We focus on the inference problem of computing the posterior of query variables given observations. It is important to efficiently integrating out relational atoms (e.g. Revenue[B] = {*Revenue*(b_1), $\cdots$, *Revenue*(b_m)}) for solving this inference problem. Eliminating variables is not trivial because many potentials over continuous domains are not analyticaly integrable, or there are polynomial number of parameters for each variable.

---

[37] *Revenue*() refers a random variable. *Revenue*[] refers a relational atom.

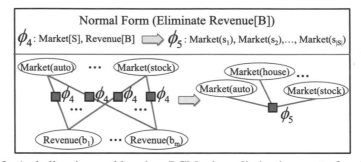

Figure 14.2: A challenging problem in a RCM when eliminating a set of variables (*Revenue[B]*). Eliminating *Revenue[B]* in ϕ_4 generates an integral ϕ_5 that makes all variables in *Market[S]* ground. Thus, the elimination makes the RCM into a ground network.

In the following description, we omit the (inequality between logical variables and objects) constraints from parfactors. This allows us to focus on the potential functions inside those parfactors. The treatment below holds with little change for parfactors with such constraints.

Relational Pairwise Potentials: This section focuses on the product of potentials which we call **Relational Normals (RNs)**. A RN is the following function with arity 2 (Section 14.2.2 provides a generalization for arbitrary potentials).:

$$\phi_{RN}(X, Y) = \prod_{x \in X, y \in Y} \frac{1}{\sigma \sqrt{2\pi}} exp\left(-\frac{(x-y)^2}{2\sigma^2}\right)$$

This potential indicates that the difference between two random variables follows Gaussian distributions. An interesting characteristic of the RN is that marginals of joint probabilities that are represented with pairwise potentials can also be represented with pairwise potentials.

Consider the models shown in Figure 14.2 and 14.3. The models represent the relationships between each market change and revenue of each bank. To simplify notations, we respectively shorten *Market(s)*, *Gain(s, b)* and *Revenue(b)* to $M(s)$, $G(s, b)$ and $R(b)$. The potential ϕ_4 in these figures is $\phi_{RN}(M(s), R(b))$, and the product of potential is $\prod_{s \in S, b \in B} \phi_{RN}(M(s), R(b))$

Figure 14.3 shows that integrating out a random variable $R(b_i)$ from the joint density results in the product of RNs again (c and c' are constants) as follow.

$$\int_{R(b_i)} \prod_{s \in S} \phi_4(M(s), R(b_i)) = c \cdot \exp\left(\frac{(\sum_{s \in S} M(s))^2}{2\sigma^2 \cdot |S|} - \frac{\sum_{s \in S} M(s)^2}{2\sigma^2}\right)$$

$$= c \cdot \prod_{1 \leq i < j \leq |S|} exp\left(-\frac{(M(s_i) - M(s_j))^2}{2\sigma^2 \cdot |S|}\right) = c' \cdot \prod_{1 \leq i < j \leq |S|} \phi_5'(M(s_i), M(s_j)) \quad (14.2)$$

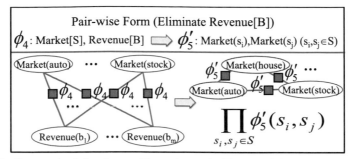

Figure 14.3: Our method for the problem shown in Figure 14.2. When eliminating *Revenue[B]*, we do not generate a ground network. Instead, we directly generate the pairwise form which allows the inference at the lifted level.

Note that, the following equation holds for the integration of a quadratic exponential function.

$$\int_{\mathbf{R(b_i)}} \exp\left(-a\mathbf{R(b_i)}^2 + 2b\mathbf{R(b_i)} + c\right) = \sqrt{\frac{\pi}{a}} exp\left(\frac{b^2}{a} + c\right) \qquad (14.3)$$

Here, the terms a and b can include random variables except $R(b_i)$.

Definition 14.1 (Connected Relational Normal) *The product of RNs is connected, when the connectivity graph is a connected component. Each vertex of the connectivity graph is a random variable or a constant in RNs, and each edge is a potential (RN).* ■

Remark 14.1 *The product of RNs is a probability density function when it is connected, and at least a RN includes a constant argument.*

14.2.1.2   A Linear Time Relational Atom Elimination   This section provides a linear time variable elimination algorithm $O(|U|)$ which can be applied to any product of RNs. This algorithm is used when the constant time algorithms of the previous sections are not applicable.

14.2.1.3   Elimination of multiple atoms in $\prod \phi_{RN}(X_i, X_j)$ This problem is to marginalize some variables in U, $(U = \{X_1, X_2, \cdots, X_{|N|}\})$ in the product of RNs between two relational atoms: $\prod \phi_{RN}(X_i, X_j)$. If all relational atoms have pairwise relationships among each other, there are $\frac{|N| \cdot |N-1|}{2}$ pairwise RNs.

Lemma 14.1   *For $|U|$ variables in $|N|$ relational atoms ($U = \{X_1, X_2, \cdots, X_{|N|}\}$) and RN potentials, marginalizing n variables in a ground model takes $O(n \cdot |U|^2)$.*

Proof   Suppose we eliminate a variable $x \in U$. Eliminating a variable x in RN needs to update coefficients of terms $(x_i x_j)$ where x_i and x_j have relations with the variable x. When x has relations with all other variables in U, the number of terms is bounded by $O(|U|^2)$.

Thus, eliminating n variables takes $O(n \cdot |U|^2)$ because it needs n iterations. ∎ Thus, any inference algorithm in a ground model has an order of $O(|U|^3)$ time complexity, when it eliminates all ground variables except a few query variables.

To reduce the time complexity, our lifted inference algorithm uses the following notations which refer ground variables in an atom X compactly: $X_{[m]} = \sum_{1 \le i \le m} x_i$; $X_{[m]^2} = \sum_{1 \le i \le m} x_i^2$; and $X_{[m][m]} = \sum_{1 \le i < j \le m} x_i \cdot x_j$. The notations give the following properties (when $|X| = m$ and $|Y| = n$):

$$(X_{[m]})^2 = X_{[m]^2} + 2X_{[m][m]}$$

$$exp\left(2X_{[m][m]} - (m-1)X_{[m]^2}\right) = \prod_{x_i, x_j \in X} exp\left(-(x_i - x_j)^2\right) = \phi'_{RN}(X, X)$$

$$exp\left(2X_{[m]}Y_{[n]} - nX_{[m]^2} - mY_{[n]^2}\right) = \prod_{x_i \in X, y_k \in Y} exp\left(-(x_i - y_k)^2\right) = \phi''_{RN}(X, Y)$$

For the product of potentials over X, Y, and $\{x'\}$, our algorithm marginalizes x':

$$\int_{x'} \phi_{RN}(X, x') \cdot \phi_{RN}(Y, x')$$

$$= \int_{x'} exp\left(-(m+n)x'^2 + 2(X_{[m]} + Y_{[n]})x' - (X_{[m]^2} + Y_{[n]^2})\right)$$

$$= \sqrt{\frac{\pi}{m+n}} \cdot exp\left(\frac{(X_{[m]} + Y_{[n]})^2}{m+n} - (X_{[m]^2} + Y_{[n]^2})\right)$$

$$= c \cdot exp\left(\frac{2X_{[m][m]} + 2X_{[m]}B_{[n]} + 2Y_{[n][n]} - (m+n-1)(X_{[m]^2} + Y_{[n]^2})}{m+n}\right)$$

$$= c \cdot \phi'_{RN}(X, X) \cdot \phi''_{RN}(X, Y) \cdot \phi'''_{RN}(Y, Y) \qquad (14.4)$$

It iterates until all n variables are eliminated.

Theorem 14.1 *For $|U|$ variables in $|N|$ relational atoms ($U = \{X_1, X_2, \cdots, X_{|N|}\}$) and potentials in RN, 'Pairwise Linear' eliminates n variables in $O(n \cdot |N|^2)$.*

That is, 'Pairwise *Linear*' has linear time complexity $O(|U|)$ with respect to the number of ground variables.

14.2.2   Exact Lifted Inference with RCMs

This section presents our algorithm, *ELIMINATE-CONTINUOUS*, which generates a new parfactor after eliminating a set of relational atoms given a set of parfactors. A potential of each parfactor is the product of *Relational Pairwise Potentials (RPPs)*:

$$\phi_{RPP}(X, Y) = \prod_{x \in X, y \in Y} \phi_{RPP}(x, y)$$

A *relational pairwise model* is a RCM whose potentials are RPPs.

14.2.3 Conditions for Exact Lifted Inference

The lifted *ELIMINATE CONTINUOUS* algorithm provides the exact solution for potentials of parfactors when the potentials satisfy three conditions: *Condition (I)*, analytically integrable; *Condition (II)*, closed under the product operations; and *Condition (III)*, closed under marginalizations of variables, thus represented with the product of *relational pairwise potentials* again. The RNs are an example that satisfies the conditions. Here, we introduce another potential, a linear Gaussian, which satisfies the conditions.

Remark 14.2 *The product of RNs with non-zero Means (RNMs) satisfies the three conditions. A RNM has the following form (d is a constant).*

$$\phi_{RN}(X, Y) = \prod_{x \in X, y \in Y} \frac{1}{\sigma \sqrt{2\pi}} exp\left(-\frac{(x - y - d)^2}{2\sigma^2}\right)$$

14.2.3.1 Inversion-Elimination
Inversion elimination is applicable when the set of logical variables in g is same with the set of logical variables in e, $LV(e) = LV(g)$. Let $\theta_1, ..., \theta_n$ be enumeration of Θ_g.

$$\int_{RV(e)} \phi(g) = \int_{RV(e)} \prod_{\theta \in \Theta_g} \phi_g(A_g\theta) = \int_{e[\theta_1]} \cdots \int_{e[\theta_n]} \phi_g(A_g\theta_1) \cdots \phi_g(A_g\theta_n)$$

$$= \prod_{\theta \in \Theta_g} \int_{e[\theta]} \phi_g(A_g\theta)(\because split \text{ (Section 14.2.1)}) = \prod_{\theta \in \Theta_g} \int_{e} \phi_g(A'\theta, e)$$

$$= \prod_{\theta \in \Theta_g} \phi'(A'\theta) = \phi_{g'}$$

Return to the econometric market example, inversion elimination can be applied to $G[S, B]$. Before an elimination, it combines two parfactors which include ϕ_2 and ϕ_3 respectively. The combined parfactor is $g = (\{S, B\}, \top, (M[S], G[S, B], R[B]), \phi_2 \cdot \phi_3)$. Then, the elimination procedure is as follow,

$$
\int_{RV(G)} \phi(g) = \int_{RV(G)} \prod_{s \in S, b \in B} \phi_g(M(s), G(s, b), R(b))
$$

$$
= \prod_{s \in \{auto, \cdots, stock\}, b \in \{b_1, \cdots, b_m\}} \left(\int_{G(s,b)} \phi_g(M(s), G(s, b), R(b)) \right)
$$

$$
= \prod_{s \in \{auto, \cdots, stock\}, b \in \{b_1, \cdots, b_m\}} \phi_{new}(M(s), R(b)) = \phi_{new}(M[S], R[b]) = \phi_{g'}
$$

Note that, the number of substitutions ($|\Theta_g|$) is the number of market sectors ($|S|$) times the number of banks ($|B|$). Regardless the number of substitutions, we can apply the same integration to eliminate $|S| \cdot |B|$ random variables ($G(s,b)$). Thus, it calculates the integral ($= \int_L \phi_g(M(s), G(s, b), R(b))$) once regardless of how many variables are generated by $s \in S$ and bB. The marginal ($\phi_{new}(M[S], R[B])$) becomes the potential of the output parfactor (g').

14.2.3.2   Relational-Atom-Elimination   *Relational-Atom-Elimination* marginalizes atoms when *Inversion-Elimination* is not applicable. It is a generalized algorithm of those for *RN* shown in Section 14.2.1.1. It marginalizes each relational atom of a parfactor g according to three cases: (1) variables in the atom e form a relationship with an atom (i.e. '$\phi(X, Y)$'); (2) variables in the atom e form relationships with each other (i.e. '$\phi(X, X)$') only; and (3) other general cases (i.e. '$\prod \phi(X_i, X_j)$').

For the case (1), a modified '*Pairwise Constant*$_1$' eliminates an atom e. In this case, integrating out a random variable in the atom does not affect integrating another variable in the atom. That is, $\int_{RV(e)} \prod_{\theta \in \Theta_g} \phi_g(.) = \prod_{\theta_e \in \Theta_e} \int_{e(\theta_e)} \prod_{\theta \in \Theta_{g \setminus \{e\}}} \phi_g(.)$. Here, E is the set of atoms in g, and $\overline{E} = E \setminus \{e\}$, and Θ_E is the set of all substitutions for E.

$$
\int_{RV(e)} \phi(g) = \int_{RV(e)} \prod_{\theta \in \Theta_E} \phi_g(A_g \theta) = \int_{RV(e)} \prod_{\theta_e \in \Theta_{\{e\}}} \prod_{\theta \in \Theta_{E \setminus \{e\}}} \phi_g(A_g \theta_e, A_g \theta)
$$

$$
= \prod_{\theta_e \in \Theta_{\{e\}}} \int_{e[\theta_e]} \prod_{\theta \in \Theta_{E \setminus \{e\}}} \phi_g(A_g \theta_e, A_g \theta) = \prod_{\theta_e \in \Theta_{\{e\}}} \phi'(RV(\overline{E}))(\because Condition(I))
$$

$$
= \phi'(RV(\overline{E}))^{|RV(e)|} = \phi''(RV(\overline{E}))(\because Condition(II))
$$

Normally, the marginal $\phi''(RV(\overline{E}))$ is not a *relational pairwise potential* because all random variables in $\overline{E}$ are arguments of the potential. However, when *Condition (III)* is satisfied, the marginal can be converted into the product of *relational pairwise potentials*: $\phi''(RV(\overline{E})) = \prod_{X_i, X_j \in RV(\overline{E})} \phi_{RPP}(X_i, X_j)$.

In the financial example, it eliminates $R[B]$ as follow.

$$\int_{RV(R)} \phi(g') = \int_{RV(R)} \prod_{s \in S, b \in B} \phi_{new}(M(s), R(b))$$

$$= \prod_{b \in B} \int_{R(b)} \prod_{s \in S} \phi_{new}(M(s), R(b)) = \prod_{b \in B} \phi'_{new}(M(auto), \cdots, M(stock))$$

$$= \phi'_{new}(M(auto), \cdots, M(stock))^{|RV(R)|} = \phi''_{new}(M(auto), \cdots, M(stock))$$

Beyond the Relational Gaussian, any potential function satisfying the Condition III) can convert the potential ϕ''_{new} into the pairwise form $\prod \phi'''_{new}$.

$$\phi''_{new}(M(auto), \cdots, M(stock)) = \prod_{s_1, s_2 \in S} \phi'''_{new}(M(s_1), M(s_2))$$

Likewise, for the cases (2) and (3), generalized algorithms of '*Pairwise Constant$_2$*' and '*Pairwise Linear*' are also applied respectively.

14.3   Relational Kalman Filtering

Many real-world systems can be modeled by continuous variables and relationships (or dependences) among them. The Kalman filter (KF) (Kalman, 1960) accurately estimates the state of a dynamic system given a sequence of control-inputs and observations. It has been applied in a broad range of domains which include weather forecasting (Burgers et al., 1998), localization and tracking in robotics (Limketkai et al., 2005), economic forecasting in finance (Bahmani-Oskooee and Brown, 2004) and many others. Given a sequence of observations and Gaussian dependences between variables, the filtering problem is to calculate the conditional probability density of the state variables at each timestep. Unfortunately, the KF computations are cubic in the number of random variables which limits current exact methods to domains with limited number of random variables. This has led to the combination of approximation and sampling (e.g. the Ensemble Kalman filter (Evensen, 1994)).

This chapter leverages the ability of relational languages (Friedman et al., 1999; Poole, 2003; Milch et al., 2005; Richardson and Domingos, 2006) to specify models with size of representation independent of the size of populations involved. Various lifted inference algorithms for relational models have been proposed (Poole, 2003; de Salvo Braz et al., 2005; Milch and Russell, 2006; Singla and Domingos, 2008; Wang and Domingos, 2008; Choi et al., 2010). These seek to carry computations in time independent of the size of the populations involved. However, the key challenge in relational filtering (of dynamic systems) is ensuring that the representation does not degenerate to the ground case when multiple observation are made. As more observations are received, an increasing number of objects become distinguished. This precludes the application of previously known algorithms unless approximately equivalent objects are grouped with expensive clustering algorithms.

We present Relational Gaussian Models (RGMs) to model dynamic systems of large number of variables in a relational fashion. RGMs have as their main building block the pairwise linear Gaussian potential as detailed in Section 14.3.1. Further, we propose a new lifted filtering algorithm that is able to marginalize out random variables of the previous timestep efficiently (in time linear in the number of random variables) while maintaining the relational (RGM) representation. This prevents the model from being increasingly grounded even when individual observations are made for all random variables. Moreover, updating the relational representation takes only quadratic time in the number of relational atoms (sets of random variables). One key insight is that, given identical observation models, even when the means of the random variables are dispersed their variances remain identical. This is sufficient to maintain a relational representation.

14.3.1   Model and Problem Definitions

This section defines **Relational Gaussian Models (RGMs)** and introduce the filtering problem for dynamic relational models.

14.3.1.1   Relational Gaussian Models (RGMs)   are a subset of Relational Continuous Models (RCMs) introduced in Section 14.2 where potentials are restricted to be Gaussian distributions. RGMs are composed of three types of parfactor models: (1) Relational Transition Models (RTMs); (2) Relational Pairwise Models (RPMs); and (3) Relational Observation Models (ROMs). Suppose that we have n relational atoms: $X_t^1(L), \ldots, X_t^n(L)$ where L is a list of logical variables. In a relational linear dynamic model, relational atoms are linearly influenced by control-inputs $U_t^1(L), \ldots, U_t^n(L)$. Similarly, a linear observation model specifies the relationship between observation variables $O_t^1(L), \ldots, O_t^n(L)$ and other relational atoms. Control inputs and observations are associated with relational atoms in two ways: (1) direct association; and (2) indirect association. We provide further details in Section 14.3.1.2.

Relational Transition Models (RTMs)   represent the dependence of relational atoms at the next timestep, $X_{t+1}^j(a')$, on relational atoms at the current timestep, $X_t^i(a)$, and (when available) control-input information. They take the following form,

$$X_{t+1}^j(a') = B_X^{i,j} \cdot X_t^i(a) + B_U^{i,j} \cdot U_t^i(a) + G_{RTM}^{i,j}, \tag{14.5}$$

where $G_{RTM}^{i,j} \sim N(0, \sigma_{RTM}^{i,j})$ and $N(m, \sigma^2)$ is the normal distribution with mean m and variance σ^2. $B_X^{i,j}$ and $B_U^{i,j}$ are the transition models, matrices or a constants, corresponding to two relational atoms.

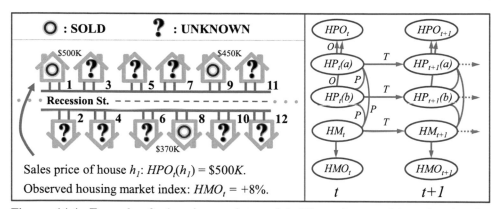

Sales price of house h_j: $HPO_t(h_1) = \$500K$.
Observed housing market index: $HMO_t = +8\%$.

Figure 14.4: Example of a housing market model. We are interested in estimating the hidden value of houses given observations of house sales prices (e.g. $HPO_t(1) = \$500K$). Both, the hidden value of a house and the observed sales prices are affected by several factors, e.g., house values increase by a certain rate every year and are also influenced by a housing market index (HM_t).

For univariate state variables, we can represent the transition model with a linear Gaussian,

$$\phi_{RTM}(X_{t+1}^j(a')|X_t^i(a), U_t^i(a)) \propto \exp\left(-\frac{(X_{t+1}^j(a') - B_X^{ij} \cdot X_t^i(a) - B_U^{ij} \cdot U_t^i(a))^2}{2 \cdot \sigma_{RTM}^{ij}{}^2}\right). \qquad (14.6)$$

The most common transition is the one from the current state $X_t^i(a)$ to the next $X_{t+1}^i(a)$. It is represented as follows,

$$X_{t+1}^i(a) = B_X^i \cdot X_t^i(a) + B_U^i \cdot U_t^i(a) + G_{RTM}^i. \qquad (14.7)$$

Relational Observation Models (ROMs)   represent the relationships between the hidden (state) variables, $X_t^i(a)$, and the observations made at the corresponding timestep, $O_t^i(a)$,

$$O_t^i(a) = H_t^i \cdot X_t^i(a) + G_{ROM}^i, \qquad (14.8)$$

where $G_{ROM}^i \sim N(0, \sigma_{ROM}^i)$. H_t^i is the observation model, a matrix or a constant, between the hidden variables and the observations.

In the linear Gaussian representation, they take the following form,

$$\phi_{ROM}(O_t^i(a)|X_t^i(a)) \propto \exp\left(-\frac{(O_t^i(a) - H_t^i \cdot X_t^i(a))^2}{2 \cdot \sigma_{ROM}^i{}^2}\right). \qquad (14.9)$$

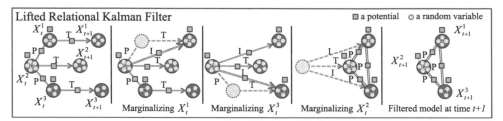

Figure 14.5: This model has three relational atoms, X_i, which may *represent* any number of random variables. The relational representation dramatically eliminates the need for redundant potentials. Hence, representation and filtering become much more efficient than filtering in the propositional case. Note that the conventional KF representation is not suited for efficient (i.e. lifted) inference.

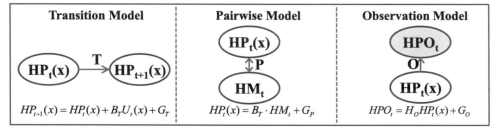

Figure 14.6: Three relational models: *Relational Transition Models (RTMs)*; *Relational Pairwise Models (RPMs)*; and *Relational Observation Models (ROMs)*. *RTMs* represent the linear relation of random variables in subsequent timesteps (e.g. $HP_t(x)$ and $HP_{t+1}(x)$). *RTMs* comprise control-inputs $U_t(x)$ multiplied by a coefficient (or a matrix) B_M and a Gaussian noise G_M component. *RPMs* represent a Gaussian relation between two random variables at a given timestep (e.g. $HP_t(x)$ and HM_t). *RPMs* comprise a coefficient (or a matrix) B_T and Gaussian noise G_T. *ROMs* represent the linear relation between one or more observations variables (e.g. HPO_t) and a group of hidden variables ($HP_t(x)$). *ROMs* also comprise a coefficient (or a matrix) H_t and Gaussian noise G_O.

Relational Pairwise Models (RPMs)   represent Gaussian dependences between pairs of relational atoms within the same timestep as follows,

$$X_t^i(a) = R_t^{i,j} \cdot X_t^j(a') + G_{RPM}^{i,j}, \tag{14.10}$$

where $G_{RPM}^{i,j} \sim N(0, \sigma_{RPM}^{i,j})$. $R_t^{i,j}$ is the pairwise coefficient, a matrix or a constant, between the two relational atoms.

Note that RTMs and ROMs are directed models while RPMs are undirected. The directed models represent the nature of dynamic systems (e.g. the state at the next timestep depends on the current timestep). The product of RPMs is an efficient way to represent a multivariate Gaussian density over all the state variables.[38]

14.3.1.2   A Relational Filtering Problem   Given a prior (or current belief) over the state variables, the filtering problem is to compute the posterior after a sequence of timesteps. The input to the problem is: (1) Relational Gaussian Model (RTMs, RPMs and ROMs); (2) current belief over the relational atoms (X_0^i) represented by a product of relational Gaussian potentials; (3) sequence of control-inputs $(U_1^i, \ldots, U_T^i)$; and (4) sequence of observations $(O_1^i, \ldots, O_T^i)$. The output is the relational Gaussian posterior distribution over the relational atoms (X_T^i) at timestep T.

Input and Observation Association   At every timestep the control-inputs and observations must be associated with the random variables they affect. The ideas in this section apply to control-inputs and observations but we illustrate them for observations.

We distinguish two types of observations: direct and indirect. Direct observations are those made for a specific random variable. For instance, if we make an observation for each random variable in a subset $A_t^i \subseteq X_t^i$ of the ground substitutions of relational atom X_t^i, we are looking at the following model,

$$\prod_{a_j \in A_t^i} \phi_{ROM}\left(o_t^i(a_j) | X_t^i(a_j)\right). \tag{14.11}$$

In the example of Figure 14.4, observing the selling price of a house would dramatically reduce the variance of the hidden variable that represents the true value of that house.

Similarly, multiple direct observations, $O_t^i = o_t^{i,1}, o_t^{i,2}, \ldots, o_t^{i,|O_t^i|}$, could be made for each variable in some sets of random variables,

$$\prod_{a_j \in A_t^i} \prod_{o_t^{i,k} \in O_t^i} \phi_{ROM_k}\left(o_t^{i,k}(a_j) | X_t^i(a_j)\right). \tag{14.12}$$

Given a notion of neighborhood (e.g. a residential neighborhood or a block of houses), indirect observation allows the possibility that observations made for a random variable, $o_t^i(a')$, would influence nearby random variables, $X_t^i(a_j)$, $a' \neq a_j$,

$$\prod_{a_j \in A_t^i} \phi_{ROM}\left(o_t^i(a') | X_t^i(a_j)\right). \tag{14.13}$$

For example, this allows the possibility that the observation of the selling price of a house would reduce the variance of the true values of neighboring houses.

Many exact lifted inference algorithms (e.g. (Kersting et al., 2006; Choi et al., 2010)) handle observations by partitioning the relational atoms into groupings of groundings for

---

[38] Note that a multivariate Gaussian density (of state variables) is a quadratic exponential form. The quadratic exponential form can always be decomposed into terms involving only single variables and pairs of variables.

which identical observations and observation models apply. In contrast, our approach partitions a relational atom into sets according to the number of different types of observations associated with each random variable. For instance, if an individual observation of the same ROM type is made for each random variable then no partitioning is necessary at all. The intuition for this is that the filtering process will assign the same variance to any two hidden variables for which the same number of observations is made at the current timestep.

Here, the partition will determine new RPMs, the pairwise parfactors which maintain the variances and covariances. In particular, the number of new RPMs is quadratic in the size of the partition. Since individual observations cause the means of the random variables to differ we store the mean information in the prior and posterior (P and P_{new} in Section 14.3.2). Hence, the number of priors and posteriors is linear in the number of random variables. However, this will not affect the computational complexity of inference as long as the RPMs do not degenerate. Further details are given in Sections 14.3.2 and 14.3.3.

Formally, given a partition $\Pi^i = (M_1^i, M_2^i, \ldots, M_{|\Pi^i|}^i)$ of a relational atom, X^i, the observation model takes the form,

$$\prod_{M_l^i \in \Pi^i} \prod_{a_j \in M_l^i} \prod_{o^{i,k} \in O_l^i} \phi_{ROM_k}\left(o^{i,k}(a_j)|X^i(a_j)\right), \tag{14.14}$$

where we omit the time subscript and where O_l^i is the set of observations relevant to part l.

14.3.2   Lifted Relational Kalman Filter

The **Lifted Relational Kalman filter (LRKF)**, just like the conventional Kalman filter, carries two recursive computations: the prediction step and the update/correction step.

Lifted Prediction   In the prediction step, our current belief over the states of the relational atoms together with the RTMs, RPMs and control-inputs are used to make the best estimate of state without observation information. First, the product of potentials in the RTMs and RPMs is built. Second, the variables from the previous timestep are marginalized to form new RPMs which estimates the relational atoms in the current timestep. We call these estimates the *intermediate posterior*, the input to the update step.

$$\int_{X_t^1, \ldots, X_t^n} \prod_{\substack{1 \le i < j \le n}} \prod_{\substack{a \in A^i \\ a' \in A^j}} \phi_{RTM}^{i,j}\left(X_{t+1}^j(a')|X_t^i(a), U_t^i(a)\right) P^i\left(X_t^i(a)\right) \phi_{RPM}^{i,j}\left(X_t^i(a), X_t^j(a')\right)$$

$$= \prod_{\substack{1 \le i < j \le n}} \prod_{\substack{a \in A^i \\ a' \in A^j}} {\phi'}_{RPM}^{i,j}\left(X_{t+1}^i(a), X_{t+1}^j(a')\right) \cdot \prod_{\substack{1 \le i \le n}} \prod_{a \in A^i} P'^i\left(X_{t+1}^i(a)\right). \tag{14.15}$$

Here, ${\phi'}_{RPM}^{i,j}$, P^i and P'^i are respectively the updated RPMs, the priors and the intermediate posteriors.

Lifted Update   In the update step, the intermediate posterior P'^i and ROMs are used to correct our estimate of the relational atoms.

When a single observation, o^i_{t+1}, is associated with all variables in a relational atom, we calculate the posterior for one random variable $X^i_{t+1}(a)$ and use the result for the rest of the groundings of the same relational atom,

$$P'^i(X^i_{t+1}(a)) \cdot \phi_{ROM}\left(o^i_{t+1}|X^i_{t+1}(a)\right) \propto \exp\left(-\frac{(X^i_{t+1}(a) - \mu^i_{P_{new}})^2}{\sigma^i_{P_{new}}{}^2}\right) = P^i_{new}(X^i_{t+1}(a)). \quad (14.16)$$

In the case of multiple observations $O^i_{t+1} = o^{i,1}_{t+1}, o^{i,2}_{t+1}, \ldots, o^{i,|O^i_t|}_{t+1}$ we may also do the computation of the posterior for a single random variable $X^i_{t+1}(a)$ and use the resulting posterior for all other groundings of the relational atom (to which the same set of observations applies). The calculation is similar to the above, except that multiple observations need to be considered,

$$P'^i\left(X^i_{t+1}(a)\right) \prod_{o\in O^i_{t+1}} \phi_{ROM}\left(o|X^i_{t+1}(a)\right) \propto \exp\left(-\frac{(X^i_{t+1}(a) - \mu^i_{P_{new}}(a))^2}{\sigma^i_{new}{}^2}\right) = P^i_{new}\left(X^i_{t+1}(a)\right).$$

$$(14.17)$$

Lifted Inference with Individual Observations   One of the key challenges in lifted inference is to handle individual observations. Current methods ground a relational atom when different observations are made for its random variables. It is usually the case that models shatter combinatorially fast and thus forfeit the benefits of a relational representation and the applicability of lifted inference.

We solve this problem in the LRKF by noting that the variances and covariances in the model are not affected by individual observations. We are thus able to represent the variances and covariances in a relational way while allowing variables to carry individual means. Further, the lifted prediction operation applies unmodified to this representation.

Lemma 14.2 *The variances of two random variables $X(a)$, $X(b)$ in an RGM are identical after a filtering step (Lifted Prediction and Lifted Update) if the following conditions hold before the filtering step: (1) both random variables are in the same relational atom; (2) the variance of both variables is the same; (3) observations are made for both variables or none of them.*

Proof   Given conditions (1) and (2), we first prove that the variance of both random variables is the same after the Lifted Prediction step. Note that condition (3) is not relevant to this step.

WLOG we assume $X_t(a)$ and $X_t(b)$ have different means, $\mu_t(a)$ and $\mu_t(b)$. Moreover, it is easy to see that the variance of $X^i_{t+1}(a)$ and $X^i_{t+1}(b)$ is the same after marginalizing all random variables of timestep t due to the following reasons: (i) $X(a)$ and $X(b)$ are in the same relational atom and thus share the same relationships with other random variables; (ii) the means are not involved in the marginalizations (see Section 14.3.2). It follows that we can represent the potentials relevant to the marginalization of $X_t(a)$ and $X_t(b)$ as follows:

$$\exp\left(-\frac{(X_t(a)-\mu_t(a))^2}{\sigma^2_{X_t(a)}}\right)\phi_{RTM}\left(X_{t+1}(a)|X_t(a),U_t(a)\right)\phi_{RPM}\left(X_t(a),X_t(b)\right)$$

$$\exp\left(-\frac{(X(b)_t-\mu_t(b))^2}{\sigma^2_{X_t(b)}}\right)\phi_{RTM}\left(X_{t+1}(b)|X_t(b),U_t(b)\right)\phi_{RPM}\left(X(a)_{t+1},X(b)_{t+1}\right)$$

$$= \exp\left(c_{X_t(a)^2}X_t(a)^2 + c_{X_t(a)}X_t(a)\right)\exp\left(\frac{2B_X^i}{\sigma^2_{RTM}}X_t(a)X_{t+1}(a)\right)\exp\left(\frac{X_t(a)X_t(b)}{\sigma^2_{RPM}}\right)$$

$$\exp\left(c_{X_t(b)^2}X_t(b)^2 + c_{X_t(b)}X_t(b)\right)\exp\left(\frac{2B_X^i}{\sigma^2_{RTM}}X_t(b)X_{t+1}(b)\right)\phi_{other}\left(X_{t+1}(a),X_{t+1}(b)\right),$$

where c_X refers to the coefficient of the term X.[39]

After $X_t(a)$ and $X_t(b)$ are marginalized we get a potential of $X_{t+1}(a)$ and $X_{t+1}(b)$. The variances of the random variables are the inverses of the coefficients of their squares in the resulting potential. Thus, all we need to show is that the coefficients of the square of the random variables, $X_{t+1}(a)^2$ and $X_{t+1}(b)^2$, are identical after the marginalization. The two coefficients can be represented as follows,

$$c_{X_{t+1}(a)^2} = \frac{-c_{X_t(b)^2}\left(\frac{B_X^i}{\sigma^2_{RTM}}\right)^2}{\left(\frac{1}{\sigma^2_{RTM}}\right)^2 - c_{X_t(a)}\,c_{X_t(b)}}, \quad c_{X_{t+1}(b)^2} = \frac{-c_{X_t(a)^2}\left(\frac{B_X^i}{\sigma^2_{RTM}}\right)^2}{\left(\frac{1}{\sigma^2_{RTM}}\right)^2 - c_{X(a)_t}\,c_{X_t(b)}}$$

where, $c_{X_t(\cdot)^2} = -\left(\frac{1}{\sigma^2_{X_t(\cdot)}} + \frac{1}{\sigma^2_{RTM}} + \frac{1}{\sigma^2_{RPM_t}}\right)$.

Condition (2) ($\sigma^2_{X_t(a)}=\sigma^2_{X_t(b)}$) implies $c_{X_t(a)^2} = c_{X_t(b)^2}$ which in turn implies $c_{X_{t+1}(a)^2} = c_{X_{t+1}(b)^2}$. This is enough to prove that the variance of two random variables $X(a)$ and $X(b)$ with different means is the same after the Lifted Prediction step.

We now prove the result for the Lifted Update step. Regarding Condition (3) there are two cases: (a) observations were made for both variables; or (b) no observations were made for either variable. In the case of (b) the proof is complete. In the case of (a), the update step for $X(a)$ can be represented by,

$$\exp\left(-\frac{(X_{t+1}(a)-\mu_{X_{t+1}}(a))^2}{\sigma^2_{X_{t+1}(a)^i}} - \frac{(X_{t+1}(a)-o_{a_t})^2}{\sigma_{X_{ROM}}{}^2}\right) = \exp\left(-\frac{(X_{t+1}(a)-\mu^+{}_{X_{t+1}}(a))^2}{\sigma^+{}_{X_{t+1}}(a)^2}\right)$$

---

[39] For the sake of exposition the RTMs here represent dependences from state variables at time t to the same state variable at time $t+1$ (e.g. from $X_t(a)$ to $X_{t+1}(a)$). However, the general RTMs (e.g dependences from $X_t(a)$ to $X_{t+1}(b)$) produce similar forms.

where,

$$\sigma^{+2}_{X_{t+1}(a)} = \frac{\sigma^2_{X_{t+1}(a)}\sigma^2_{X_{ROM}}}{\sigma^2_{X_{t+1}(a)^i} + \sigma^2_{X_{ROM}}}, \quad \mu^+_{X_{t+1}(a)} = \frac{\sigma^2_{X_{ROM}}\mu_{X_{t+1}(a)} + \sigma^2_{X_{t+1}(a)}O_{a_t}}{\sigma^2_{X_{ROM}} + \sigma^2_{X_{t+1}(a)}}$$

Likewise, after the update step the variance of $X(b)$ is,

$$\sigma^{+2}_{X_{t+1}(b)} = \frac{\sigma^2_{X_{t+1}(b)^i}\sigma^2_{X_{ROM}}}{\sigma^2_{X_{t+1}(b)^i} + \sigma^2_{X_{ROM}}}$$

By Condition (2) and the proof for the prediction step, $\sigma_{X_{t+1}(a)^i} = \sigma_{X_{t+1}(b)^i}$. Thus, $\sigma^+_{X_{t+1}(a)} = \sigma^+_{X_{t+1}(b)}$. ∎

Lemma 14.3 *The covariances of two pairs of variables $(X(a), X(b))$ and $(X(a), X(c))$ in an RGM are identical after a filtering step (Lifted Prediction and Lifted Update) if the following conditions hold before the filtering step: (1) the three random variables are in the same relational atom; (2) the covariances of both pairs of variables are the same; (3) observations are made for the three variables or none of them.*

Proof   The method used in the proof of Lemma 14.2 can be employed in this proof: The terms involving the individual observations do not affect terms which determine the covariance of two random variables. ∎

14.3.3 Algorithms and Computational Complexity

Let $\mathbb{X}$ ($|\mathbb{X}|$) be the set (number) of all random variables in the model and $X = (X^1, \ldots, X^{|X|})$ be the set of relational atoms (also, a partition of $\mathbb{X}$).

Algorithm 10 presents our Lifted Relational Kalman filtering algorithm. The inputs to the algorithm are: relational atoms, X; the RGM, RTMs M_X, RPMs M_P and ROMs M_O; the prior over the relational atoms, P_0; and the control-inputs, $U_{[1,\ldots,T]}$, and observations, $O_{[1,\ldots,T]}$, for each timestep.

The algorithm computes the posterior recursively. **Split** partitions the domains of each relational atom X^i as induced by the control-inputs U_t. **Lifted_Predict** calculates new RPMs, $M_P{}^{40}$, and intermediate posterior, P_{int}, based on the transition models, M_X, and the control-inputs, U_t. Then, **Split_Obs** partitions the domains of each relational atom X^i as induced by the observations, O^i_t. **Lifted_Update** calculates the new posterior, P_{cur}, based on the intermediate posterior, P_{int}, the observation models, M_O, and the observations, O^i_t.

Given the control-inputs, **Split** partitions relational atoms as done in previous work: e.g. *Split* (Poole, 2003) and *SHATTER* (de Salvo Braz et al., 2005). If the control-inputs are allowed to differ for the variables in a relational atom, the model will be propositionalized.

---

[40] In our representation the number of relational atoms determines the number of RPMs which is equal to $E(|X|, 2)$ (the number of 2-combinations of $|X|$ with repetition).

Algorithm 10 Lifted_Relational_Kalman_Filter(LRKF)

Require: Atoms, $X = (X^1, \ldots, X^{|X|})$; RTM, M_X, RPM, M_P, and ROM M_O; prior, P_0;
    control-inputs, $U_{[1,\ldots,T]}$; observations, $O_{[1,\ldots,T]}$.

    $P_{cur} \leftarrow P_0, X_{cur} \leftarrow X$

    for $t = 1$ to T **do**

        $[X_{cur}, M_X, M_P, M_O] \leftarrow$ **Split**$(X_{cur}, U_t, M_X, M_P, M_O)$

        $[P_{int}, M_p] \leftarrow$ **Lifted_Predict**$(X_{cur}, P_{cur}, M_X, M_P, U_t)$ (§14.3.2)

        $[X_{cur}, M_O] \leftarrow$ **Split_Obs**(X_{cur}, O_t, M_O) (§14.3.2)

        $[P_{cur}] \leftarrow$ **Lifted_Update**$(X_{cur}, M_O, P_{int}, O_t)$ (§14.3.2)

    end for

    return X_{cur}, P_{cur}

Hence, there is little advance in how we handle individual control-inputs with respect to previous algorithms (Choi et al., 2010).[41]

Algorithm **Split_Obs** partitions a relational atom X^i based on the observations. However, **Split_Obs** will only partition a relational atom in case the conditions of Lemmas 14.2 and 14.3 do not hold, i.e., when different number of observations are made for the relational variables. If the conditions of Lemmas 14.2 and 14.3 hold, the efficiency of the relational representation will be preserved even if multiple observations are made for all variables in some or all of the relational atoms.

Lemma 14.4 *The complexity of **Lifted_Predict** is $O(|\mathbb{X}| \cdot |X_{t^+}|^2)$. Where X_{t^+} is the set of relational atoms output by **Split**.*

Proof   This step corresponds to the marginalization (Equation (14.15)) of the variables in $\mathbb{X}$. For every variable that is integrated the parameters of all, $E(|X_{t^+}|, 2)$, pairwise interactions between relational atoms must be updated. ∎

Lemma 14.5 *The complexity of **Lifted_Update** is $O(|X_{t^{+o}}| \cdot |O_{max}|)$. Where $X_{t^{+o}}$ is the set of relational atoms output by **Split_Obs** and O_{max} is the largest set of observations associated with a relational atom.*

Proof   For each relational atom in $X_{t^{+o}}$ the computation in Equation (14.17) iterates over all relevant observations. ∎

The result is as follows,

---

[41] However, we conjecture that techniques similar to the ones we used for ROMs can be applied to RTMs. Any two random variables in the same atom will have the same variance after the Lifted_Predict step if they receive the same types of control inputs. That is, RTMs of the same type will increase the variances of the random variables by the same amount.

Theorem 14.2 *The computational complexity of* **LRKF** *is* $O(T \cdot (|\mathbb{X}| \cdot |X^*_{t^+}|^2 + |X^*_{t^+o}| \cdot |O^*_{max}|))$ *where T is the number of timesteps,* $\mathbb{X}$, $X^*_{t^+}$, $X^*_{t^+o}$ *and* O^*_{max} *are as above with the* $*$ *indicating the largest set across all timesteps.*

14.4 Lifted Message Passing for Hybrid Relational Models

Lifted message passing algorithms exploit relational structures in a probabilistic graphical model to answer queries efficiently. Lifted message passing algorithms construct a lifted hypergraph which is composed of a set of nodes (supernodes) and potentials (superpotential) those are indistinguishable given the evidence. In general cases, a lifted algorithm need to construct a separate lifted network for each evidence case and run a modified message passing algorithm separately on each lifted network. In this section, we present a lifted message passing algorithm which exploits symmetries across multiple evidence cases for relational Linear Gaussian models.

14.4.1 Lifted Gaussian Belief Propagation

This section presents a unified review of **Gaussian Belief Propagation (GaBP)** and lifted belief propagation (LBP) leading to **Lifted Gaussian Belief Propagation (LGaBP)**. We develop LGaBP in the context of solving linear systems that are key to our lifted PageRank and Kalman filtering applications presented later.

Many real world applications such as environmental sensor networks, information diffusion in social networks, and localization in robotics involve systems of continuous variables. One of the most fundamental problems encountered in these applications is solving linear systems of the form $\mathbf{Ax} = \mathbf{b}$ where $\mathbf{A} \in \mathbb{R}^{n \times n}$ is a real-valued square matrix, and $\mathbf{b} \in \mathbb{R}^n$ is real-valued column vector, and we seek the column vector $\mathbf{x}$ such that equality holds. As a running example, consider $\mathbf{b} = (0\ 0\ 1)^t$ (where ^{t} denotes transpose) and

$$\mathbf{A} = \begin{pmatrix} 10 & 4 & 3 \\ 4 & 10 & 3 \\ 5 & 5 & 11 \end{pmatrix}. \tag{14.18}$$

Shental *et al.* (Shental et al., 2008) have shown how to translate this problem into a probabilistic inference problem, i.e., to solve a linear system of equations of size n we compute the marginals of the Gaussian variables $x_1, \ldots, x_n$ in an appropriately defined graphical model. Given the matrix $\mathbf{A}$ and the observation matrix $\mathbf{b}$, the Gaussian density function $p(\mathbf{x}) \sim \exp(-\frac{1}{2}\mathbf{x}^t \mathbf{Ax} + \mathbf{b}^t\mathbf{x})$ can be factorized according to the graph consisting of edge potentials ψ_{ij} and self potentials ϕ_i as follows: $p(\mathbf{x}) \propto \prod_{i=1}^{n} \phi_i(x_i) \prod_{i,j} \psi_{ij}(x_i, x_j)$, where the potentials are $\psi_{ij}(x_i, x_j) := \exp(-\frac{1}{2}x_i A_{ij} x_j)$ and $\phi_i(x_i) := \exp(-\frac{1}{2}A_{ii}x_i^2 + b_i x_i)$. The edge potentials ψ_{ij} are defined for all (i, j) s.t. $\mathbf{A}_{ij} > 0$.

To solve the inference task, Shental *et al.* proposed to use Weiss *et al.*'s (Weiss and Freeman, 2001) Gaussian BP (GaBP) which is a special case of continuous BP, where the underlying distribution is Gaussian. BP in Gaussian models gives simpler update formulas than the general continuous case and the message updates can directly be written in terms of the mean vector and the precision matrix (the inversion of the covariance matrix). Since $p(\mathbf{x})$ is jointly Gaussian, the messages are proportional to Gaussian distributions $\mathcal{N}(\mu_{ij}, P_{ij}^{-1})$ with precision $P_{ij} = -A_{ij}^2 P_{i\backslash j}^{-1}$ and mean $\mu_{ij} = -P_{ij}^{-1} A_{ij}\mu_{i\backslash j}$ where

$$P_{i\backslash j} = \tilde{P}_{ii} + \sum\nolimits_{k \in N(i)\backslash j} P_{ki}$$

$$\mu_{i\backslash j} = P_{i\backslash j}^{-1}(\tilde{P}_{ii}\tilde{\mu}_{ii} + \sum\nolimits_{k \in N(i)\backslash j} P_{ki}\mu_{ki})$$

for $i \neq j$ and $\tilde{P}_{ii} = A_{ii}$ and $\tilde{\mu}_{ii} = b_i/A_{ii}$. Here, $N(i)$ denotes the set of all the nodes neighboring the ith node and $N(i)\backslash j$ excludes the node j from $N(i)$. All messages parameters P_{ij} and μ_{ij} are initially set to zero. The marginals are Gaussian probability density functions $\mathcal{N}(\mu_i, P_i^{-1})$ with precision $P_i = \tilde{P}_{ii} + \sum_{k \in N(i)} P_{ki}$ and mean $\mu_i = P_{i\backslash j}^{-1}(\tilde{P}_{ii}\tilde{\mu}_{ii} + \sum_{k \in N(i)} P_{ki}\mu_{ki})$. If the spectral radius of the matrix $\mathbf{A}$ is smaller than 1 then GaBP converges to the true marginal means ($\mathbf{x} = \mu$). We refer to (Shental et al., 2008) for details.

Although already quite efficient, many graphical models produce inference problems with symmetries not reflected in the graphical structure. LBP can exploit this structure by automatically grouping nodes (potentials) of the graphical model G into supernodes (superpotentials) if they have identical *computation trees* (i.e., the tree-structured unrolling of the graphical model computations rooted at the nodes). This compressed graph $\mathcal{G}$ is computed by passing around color signatures in the graph that encode the message history of each node. The signatures are initialized with the color of the self potentials, i.e., $cs_i^0 = \phi_i$ and iteratively updated by $cs_i^k = \{cs_i^{k-1}\} \cup \{[\psi_{ij}, cs(\psi_j^{k-1})] | j \in N(i)\}$. For the algorithmic details of color passing, please refer to (Kersting et al., 2009). The key point to observe is that this very same process also applies to GaBP (viewing "identical" for potentials only up to a finite precision), thus leading to a novel LGaBP algorithm.

To continue our running example, let us examine computing the inverse of matrix $\mathbf{A}$ in Equation (14.18) using LGaBP. We note that $\mathbf{A}^{-1} = [\mathbf{x}_1, \ldots, \mathbf{x}_n]$ can be computed by solving $\mathbf{A}\mathbf{x}_i = \mathbf{e}_i$ for $i = 1 \ldots n$, where $\mathbf{e}_i$ is the ith *basis vector* — $\mathbf{I} = [\mathbf{e}_1, \ldots, \mathbf{e}_n]$ for the $n \times n$ identity matrix $\mathbf{I}$. In our running example, this yields the respective lifted networks in Figures 14.7(a–c). As one can see, for the evidence cases e_1 and e_2 there is no compression and lifted inference is essentially ground. All nodes get different colors for these cases (Figures 14.7(a) and (b)). For the evidence case $\mathbf{e}_3$, however, variables x_1 and x_2 are assigned the same color by LGaBP (Figure 14.7(c)). The final lifted graph $\mathcal{G}$ is constructed by grouping all nodes (potentials) with the same color (signatures) into *supernodes (superpotentials)*, which are sets of nodes (potentials) that behave identical at each step of carrying out GaBP on $\mathcal{G}$ (Figure 14.7(d)). On the lifted graph $\mathcal{G}$, LGaBP then runs a modified GaBP. The modified messages simulate running GaBP on the original graph G. Following LBP, we have to pay special attention to the self-loops introduced by lifting

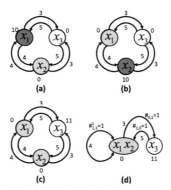

Figure 14.7: Lifted graphical models produced when inverting $\mathbf{A}$ in Equation (14.18) using LGaBP. An edge from i to j encodes potential ψ_{ij}. The ϕ potentials are associated with the nodes. (a) Colored network when computing $\mathbf{A}x_1 = \mathbf{e}_1$. All nodes get different colors; no compression and lifted inference is essentially ground. (b) Colored network for $\mathbf{A}x_2 = \mathbf{e}_2$. Again no compression. Note, however, the symmetries between (a) and (b). (c) Colored network for $\mathbf{A}x_3 = \mathbf{e}_3$. Nodes x_1 and x_2 get the same color and are grouped together (d).

that correspond to messages between different nodes of the same supernode. Reconsider our running example. As shown in Figure 14.7(d), there is a self-loop for the supernode $\{x_1, x_2\}$. In general, there might be several of them for each supernode and we assume that they are indexed by k. To account for the self-loops and in contrast to GaBP, we introduce "self-messages" $P_{ii}^k = -A_{ii}^2 (P_{i\backslash i}^k)^{-1}$ and $\mu_{ii}^k = -P_{ij}^{-1} A_{ii}\mu_{i\backslash i}^k$, with

$$P_{i\backslash i}^k = \tilde{P}_{ii} + \Big(\sum_{l\in S(i)\backslash k} \#_{ii}^l P_{ii}^l \Big) + \Big(\sum_{l\in N(i)\backslash i} \#_{li} P_{li} \Big)$$

$$\mu_{i\backslash i}^k = P_{i\backslash j}^{-1}\Big[\tilde{P}_{ii}\tilde{\mu}_{ii} + \sum_{l\in S(i)\backslash k} \#_{ii}^l P_{ii}^l \mu_{ii}^l + \sum_{l\in N(i)\backslash i} \#_{li} P_{li}\mu_{li} \Big] .$$

As i is now a neighbor of itself, the term $N(i) \setminus i$ is required. Furthermore, $\#_{ij}^l$ — also given in Figure 14.7(d) — are counts that encode how often the message (potential) would have been used by GaBP on the original network G. Using these counts we can exactly simulate the messages that would have been sent in the ground network. Messages between supernode i and j, $i \neq j$, are modified correspondingly:

$$P_{i\backslash j} = \tilde{P}_{ii} + \sum_{k\in S(i)} \#_{ii}^k P_{ii}^k + \Big(\sum_{k\in N(i)\backslash i,j} \#_{ki} P_{ki} \Big)$$

$$\mu_{i\backslash j} = P_{i\backslash j}^{-1}\Big[\tilde{P}_{ii}\tilde{\mu}_{ii} + \sum_{k\in S(i)} \#_{ii}^k P_{ii}^k \mu_{ii}^k + \sum_{k\in N(i)\backslash i,j} \#_{ki} P_{ki}\mu_{ki} \Big]$$

where $N(i) \setminus i,j$ denotes all neighbours of node i without i and j. Adapting the arguments from (Kersting et al., 2009), the following LGaBP correctness theorem can be proved:

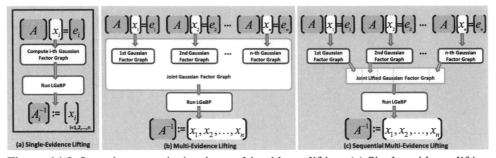

Figure 14.8: Inverting a matrix $\mathbf{A}$ using multi-evidence lifting. (a) Single-evidence lifting runs LGaBP solving $\mathbf{A}x_i = \mathbf{e}_i$ for each $i = 1, 2, \ldots, n$ separately; $\mathbf{e}_i$ denotes the ith basis vector; thus LGaBP computes each column vector of $\mathbf{A}^{-1}$ separately. (b) Multi-evidence lifting runs LGaBP on the joint graphical model of $\mathbf{A}x_i = \mathbf{e}_i$ for $i = 1, 2, \ldots, n$; thus LGaBP computes $\mathbf{A}^{-1}$ for all $\mathbf{e}_i$ at once. (c) Sequential multi-evidence lifting directly builds the lifted joint graphical model in a sequential fashion, avoiding cubic space complexity

Theorem 14.3 *Given a Gaussian model G, LGaBP computes the minimal compressed lifted model, and running modified GaBP on $\mathcal{G}$ produces the same marginals as GaBP on G.*

14.4.2   Multi-evidence Lifting

Both applications considered in this section — lifted PageRank and Kalman filtering — require the inversion of very large, structured matrices. As shown previously, this task can be reduced to the problem of solving several linear systems with GaBP — more precisely, calling GaBP multiple times, each time with a different evidence case.

Returning to our running example, inverting $\mathbf{A}$ from Equation (14.18) using GaBP results in three graphical models of size 3/9 (nodes/potentials). Given the recent success of LBP approaches, we ask "can we do better?" A first attempt to affirmatively answer the question is to simply replace GaBP with LGaBP for each evidence case $\mathbf{e}_i$ as illustrated in the three lifted graphical models of Figures 14.7(a–d). In general, this *single-evidence lifting* approach is illustrated in Figure 14.8(a). Due to lifting, we can hope to greatly reduce the cost of inference in each iteration. For our running example, this results in two ground networks both of size 3/9 and one lifted network of size 2/5 shown respectively in Figures 14.7(a,b) and (d). Thus, the total size (sum of individual sizes) drops to 8/23 This LGaBP approach — as we will show in our experiments — can already result in considerable efficiency gains. However, we can do even better.

Repeatedly constructing the lifted network for each new evidence case can be avoided when symmetries across multiple evidence cases are not exploited (e.g., the common 0's amongst all $\mathbf{e}_i$). As an example, let's compare for example the graphs depicted in Figures 14.7(a) and (b). These two networks are basically symmetric which is not exploited in the single-evidence case. They stay essentially ground if they are processed seperately. To overcome this, we propose *multi-evidence lifting* (as illustrated in Figure 14.8(b) for inverting a matrix):

> *Compute the graphical models of each evidence case, form their union, and run LGaBP on the resulting joint graphical model.*

This automatically employs the symmetries within and across the evidence cases due to Theorem 14.3. In our running example, we get a *single* lifted graph of size 5/14. Intuitively, multi-evidence lifting only produces (in this example) one of the two ground networks and hence consists of the union of the networks shown in Figures 14.7(a) and (d). This is clearly a reduction compared to LGaBP's size of 8/23.

However, there is no free lunch. Multi-evidence lifting sacrifices space complexity for a lower time complexity. Inverting a $n \times n$ matrix may result in a joint ground graphical model with n^2 nodes (n nodes for each of the n systems of linear equations) and $O(n^3)$ edges ($O(n^2)$ edges for each of the n systems of linear equations; edges are omitted if $\mathbf{A}_{ij} = 0$). This makes multi-evidence essentially intractable for large n. For instance, already for $n > 100$ we have to deal with millions of edges, easily canceling the benefits of lifted inference. We develop an efficient sequential multi-evidence lifting approach that computes the joint lifted network by considering one evidence case after the other (Figure 14.8(c)).

Indeed, one is tempted to employ one of the efficient algorithms recently presented for updating the structure of an existing lifted network with incremental changes to the evidence to solve the problem (Nath and Domingos, 2010; Ahmadi et al., 2010). While employing the symmetries in the graphical model across multiple evidence cases for the lifting, in the inference stage they need to construct a separate lifted network for each evidence case and run a modified message passing algorithm on each lifted network separately. Thus, symmetries across evidence cases are missed.

14.4.3   Sequential Multi-Evidence Lifting

We seek a way to efficiently construct the joint lifted network while still being able to lift across multiple evidence cases. To do this, we can modify the sequential single-evidence lifting (Ahmadi et al., 2010) to the multi-evidence case. Ahmadi *et al.* gives a clear characterization of the core information required for sequential clamping for lifted message passing, namely the shortest-paths connecting the variables in the network which resemble the computation paths along which the nodes communicate in the network.

Thus, to be able to adapt the lifted network for incoming evidence in the single-evidence case, we compute in a first step the set of shortest paths connecting any two nodes in the graph. Now, when there is new evidence for a node, the adapted lifted network is computed as a combination of the nodes' initial coloring, i.e. the lifting without evidence, and the set of shortest paths to the nodes in our evidence.

Algorithm 11 Sequential Multi-Evidence Lifted GaBP

Require: Matrix $\mathbf{A}$
  Construct network G_0 for $\mathbf{Ax} = \mathbf{0}$
  Compute path color matrix $\mathbf{PC}$ on G_0
  Lift G_0 to obtain lifted network H_0
  for each unit vector e_i **do**
      $colors(H_i) = colors(H_0)$
      $evidence = \{i{:}A_{ii}\}$
      repeat
          $colors(H_i) = \mathrm{newColor}(H_0, PC, evidence)$
          Add nodes that have changed color to $evidence$
      until $colors(H_i)$ does not change
      if H_i is previously unseen **then**
          Add H_i to joint lifted network $H_{\{1,...,i-1\}}$
      else
          Bookmark the corresponding index j
  end for
  Run modified GaBP on joint lifted network $H_{\{1,...,n\}}$
  return $\mathbf{X} = (x_1, x_2, \ldots, x_n)$, the inverse of $\mathbf{A}$

Algorithm 11 describes how we can adapt this to the multi-evidence case. Intuitively, we would like to view each $\mathbf{Ax}_i = \mathbf{e}_i$ ($i = 1, 2, \ldots, n$) as conditioning an initial network on node i and efficiently computing its contribution to the resulting lifted joint network directly from the initial one. But what should be the initial network? It is not provided by the task itself. Thus, to be able to efficiently find the lifted network structure, we propose to introduce an additional system of linear equations, namely the one with no evidence: $\mathbf{Ax} = \mathbf{0}$ (Algorithm 11, line 1). Indeed, it is not needed in the inversion task per se but it allows us to significantly speed up the multi-evidence lifting process. Intuitively, each $\mathbf{e}_i$ only differs from $\mathbf{0}$ in exactly one element so it serves as the basis for lifting all subsequent networks. To do this, we compute the path colors for all pairs of nodes (line 2) to know how it affects the other nodes when conditioning on a variable i and the initial lifting without evidence (line 3). The lifted graph H_i for the ith system of linear equations $\mathbf{Ax}_i = \mathbf{e}_i$ can now be adaptively computed by combining the initial lifting and the path colors to node i (line 8). The combination is essentially an elementwise concatenation of the two respective vectors.

Consider computing the lifted network for $\mathbf{Ax}_3 = \mathbf{e}_3$ for our running example. When we have no evidence the nodes x_1 and x_2 are clustered together, and x_3 is in a separate cluster. Thus, we obtain an initial color vector $C = (0, 0, 1)$. Since we want to compute the network conditioned on x_3 we have to combine this initial clustering C with the path colors with respect to x_3, $PC_{x_3} = (\{3,5\},\{3,5\},\emptyset) = (0,0,1)$. To obtain the lifted network conditioned on x_3 we have to (1) do an elementwise concatenation (in the following depicted by $\oplus$) of the two vectors and (2) interpret the result as a new color vector: $H_3 = (0,0,1) \oplus (0,0,1) =_{(1)} (00,00,11) =_{(2)} (2,2,3)$. Since only the shortest paths are computed, adapting the colors has to be performed iteratively to let the evidence propagate. For further details on adapting the color vector (line 8) we refer to (Ahmadi et al., 2010).

Moreover, we can implement a type of memoization when we perform the lifting in this fashion. Because we know each resulting lifted network H_i in advance, we can check whether an equivalent lifted network was already constructed: if the same color pattern exists already, we simply do not add H_i and instead only bookmark the correspondence of nodes (line 11-14) [42]. This does not affect the counts at all and, hence, still constructs the correct joint lifted network. This argument together with the correctness of the sequential single-evidence lifting and multi-evidence lifting effectively proves the correctness of sequential multi-evidence lifting:

Theorem 14.4 *Sequential multi-evidence lifting computes the same joint lifted model as in the batch case. Hence, running the modified GaBP on it produces the same marginals.*

14.5 Approximate Lifted Inference for General RCMs

In continuous relational models with limited potential functions, Gaussian for example, inference can be performed efficiently. However, in real world applications, flexible potential functions might be needed for better modeling the true distribution, and in the case of having arbitrary potential functions, exact inference in RCM is intractable. Hybrid Lifted Belief Propagation (HLBP) Chen et al. (2019), extends lifted belief propagation to continuous domains by using particle message passing and coarse-to-fine approximate lifting to handle continuous evidence. This provides a way for approximate lifted inference in general RCM. Similar to Counting BP Kersting et al. (2009), it first applies color passing on the grounded graph of the RCM, clustering similar nodes and factors, then performs inference on the compressed graph. Different from discrete case lifted BP, lifted BP with continuous domain requires integration over variables while computing the message from factor node to variable node. Instead of calculating the message exactly, in HLBP, it is approximated by a set of sampling points. The algorithm adopts important sampling technique to draw samples from a proposal distribution q, and multiple each sampling value with important weight. Specifically, the proposal q could be a Gaussian distribution obtained by expectation propagation.

In situations with a large number of continuous evidence variables, the number of exact symmetries in the factor graph is severely limited. To tackle this problem, a coarse-to-

---

[42] This is not as hard as solving (sub)graph-isomorphisms. We only have to check whether the color pattern of H_i was previously seen. If so, the result has already been memoized.

fine approximate lifting is applied. It clusters evidences that have close value together in the beginning and treat all of the evidence variables in the same cluster as having the same value, given by the cluster average, resulting a compressed graph with coarse evidence. The algorithm refines the compressed graph by splitting evidence clusters and running color passing algorithm. In HLBP, we interleave evidence refinement and message passing. This procedure guarantees the improvement of message accuracy in each iteration, and reduces the total computational cost in the earlier round of BP iterations.

For HLBP, the quality of inference largely depends on sampling. While it performs well on models with simple beliefs, it can fail to accurately capture the shape of multi-modal distributions and may have numerical and convergence issues. An alternative would be using variational inference instead of BP, namely Hybrid Lifted Variational Inference (HLVI) Chen et al. (2020). This method approximates the marginal distribution with mixture of components. Specifically,

$$q(x) = \sum_{k=1}^{K} w_k q^k(x) = \sum_{k=1}^{K} w_k \prod_i q_i^k(x_i) \tag{14.19}$$

where K is the number of mixture components, $w_k \geq 0$ is the weight of the kth mixture (a shared parameter across all marginal distributions), and $\sum_{k=1}^{K} w_k = 1$. Each component $q_i^k(x_i)$ is some valid distribution with parameters η_i^k. For continuous variable, we could use Gaussian as components, and categorical distribution for the discrete case. Inference is performed by minimizing the lifted Bethe free energy objective function, which is the same as minimizing the Kullback–Leibler divergence between the approximation q and the true distribution. Similar to HLBP, we could adopt coarse-to-fine lifting which interleaves evidence refinement and gradient descent update.

14.6   Conclusions

In this chapter, we reviewed lifted inference algorithms for hybrid relational models with continuous and discrete domains. We presented efficient lifted inference algorithm for large-scale probabilistic graphical models with continuous variables and interesting insights which transform probabilistic first-order language into compact approximations which allow efficient inference in relational linear dynamical systems, Kalman filtering, and in the PageRank problem of personalize recommendations. We also presented an approximate lifted inference for general relational hybrid models.

Acknowledgments

This work was partly supported by NSF award IIS-09-17123 – RI: Scaling Up Inference in Dynamic Systems with Logical Structure; Basic Science Research Program through the National Research Foundation of Korea (NRF) grant funded by the Korea government (MSIT: the Ministry of Science and ICT) (NRF-2017R1A1A1A05001456); and Institute for Information Communications Technology Planning and Evaluation (IITP) grant funded by the MSIT (No. 2017-0-01779, a machine learning and statistical inference framework for explainable artificial intelligence)

IV   BEYOND PROBABILISTIC INFERENCE

15 Color Refinement and Its Applications

Martin Grohe, Kristian Kersting, Martin Mladenov, and Pascal Schweitzer

Abstract. Color refinement is a simple algorithm that partitions the vertices of a graph according to their "iterated degree sequence." It has very efficient implementations, running in quasilinear time, and a surprisingly wide range of applications. The algorithm has been designed in the context of graph isomorphism testing, and it is used as a subroutine in almost all practical graph isomorphism tools. Somewhat surprisingly, other applications, among them lifted inference, have been discovered recently.

In this chapter, we introduce the basic color refinement algorithm and explain how it can be implemented in quasilinear time. We discuss variations of this basic algorithm and its somewhat surprising characterizations in terms of logic and linear programming, and we survey three important applications of color refinement: graph isomorphism testing, linear programming, and graph kernels.

15.1 Introduction

Color refinement, also known as *naive vertex classification* and *1-dimensional Weisfeiler-Leman algorithm*, is a combinatorial algorithm that aims to classify the vertices of a graph by similarity. It iteratively partitions, or *colors*, the vertices in a sequence of *refinement rounds*. Initially, all vertices get the same color. Then in each refinement round, any two vertices v, w that still have the same color get different colors if there is some color c such that v and w have a different number of neighbors of color c; otherwise they keep the same color. Thus, after the first refinement round, two vertices have the same color if and only they have the same degree (number of incident edges). After the second round, they have the same color if and only if for each k they have the same number of neighbors of degree k. The refinement process stops if in some refinement round the partition induced by the colors is no longer refined, that is, all pairs of vertices that have the same color before the refinement round still have the same color after the round. We call such a coloring that can no longer be refined a *stable coloring*.

Figure 15.1 shows how the coloring evolves on a randomly chosen graph with 30 vertices and 30 edges. A stable coloring is already reached after three refinement rounds. Observe that after the first refinement round, the colors reflect degrees: the (small) blue vertices have degree 0, the purple vertices have degree 1, the orange vertices degree 2, et cetera.

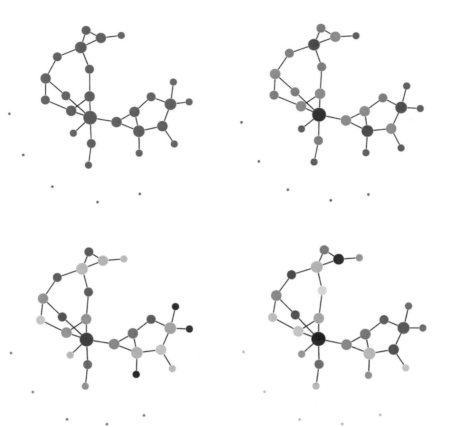

Figure 15.1: Color refinement on a random graph (from left to right, from top to bottom). A stable coloring is already reached after three refinement rounds. Observe that after the first refinement round, the colors reflect degrees: the (small) blue vertices have degree 0, the purple vertices have degree 1, the orange vertices degree 2, et cetera.

A naive implementation of color refinement will have a quadratic running time of $O(nm)$ steps on a graph with n edges and m vertices: we need at most n refinement rounds, and each round requires $O(m)$ steps. However, there is a more efficient "asynchronous" refinement strategy, akin to asynchronous belief propagation, which runs in quasi-linear time $O(m \log n)$ (Cardon and Crochemore, 1982; Paige and Tarjan, 1987) (see Section 15.2).

15.1.1   Applications

Before we continue with variants of the basic algorithm, let us briefly describe its main applications. The original application of color refinement is *graph isomorphism testing* (see (Read and Corneil, 1977)). To test if two graphs, G and H are isomorphic, we can run color refinement on their disjoint union $G \uplus H$ and then compare the color patterns on the two graphs. If they are different, that is, there is a color such that G and H have different number of vertices of this color, then the graphs cannot be isomorphic. We say that color refinement *distinguishes* G and H. If the color patterns of G and H are the same, we still do not know if G and H are isomorphic or not. However, very often this simple isomorphism test succeeds. In fact, Babai, Erdős, and Selkow (Babai et al., 1980) proved that color refinement distinguishes almost all non-isomorphic graphs. More advanced graph isomorphism tests and almost all practical isomorphism tools use color refinement as a subroutine (see Section 15.3 for details).

An application that has surfaced much more recently is the use of color refinement as a *pre-processing routine for solving mathematical programs or probabilistic inference tasks.* The goal is to "compress" the problem instances by identifying objects, for example, nodes of a graphical model or rows and columns of the matrix of a linear program, that receive the same color in a suitable adaptation of color refinement to the instances of the problem at hand (see below). The idea is that objects of the same color are sufficiently similar so that we can combine them to form one "super object". This way, we obtain a smaller compressed instance on which we solve our original task. Then we transform the solution of the compressed instance to a solution of the original instance (see Section 15.6 for details).

A third application is the design of graph kernels based on color refinement. We need a variant of color refinement where we use $1, 2, \ldots$ as color names. It is crucial that the color names are assigned *canonically*, that is, do not depend on the representation of the input graph, but only on its isomorphism class. Then we associate integer vectors with the colorings: with a coloring that assigns colors $1, \ldots, k$ to the vertices of a graph, we associate the vector $(n_1, \ldots, n_k)$ where n_i is the number of vertices of color i. Thus, color refinement yields such a vector for each graph, and we can use the scalar product on these vectors as our graph kernel. To avoid overfitting it has turned out to be best to only use the colorings obtained after the first few iterations of color refinement (see Section 15.7 for details).

15.1.2 Variants

Color refinement has several interesting variants, including a hierarchy of algorithms known as the Weifeiler-Leman algorithm(s), and generalizations to other structures such as vertex- and edge-labeled graphs, directed graphs and arbitrary relational structures, weighted graphs and even matrices. Many of these variations are needed in the applications of color refinement.

The simplest extensions of color refinement is the one to vertex-labeled graphs: instead of starting the refinement procedure from the coloring that assigns the same color to all vertices, we start from the given vertex labeling as our initial coloring. The extension to graphs that (also) have labeled edges is not much harder; all we need to do in the refinement step is to consider the degrees with respect to different edge labels separately. Thus in a refinement round, two vertices v, w that still have the same color get different colors if there is some edge label e and some color c of the current coloring such that v and w have a different number of e-labeled edges into the color class c. To extend color refinement to directed graphs, we consider in-degrees and out-degrees separately.

Now consider weighted graphs with edge weights in the reals (in fact, the edge weights can be elements of an arbitrary commutative monoid). One way of applying color refinement would be to just view the weights as edge labels and apply the algorithm for edge-labeled graphs. But for the "pre-processing" applications described in Section 15.6, the following variant is better: instead of refining by the number of edges into some color class, we refine by the sum of the weights of these edges. Thus in a refinement round, two vertices v, w that still have the same color get different colors if there is some color c of the current coloring such that the sum of the weights of the edges from v into color class c is different from the sum of the weights of the edges from w into color class c. Since we can interpret matrices as weighted bipartite graphs, this also gives us a natural color refinement procedure for matrices (see Section 15.5.1 for details).

Let us return to simple graphs (unweighted, undirected graphs containing no loops or multiple edges) and think about other ways of applying the iterative-coloring idea. What if we color edges instead of vertices? But wait, why not color pairs of non-adjacent vertices as well? Or even color triples of vertices? This leads to the idea of the **Weisfeiler-Leman algorithm**. In the k-dimensional version of this algorithm, we color k-tuples of vertices; the 1-dimensional version is just color refinement. We shall describe the algorithm in Section 15.4.

There are also extension of color refinement and Weisfeiler-Leman algorithm to relational structures and to hypergraphs (see (Böker, 2019)).

15.1.3   A Logical Characterization of Color Refinement

There is an interesting connection between color refinement and logic: the algorithm can be viewed as an equivalence test for the logic C^2, the fragment of first order logic extended by counting quantifiers of the form $\exists^{\geq i} x$ ("there are at least i values for x") consisting of all formulas with just at most two variables. This means that two graphs satisfy the same C^2-sentences if and only if they cannot be distinguished by color refinement. The correspondence can be lifted to the higher-dimensional Weisfeiler-Leman algorithm: the k-dimensional Weisfeiler-Leman algorithm is an equivalence test for the logic C^{k+1}, defined like C^2 with $(k + 1)$ variables per formula. This connection goes back to (Immerman and Lander, 1990; Cai et al., 1992). It places color refinement and the Weisfeiler-Leman algorithm in the context of other logical equivalence tests, for example, *bisimilarity*, which may be viewed as an equivalence test for modal logic on finite transitions systems (that is, colored directed graphs). We shall not explore the connection between color refinement and logic in this chapter and refer the reader to (Cai et al., 1992) and (Grohe, 2017, Chapter 3) for details.

15.2   Color Refinement in Quasilinear Time

We mentioned in the introduction that a naive implementation of color refinement has (at least) a quadratic running time; each of the n refinement rounds requires time $O(m)$, because all vertices and edges of the input graph need to be inspected to update the coloring. In this section, we shall describe a more efficient implementation running in time $O(m \log n)$. It uses a trick sometimes described as "processing the smaller half", which goes back to Hopcroft's quasilinear time algorithm for DFA-minimization (Hopcroft, 1971). Actually, we present a simplified version of the algorithm (due to McKay (McKay, 1981)) running in time $O(n^2 \log n)$ and only hint at the improvements (due to Cardon and Crochemore (Cardon and Crochemore, 1982), also see (Paige and Tarjan, 1987; Berkholz et al., 2013)) reducing the time to $O(m \log n)$.

Let $G = (V, E)$ be a graph. For each vertex-coloring C of G, we let C' be the coloring obtained from C by applying one refinement round (as described in the introduction). At the moment, we do not care about the actual colors used, but only about the partition of the vertices into color classes that the coloring induces. We call a coloring C **stable** if C and C' induce the same partition. We say that a coloring C is *coarser* than a coloring D if the partition induced by D refines the partition induced by C. We observe that the refinement operation is monotone with respect to this partial order: if C is at least as coarse as D then C' is at least as coarse as D'. A consequence of this observation is that color refinement computes the **coarsest stable coloring**.

Now consider the algorithm displayed in Figure 12 for computing the coarsest stable coloring of a given graph G. Starting from the constant coloring, it repeatedly picks a color q and *refines all other colors with respect to q*, that is, for each color c it splits the vertices v of color $C(v) = c$ according to the number $D(v)$ of neighbors of color q they have, and then assigns new colors to these vertices according to these numbers. The colors q that are used for splitting are stored in a queue Q, and the algorithm stops once this queue is empty. Whenever a color class c is split into new classes $B_{k_1}, \ldots, B_{k_2}$, then the new colors of the vertices (assigned in line 17) will be $c_{\max+i}$ for $k_1 \leq i \leq k_2$, where $c_{\max}$ is the last color used in the previous step. We add all of these new colors to the queue Q *except for the i^* such that the size B_{i^*} is maximum* (in line 12). If there are several B_i of maximum size, we take the first as B_{i^*}.

To see that the algorithm is correct, imagine first that we add all new colors to the queue Q, that is, we replace line 12 by $Q \leftarrow Q \cup \{c_{\max} + i \mid k_1 \leq i \leq k_2\}$. Then the algorithm would do exactly the same as the "normal" color refinement. But why does it also work if we do not add the largest new colors to the queue? To understand this, suppose that we split color c into new colors $d_1, \ldots, d_\ell$ and that, without loss of generality, d_ℓ is the largest one, which we omit from the queue. Now consider some other color c'. If we first refine c' with c and then with $d_1, \ldots, d_{\ell-1}$, then we have already taken d_ℓ into account, because

---

Algorithm 12 Efficient color refinement algorithm in pseudocode.

COLOR-REFINEMENT(G)

Require: Graph $G = (V, E)$

Ensure: Coarsest stable coloring C of G

1: $C(v) \leftarrow 1$ for all $v \in V$ ▷ initial coloring

2: $P(1) \leftarrow V$ ▷ P associates with each color the vertices of this color

3: $c_{\min} \leftarrow 1; c_{\max} \leftarrow 1$ ▷ color names are always between $c_{\min}$ and $c_{\max}$

4: initialize queue $Q \leftarrow \{1\}$ ▷ Q contains colors that will be used for refinement

5: **while** $Q \neq \emptyset$ **do**

6:     $q \leftarrow$ DEQUEUE(Q)

7:     $D(v) \leftarrow |N_G(v) \cap P(q)|$ for all $v \in V$ ▷ number of neighbors of v of color q

8:     $(B_1, \ldots, B_k)$ ordered partition of vertices v sorted lexicographically by $(C(v), D(v))$

9:     **for all** $i = c_{\min}$ to $c_{\max}$ **do**

10:        let $k_1 \leq k_2$ such that $P(c) = B_{k_1} \cup \ldots \cup B_{k_2}$ ▷ color class of c will be split

11:   ▷ into classes $B_{k_1}, \ldots, B_{k_2}$

12:        $i^* \leftarrow \arg\max_{k_1 \leq i \leq k_2} |B_i|$

13:        $Q \leftarrow Q \cup \{c_{\max} + i \mid k_1 \leq i \leq k_2, \, i \neq i^*\}$ ▷ add all colors except the one

14:   ▷ with the largest class to queue Q

15:     **end for**

16:     $c_{\min} \leftarrow c_{\max} + 1; c_{\max} \leftarrow c_{\max} + k$ ▷ new color range

17:     **for** $b = c_{\min}$ to $c_{\max}$ **do**

18:        $P(b) \leftarrow B_{b+1-c_{\min}}$ ▷ new coloring

19:        $C(v) \leftarrow b$ for all $v \in P(b)$

20:     **end for**

21: **end while**

22: **return** C

---

the number of neighbors of a vertex in v of color d_ℓ is the number of neighbors of color c minus the sum of the number of neighbors of colors $d_1, \ldots, d_{\ell-1}$.

As described, the algorithm runs in time $O(n^2 \log n)$. Obviously, the running time is dominated by the time spent in the main loop in lines 5–19. To compute the degrees $D(v)$ in line 7, we go through the neighbors of all vertices v in the color class $P(q)$, this requires time $O(n \cdot |P(v)|)$. It turns out that this is the most expensive step of the whole loop: lines 8–19 only require time $O(n)$; to create the ordered partition in line 8 we can use bucket sort. To bound the overall running time, we charge a time of $O(n)$ to each vertex in $P(q)$.

Now the crucial observation is that each vertex appears at most $O(\log n)$ times in a color class $P(q)$ for some q taken from the q in line 6. To see this, note that whenever a new color $b_i = c_{\max} + i$ is added to the queue Q in line 12 its color class B_i has at most half the size of the class $P(c)$. Now suppose a vertex v that appears on the queue ℓ times, say, in the colors $q_1, q_2, \ldots, q_\ell$. Then $q_{i+1} \leq q_i/2$ and thus $\ell \leq \log n$.

This means that the overall running time is $O(n^2 \log n)$, because each iteration of the loop requires time $O(n)$ per vertex in q, and each of the n vertices appears at most $\log n$ times.

To improve the running time to $O(m \log n)$, or more precisely $O(n + m \log n)$, we need to implement the main loop in such a way that one iteration requires time proportional to the number of edges incident with the vertices in the color class $P(q)$. Then we can charge time proportional to its degree to every vertex in $P(q)$, and essentially the same analysis as above yields the desired overall running time of $O(m \log n)$. To be able to execute lines 6–18 in time proportional to the number of edges out of $P(q)$ we need to make one significant change: all vertices that are not adjacent to vertex in $P(q)$ keep their old color, so we never need to touch these vertices. The rest mainly amounts to a careful choice of data structures. We refer the reader to (Berkholz et al., 2016) for details.

We remark that Berkholz, Bonsma, and Grohe (Berkholz et al., 2016) proved that no faster color refinement algorithm is possible under some reasonable assumption about the type of algorithm, which includes all known approaches.

15.3 Application: Graph Isomorphism Testing

The **graph isomorphism problem (GI)** is the algorithmic problem of deciding whether two given graphs are isomorphic. Recall that an *isomorphism* between two graphs is a bijective mapping that preserves adjacencies and non-adjacencies. The question for the complexity of the isomorphism problem is one of the best known open problem in computer science; it was already mentioned in Karp's seminal paper on NP-completeness (Karp, 1972) more than 40 years ago. In a recent breakthrough, Babai (Babai, 2016) proved that GI is solvable in quasipolynomial time $n^{(\log n)^{O(1)}}$.

The most straightforward way to use color refinement as an isomorphism test for two graphs G, H is to run it on the disjoint union $G \uplus H$ (the two graphs are "added together", with no new edges) and then compare the color patterns obtained on the graphs. If they differ, then we know that the graphs are not isomorphic; we say that color refinement *distinguishes* the two graphs. However, if the color pattern on the two graphs is the same, this does not mean that they are isomorphic. Thus color refinement (used this way) is an *incomplete* isomorphism test. Let us say that color refinement *identifies* a graph G if it distinguishes G from all graphs H that are not isomorphic to G. Babai, Erdös, and Selkow (Babai et al., 1980) proved that color refinement identifies almost all graphs, in the sense that the fraction of graphs of order n that color refinement identifies converges to 1 as n goes to infinity. The convergence is exponentially fast (Babai and Kučera, 1979). Color refinement also identifies all trees and forests (Immerman and Lander, 1990).

However, there are also very simple non-isomorphic graphs not distinguished by color refinement. The simplest example is a cycle of length 6 vs the disjoint union of two triangles. Clearly, these two graphs are not isomorphic, but color refinement assigns the same color to all vertices and effectively does not do anything. In fact, color refinement cannot distinguish any two regular graphs (every vertex has the same degree) of the same size and the same degree.

Even though color refinement fails to be a complete isomorphism test as a standalone algorithm, it is a key component of almost all practical graph isomorphism tools, which built on the so called **individualization/refinement** (for short: *IR*) paradigm. The prototype for all these isomorphism tools is McKays "nauty" (McKay, 1981), which is still one of the best isomorphism tools.

IR-algorithms are usually implemented as canonization algorithms. They take a single graph G as input and return a graph G^* such that G^* is isomorphic to G and for isomorphic input graphs G and H the graphs G^* and H^* are identical. Typically, G^* will have an initial segment of the positive integers as its vertex set. Obviously, a canonization algorithm yields an isomorphism test: given two graphs G and H we simply compute G^* and H^* and compare them.

We call a coloring of a graph *discrete* if all vertices have different colors. We always assume that the colors of our colorings come from some ordered set, typically the positive integers. Then from a discrete coloring D of a graph G we obtain an isomorphic copy $G^*(D)$ with vertex set $1, \ldots, n$ by simply renaming the vertices to $1, \ldots, n$ in the order of their colors. Recall that color refinement is a *canonical* coloring algorithm, that is, if f is an isomorphism from a graph G to a graph H and C_G, C_H are the stable colorings that color refinement produces, then $C_G(v) = C_H(f(v))$ for all vertices v of G.

The most basic version of an IR algorithm proceeds as follows. It runs color refinement on the input graph G. If the stable coloring C_G is discrete, it returns an isomorphic copy $G^*(C_G)$ corresponding to this coloring. If the coloring is not yet discrete, it *individualizes* a vertex in a color class with more than one vertex, that is, assigns a fresh color to the vertex. Then it runs color refinement again. If the coloring is still not discrete, it individualizes another vertex, and it repeats the individualization and refinement steps until eventually a discrete coloring is reached. This way we obtain a discrete coloring, but not in a canonical way. To fix this, whenever we individualize a vertex we have to branch on all vertices of the same color, individualizing one vertex in each branch. We build the *tree* of all these colorings. This tree is isomorphism-invariant. Associated with each leaf ℓ of the tree is a discrete coloring D_ℓ of the input graph G. We let $G_\ell^* = G^*(D_\ell)$ be the isomorphic copy of G corresponding to this discrete coloring. Let S_ℓ be the string of length n^2 that consists of the rows of the adjacency matrix of G_ℓ^*. Recall that the vertices of G_ℓ^* are $1, \ldots, n$; hence we can order the rows and columns of the adjacency matrix by the natural order on the vertices. Then our algorithm returns G_ℓ^* for a leaf ℓ with the lexicographically minimal string S_ℓ.

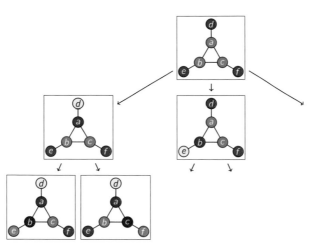

Figure 15.2: Individualization/Refinement. Shown is a part of the tree of colorings built on a simple, six-vertex graph. The root is the coloring obtained by color refinement. The leafs are discrete colorings.

Example 15.1 Figure 15.2 shows part of the tree of colorings built on a simple, six-vertex graph. At the root, we have the coloring obtained by color refinement. At the left child of the root, we have individualized vertex d and run color refinement again. At the middle child we have individualized e, and at the right child (not shown) we have individualized f. On the next level, on the leftmost branch we have individualized e, et cetera.

If the colors are ordered red $<$ green $<$ yellow $<$ blue $<$ orange $<$ dark red, then the graph G_ℓ^* associated with the leftmost leaf ℓ is obtained from G by renaming the vertices as follows: $f \mapsto 1$, $c \mapsto 2$, $d \mapsto 3$, $a \mapsto 4$, $e \mapsto 5$, $b \mapsto 6$. The string S_ℓ is 010000 100101 000100 011001 000001 010110. Curiously, for this graph the string is the same at all leaves.

The running time of our algorithm is exponential in the worst case, but we hope that usually most of the work is done by color refinement. Unfortunately, this is not always the case, in particular if the input graph has many symmetries. A key idea that makes the algorithm efficient in practice is that we can exploit symmetries to prune the tree of colorings. Whenever we arrive at a leaf ℓ of the tree such that $S_k = S_\ell$ for some leaf k processed before, or equivalently, $G_k^* = G_\ell^*$, then the colored graphs (G, D_k) and (G, D_ℓ) are isomorphic. The (unique) mapping $\gamma : V \to V$ that preserves the colors is an isomorphism between the colored graphs and hence an automorphism of G. We can exploit this information in the following way. Consider the lowest common ancestor t of the leaves k, ℓ, and let T_k and T_ℓ be the subtrees rooted at the children of t that contain k, ℓ, respectively. Let v_k, v_ℓ be the vertices individualized in the respective branches. Then we know that the whole trees T_k and T_ℓ of colorings are isomorphic, because the automorphism γ maps v_k to v_ℓ and fixes all vertices that have been individualized on the path from the root to t. Thus, there is no need to explore the tree T_k any further. There is a second way in which we use the automorphisms γ found this way. We store the automorphisms and gradually build up a subgroup Γ of the automorphism group of G. Then, whenever we are about to enter a new branch of the tree, say, by individualizing a vertex v_ℓ of the same color as a vertex v_k that has been individualized before (that is, v_k is a sibling of v_k on the left of v_ℓ), and we find an automorphism γ in our group Γ such that $\gamma(v_k) = v_\ell$ and γ fixes all vertices that have been individualized on the path from the root to the current node, then there is no need to enter the new branch where v_ℓ is individualized at all, because the subtree will look exactly like the subtree for v_k.

Example 15.2 Figure 15.3 shows the full tree of colorings for the same graph G as in Example 15.1 if we apply the pruning techniques just described. Let ℓ_1, ℓ_3, ℓ_3 be the three leaves of the tree. Once we arrive at ℓ_2, we find that $S_{\ell_1} = S_{\ell_2}$ and hence that the permutation $\gamma = (b\ c)(e\ f)$ (in cycle notation, more explicitly, $g = \{a \mapsto a, b \mapsto c, c \mapsto b, d \mapsto d, e \mapsto f, f \mapsto e\}$) is an automorphism of G. Then, when we arrive at the leaf ℓ_3, we find that $S_{\ell_1} = S_{\ell_3}$ and hence that the permutation $\delta = (a\ b)(d\ e)$ is an automorphism. The lowest common ancestor of ℓ_1 and ℓ_3 is the root, and this means we do not have to explore the subtree rooted in the second child of the root any further. Hence we immediately jump back to the root, where in the next branch we individualize the node f (after we have individualized d, e at the first two children). However, we have already found the automorphism γ which maps e to f, and since we have already explored the e-branch, there is no need to do so for the f-branch. Thus, the tree is complete.

15.4   The Weisfeiler-Leman Algorithm

The k-dimensional Weisfeiler-Leman algorithm (for short: k-WL) is a version of color refinement that colors k-tuples of vertices instead of single vertices. This makes the algorithm more powerful. For example, with high probability 2-WL distinguishes two randomly chosen d-regular graphs of the same size (Bollobás, 1982), and 3-WL identifies all planar graphs (Grohe, 1998; Kiefer et al., 2017). The price we pay for this additional power is a significantly higher runtime: the best-known implementation of k-WL runs in time $O(k^2 n^{k+1} \log n)$ (Immerman and Lander, 1990).

To describe the k-dimensional Weisfeiler-Leman algorithm, let us start with a different, somewhat more formal, view on the color refinement algorithm. Let $G = (V, E)$ be a graph. Let C_0, defined by $C_0(v) = 1$ for all $v \in V$, be the initial coloring. Denoting the coloring obtained after the ith refinement round by C_i, the new coloring after the $(i + 1)$th round can

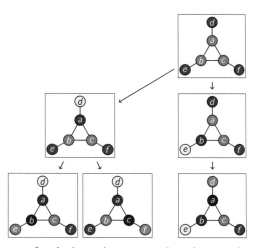

Figure 15.3: Full tree of colorings due to pruning the search tree of individualiza-tions/refinements.

be defined by assigning to each vertex the pair consisting of its old color and the multiset[43] of colors of its neighbors:

$$C_{i+1} := \Big(C_i(v), \{\!\{C_i(w) \mid vw \in E\}\!\}\Big).$$

To avoid the awkwardness of handling nested multisets, in an implementation we would usually sort the multisets lexicographically and label them by integers, which we then use as the new colors. Note that the coarsest stable coloring is $C_\infty = C_i$ for the smallest i such that for all $v, w \in V$ it holds that $C_i(v) = C_i(w) \iff C_{i+1}(v) = C_{i+1}(w)$.

We need one more definition. The *atomic type* atp$(\bar{v})$ of a k-tuple $\bar{v} = (v_1, \dots, v_k)$ of vertices of a graph G describes the labeled subgraph induced by G on this tuple; formally we may describe it by a $(k \times k)$-matrix A with entries $A_{ij} = 2$ if $v_i = v_j$ and $A_{ij} = 1$ if $v_i v_j \in E$ and $A_{ij} = 0$ otherwise; the crucial property of atomic types is that for tuples $\bar{v} = (v_1, \dots, v_k), \bar{w} = (w_1, \dots, w_k)$ we have atp$(\bar{v})$ = atp$(\bar{w})$ if and only if the mapping $v_i \mapsto w_i$ is well-defined and an isomorphism from the induced subgraph $G[\{v_1, \dots, v_k\}]$ to the induced subgraph $G[\{w_1, \dots, w_k\}]$.

Now we can describe k-WL as the algorithm that, given a graph $G = (V, E)$, computes the following sequence of "colorings" C_i^k of V^k for $i \geq 0$ until it returns $C_\infty^k = C_i^k$ for the smallest i such that for all $\bar{v}, \bar{w}$ it holds that $C_i^k(\bar{v}) = C_i^k(\bar{w}) \iff C_{i+1}^k(\bar{v}) = C_{i+1}^k(\bar{w})$. The initial coloring C_0^k assigns to each tuple its atomic type: $C_0^k(\bar{v}) := \text{atp}(\bar{v})$. In the $(i+1)$th refinement round, the coloring C_{i+1}^k is defined by $C_{i+1}^k(\bar{v}) := (C_i^k(\bar{v}), M_i(\bar{v}))$, where $M_i(\bar{v})$ is

---

[43] A *multiset* is an unordered collection of not necessarily distinct elements (whose multiplicities are counted). Formally, we may view a multiset as a mapping from a set to the positive integers.

the multiset

$$\big\{\!\!\big\{ \big(\mathrm{atp}(v_1,\ldots,v_k,w),\, C_i^k(v_1,\ldots,v_{k-1},w),\, C_i^k(v_1,\ldots,v_{k-2},w,v_k),\ldots,C_i^k(w,v_2,\ldots,v_k)\big) \,\big|\, w \in V \big\}\!\!\big\}$$

for $\bar{v} = (v_1,\ldots,v_k)$. If $k \geq 2$, we can omit the entry $\mathrm{atp}(v_1,\ldots,v_k,w)$ from the tuples in the multiset, because all the information it contains is also contained in the entries $C_i^k(\ldots)$. It is easy to see that the coloring C_i^1 computed by 1-WL coincides with the colorings C_i computed by color refinement, in the sense that $C_i(v) = C_i(w) \iff C_i^1(v) = C_i^1(w)$ for all vertices v, w.

A fairly straightforward modification of the algorithm discussed in Section 15.2 computes the coloring C_∞^k in time $O(k^2 n^{k+1} \log n)$ (Immerman and Lander, 1990).

Example 15.3 Let G be the graph consisting of the disjoint union of a cycle of length 3 and a cycle of length 4, say, with vertex set $V = \{1,\ldots,7\}$, where vertices $1, 2, 3$ form a triangle and vertices $4,\ldots,7$ form a 4-cycle. If we run color refinement on G, all vertices get the same color: $C_\infty(v) = C_0(v) = 1$ for all $v \in V$.

Now let us run 2-WL on G. There are three atomic types of pairs of elements, let us call them equal, adjacent, and non-adjacent. For example, $C_0^2(4,4) = \mathrm{e}$, $C_0^2(4,5) = \mathrm{a}$, and $C_0^2(4,3) = C_0^2(4,6) = \mathrm{n}$. For the coloring C_1^2 after the first refinement round we have, for example,

$$C_1^2(4,3) = \big(\mathrm{n}, \{\!\!\{(\mathrm{n,a}),(\mathrm{n,a}),(\mathrm{n,e}),(\mathrm{e,n}),(\mathrm{a,n}),(\mathrm{n,n}),(\mathrm{a,n})\}\!\!\}\big),$$

$$C_1^2(4,6) = \big(\mathrm{n}, \{\!\!\{(\mathrm{n,n}),(\mathrm{n,n}),(\mathrm{n,n}),(\mathrm{e,n}),(\mathrm{a,a}),(\mathrm{n,e}),(\mathrm{a,a})\}\!\!\}\big),$$

where we omit the atomic types from the tuples in the multiset.

The coloring C_1^2 is not yet the stable coloring. C_2^2 does not refine C_1^2 on the pairs of distinct vertices, but it does on the pairs of equal vertices. The reader can easily verify that C_1^2 takes the same value on all pairs (v,v), but we have $C_2^2(1,1) \neq C_2^2(4,4)$. The coloring C_2^2 is stable. See Figure 15.4(a) for an illustration. The coloring is symmetric, that is, $C_2^2(v,w) = C_2^2(w,v)$ for all v, w, hence we can display the colors as undirected edges.

By comparison, Figure 15.4(b) shows the coloring obtained by running 2-WL on a cycle of length 7. Since the color pattern is different from the one on the graph G in Figure 15.4(a), 2-WL distinguishes the two graphs. Color refinement does not, because both graphs are 2-regular.

15.5   Fractional Isomorphism

Suppose we try to solve the graph isomorphism problem by linear programming. We want to describe the isomorphisms between G and H as the solutions of an integer linear

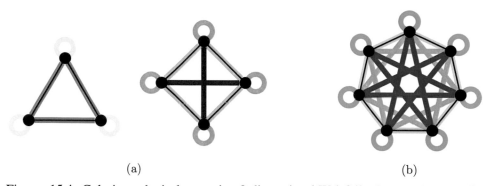

(a)   (b)

Figure 15.4: Colorings obtain by running 2-dimensional Weisfeiler-Leman (a) on a disjoint union of a cycle of length 3 and a cycle of length 4 as well as (b) on a cycle of length 7.

program, or rather, a system of linear equalities and inequalities.[44] Quite surprisingly, there is a close correspondence between the real solutions of this system and the stable colorings of the disjoint union $G \uplus H$.

For the rest of the section, let G and H be two graphs with vertex sets V, W of the same size n, and let A and B be their adjacency matrices. Without loss of generality we assume that V and W are disjoint. A and B are $n \times n$ matrices with $\{0, 1\}$-entries $A_{vv'}$ and $B_{ww'}$. We think of the rows and columns of A as being indexed by vertices of G and the rows and columns of B as being indexed by vertices of H. We write $A \in \{0, 1\}^{V \times V}$ and $B \in \{0, 1\}^{W \times W}$.

The graphs G and H are isomorphic if and only if there is a permutation matrix $X \in \{0, 1\}^{V \times W}$ such that $X^\top A X = B$. Recall that a *permutation matrix* is a $\{0, 1\}$-matrix with exactly one 1-entry in each row and in each column. A permutation matrix $X \in \{0, 1\}^{V \times W}$ encodes the bijective mapping $\pi_X : V \to W$ defined by $\pi_X(v) = w$ for the unique w with $X_{vw} = 1$. Then $X^\top A X$ is the adjacency matrix of the graph $\pi_X(G)$. To turn it into a system of of linear equations, we equivalently rewrite the equation $X^\top A X = B$ as

$$AX = XB, \tag{15.1}$$

using the fact that permutation matrices X are orthogonal, that is, the transpose $X^\top$ is the inverse X^{-1}. Equation (15.1) combined with equations forcing X to be a permutation matrix leads to the following system of linear equations and inequalities in the variables

---

[44] Graph isomorphism is not an optimization problem, hence our "linear program" has no objective function, or rather, we are only interested in the feasible solutions. Still, it will sometimes be convenient to use linear programming terminology.

X_{vw} representing the entries of X:

$$\text{ISO}(G, H) \qquad \sum_{v' \in V} A_{vv'} X_{v'w} = \sum_{w' \in W} X_{vw'} B_{w'w} \qquad (15.2)$$

$$\sum_{w' \in W} X_{vw'} = \sum_{v' \in V} X_{v'w} = 1 \qquad (15.3)$$

$$X_{vw} \geq 0 \qquad\qquad \text{for all } v \in V, w \in W. \qquad (15.4)$$

Our discussion shows that the integer solutions to this system correspond to the isomorphisms between G and H. But what about the real (or rational) solutions? Let us call them **fractional isomorphisms**. Note that fractional isomorphism are doubly stochastic matrices X satisfying equation (15.1). A matrix is *doubly stochastic* if its row sums and column sums are 1, that is, if it satisfies equations (15.3). Tinhofer (Tinhofer, 1991) proved the following beautiful theorem.

Theorem 15.1 ((Tinhofer, 1991)) *There is a fractional isomorphism from G to H if and only if color refinement does not distinguish G and H.*

Proof   We first prove the easier backward direction. Suppose that color refinement does not distinguish G and H. Let $U_1, \ldots, U_\ell \subseteq V \cup W$ be the color classes of the coarsest stable coloring C_∞, and let $V_i = V \cap U_i$ and $W_i = W \cap U_i$. Since color refinement does not distinguish G and H, we have $|V_i| = |W_i|$. We define a matrix $X \subseteq \mathbb{R}^{V \times W}$ by

$$X_{vw} = \begin{cases} \frac{1}{|V_i|} & \text{if } v \in V_i, w \in W_i \text{ for some } i, \\ 0 & \text{otherwise.} \end{cases}$$

Since $|V_i| = |W_i|$, this matrix is doubly stochastic and thus satisfies equations (15.3) and (15.4). To see that it satisfies (15.2), let $v \in V_i$ and $w \in W_j$. Then

$$\sum_{v' \in V} A_{vv'} X_{v'w} = \sum_{v' \in V_j} \frac{A_{vv'}}{|V_j|} = \frac{|N_G(v) \cap V_j|}{|V_j|}$$

$$= \frac{|N_H(w) \cap W_i|}{|W_i|} = \sum_{w' \in W_i} \frac{B_{w'w}}{|W_i|} = \sum_{w' \in W} X_{vw'} B_{w'w}.$$

Here the first and last equalities follows from the definition of X and the second and fourth from the definition of the adjacency matrices and the fact that $|V_i| = |W_i|$ and $|V_j| = |W_j|$. The crucial third equality follows from the fact that the coloring is stable, which implies that there are numbers n_{ij} and n_{ji} such that every vertex in color class U_i has n_{ij} neighbors in class U_j and every vertex in class U_j has n_{ji} neighbors in class U_i. Then the number of edges between the classes V_i and V_j and between W_i and W_j $n_{ij}|V_i| = n_{ji}|V_j| = n_{ij}|W_i| = n_{ji}|W_j|$. As $n_{ij} = |N_G(v) \cap V_j|$ and $n_{ji} = |N_H(w) \cap W_i|$, this implies the equality.

To prove the forward direction, suppose that X is a fractional isomorphism from G to H. Let K be the graph with vertex set $V \cup W$ and edge set $\{vw \mid X_{vw} > 0\}$. Let $K_1, \ldots, K_\ell$ be the connected components of K, and let U_i be the vertex set of K_i. We define a coloring $C : V \cup W \rightarrow \{1, \dots, \ell\}$ by $C(u) = i$ if $u \in U_i$. We shall prove that C is a stable coloring satisfying

$$|V \cap C^{-1}(i)| = |W \cap C^{-1}(i)| \qquad \text{for all } i. \tag{15.5}$$

Once we have proved this, we are done, because the satbel coloring C refines the coarsest stable coloring C_∞, and thus by (15.5) C_∞ satisfies $|V \cap C_\infty^{-1}(c)| = |W \cap C_\infty^{-1}(c)|$ for all colors c. Hence color refinement does not distinguish G and H.

Equation (15.5), or equivalently $|V \cap U_i| = |W \cap U_i|$ is proved by a straightforward calculation:

$$|V \cap U_i| = \sum_{v \in V \cap U_i} \sum_{w \in W} X_{vw} = \sum_{v \in V \cap U_i} \sum_{w \in W \cap U_i} X_{vw} = \sum_{w \in W \cap U_i} \sum_{v \in V} X_{vw} = |W \cap U_i|.$$

where the first and fourth equalities follow from (15.3) and the second and third equalities from the fact that $X_{vw} = 0$ unless v and w belong to the same set U_j.

To prove that the coloring C is stable, we consider color classes U_i, U_j. We need to prove that for all $v, w \in U_i$ we have $|N(v) \cap U_j| = |N(w) \cap U_j|$. Here by $N(u)$ we mean $N_G(u)$ if $u \in V$ and $N_H(u)$ if $u \in W$. It will be important to distinguish $N(u)$ from $N_K(u) = \{u' \mid X_{uu'} > 0 \text{ or } X_{u'u} > 0\}$. So let $v \in U_i$, where without loss of generality we assume that $v \in V$. Then

$$|N(v) \cap U_j| = \sum_{v' \in V \cap U_j} A_{vv'} = \sum_{v' \in V \cap U_j} A_{vv'} \sum_{w \in W} X_{v'w} = \sum_{w \in W \cap U_j} \sum_{v' \in V} A_{vv'} X_{v'w}$$

$$= \sum_{w \in W \cap U_j} \sum_{w' \in W} X_{vw'} B_{w'w} = \sum_{w' \in W} X_{vw'} \sum_{w \in W \cap U_j} B_{w'w} = \sum_{w' \in N_K(v)} X_{vw'} |N(w') \cap U_j|.$$

Here the second equality follows from (15.3) and the third from the fact that $X_{v'w} \neq 0$ only if v' and w are from the same color class U_j. The fourth equality follows from (15.2), and the last equality follows from the fact that $X_{vw'} \neq 0$ only if w' is neighbor of v in the graph K. Note that we have written $|N(v) \cap U_j|$ as a *positive convex combination* of the numbers $|N(w) \cap U_j|$ for $w \in N_K(v)$, that is, a linear combination with positive coefficients whose sum is 1.

Now we use the following simple fact: *if f is a real valued function defined on the vertex set of a connected graph, and for all v the value $f(v)$ is a positive convex combination of the values $f(w)$ for the neighbors w of v, then f is constant.* We apply this fact to the graph K_i and the function $f(u) = |N(u) \cap U_j|$. Thus, $|N(u) \cap U_j| = |N(u') \cap U_j|$ for all $u, u' \in U_i$, which proves that the coloring C is stable. □

Tinhofer's Theorem has a nice generalization to the higher dimensional Weisfeiler-Leman algorithm. To put it in context, let us briefly review some ideas from combinatorial optimization. Suppose we want to solve an integer linear program L. We may start by looking at the real solutions to L, that is, drop the integrality constraints, and then maybe round them to integral solutions. But chances are that the polytope $P_\mathbb{R}$ of real solutions is too far from the polytope $P_\mathbb{Z}$ generated by the integer solutions, the latter being the one we are interested in. Then what we can do is add additional linear constraints in such a way that the resulting linear program L' has the same integer solutions. Let $P'_\mathbb{R}, P'_\mathbb{Z}$ be the polytopes corresponding to L'. Then $P_\mathbb{Z} = P'_\mathbb{Z} \subseteq P'_\mathbb{R} \subseteq P_\mathbb{R}$. Maybe $P'_\mathbb{R}$ is sufficiently close to $P'_\mathbb{Z}$ for our purposes. If it is not, we may repeat the process and add further linear constraints so the we get a linear program L'', then maybe a linear program L''', et cetera. There are systematic ways of doing this, leading to so-called "lift-and-project"[45] hierarchies (Balas et al., 1993; Lovász and Schrijver, 1991; Sherali and Adams, 1990; Lasserre, 2002).

One of the most important of these hierarchies is the **Sherali-Adams hierarchy** (Sherali and Adams, 1990). Atserias and Maneva (Atserias and Maneva, 2013) established a close connection between the Sherali-Adams hierarchy over the linear program $\mathrm{ISO}(G, H)$ and distinguishability by k-WL. The following linear program $\mathrm{ISO}^{(k)}(G, H)$ is a combination of of the equations of the kth level and the $(k + 1)$th level of the Sherali-Adams hierarchy over $\mathrm{ISO}(G, H)$. The variables of this linear program are X_p for $p \subseteq V \times W$ with $|p| \le k + 1$. To understand the equations, we should view the indices p as relations between $(k + 1)$ vertices of G and of H. It can be shown that the equations and inequalities imply that $X_p > 0$ only if p is a *local isomorphism*, that is, an injective mapping that preserves adjacencies and non-adjacencies.

$$\mathrm{ISO}^{(k)}(G, H) \qquad \sum_{v' \in V} A_{vv'} X_{p \cup \{(v', w)\}} = \sum_{w' \in W} X_{p \cup \{(v, w')\}} B_{w'w}$$

$$\text{for all } p \text{ of size } |p| \le k - 1 \text{ and all } v, w$$

$$\sum_{w \in W} X_{p \cup \{(v,w)\}} = \sum_{v \in V} X_{p \cup \{(v,w)\}} = X_p$$

$$\text{for all } p \text{ of size } |p| \le k \text{ and all } v, w$$

$$X_\emptyset = 1$$

$$X_p \ge 0 \qquad \text{for all } p \text{ of size } |p| \le k + 1$$

Theorem 15.2 ((Atserias and Maneva, 2013; Grohe and Otto, 2015; Malkin, 2014))
$\mathrm{ISO}^{(k)}(G, H)$ has a rational solution if and only if k-WL does not distinguish G and H.

15.5.1   Color Refinement and Fractional Isomorphism of Matrices

There is an interesting and, as we shall see in the next section, quite useful generalization of Tinhofer's theorem to weighted graphs, which we may also view as symmetric real matrices. We adopt color refinement to weighted graphs by refining not by degree, but by the sum of the edge weights. So let $A \in \mathbb{R}^{V \times V}$ be a symmetric matrix. We may view A as

---

[45] Please note that "lift" here is used in a different way then in "lifted" probabilistic inference.

the adjacency matrix of a weighted graph G with vertex set V and an edge with weight A_{vw} between v and w for all v, w such that $A_{vw} \neq 0$. By a *coloring of A* we mean a coloring of the index set V. Color refinement computes colorings $C_0, C_1, \dots$ of A in a sequence of refinement rounds. The initial coloring C_0 assigns the same color to all $v \in V$. In the $(i + 1)$th refinement round, the algorithm assigns the same color to vertices v, w if and only if $C_i(v) = C_i(w)$ and for all colors c in the range of C_i,

$$\sum_{w \in V \cap C_i^{-1}(c)} A_{vw} = \sum_{w' \in V \cap C_i^{-1}(c)} A_{v'w'}. \tag{15.6}$$

We let $C_\infty = C_i$ for the smallest i such that C_{i+1} induces the same partition of V into color classes as C_i.

We call a coloring C of A *stable* if it satisfies (15.6) (with C instead of C_i) for all $v, w \in V$ with $C(v) = C(w)$. It is easy to see that C_∞ is the coarsest stable coloring of A. Adopting the algorithm described in Section 15.2, we can compute C_∞ in time $O(n^2 \log n)$, where $n = |V|$, and even $O(m \log n)$, where m is the number of nonzero entries of A, assuming the matrix is suitably represented.

Now let $A \in \mathbb{R}^{V \times V}$ and $B \in \mathbb{R}^{W \times W}$, where $|V| = |W|$ and V and W are disjoint, be two symmetric matrices. The following matrix corresponds to the disjoint union of the weighted graphs:

$$A \uplus B = \begin{pmatrix} A & 0 \\ 0 & B \end{pmatrix} \in \mathbb{R}^{(V \cup W) \times (V \cup W)}.$$

Let C_∞ be the coarsest stable coloring of $A \uplus B$. We say that color refinement *distinguishes* A and B if for all colors c in the range of C_∞ we have $|V \cap C_\infty^{-1}(c)| = |W \cap C_\infty^{-1}(c)|$.

A **fractional isomorphism** from A to B is a doubly stochastic matrix $X \in \mathbb{R}^{V \times W}$ such that $AX = XB$. We obtain a direct generalization of Tinhofer's Theorem, with essentially the same proof.

Theorem 15.3 ((Grohe et al., 2014))   *There is a fractional isomorphism from A to B if and only if color refinement does not distinguish A and B.*

We can further generalize this to arbitrary, not necessarily symmetric matrices. Let $A \in \mathbb{R}^{V_1 \times V_2}$, where V_1 and V_2 are disjoint (but not necessarily of the same size). We say that a **stable bicoloring** of A is a pair (C_1, C_2) of colorings of V_1, V_2, respectively, such that

$$\sum_{v_{3-i} \in V_{3-i} \cap C_{3-i}^{-1}(c)} A_{v_1 v_2} = \sum_{v'_{3-i} \in V_{3-i} \cap C_{3-i}^{-1}(c)} A_{v'_1 v'_2}$$

for $i = 1, 2$, all $v_i, v'_i \in V_i$ with $C_i(v_1) = C_i(v'_1)$, and all c in the range of C_{3-i}. Equivalently, we may view a stable bi-coloring as a stable coloring C of the matrix

$$\begin{pmatrix} 0 & A \\ A^\top & 0 \end{pmatrix} \in \mathbb{R}^{(V_1 \cup V_2) \times (V_1 \cup V_2)}$$

that refines the initial coloring with two color classes V_1 and V_2. This shows that we can compute the coarsest stable bicoloring of a matrix in time $O(m \log(n_1 + n_2))$, where $n_i = |V_i|$ and m is the number of nonzero entries of A.

We can then define a notion of a *fractional bi-isomorphism* between two matrices $A \in \mathbb{R}^{V_1 \times V_2}$, $B \in \mathbb{R}^{W_1 \times W_2}$ as a pair of doubly stochastic matrices $X_1 \in \mathbb{R}^{V_1 \times W_1}$, $X_2 \in \mathbb{R}^{V_2 \times W_2}$ such that $AX_2 = X_1B$. It is not difficult to prove the analogue of Theorem 15.3 for fractional bi-isomorphism. We only state a technical lemma (for later reference) that follows from the easy direction of the proof.

Lemma 15.1 ((Grohe et al., 2014)) *Let $A \in \mathbb{R}^{V_1 \times V_2}$, and let (C_1, C_2) be a stable bi-coloring of A. For $i = 1, 2$, let $X_i \in \mathbb{R}^{V_i \times V_i}$ be the matrix with entries*

$$(X_i)_{vv'} = \begin{cases} \frac{1}{|C_i^{-1}(c)|} & \text{if } C_i(v) = C_i(v') = c \text{ for some } c \text{ in the range of } C_i, \\ 0 & \text{if } C_i(v) \neq C_i(v'). \end{cases}$$

*Then (X_1, X_2) is a **fractional automorphism** of A, that is, X_1, X_2 are doubly stochastic and $AX_2 = X_1A$.*

15.6   Application: Linear Programming

The results of the previous section establish a surprising connection between color refinement and linear algebra. In this section, we will show how this connection can be used for pre-processing linear programs in order to "compress" ("lift", in the sense of lifted probabilistic inference) them to smaller equivalent linear programs.

To explain the method, we start with the problem of solving a system $Ax = \mathbf{1}$, of linear equations, where $A \in \mathbb{R}^{m \times n}$ and $\mathbf{1} = (1, \ldots, 1)^\top \in \mathbb{R}^m$. Let (C, D) be a stable bi-coloring of A. Suppose that the range of C is $\{1, \ldots, k\}$ and the range of D is $\{1, \ldots, \ell\}$. We define two matrices $P \in \mathbb{R}^{m \times k}$ and $P^* \in \mathbb{R}^{k \times m}$ by

$$P_{ip} = \begin{cases} 1 & \text{if } C(i) = p, \\ 0 & \text{otherwise,} \end{cases} \qquad P^*_{pi} = \begin{cases} \frac{1}{|C^{-1}(p)|} & \text{if } C(i) = p, \\ 0 & \text{otherwise.} \end{cases} \qquad (15.7)$$

Observe that $X = PP^* \in \mathbb{R}^{m \times m}$ is the matrix with entries $X_{ii'} = 1/|C^{-1}(p)|$ if for i, i' with $C(i) = C(i') = p$ and $X_{ii'} = 0$ for i, i' with $C(i) \neq C(i')$. Similarly, we define matrices $Q \in \mathbb{R}^{n \times \ell}$ and $Q^* \in \mathbb{R}^{\ell \times n}$ by

$$Q_{jq} = \begin{cases} 1 & \text{if } D(j) = q, \\ 0 & \text{otherwise,} \end{cases} \qquad Q^*_{qj} = \begin{cases} \frac{1}{|D^{-1}(q)|} & \text{if } D(j) = q, \\ 0 & \text{otherwise} \end{cases} \qquad (15.8)$$

and let $Y = QQ^* \in \mathbb{R}^{n \times n}$. Observe that P, P^*, Q and Q^* are all stochastic matrices and that X and Y are doubly stochastic. By Lemma 15.1, we have

$$AY = XA. \qquad (15.9)$$

Moreover, a simple calculation shows

$$P^* = P^*X, \qquad\qquad Q = YQ. \qquad\qquad (15.10)$$

Now we let $A' = P^*AQ \in \mathbb{R}^{k \times \ell}$ and $\mathbf{1}' = (1, \ldots, 1)^\top \in \mathbb{R}^k$. Note that the matrix A' may be substantially smaller than A; this happens if A has many symmetries or at least "regularities". We claim that the system $A'x' = \mathbf{1}'$ is equivalent to our original system in the sense that the solution spaces can be mapped into each other by simple linear transformation. Indeed, if x is a solution to $Ax = \mathbf{1}$, then $x' = Q^*x$ is a solution to $A'x' = \mathbf{1}'$:

$$A'x' = P^*AQQ^*x = P^*AYx = P^*XAx = P^*X\mathbf{1} = P^*PP^*\mathbf{1} = \mathbf{1}', \qquad (15.11)$$

where the last equality holds because the matrices P and P^* are stochastic and thus $P^*\mathbf{1} = \mathbf{1}'$ and $P\mathbf{1}' = \mathbf{1}$. Conversely, suppose that x' is a solution to $A'x' = \mathbf{1}'$, and let $x = Qx'$. Then

$$Ax = AQx' = AYQx' = XAQx' = PA'x' = P\mathbf{1}' = \mathbf{1}. \qquad (15.12)$$

The method easily extends to linear programs. Consider a linear program in standard form:

$$\text{L} \qquad\qquad \min \quad c^\top x$$

$$\text{subject to} \quad Ax = b, \quad x \geq \mathbf{0},$$

where $A \in \mathbb{R}^{m \times n}$, $b = (b_1, \ldots, b_m)^\top \in \mathbb{R}^m$, and $c = (c_1, \ldots, c_n)^\top \in \mathbb{R}^n$. We define an initial bi-coloring (C_0, D_0) of the matrix A by $C_0(i) = b_i$ and $D_0(j) = c_j$, and we let (C, D) be the coarsest stable bi-coloring of A that refines (C_0, D_0). We define the matrices P, P^*, Q, Q^*, X, Y as above and note that (15.9) and (15.10) still hold. We let $A' = P^*AQ$ and $b' = P^*b$ and $c' = Q^*c$. We obtain a new linear program:

$$\text{L}' \qquad\qquad \min \quad (c')^\top x'$$

$$\text{subject to} \quad A'x' = b', \quad x' \geq \mathbf{0}.$$

Let x be a feasible solution to L and $x' = Q^*x$. Then x' is nonnegative, and the same calculation as in (15.11) shows that $A'x' = b'$. Hence x' is a feasible solution to L'.

Conversely, let x' be a feasible solution to L', and let $x = Qx'$. Then x is nonnegative, and the calculation (15.12) shows that $Ax = Pb'$. As the stable coloring C refines our initial coloring C_0, for all i, i' with $C(i) = C(i')$ we have $b_i = b_{i'}$. Thus,

$$(Pb')_i = (Xb)_i = \sum_{i'=1}^m X_{ii'}b'_i = \sum_{i' \in C^{-1}(i)} \frac{1}{|C^{-1}(i)|} b_{i'} = b_i \sum_{i' \in C^{-1}(i)} \frac{1}{|C^{-1}(i)|} = b_i$$

for all i. Hence $Pb' = b$, and therefore x is a feasible solution to L.

Further calculations show that if x is an optimal solution to L then $x' = Q^*x$ is an optimal solution to L' and, conversely, if x' is an optimal solution to L' then $x' = Q^*x$ is an optimal solution to L (see (Grohe et al., 2014) for details).

What this all means is that for a given linear program L, color refinement gives us a potentially much smaller linear program L′ and two linear mappings Q^* and Q transforming feasible solutions to L into feasible solutions to L′ and vice versa and preserving optimality. Thus, instead of solving L directly, we can first compute L′, then solve it and transform the solution back to a solution to L using Q. Experiments (Grohe et al., 2014) have shown that for linear programs with many regularities, this often yields a substantial speed-up in total processing time.

15.7   Application: Weisfeiler-Leman Graph Kernels

Kernels are a concept that can be used to perform pattern analysis, a basic task in machine learning. Normally, in pattern analysis algorithms, each object is assigned a vector in a high-dimensional feature space and the similarity of two arising vectors is interpreted as measuring the similarity of the objects the vectors were generated from. In contrast to this general approach, a kernel is an implicitly given similarity measure, typically as a scalar product of the corresponding feature vectors. A crucial point (often referred to as the kernel trick) is that kernel methods can avoid the explicit computation of the feature vector and directly evaluate the similarity value of a pair of objects. With such a similarity measure at hand, it is then possible to perform tasks such as classification and clustering using the by now well developed kernel methods from machine learning (see (Hofmann et al., 2008)). Among these methods are techniques like support vector machines, kernel regression, and principal component analysis.

An important and growing branch of machine learning deals with *graph-structured data*. Here data comes in form of annotated (that is edge- and/or vertex-labeled) graphs. **Graph Kernels** constitute the application of said kernel methods to graph structure data. While early research focused on similarity of vertices within one given graph (Kondor and Lafferty, 2002), the focus moved to comparing graphs to each other (Gartner et al., 2003; Vishwanathan et al., 2010). Over time, numerous graph kernels have been considered. We refer to (Vishwanathan et al., 2010) for a unified treatment. Most of these kernels have in common that the counts of some form of substructure (e.g., subgraphs, walks between paths, etc.) are used to compile the feature vector. Of course, whether two graphs should be considered similar always depends on the specific application. However, experiments show that counting such substructures yields satisfactory results for the desired pattern analysis tasks.

A major drawback of the kernel methods that count such substructures is their excessive running time. In fact, they do not scale particularly well to larger inputs. This is where color-refinement and the Weisfeiler-Leman Graph Kernel come into play. For an integer h, the *number of iterations*, the **Weisfeiler-Leman Graph Kernel** assigns to every pair of graphs G and $G′$ a number $k^h(G, G′)$ as follows. As before we let C_i be the coloring after the i-th round of color refinement performed on the disjoint union $G \uplus G'$. Then we define

$$k^h(G, G') = \sum_{v \in V(G)} \sum_{v' \in V(G)} \delta(C_0(v), C_0(v')) + \delta(C_1(v), C_1(v')) + \cdots + \delta(C_h(v), C_h(v')),$$

where $\delta(a, b)$ is 1 if $a = b$ and 0 otherwise. Thus, we count the number of vertex pairs from the two graphs of equal color and do so for each of the first h iterations. Figure 15.5 shows an example. In this example, in order to keep color names short, colors in the later iterations are repeatedly renamed to previously unused integers.

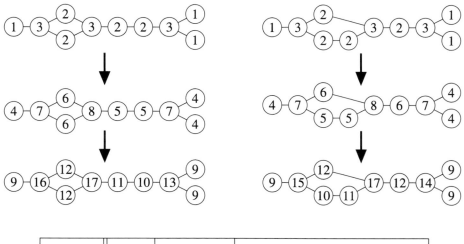

histogram	1	2	3	4	5	6	7	8	9	10	11	12	13	14	15	16	17
G	3	4	3	3	2	2	2	1	3	1	1	2	1	0	0	1	1
G'	3	4	3	3	2	2	2	1	3	1	1	2	0	1	1	0	1

Figure 15.5: The computation of the implicit feature vectors of two graphs G (left) and G' (right) in the Weisfeiler-Leman Graph Kernel. The figure also shows the histograms of colors that appear in the two graphs highlighting their similarity. We obtain $k^0(G, G') = 10^2 = 100$, $k^1(G, G') - 100 + 9 + 16 + 9 = 134$, $k^2(G, G') = 134 + 9 + 4 + 4 + 4 + 1 = 156$ and $k^3(G, G') = 156 + 9 + 1 + 1 + 4 + 0 + 0 + 0 + 0 + 1 - 172$.

Experiments show that the information collected by the kernel is sufficient to adequately perform desired classification tasks (Shervashidze et al., 2011). These experiments also indicate that choosing the number of iterations as $h \approx 5$ is best. Indeed, since almost all graphs are distinguished by color refinement, performing too many iterations means that the colors contain too much (global) information about the graphs they are situated in and cannot be used as a meaningful measure of similarity. (In fact, two graphs will most likely not share any vertices of the same color if the number of rounds is too large.)

As with the other applications for color refinement, a major advantage of the methods is its running time. Since an iteration of color refinement can be performed in time $O(m)$, to compute the Weisfeiler-Leman Graph Kernel that uses h iterations on a pair of graphs on m edges, we require time $O(mh)$. However, in applications, we intend to apply the kernel between all pairs of graphs coming from a large set of examples, of size N say. A crucial observation is that instead of performing color refinement on the disjoint union of the two graphs we can also perform it separately on each graph. Doing so, we cannot simply rename the new arising colors, since this might be inconsistent between graphs. However, using a suitable hash function we can remedy the situation. We obtain a running time of $O(Nhm + N^2 hn)$ to compute the kernels between every pair of graphs from a set of size N. Here the first summand arises from color refinement on N graphs, while the second summand is the computation of the scalar product for each of the N^2 pairs of graphs.

It is possible to combine the concept of the Weisfeiler-Leman Graph Kernel with other kernels in a more general framework. We refer to (Shervashidze et al., 2011) for details. With this general applicability and due to its fast computation time, the Weisfeiler-Leman Graph Kernel has found its application in various machine learning areas. For example it has been applied for Malware detection (Sahs and Khan, 2012) in the context of code similarity (Li et al., 2016), fMRI analysis (Gkirtzou et al., 2013) and Resource Description Framework data (Ristoski and Paulheim, 2016). While the Kernel itself hinges on the annotation of given the data (i.e., the labels) coming from a discrete domain, recent work (Morris et al., 2016) shows how randomization and hashing can be applied to transform the seemingly intrinsic discrete kernel into a kernel suitable for continuous data.

While kernel methods are still among the best performing machine learning methods for graphs, there has been an increasing interest in deep learning approaches. Many of these approaches fit into a framework of **graph convolutional neural networks (GNNs)** (Kipf and Welling, 2017; Hamilton et al., 2017). GNNs can be seen as a computing a "differentiable" version of color refinement. Indeed, it was proved in (Morris et al., 2019) that GNNs have exactly the same expressiveness as color refinement when it comes to distinguishing graphs or nodes within a graph.

15.8   Conclusions

In this chapter, we reviewed the basic color refinement algorithm and explained how it can be implemented in quasilinear time. In particular, we connected it to lifted inference and graph kernels.

Acknowledgments

This work was partly supported by the German Science Foundation (DFG) within the Koselleck Grant "Logic, Structure, and the Graph Isomorphism Problem" and the Collaborative Research Center SFB 876 "Providing Information by Resource-Constrained Data Analysis", project A6 "Resource-efficient Graph Mining".

16 Stochastic Planning and Lifted Inference

Roni Khardon and Scott Sanner

Abstract. Lifted probabilistic inference (Poole, 2003) and symbolic dynamic programming for lifted stochastic planning (Boutilier et al, 2001) were introduced around the same time as algorithmic efforts to use abstraction in stochastic systems. Over the years, these ideas evolved into two distinct lines of research, each supported by a rich literature. Lifted probabilistic inference focused on efficient arithmetic operations on template-based graphical models under a finite domain assumption while symbolic dynamic programming focused on supporting sequential decision-making in rich quantified logical action models and on open domain reasoning. Given their common motivation but different focal points, both lines of research have yielded highly complementary innovations. In this chapter, we aim to help close the gap between these two research areas by providing an overview of lifted stochastic planning from the perspective of probabilistic inference, showing strong connections to other chapters in this book. This also allows us to define *generalized lifted inference* as a paradigm that unifies these areas and elucidates open problems for future research that can benefit both lifted inference and stochastic planning.

16.1 Introduction

In this chapter we illustrate that stochastic planning can be viewed as a specific form of probabilistic inference and show that recent symbolic dynamic programming (SDP) algorithms for the planning problem can be seen to perform "generalized lifted inference", thus making a strong connection to other chapters in this book. As we discuss below, although the SDP formulation is more expressive in principle, work on SDP to date has largely focused on algorithmic aspects of reasoning in open domain models with rich quantified logical structure whereas lifted inference has largely focused on aspects of efficient arithmetic computations over finite domain (quantifier free) template-based models. The contributions in these areas are therefore largely along different dimensions. However, the intrinsic relationships between these problems suggest a strong opportunity for cross-fertilization where the true scope of generalized lifted inference can be achieved. This chapter intends to highlight these relationships and lay out a paradigm for generalized lifted inference that subsumes both fields and offers interesting opportunities for future research.

Figure 16.1: A formal desciption of the BOXWORLD adapted from Boutilier et al. (2001). We use a simple STRIPS-like (Fikes and Nilsson, 1971) add and delete list representation of actions and, as a simple probabilistic extension in the spirit of PSTRIPS (Pednault, 1989), we assign probabilities that an action successfully executes conditioned on various state properties.

- *Domain Object Types (i.e., sorts)*: Box, Truck, City = {paris, . . .}
- *Relations (with parameter sorts)*:
  BoxIn: *BIn(Box, City)*, TruckIn: *TIn(Truck, City)*, BoxOn: *On(Box, Truck)*
- *Reward*: if $\exists B, BIn(B, paris)$ then 10 else 0
- *Actions (with parameter sorts)*:

  − *load(Box : B, Truck : T, City : C)*:
    * Success Probability: if $(BIn(B, C) \wedge TIn(T, C))$ then .9 else 0
    * Add Effects on Success: $\{On(B, T)\}$
    * Delete Effects on Success: $\{BIn(B, C)\}$

  − *unload(Box : B, Truck : T, City : C)*:
    * Success Probability: if $(On(B, T) \wedge TIn(T, C))$ then .9 else 0
    * Add Effects on Success: $\{BIn(B, C)\}$
    * Delete Effects on Success: $\{On(B, T)\}$

  − *drive(Truck : T, City : C_1, City : C_2)*:
    * Success Probability: if $(TIn(T, C_1))$ then 1 else 0
    * Add Effects on Success: $\{TIn(T, C_2)\}$
    * Delete Effects on Success: $\{TIn(T, C_1)\}$

  − *noop*
    * Success Probability: 1
    * Add Effects on Success: $\emptyset$
    * Delete Effects on Success: $\emptyset$

To make the discussion concrete, let us introduce a running example for stochastic planning and the kind of generalized solutions that can be achieved. For illustrative purposes, we borrow a planning domain from Boutilier et al. (2001) that we refer to as BOXWORLD. In this domain, outlined in Figure 16.1, there are several cities such as *london*, *paris* etc., trucks *truck$_1$*, *truck$_2$* etc., and boxes *box$_1$*, *box$_2$* etc. The agent can load a box onto a truck or unload it and can drive a truck from one city to another. When any box has been delivered to a specific city, *paris*, the agent receives a positive reward. The agent's planning task is to find a policy for action selection that maximizes this reward over some planning horizon.

Figure 16.2: A decision-list representation of the optimal policy and expected discounted reward for the BOXWORLD problem. The optimal action parameters in the *then* conditions correspond to the existential bindings that made the *if* conditions true.

---

if $(\exists B, BIn(B, paris))$ then do *noop* (value = 100.00)

else if $(\exists B, T, TIn(T, paris) \land On(B, T))$ then do *unload*(B, T, paris) (value = 89.0)

else if $(\exists B, C, T, On(B, T) \land TIn(T, C))$ then do *drive*(T, C, paris) (value = 80.0)

else if $(\exists B, C, T, BIn(B, C) \land TIn(T, C))$ then do *load*(B, T, C) (value = 72.0)

else if $(\exists B, C_1, T, C_2, BIn(B, C_1) \land TIn(T, C_2))$ then do *drive*(T, C_2, C_1) (value = 64.7)

else do *noop* (value = 0.0)

Our objective in lifted stochastic planning is to obtain an abstract policy, for example, like the one shown in Figure 16.2. In order to get some box to *paris*, the agent should drive a truck to the city where the box is located, load the box on the truck, drive the truck to *paris*, and finally unload the box in *paris*. This is essentially encoded in the symbolic value function shown in Fig. 16.2, which was computed by discounting rewards t time steps into the future by 0.9^t.

Similar to this example, for some problems we can obtain a solution which is described abstractly and is independent of the specific problem instance or even its size — for our example problem the description of the solution does not depend on the number of cities, trucks or boxes, or on knowledge of the particular location of any specific truck. Accordingly, one might hope that computing such a solution can be done without knowledge of these quantities and in time complexity independent of them. This is the computational advantage of symbolic stochastic planning which we associate with lifted inference in this chapter.

The next two subsections expand on the connection between planning and inference, identify opportunities for lifted inference, and use these observations to define a new setup which we call *generalized lifted inference* which abstracts some of the work in both areas and provides new challenges for future work.

16.1.1   Stochastic Planning and Inference

Planning is the task of choosing what actions to take to achieve some goals or maximize long-term reward. When the dynamics of the world are deterministic, that is, each action has exactly one known outcome, then the problem can be solved through logical inference. That is, inference rules can be used to deduce the outcome of individual actions given the current state, and by combining inference steps one can prove that the goal is achieved. In this manner a proof of goal achievement embeds a plan. This correspondence was at the heart of McCarthy's seminal paper (McCarthy, 1958) that introduced the topic of AI and viewed planning as symbolic logical inference. Since this formulation uses first-order logic, or the closely related situation calculus, lifted logical inference can be used to solve deterministic planning problems.

When the dynamics of the world are non-deterministic, this relationship is more complex. In particular, in this chapter we focus on the stochastic planning problem where an action can have multiple possible known outcomes that occur with known state-dependent probabilities. Inference in this case must reason about probabilities over an exponential number of state trajectories for some planning horizon. While lifted inference and planning may seem to be entirely different problems, analogies have been made between the two fields in several forms (Attias, 2003; Toussaint and Storsky, 2006; Domshlak and Hoffmann, 2006; Lang and Toussaint, 2009; Furmston and Barber, 2010; Liu and Ihler, 2012; Cheng et al., 2013; Lee et al., 2014, 2016; Issakkimuthu et al., 2015; van de Meent et al., 2016). To make the connections concrete, consider a finite domain and the finite horizon goal-oriented version of the BoxWorld planning problem of Figure 16.1, e.g., two boxes, three trucks, and four cities and a planning horizon of 10 steps where the goal is to get some box in *paris*. In this case, the value of a state, $V(S)$, corresponds to the probability of achieving the goal, and goal achievement can be modeled as a specific form of inference in a Bayesian network or influence diagram.

We start by considering the *conformant planning problem* where the intended solution is an explicit sequence of actions. In this case, the sequence of actions is determined in advance and action choice at the ith step does not depend on the actual state at the ith step. For this formulation, one can build a Dynamic Bayesian Network (DBN) model where each time slice represents the state at that time and action nodes affect the state at the next time step, as in Figure 16.3(a). The edges in this diagram capture $p(S'|S, A)$, where S is the current state, A is the current action and S' is the next state, and each of S, S', A is represented by multiple nodes to show that they are given by a collection of predicates and their values. Note that, since the world dynamics are known, the conditional probabilities for all nodes in the graph are known. As a result, the goal-based planning problem where a goal G must hold at the last step, can be modeled using standard inference. The value of conformant planning is given by **marginal MAP** (where we seek a MAP value for some variables but take expectation over the remaining variables) (Domshlak and Hoffmann, 2006; Lee et al., 2014, 2016; Cui et al., 2018):

$$
\begin{aligned}
V_{\text{conformant}}(S_0) &= \max_{A_0}, \ldots, \max_{A_{N-1}} Pr(G|S_0, A_0, \ldots, A_{N-1}) \\
&= \max_{A_0}, \ldots, \max_{A_{N-1}} \sum_{S_1, S_2, \ldots, S_N} Pr(G, S_1, \ldots, S_N | S_0, A_0, \ldots, A_{N-1}).
\end{aligned}
$$

The optimal conformant plan is extracted using argmax instead of max in the equation.

The standard MDP formulation with a reward per time time step which is accumulated can be handled similarly, by normalizing the cumulative reward and adding a binary node

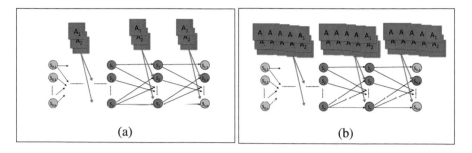

Figure 16.3: Planning as inference: conditioning on start and goal state. (a) Conformant planning – actions selected per time step without knowledge of the state. (b) An exponential size policy at each time step determines action selection. The transition depends on the current state and policy's actions for that state.

G whose probability of being true is a function of the normalized cumulative reward. Several alternative formulations of planning as inference have been proposed by defining an auxiliary distribution over finite trajectories which captures utility weighted probability distribution over the trajectories (Toussaint and Storsky, 2006; Furmston and Barber, 2010; Liu and Ihler, 2012; Cheng et al., 2013; van de Meent et al., 2016) and related formulations have been used in reinforcement learning (Levine, 2018). While the details vary, the common theme among these approaches is that the planning objective is equivalent to calculating the partition function (or "probability of evidence") in the resulting distribution. This achieves the same effect as adding a node G that depends on the cumulative reward. To simplify the discussion, we continue the presentation with the simple goal based formulation.

The same problem can be viewed from a Bayesian perspective, treating actions as random variables with an uninformative prior. In this case we can use

$$Pr(G|S_0, A_0, \ldots, A_{N-1}) = \frac{Pr(A_0, \ldots, A_{N-1}|G, S_0)Pr(G|S_0)}{Pr(A_0, \ldots, A_{N-1}|S_0)}.$$

to observe that (Attias, 2003; Toussaint and Storsky, 2006; Lang and Toussaint, 2009)

$$\operatorname*{argmax}_{A_0}, \ldots, \operatorname*{max}_{A_{N-1}} Pr(G|S_0, A_0, \ldots, A_{N-1}) = \operatorname*{argmax}_{A_0}, \ldots, \operatorname*{max}_{A_{N-1}} Pr(A_0, \ldots, A_{N-1}|G, S_0)$$

and therefore one can alternatively maximize the probability conditioned on G.

However, linear plans, as the ones produced by the conformant setting, are not optimal for probabilistic planning. In particular, if we are to optimize goal achievement then we must allow the actions to depend on the state they are taken in. That is, the action in the second step is taken with knowledge of the probabilistic outcome of the first action, which is not known in advance. We can achieve this by duplicating action nodes, with a copy for each possible value of the state variables, as illustrated in Figure 16.3(b). This represents a separate policy associated with each horizon depth which is required because finite horizon problems have non-stationary optimal policies. In this case, state transitions depend on the identity of the current state and the action variables associated with that state. The corresponding inference problem can be written as follows:

$$V(S_0) = \max_{A_0(S_0)}, \ldots, \max_{A_{N-1}(S_N-1)} Pr(G|S_0, A_0(S_0), \ldots, A_{N-1}(S_N-1)). \qquad (16.1)$$

However, the number of random variables in this formulation is prohibitively large since we need the number of original action variables to be multiplied by the size of the state space.

Alternatively, the same desideratum, optimizing actions with knowledge of the previous state, can be achieved without duplicating variables in the equivalent formulation

$$V(S_0) = \max_{A_0} \sum_{S_1} Pr(S_1|S_0, A_0) \max_{A_1} \sum_{S_2} Pr(S_2|S_1, A_1) \ldots$$

$$\max_{A_{N-2}} \sum_{S_{N-1}} Pr(S_{N-1}|S_{N-2}, A_{N-2}) \max_{A_{N-1}} \sum_{S_N} Pr(S_N|S_{N-1}, A_{N-1}) Pr(G|S_n). \qquad (16.2)$$

In fact, this formulation is exactly the same as the finite horizon application of the value iteration (VI) algorithm for (goal-based) Markov Decision Processes (MDP) which is the standard formulation for sequential decision making in stochastic environments. The standard formulation abstracts this by setting

$$V_0(S) = Pr(G|S)$$
$$V_{k+1}(S) = \max_A \underbrace{\sum_{S'} Pr(S'|S, A) V_k(S')}_{Q(S,A)}. \qquad (16.3)$$

The optimal policy (at S_0) can be obtained as before by recording the argmax values. In terms of probabilistic inference, the problem is no longer a marginal MAP problem because summation and maximization steps are constrained in their interleaved order. But it can be seen as a natural extension of such inference questions with several alternating blocks of expectation and maximization. We are not aware of an explicit study of such problems outside the planning context.

16.1.2   Stochastic Planning and Generalized Lifted Inference

Given that planning can be seen as an inference problem, one can try to apply ideas of lifted inference to planning. Taking the motivating example from Figure 16.1, let us specialize the reward to a ground atomic goal G equivalent to $BIn(b^*, paris)$ for constants b^* and *paris*. Then we can query $\max_A Pr(BIn(b^*, paris)|s_0, A)$ to compute $V(S_0 = s_0)$ where s_0 is the concrete value of the current state.

Given that Figure 16.1 implies a complex relational specification of the transition probabilities, lifted inference techniques are especially well-placed to attempt to exploit the structure of this query to perform inference in aggregate and thus avoid redundant computations. However, we emphasize that, even if lifted inference is used, this is a standard query in the graphical model where evidence constrains the value of some nodes, and the solution is a single number representing the corresponding probability (together with a MAP assignment to variables).

However, Eq 16.3 suggests an explicit additional structure for the planning problem. In particular, the intermediate expressions $V_k(S)$ include the values (the probability of reaching the goal in k steps) for all possible concrete values of S. Similarly, the final result $V_N(S)$ includes the values for all possible start states. In addition, as in our running example we can consider more abstract rewards. This suggests a first generalization of the standard setup in lifted inference. Instead of asking about a ground goal $Pr(BIn(b^*, paris))$ and expecting a single number as a response, we can abstract the setup in two ways: first, we can ask about more general conditions such as $Pr(\exists B, BIn(B, paris))$ and second we can expect to get a structured result that specifies the corresponding probability for every concrete state in the world. If we had two box instances b_1, b_2 and m truck instances $t_1, \ldots, t_m$, the answer for $V_1(S)$, i.e., the value for the goal based formulation with horizon one, might take the form:

if $(BIn(b_1, paris) \vee BIn(b_2, paris))$
      then $V_1(S) = 10$
else if $((TIn(t_1, paris) \wedge On(b_1, t_1)) \vee \ldots \vee (TIn(t_m, paris) \wedge On(b_2, t_m)))$
      then $V_1(S) = 9$
else $V_1(S) = 0.$

The significance of this is that the question can have a more general form and that the answer solves many problems simultaneously, providing the response as a case analysis depending on some properties of the state. We refer to this reasoning as *inference with generalized queries and answers*. In this context, the goal of lifted inference will be to calculate a structured form of the reply directly.

A second extension arises from the setup of generalized queries. The standard form for lifted inference is to completely specify the domain in advance. This means providing the number of objects and their properties, and that the response to the query is calculated only for this specific domain instantiation. However, inspecting the solution in the previous paragraph it is obvious that we can at least hope to do better. The same solution can be described more compactly as

if $(\exists B, BIn(B, paris))$
      then $V_1(S) = 10$
else if $(\exists B, \exists T, (TIn(T, paris) \wedge On(B, T))$
      then $V_1(S) = 9$
else $V_1(S) = 0.$

Arriving at such a solution requires us to allow open domain reasoning over all potential objects (rather than grounding them, which is impossible in open domains), and to extend ideas of lifted inference to exploit quantifiers and their structure. Following through with this idea, we can arrive at a *domain-size independent value function and policy* as the one shown in Figure 16.2. In this context, the goal of lifted inference will be to calculate an abstracted form of the reply directly. We call this problem *inference with generalized models*. As we describe in this chapter, SDP algorithms are able to perform this type of inference.

The previous example had enough structure and a special query that allowed the solution to be specified without any knowledge of the concrete problem instance. This property is not always possible. For example, consider a setting where we get one unit of reward for every box in *paris*: $\sum_{B:Box}$ [if ($BIn(B, paris)$) then 1 else 0]. In addition, consider the case where, after the agent takes their action, any box which is not on a truck disappears with probability 0.2. In this case, we can still potentially calculate an abstract solution, but it requires access to more complex properties of the state, and in some cases the domain size (number of objects) in the state. For our example this gives:

Let $n = (\#_B, BIn(B, paris))$
if $(\exists B, \exists T, (TIn(T, paris) \wedge On(B, T))$
       then $V_1(S) = n * 8 + 7.2$
else $V_1(S) = n * 8$.

Here we have introduced a new notation for count expressions where, for example, $(\#_B, BIn(B, paris))$ counts the number of boxes in Paris in the current state. To see this result note that any existing box in Paris disappears 20% of the time and that a box on a truck is successfully unloaded 90% of the time but remains and does not disappear only in 80% of possible futures leading to the value 7.2. This is reminiscent of the type of expressions that arise in existing lifted inference problems and solutions. Typical solutions to such problems involve parameterized expressions over the domain (e.g., counting, summation, etc.), and critically do not always require closed-domain reasoning (e.g., *a priori* knowledge of the number of boxes). They are therefore suitable for inference with generalized models. Some work on SDP has approached lifted inference for problems with this level of complexity, including exogenous activities (the disappearing boxes) and additive rewards. But, as we describe in more detail, the solutions for these cases are much less well understood and developed.

To recap, our example illustrates that stochastic planning potentially enables abstract solutions that might be amenable to lifted computations. SDP solutions for planning problems have focused on the computational advantages arising from these expressive generalizations. At the same time, the focus in SDP algorithms has largely been on problems where the solution is completely independent of domain size and does not require numerical properties of the state. These algorithms have thus skirted some of the computational issues that are typically tackled in lifted inference. It is the combination of these aspects, as illustrated in the last example, which we call **generalized lifted inference**. As the discussion suggests, generalized lifted inference is still very much an open problem. In addition to providing a survey of existing SDP algorithms, the goal of this chapter is to highlight the opportunities and challenges in this exciting area of research.

16.2 Preliminaries

This section provides a formal description of the representation language, the relational planning problem, and the description of the running example in this context.

16.2.1 Relational Expressions and Their Calculus of Operations

The computation of SDP algorithms is facilitated by a representation that enables compact specification of functions over world states. Several such representations have been devised and used. In this chapter we chose to abstract away some of those details and focus on a simple language of relational expressions. This is closest to the GFODD representation of (Joshi et al., 2011, 2013), but it resembles the case notation of (Boutilier et al., 2001; Sanner and Boutilier, 2009).

Syntax. We assume familiarity with basic concepts and notation in **first-order logic** (FOL) (Lloyd, 1987; Russell and Norvig, 2010; Chang and Keisler, 1990). Relational expressions are similar to expressions in FOL. They are defined relative to a relational signature, with a finite set of predicates $p_1, p_2, \ldots, p_n$ each with an associated arity (number of arguments), a countable set of variables $x_1, x_2, \ldots$, and a set of constants $c_1, c_2, \ldots, c_m$. We do not allow function symbols other than constants (that is, functions with arity ≥ 1). A term is a variable (often denoted in uppercase) or constant (often denoted in lowercase) and an atom is either an equality between two terms or a predicate with an appropriate list of terms as arguments. Intuitively, a term refers to an object in the world of interest and an atom is a property which is either true or false.

We illustrate relational expressions informally by some examples. In FOL we can consider open formulas that have unbound variables. For example, the atom $color(X, Y)$ is such a formula and its truth value depends on the assignment of X and Y to objects in the world. To simplify the discussion, we assume for this example that arguments are typed (or sorted) and X ranges over "objects" and Y over "colors". We can then quantify over these variables to get a sentence which will be evaluated to a truth value in any concrete possible world. For example, we can write $[\exists Y, \forall X, color(X, Y)]$ expressing the statement that there is a color associated with all objects. Generalized expressions allow for more general open formulas that evaluate to numerical values. For example, $E_1 = [\text{if } color(X, Y) \text{ then } 1 \text{ else } 0]$ is similar to the previous logical expression but $E_2 = [\text{if } color(X, Y) \text{ then } 0.3 \text{ else } 0.5]$ returns non-binary values.

Quantifiers from logic are replaced with **aggregation** operators that combine numerical values and provide a generalization of the logical constructs. In particular, when the open formula is restricted to values 0 and 1, the operators max and min simulate existential and universal quantification. Thus, $[\max_Y, \min_X, \text{if } color(X, Y) \text{ then } 1 \text{ else } 0]$ is equivalent to the logical sentence $[\exists Y, \forall X, color(X, Y)]$ given above. But we can allow for other types of aggregations. For example, $[\max_Y, \text{sum}_X, \text{if } color(X, Y) \text{ then } 1 \text{ else } 0]$ evaluates to the largest number of objects associated with one color, and the expression $[\text{sum}_X, \min_Y, \text{if } color(X, Y) \text{ then } 0 \text{ else } 1]$ evaluates to the number of objects that have no color association. In this manner, a generalized expression represents a function from possible worlds to numerical values and, as illustrated, can capture interesting properties of the state.

Relational expressions are also related to work in statistical relational learning (Richardson and Domingos, 2006; Raedt et al., 2007; Van den Broeck et al., 2011). For example, if the open expression E_2 given above captures probability of ground facts for the predicate $color()$ and the ground facts are mutually independent then $[\text{product}_X, \text{product}_Y, \text{if } color(X, Y) \text{ then } 0.3 \text{ else } 0.5]$ captures the joint probability for all facts for $color()$. Of course, the open formulas in logic can include more than one atom and similarly expressions can be more involved.

In the following we will drop the cumbersome if-then-else notation and instead will assume a simpler notation with a set of mutually exclusive conditions which we refer to as *cases*. In particular, an expression includes a set of mutually exclusive open formulas in FOL (without any quantifiers or aggregators) denoted $c_1, \ldots, c_k$ associated with corresponding numerical values $v_1, \ldots, v_k$. The list of cases refers to a finite set of variables $X_1, \ldots, X_m$. A generalized expression is given by a list of aggregation operators and their variables and the list of cases $[agg_{X_1}, agg_{X_2}, \ldots, agg_{X_m}[c_1 : v_1, \ldots, c_k : v_k]]$ so that the last expression is canonically represented as $[\text{product}_X, \text{product}_Y, [color(X, Y) : 0.3; \neg color(X, Y) : 0.5]]$.

Semantics. The semantics of expressions is defined inductively exactly as in first order logic and we skip the formal definition. As usual, an expression is evaluated in an *interpretation* also known as a possible world. In our context, an interpretation specifies (1) a finite set of n domain elements also known as objects, (2) a mapping of constants to domain elements, and (3) the truth values of all the predicates over tuples of domain elements of appropriate size to match the arity of the predicate. Now, given an expression $B = (agg_X, f(X))$, an interpretation I, and a substitution ζ of variables in X to objects in I, one can identify the case c_i which is true for this substitution. Exactly one such case exists since the cases are mutually exclusive and exhaustive. Therefore, the value associated with ζ is v_i. These values are then aggregated using the aggregation operators. For example, consider again the expression $[\text{product}_X, \text{product}_Y, [color(X, Y) : 0.3; \neg color(X, Y) : 0.5]]$ and an interpretation I with objects a, b and where a is associated with colors black and white and b is associated with color black. In this case we have exactly 4 substitutions evaluating to 0.3, 0.3, 0.5, 0.3. Then the final value is $0.3^3 \cdot 0.5$.

Operations over expressions. Any binary operation op over real values can be generalized to open and closed expressions in a natural way. If f_1 and f_2 are two closed expressions, $f_1 \ op \ f_2$ represents the function which maps each interpretation w to $f_1(w) \ op \ f_2(w)$. This provides a definition but not an implementation of binary operations over expressions. For implementation, the work in (Joshi et al., 2011) showed that if the binary operation is *safe*, i.e., it distributes with respect to all aggregation operators, then there is a simple algorithm (the Apply procedure) implementing the binary operation over expressions. For example, $+$ is safe w.r.t. max aggregation, and it is easy to see that $(\max_X f(X)) + (\max_X g(X)) = \max_X \max_Y f(X)+g(Y)$, and the open formula portion of the result can be calculated directly from the open expressions $f(X)$ and $g(Y)$. Note that we need to standardize the expressions apart, as in the renaming of $g(X)$ to $g(Y)$ for such operations. When $f(x)$ and $g(y)$ are open relational expressions the result can be computed through a cross product of the cases. For example,

$$[\max_X, \min_Y [color(X,Y) : 3; \neg color(X,Y) : 5]] \oplus [\max_X, [box(X) : 1; \neg box(X) : 2]]$$

$$= [\max_Z, \max_X, \min_Y [color(X,Y) \wedge box(Z) : 4; \neg color(X,Y) \wedge box(Z) : 6;$$

$$color(X,Y) \wedge \neg box(Z) : 5; \neg color(X,Y) \wedge \neg box(Z) : 7]]$$

When the binary operation is not safe then this procedure fails, but in some cases, operation-specific algorithms can be used for such combinations.[46]

As will become clear later, to implement SDP we need the binary operations $\oplus$, $\otimes$, max and the aggregation includes max in addition to aggregation in the reward function. Since $\oplus$, $\otimes$, max are safe with respect to max, min aggregation one can provide a complete solution when the reward is restricted to have max, min aggregation. When this is not the case, for example when using sum aggregation in the reward function, one requires a special algorithm for the combination. Further details are provided in (Joshi et al., 2011, 2013).

Summary. Relational expressions are closest to the GFODD representation of (Joshi et al., 2011, 2013). Every case c_i in a relational expression corresponds to a path or set of paths in the GFODD, all of which reach the same leaf in the graphical representation of the GFODD. GFODDs are potentially more compact than relational expressions since paths share common subexpressions, which can lead to an exponential reduction in size. On the other hand, GFODDs require special algorithms for their manipulation. Relational

---

[46] For example, a product of expressions that include only product aggregations, which is not safe, can be obtained by scaling the result with a number that depends on domain size, and $[\prod_{x_1} \prod_{x_2} \prod_{x_3} f(x_1, x_2, x_3)] \otimes [\prod_{y_1} \prod_{y_2} g(y_1, y_2)]$ is euqal to $[\prod_{x_1} \prod_{x_2} \prod_{x_3} [f(x_1, x_2, x_3) \times g(x_1, x_2)^{1/n}]]$ when the domain has n objects.

expressions are also similar to the case notation of (Boutilier et al., 2001; Sanner and Boutilier, 2009). However, in contrast with that representation, cases are not allowed to include any quantifiers and instead quantifiers and general aggregators are globally applied over the cases, as in standard quantified normal form in logic.

16.2.2   Relational MDPs

In this section we define MDPs, starting with the basic case with enumerated state and action spaces, and then providing the relational representation.

MDP Preliminaries. We assume familiarity with basic notions of Markov Decision Processes (MDPs) (Russell and Norvig, 2010; Puterman, 1994). Briefly, a MDP is a tuple $\langle S, A, P, R, \gamma \rangle$ given by a set of states S, set of actions A, transition probability $Pr(S'|S,A)$, immediate reward function $R(S)$ and discount factor $\gamma < 1$. The solution of a MDP is a policy π that maximizes the expected discounted total reward obtained by following that policy starting from any state. The Value Iteration algorithm (VI) informally introduced in Eq 16.3, calculates the optimal value function by iteratively performing Bellman backups, $V_{k+1} = T[V_k]$, defined for each state $s \in S$ as,

$$V_{k+1}(s) = T[V_k](s) \leftarrow \max_{a \in A}\{R(s) + \gamma \sum_{s' \in S} Pr(s'|s, a)V_k(s')\}. \qquad (16.4)$$

Unlike Eq 16.3, which was goal-oriented and had only a single reward at the terminal horizon, here we allow the reward R(S) to accumulate at all time steps as typically allowed in MDPs. If we iterate the update until convergence, we get the optimal infinite horizon value function typically denoted by V^* and optimal stationary policy π^*. For finite horizon problems, which is the topic of this chapter, we simply stop the iterations at a specific k. In general, the optimal policy for the finite horizon case is not stationary, that is, we might make different choice in the same state depending on how close we are to the horizon.

Logical Notation for Relational MDPs (RMDPs). RMDPs are simply MDPs where the states and actions are described in a function-free first order logical language. A state corresponds to an interpretation over the corresponding logical signature, and actions are transitions between such interpretations.

A relational planning problem is specified by providing the logical signature, the start state, the transitions as controlled by actions, and the reward function. As mentioned above, one of the advantages of relational SDP algorithms is that they are intended to produce an abstracted form of the value function and policy that does not require specifying the start state or even the number of objects n in the interpretation at planning time. This yields policies that generalize across domain sizes. We therefore need to explain how one can use logical notation to represent the transition model and reward function in a manner that does not depend on domain size.

Two types of transition models have been considered in the literature:

- **Endogenous Branching Transitions:** In the basic form, state transitions have limited stochastic branching due to a finite number of action outcomes. The agent has a set of action types $\{A\}$ each parametrized with a tuple of objects to yield an action template $A(X)$ and a concrete ground action $A(x)$ (e.g. template $unload(B, T)$ and concrete action $unload(box23, truck1)$). Each agent action has a finite number of action variants $A_j(X)$ (e.g., action success vs. action failure), and when the user performs $A(X)$ in state s one of the variants is chosen randomly using the state-dependent action choice distribution $Pr(A_j(X)|A(X))$. To simplify the presentation we follow (Wang et al., 2008; Joshi et al., 2011) and require that $Pr(A_j(X)|A(X))$ are given by open expressions, i.e., they have no aggregations and cannot introduce new variables. For example, in BOXWORLD, the agent action $unload(B, T, C)$ has success outcome $unloadS(B, T, C)$ and failure outcome $unloadF(B, T, C)$ with action outcome distribution as follows:

$$P(unloadS(B, T, C)|unload(B, T, C)) = [(On(B, T) \wedge TIn(T, C)) : .9; \neg : 0]$$

$$P(unloadF(B, T, C)|unload(B, T, C)) = [(On(B, T) \wedge TIn(T, C)) : .1; \neg : 1] \quad (16.5)$$

where, to simplify the notation, the last case is shortened as $\neg$ to denote that it complements previous cases. This provides the distribution over deterministic outcomes of actions.

The deterministic action dynamics are specified by providing an open expression, capturing successor state axioms (Reiter, 2001), for each variant $A_j(X)$ and predicate template $p'(Y)$. Following (Wang et al., 2008) we call these expressions TVDs, standing for truth value diagrams. The corresponding TVD, $T(A_j(X), p'(Y))$, is an open expression that specifies the truth value of $p'(Y)$ *in the next state* (following standard practice we use prime to denote that the predicate refers to the next state) when $A_j(X)$ has been executed *in the current state*. The arguments X and Y are intentionally different logical variables as this allows us to specify the truth value of all instances of $p'(Y)$ simultaneously. Similar to the choice probabilities we follow (Wang et al., 2008; Joshi et al., 2011) and assume that TVDs $T(A_j(X), p'(Y))$ have no aggregations and cannot introduce new variables. This implies that the regression and product terms in the SDP algorithm of the next section do not change the aggregation function, thereby enabling analysis of the algorithm. Continuing our BOXWORLD example, we define the TVD for $BIn'(B, C)$ for

$unloadS(B_1, T_1, C_1)$ and $unloadF(B_1, T_1, C_1)$ as follows:

$$BIn'(B,C) \equiv T(unloadS(B_1, T_1, C_1), BIn'(B,C))$$
$$\equiv [(BIn(B,C) \vee$$
$$((B_1 = B) \wedge (C_1 = C) \wedge On(B_1, T_1) \wedge TIn(T_1, C_1))) : 1; \neg : 0]$$

$$BIn'(B,C) \equiv T(unloadF(B_1, T_1, C_1), BIn'(B,C))$$
$$\equiv [BIn(B,C) : 1; \neg : 0] \tag{16.6}$$

Note that each TVD has exactly two cases, one leading to the outcome 1 and the other leading to the outcome 0. Our algorithm below will use these cases individually. Here we remark that since the next state (primed) only depends on the previous state (unprimed), we are effectively logically encoding the Markov assumption of MDPs.

- **Exogenous Branching Transitions:** The more complex form combines the endogenous model with an exogenous stochastic process that affects ground atoms independently. As a simple example in our BOXWORLD domain, we might imagine that with some small probability, each box B in a city C ($BIn(B,C)$) may independently randomly disappear (falsify $BIn(B,C)$) owing to issues with theft or improper routing — such an outcome is independent of the agent's own action. Another more complicated example could be an inventory control problem where customer arrival at shops (and corresponding consumption of goods) follows an independent stochastic model. Such exogenous transitions can be formalized in a number of ways (Sanner, 2008; Sanner and Boutilier, 2007; Joshi et al., 2013); we do not aim to commit to a particular representation in this chapter, but rather to mention its possibility and the computational consequences of such general representations.

Having completed our discussion of RMDP transitions, we now proceed to define the reward $R(S, A)$, which can be any function of the state and action, specified by a relational expression. Our running example with existentially quantified reward is given by

$$[\max_B [BIn(B, paris) : 10; \neg BIn(B, paris) : 0]] \tag{16.7}$$

but we will also consider additive reward as in

$$[\sum_B [BIn(B, paris) : 10; \neg BIn(B, paris) : 0]]. \tag{16.8}$$

16.3   Symbolic Dynamic Programming

The SDP algorithm is a symbolic implementation of the value iteration algorithm. The algorithm repeatedly applies so-called decision-theoretic regression which is equivalent to one iteration of the value iteration algorithm.

As input to SDP we get closed relational expressions for V_k and R. In addition, assuming that we are using the *Endogenous Branching Transition* model of the previous section, we get open expressions for the probabilistic choice of actions $Pr(A_j(X)|A(X))$ and for the dynamics of deterministic action variants as TVDs. The corresponding expressions for the running example are given respectively in Eq (16.7), Eq (16.5) and Eq (16.6).

The following SDP algorithm of (Joshi et al., 2011) modifies the earlier SDP algorithm of (Boutilier et al., 2001) and implements Eq (16.4) using the following 4 steps:

i. **Regression:** The k step-to-go value function V_k is regressed over every deterministic variant $A_j(X)$ of every action $A(X)$ to produce $Regr(V_k, A_j(X))$. Regression is conceptually similar to goal regression in deterministic planning. That is, we identify conditions that need to occur before the action is taken in order to arrive at other conditions (for example the goal) after the action. However, here we need to regress all the conditions in the relational expression capturing the value function, so that we must regress each case c_i of V_k separately. This can be done efficiently by replacing every atom in each c_i by its corresponding positive or negated portion of the TVD without changing the aggregation function. Once this substitution is done, logical simplification (at the propositional level) can be used to compress the cases by removing contradictory cases and simplifying the formulas. Applying this to regress $unloadS(B_1, T_1, C_1)$ over the reward function given by Eq (16.7) we get:

$$[\max_B [(BIn(B, paris) \vee$$

$$((B_1 = B) \wedge (C_1 = paris) \wedge On(B_1, T_1) \wedge TIn(T_1, C_1))) : 10; \neg : 0]]$$

and regressing $unloadF(B_1, T_1, C_1)$ yields

$$[\max_B [BIn(B, paris) : 10; \neg : 0]]$$

This illustrates the utility of compiling the transition model into the TVDs which allow for a simple implementation of deterministic regression.

ii. **Add Action Variants:** The Q-function $Q_k^{A(X)} = R \oplus [\gamma \otimes \oplus_j(Pr(A_j(X)) \otimes Regr(V_k, A_j(X)))]$ for each action $A(X)$ is generated by combining regressed diagrams using the binary operations $\oplus$ and $\otimes$ over expressions. Recall that probability expressions do not refer to additional variables. The multiplication can therefore be done directly on the open formulas without changing the aggregation function. As argued by (Wang et al., 2008), to guarantee correctness, both summation steps ($\oplus_j$ and $R\oplus$ steps) must standardize apart the functions before adding them.

For our running example and assuming $\gamma = 0.9$, we would need to compute the following:

$Q_k{}^{unload(B_1,T_1,C_1)}(S) =$

$\quad R(S) \oplus 0.9\cdot$

$\qquad [(Regr(V_0, unloadS(B_1, T_1, C_1)) \otimes P(unloadS(B_1, T_1, C_1)|unload(B_1, T_1, C_1)))\oplus$

$\qquad (Regr(V_0, unloadF(B_1, T_1, C_1)) \otimes P(unloadF(B_1, T_1, C_1)|unload(B_1, T_1, C_1)))].$

We next illustrate some of these steps. The multiplication by probability expressions can be done by cross product of cases and simplification. For *unloadS* this yields

$$[\max_B [((BIn(B, paris) \vee ((B_1 = B) \wedge (C_1 = paris)))$$

$$\wedge On(B_1, T_1) \wedge TIn(T_1, C_1)) : 9; \neg : 0]]$$

and for *unloadF* we get

$$[\max_B [BIn(B, paris) \wedge (On(B_1, T_1) \wedge TIn(T_1, C_1)) : 1;$$

$$BIn(B, paris) \wedge \neg(On(B_1, T_1) \wedge TIn(T_1, C_1)) : 10;$$

$$\neg : 0]].$$

Note that the values here are weighted by the probability of occurrence. For example the first case in the last equation has value 1=10*0.1 because when the preconditions of *unload* hold the variant *unloadF* occurs with 10% probability. The addition of the last two equations requires standardizing them apart, performing the safe operation through cross product of cases, and simplifying. Skipping intermediate steps, this yields

$$[\max_B [BIn(B, paris) : 10;$$

$$\neg BIn(B, paris) \wedge (B_1 = B) \wedge (C_1 = paris) \wedge On(B_1, T_1) \wedge TIn(T_1, C_1) : 9;$$

$$\neg : 0]].$$

Multiplying by the discount factor scales the numbers in the last equation by 0.9 and finally standardizing apart and adding the reward and simplifying (again skipping intermediate steps) yields

$Q_0{}^{unload(B_1,T_1,C_1)}(S) =$

$$[\max_B [BIn(B, paris) : 19;$$

$$\neg BIn(B, paris) \wedge (B_1 = B) \wedge (C_1 = paris) \wedge On(B_1, T_1) \wedge TIn(T_1, C_1) : 8.1;$$

$$\neg : 0]].$$

Intuitively, this result states that after executing a concrete stochastic *unload* action with arguments (B_1, T_1, C_1), we achieve the highest value (10 plus a discounted 0.9*10) if a box was already in Paris, the next highest value (10 occurring with probability 0.9 and discounted by 0.9) if unloading B_1 from T_1 in $C_1 = paris$, and a value of zero otherwise. The main source of efficiency (or lack thereof) of SDP is the ability to perform such operations symbolically and simplify the result into a compact expression.

iii. **Object Maximization:** Note that up to this point in the algorithm the action arguments are still considered to be concrete arbitrary objects, (B_1, T_1, C_1) in our example. However, we must make sure that in each of the (unspecified and possibly infinite set of possible) states we choose the best concrete action for that state, by specifying the appropriate action arguments. This is handled in the current step of the algorithm. To achieve this, we maximize over the action parameters X of $Q_{V_k}^{A(X)}$ to produce $Q_{V_k}^{A}$ for each action $A(X)$. This implicitly obtains the value achievable by the best ground instantiation of $A(X)$ in each state. This step is implemented by converting action parameters X to variables, each associated with the max aggregation operator, and appending these operators to the head of the aggregation function. Once this is done, further logical simplification may be possible. This occurs in our running example where existential quantification (over B_1, C_1) which is constrained by equality can be removed, and the result is:

$$Q_0^{unload}(S) =$$
$$[\max_T, \max_B [BIn(B, paris) : 19;$$
$$\neg BIn(B, paris) \wedge On(B, T) \wedge TIn(T, paris) : 8.1;$$
$$\neg : 0]].$$

iv. **Maximize over Actions:** The $k+1$st step-to-go value function $V_{k+1} = \max_A Q_{V_k}^A$, is generated by combining the expressions using the binary operation max. Concretely, for our running example, this means we would compute:

$$V_1(S) = \max(Q_0^{unload}(S), \max(Q_0^{load}(S), Q_0^{drive}(S)))$$

While we have only shown $Q_0^{unload}(S)$ above, we remark that the values achievable in each state by $Q_0^{unload}(S)$ dominate or equal the values achievable by $Q_0^{load}(S)$ and $Q_0^{drive}(S)$ in the same state. Practically this implies that after simplification we obtain the following value function:

$$V_1(S) = Q_0^{unload}(S) =$$
$$[\max_T, \max_B [BIn(B, paris) : 19;$$
$$\neg BIn(B, paris) \wedge On(B, T) \wedge TIn(T, paris) : 8.1;$$
$$\neg : 0]].$$

Critically for the objectives of lifted stochastic planning, we observe that the value function derived by SDP is indeed lifted: it holds for any number of boxes, trucks and cities.

SDP repeats these steps to the required depth, iteratively calculating V_k. For example, Figure 16.2 illustrates V_∞ for the BOXWORLD example, which was computed by terminating the SDP loop once the value function converged.

The basic SDP algorithm is an exact calculation whenever the model can be specified using the constraints above and the reward function can be specified with max and min aggregation (Joshi et al., 2011). This is satisfied by classical models of stochastic planning. As illustrated, in these cases, the SDP solution conforms to our definition of generalized lifted inference.

Extending the Scope of SDP. The algorithm above cannot handle models with more complex dynamics and rewards as motivated in the introduction. In particular, prior work has considered two important properties that appear to be relevant in many domains. The first is additive rewards, illustrated for example, in Eq 16.8. The second property is exogenous branching transitions illustrated above by the disappearing blocks example. These represent two different challenges for the SDP algorithm. The first is that we must handle sum aggregation in value functions, despite the fact that this means that some of the operations are not *safe* and hence require a special implementation. The second is in modeling the exogenous branching dynamics which requires getting around potential conflicts among such events and between such events and agent actions. The introduction illustrated the type of solution that can be expected in such a problem where counting expressions, that measure the number of times certain conditions hold in a state, determine the value in that state.

To date, exact abstract solutions for problems of this form have not been obtained. The work of Sanner and Boutilier (2007) and Sanner (2008) (Ch. 6) considered additive rewards and has formalized an expressive family of models with exogenous events. This work has shown that some specific challenging domains can be handled using several algorithmic ideas, but did not provide a general algorithm that is applicable across problems in this class. The work of Joshi et al. (2013) developed a model for "service domains" which significantly constrains the type of exogenous branching. In their model, a transition includes an agent step whose dynamics use endogenous branching, followed by "nature's step" where each object (e.g., a box) experiences a random exogenous action (potentially disappearing). Given these assumptions, they provide a generally applicable approximation algorithm as follows. Their algorithm treats agent's actions exactly as in SDP above. To regress nature's actions we follow the following three steps: (1) the summation variables are first ground using a Skolem constant c, then (2) a single exogenous event centered at c is regressed using the same machinery, and finally (3) the Skolemization is reversed to yield another additive value function. The complete details are beyond the scope of this chapter. The algorithm yields a solution that avoids counting formulas and is syntactically close to the one given by the original algorithm. Since such formulas are necessary, the result is an approximation but it was shown to be a conservative one in that it provides a monotonic lower bound on the true value. Therefore, this algorithm conforms to our definition of *approximate generalized lifted inference*.

In our example, starting with the reward of Eq (16.8) we first replace the sum aggregation with a scaled version of average aggregation (which is safe w.r.t. summation)

$$[n \cdot \text{avg}_B[BIn(B, paris) : 10; \neg : 0]]$$

and then ground it to get

$$[n \cdot [BIn(c, paris) : 10; \neg : 0]].$$

The next step is to regress through the exogenous event at c. The problem where boxes disappear with probability 0.2 can be cast as having two action variants where "disappearing-block" succeeds with probability 0.2 and fails with probability 0.8. Regressing the success variant we get the expression [0] (the zero function) and regressing the fail variant we get $[n \cdot [BIn(c, paris) : 10; \neg : 0]]$. Multiplying by the probabilities of the variants we get: [0] and $[n \cdot [BIn(c, paris) : 8; \neg : 0]]$ and adding them (there are no variables to standardize apart) we get

$$[n \cdot [BIn(c, paris) : 8; \neg : 0]].$$

Finally lifting the last equation we get

$$[n \cdot \text{avg}_B[BIn(B, paris) : 8; \neg BIn(B, paris) : 0]].$$

Next we follow with the standard steps of SDP for the agent's action. The steps are analogous to the example of SDP given above. Considering the discussion in the introduction (recall that in order to simplify the reasoning in this case we omitted discounting and adding the reward) this algorithm produces

$$[n \cdot \max_T, \text{avg}_B, [BIn(B, paris) : 8;$$
$$(\neg BIn(B, paris) \wedge On(B, T) \wedge TIn(T, paris)) : 7.2; \neg : 0]],$$

which is identical to the exact expression given in the introduction. As already mentioned, the result is not guaranteed to be exact in general. In addition, the maximization in step iv of SDP requires some ad-hoc implementation because maximization is not safe with respect to average aggregation.

It is clear from the above example that the main difficulty in extending SDP is due to the interaction of the counting formulas arising from exogenous events and additive rewards with the first-order aggregation structure inherent in the planning problem. Relational expressions, their GFODD counterparts, and other representations that have been used to date are not able to combine these effectively. A representation that seamlessly supports both relational expressions and operations on them along with counting expressions might allow for more robust versions of generalized lifted inference to be realized.

16.4   Discussion and Related Work

As motivated in the introduction, SDP has explored probabilistic inference problems with a specific form of alternating maximization and expectation blocks. The main computational advantage comes from lifting in the sense of lifted inference in standard first order logic. Issues that arise from conditional summations over combinations random variables, common in probabilistic lifted inference, have been touched upon but not extensively. In cases where SDP has been shown to work it provides *generalized lifted inference* where the complexity of the inference algorithm is completely independent of the domain size (number of objects) in problem specification, and where the response to queries is either independent of that size or can be specified parametrically. This is a desirable property but to our knowledge it is not shared by most work on probabilistic lifted inference. The most closely related work we are aware of is the notion of domain lifted inference (Van den Broeck, 2011; Van den Broeck, 2013; Kazemi et al., 2016, 2017) also discussed in Chapter 8. Here, a model is compiled into an alternative form parametrized by the domain D and where responses to queries can be obtained in polynomial time as a function of D. The emphasis in that work is on being *domain lifted* (i.e., being polynomial in domain size). Generalized lifted inference requires an algorithm whose results can be computed once, in time independent of that size, and then reused to evaluate the answer for specific domain sizes. This analogy also shows that SDP can be seen as a compilation algorithm, compiling a domain model into a more accessible form representing the value function, which can be queried efficiently. This connection provides an interesting new perspective on both fields.

 In this chapter we focused on one particular instance of SDP. Over the last 15 years SDP has seen a significant amount of work expanding over the original algorithm by using different representations, by using algorithms other than value iteration, and by extending the models and algorithms to more complex settings. In addition, several "lifted" inductive approaches that do not strictly fall within the probabilistic inference paradigm have been developed. We review this work in the remainder of this section.

16.4.1   Deductive Lifted Stochastic Planning

As a precursor to its use in lifted stochastic planning, the term SDP originated in the propositional logical context (Dearden and Boutilier, 1997; Boutilier et al., 1999) when it was realized that propositionally structured MDP transitions (i.e., dynamic Bayesian networks (Dean and Kanazawa, 1989)) and rewards (e.g., trees that exploited context-specific independence (Boutilier et al., 1996)) could be used to define highly compact *factored MDPs*; this work also realized that the factored MDP structure could be exploited for representational compactness and computational efficiency by leveraging symbolic representations (e.g., trees) in dynamic programming. Two highly cited (and still used algorithms) in this area of work are the SPUDD (Hoey et al., 1999) and APRICODD (St-Aubin et al., 2000) algorithms that leveraged algebraic decision diagrams (ADDs) (Bahar et al., 1993) for, respectively, exact and approximate solutions to factored MDPs. Recent work in this area (Lesner and Zanuttini, 2011) shows how to perform propositional SDP directly with ground representations in PPDDL (Younes et al., 2005), and develops extensions for factored action spaces (Raghavan et al., 2012, 2013).

Following the seminal introduction of *lifted* SDP (Boutilier et al., 2001), several early papers on SDP approached the problem with existential rewards with different representation languages that enabled efficient implementations. This includes the First-order value iteration (FOVIA) (Karabaev and Skvortsova, 2005; Hölldobler et al., 2006), the Relational Bellman algorithm (ReBel) (Kersting et al., 2004), and the FODD based formulation of (Wang et al., 2008; Joshi and Khardon, 2008; Joshi et al., 2010).

Along this dimension two representations are closely related to the relational expression of this chapter. As mentioned above, relational expressions are an abstraction of the GFODD representation (Joshi et al., 2011, 2013; Hescott and Khardon, 2015) which captures expressions using a decision diagram formulation extending propositional ADDs (Bahar et al., 1993). In particular, paths in the graphical representation of the DAG representing the GFODD correspond to the mutually exclusive conditions in expressions. The aggregation in GFODDs and relational expressions provides significant expressive power in modeling relational MDPs. The GFODD representation is more compact than relational expressions but requires more complex algorithms for its manipulation. The other closely related representation is the case notation of (Boutilier et al., 2001; Sanner and Boutilier, 2009). The case notation is similar to relational expressions in that we have a set of conditions (these are mostly in a form that is mutually exclusive but not always so) but the main difference is that quantification is done within each case separately, and the notion of aggregation is not fully developed. First-order algebraic decision diagrams (FOADDs) (Sanner, 2008; Sanner and Boutilier, 2009) are related to the case notation in that they require closed formulas within diagram nodes, i.e., the quantifiers are included within the graphical representation of the expression. The use of quantifiers inside cases and nodes allows for an easy incorporation of off the shelf theorem provers for simplification. Both FOADD and GFODD were used to extend SDP to capture additive rewards and exogenous events as already discussed in the previous section. While the representations (relational expression and GFODDs vs. case notation and FOADD) have similar expressive power, the difference in aggregation makes for different algorithmic properties that are hard to compare in general. However, the modular treatment of aggregation in GFODDs and the generic form of operations over them makes them the most flexible alternative to date for directly manipulating the aggregated case representation used in this chapter.

The idea of SDP has also been extended in terms of the choice of planning algorithm, as well as to the case of partially observable MDPs. Case notation and FOADDs have been used to implement approximate linear programming (Sanner and Boutilier, 2005, 2009) and approximate policy iteration via linear programming (Sanner and Boutilier, 2006) and FODDs have been used to implement relational policy iteration (Wang and Khardon, 2007). GFODDs have also been used for open world reasoning and applied in a robotic context (Joshi et al., 2012). The work of (Wang and Khardon, 2010) and (Sanner and Kersting, 2010) explore SDP solutions, with GFODDs and case notation respectively, to relational partially observable MDPs (POMDPs) where the problem is conceptually and algorithmically much more complex. Related work in POMDPs has not explicitly addressed SDP, but rather has implicitly addressed lifted solutions through the identification of (and abstraction over) symmetries in applications of dynamic programming for POMDPs (Doshi and Roy, 2008; Kang and Kim, 2012).

Finally, the work of (Cui et al., 2019, 2018) develops a solution for MDPs and Marginal MAP problems by combining a symbolic variant of **lifted belief propagation**, as discussed in Chapter11, with gradient based search.

16.4.2   Inductive Lifted Stochastic Planning

Inductive methods can be seen to be orthogonal to the inference algorithms in that they mostly do not require a model and do not reason about that model. However, the overall objective of producing lifted value functions and policies is shared with the previously discussed deductive approaches. We therefore review these here for completeness. As we discuss, it is also possible to combine the inductive and deductive approaches in several ways.

The basic inductive approaches learn a policy directly from a teacher, sometimes known as behavioral cloning. The work of (Khardon, 1999b,a; Yoon et al., 2002) provided learning algorithms for relational policies with theoretical and empirical evidence for their success. Relational policies and value functions were also explored in reinforcement learning. This was done with pure reinforcement learning using relational regression trees to learn a Q-function (Dzeroski et al., 2001), combining this with supervised guidance (Driessens and Dzeroski, 2002), or using Gaussian processes and graph kernels over relational structures to learn a Q-function (Gartner et al., 2006). A more recent approach uses functional gradient boosting with lifted regression trees to learn lifted policy structure in a policy gradient algorithm (Kersting and Driessens, 2008).

Finally, several approaches combine inductive and deductive elements. The work of Gretton and Thiebaux (2004) combines inductive logic programming with first-order decision-theoretic regression, by first using deductive methods (decision theoretic regression) to generate candidate policy structure, and then learning using this structure as features. The work of Yoon et al. (2006) shows how one can implement relational approximate policy iteration where policy improvement steps are performed by learning the intended policy from generated trajectories instead of direct calculation. Although these approaches are partially deductive they do not share the common theme of this chapter relating planning and inference in relational contexts.

16.5 Conclusions

This chapter provides a review of SDP methods, that perform abstract reasoning for stochastic planning, from the viewpoint of probabilistic inference. We have illustrated how the planning problem and the inference problem are related. Specifically, finite horizon optimization in MDPs is related to an inference problem with alternating maximization and expectation blocks and is therefore more complex than marginal MAP queries that have been studied in the literature. This analogy is valid both at the propositional and relational levels and it suggests a new line of challenges for inference problems in discrete domains. We have also identified the opportunity for generalized lifted inference, where the algorithm and its solution are agnostic of the domain instance and its size and are efficient regardless of this size. We have shown that under some conditions SDP algorithms provide generalized lifted inference. In more complex models, especially ones with additive rewards and exogenous events, SDP algorithms are yet to mature into an effective and widely applicable inference scheme. On the other hand, the challenges faced in such problems are exactly the ones typically seen in standard lifted inference problems. Therefore, exploring generalized lifted inference more abstractly has the potential to lead to advances in both areas.

Acknowledgments

This work was partly supported by NSF grants 0964457, 1616280 and 2002393. The work was partly done while RK was at Tufts University and SS was at Oregon State University.

Bibliography

Ahmadi, B., Kersting, K., and Hadiji, F. (2010). Lifted belief propagation: Pairwise marginals and beyond. In *Proceedings of the 5th European Workshop on Probabilistic Graphical Models (PGM)*. 232, 275, 341, 343

Ahmadi, B., Kersting, K., and Natarajan, S. (2012). Lifted online training of relational models with stochastic gradient methods. In *ECML PKDD*, 585–600. 89

Ahmadi, B., Kersting, K., and Sanner, S. (2011). Multi-Evidence Lifted Message Passing, with Application to PageRank and the Kalman Filter. In *Proceedings of the 22nd International Joint Conference on Artificial Intelligence, IJCAI 2011*. 275, 318

Aji, S. M. and McEliece, R. J. (2001). The generalized distributive law and free energy minimization. In *Proceedings of the 39th Allerton Conference on Communication, Control and Computing*, 672–681. 273

Aloul, F. A., Sakallah, K. A., and Markov, I. L. (2006). Efficient symmetry breaking for boolean satisfiability. *IEEE Transactions on Computers*, 55(5):549–558. 226

Anand, A., Grover, A., Mausam, M., and Singla, P. (2016). Contextual symmetries in probabilistic graphical models. In *Proceedings of the Twenty-Fifth International Joint Conference on Artificial Intelligence*, 3560–3568. 198

Andrieu, C., de Freitas, N., Doucet, A., and Jordan, M. I. (2003). An introduction to MCMC for machine learning. *Machine Learning*, 50(1-2):5–43. 190

Apsel, U. and Brafman, R. I. (2011). Extended lifted inference with joint formulas. In *Proceedings of the 27th Conference on Uncertainty in Artificial Intelligence (UAI)*, 11–18. 57

Apt, K. R. and Bezem, M. (1991). Acyclic programs. *New Generation Computing*, 9(3-4):335–363. 6

Arnborg, S., Corneil, D. G., and Proskurowski, A. (1987). Complexity of finding embeddings in a k-tree. *SIAM Journal on Algebraic Discrete Methods*, 8(2):277–284. 99

Atserias, A. and Maneva, E. (2013). Sherali-Adams relaxations and indistinguishability in counting logics. *SIAM Journal of Computation*, 42(1):112–137. 364

Attias, H. (2003). Planning by probabilistic inference. In *Proceedings of the Ninth International Workshop on Artificial Intelligence and Statistics, AISTATS 2003, Key West, Florida, USA, January 3-6, 2003*. 376, 377

Babai, L. (2016). Graph isomorphism in quasipolynomial time. In *Proceedings of the 48th Annual ACM Symposium on Theory of Computing (STOC '16)*, 684–697. 355

Babai, L., Erdős, P., and Selkow, S. (1980). Random graph isomorphism. *SIAM Journal on Computing*, 9:628–635. 351, 355

Babai, L. and Kučera, L. (1979). Canonical labelling of graphs in linear average time. In *Annual Symposium on Foundations of Computer Science (FOCS)*, 39–46. 355

Bach, S. H., Broecheler, M., Huang, B., and Getoor, L. (2017). Hinge-loss markov random fields and probabilistic soft logic. *Journal of Machine Learning Research (JMLR)*, 18:1–67. 28, 53

Bahar, R. I., Frohm, E., Gaona, C., Hachtel, G., Macii, E., Pardo, A., and Somenzi, F. (1993). Algebraic Decision Diagrams and their applications. In *IEEE /ACM International Conference on CAD*, 428–432. 393

Bahmani-Oskooee, M. and Brown, F. (2004). Kalman filter approach to estimate the demand for international reserves. *Applied Economics*, 36(15):1655–1668. 327

Balas, E., Ceria, S., and Cornuéjols, G. (1993). A lift-and-project cutting plane algorithm for mixed 0-1 programs. *Mathematical Programming*, 58:295–324. 364

Battaglia, D., Kolář, M., and Zecchina, R. (2004). Minimizing energy below the glass thresholds. *Physical Review E*, 70:036107. 226

Beame, P., Van den Broeck, G., Gribkoff, E., and Suciu, D. (2015). Symmetric weighted first-order model counting. In *PODS*, 313–328. 104, 110, 155, 156, 158, 159, 164

Belle, V., Passerini, A., and Van den Broeck, G. (2015a). Probabilistic inference in hybrid domains by weighted model integration. In *Proceedings of 24th International Joint Conference on Artificial Intelligence (IJCAI)*. 101

Belle, V., Van den Broeck, G., and Passerini, A. (2015b). Hashing-based approximate probabilistic inference in hybrid domains. In *Proceedings of the 31st Conference on Uncertainty in Artificial Intelligence (UAI)*. 101

Bellodi, E., Lamma, E., Riguzzi, F., Santos Costa, V., and Zese, R. (2014). Lifted variable elimination for probabilistic logic programming. *Theory and Practice of Logic Programming (TPLP)*, 14(4–5):681–695. 101

Berkholz, C., Bonsma, P., and Grohe, M. (2013). Tight lower and upper bounds for the complexity of canonical colour refinement. In *Proc. of the 21st Annual European Symposium on Algorithms, Sophia Antipolis, France, September 2-4, 2013.*, 145–156. Springer. 353

Berkholz, C., Bonsma, P., and Grohe, M. (2016). Tight lower and upper bounds for the complexity of canonical colour refinement. *Theory of Computing Systems*, doi:10.1007/s00224-016-9686-0. 355

Berry, A., Blair, J. R., Heggernes, P., and Peyton, B. W. (2004). Maximum cardinality search for computing minimal triangulations of graphs. *Algorithmica*, 39(4):287–298. 99

Besag, J. (1974). Spatial interaction and the statistical analysis of lattice systems. *Journal of the Royal Statistical Society. Series B (Methodological)*, 36(2):192–236. 43

Bishop, Y. M., Fienberg, S. E., and Holland, P. W. (2007). *Discrete multivariate analysis: theory and practice*. Springer Science & Business Media. 92, 96

Blockeel, H. and De Raedt, L. (1998). Top-down induction of first-order logical decision trees. *Artificial intelligence*, 101(1):285–297. 45

Bodlaender, H. L. (1993). A tourist guide through treewidth. *Acta Cybernetica*, 11(1–2):1–21. 12

Boixo, S., Rønnow, T. F., Isakov, S. V., Wang, Z., Wecker, D., Lidar, D. A., Martinis, J. M., and Troyer, M. (2013). Quantum annealing with more than one hundred qubits. *Nature Physics*, 10(3):218–224. 197

Böker, J. (2019). Color refinement, homomorphisms, and hypergraphs. *ArXiv (CoRR)*, arXiv:1903.12432 [cs.DM]. 352

Bollobás, B. (1982). Distinguishing vertices of random graphs. *Annals of Discrete Mathematics*, 13:33–50. 358

Borgwardt, S., Ceylan, I. I., and Lukasiewicz, T. (2017). Ontology-mediated queries for probabilistic databases. In *Thirty-First AAAI Conference on Artificial Intelligence*. 101

Boutilier, C., Dean, T., and Hanks, S. (1999). Decision-theoretic planning: Structural assumptions and computational leverage. *Journal of Artificial Intelligence Research (JAIR)*, 11:1–94. 392

Boutilier, C., Friedman, N., Goldszmidt, M., and Koller, D. (1996). Context-specific independence in Bayesian networks. In *Proceedings of the 12th Conference on Uncertainty in Artificial Intelligence (UAI)*, 115–123. 89, 161, 392

Boutilier, C., Reiter, R., and Price, B. (2001). Symbolic dynamic programming for first-order MDPs. In *International Joint Conference on Artificial Intelligence (IJCAI-01)*, 690–697. Seattle. xix, 374, 381, 384, 387, 393

Braun, T. and Möller, R. (2016). Lifted junction tree algorithm. In *Joint German/Austrian Conference on Artificial Intelligence (Künstliche Intelligenz)*, 30–42. Springer. 147

Braun, T. and Möller, R. (2017a). Counting and conjunctive queries in the lifted junction tree algorithm. In *International Workshop on Graph Structures for Knowledge Representation and Reasoning*, 54–72. Springer. 147

Braun, T. and Möller, R. (2017b). Preventing groundings and handling evidence in the lifted junction tree algorithm. In *Joint German/Austrian Conference on Artificial Intelligence (Künstliche Intelligenz)*, 85–98. Springer. 147

Braunstein, A., Mézard, M., and Zecchina, R. (2005). Survey propagation: An algorithm for satisfiability. *Random Structures and Algorithms*, 27(2):201–226. 217, 220, 221, 222

Braunstein, A. and Zecchina, R. (2004). Survey propagation as local equilibrium equations. *Journal of Statistical Mechanics: Theory and Experiment*, P06007:812–815. 213, 227

Bröcheler, M., Mihalkova, L., and Getoor, L. (2010). Probabilistic similarity logic. In Grünwald, P. and Spirtes, P. (eds.), *Proc. Twenty Sixth Conference on Uncertainty in Artificial Intelligence*, 73–82. AUAI Press. 53

Bruynooghe, M., De Cat, B., Drijkoningen, J., Fierens, D., Goos, J., Gutmann, B., Kimmig, A., Labeeuw, W., Langenaken, S., Landwehr, N., Meert, W., Nuyts, E., Pellegrims, R., Rymenants, R., Segers, S., Thon, I., Van Eyck, J., Van den Broeck, G., Vangansewinkel, T., Van Hove, L., Vennekens, J., Weytjens, T., and De Raedt, L. (2009). An exercise with statistical relational learning systems. In Domingos, P. and Kersting, K. (eds.), *Proceedings of the International Workshop on Statistical Relational Learning (SRL-09)*, 1–3. 16

Buchholz, P. (1994). Exact and ordinary lumpability in finite Markov chains. *Journal of Applied Probability*, 31(1):59–75. 187

Buchman, D. and Poole, D. (2015). Representing aggregators in relational probabilistic models. In *Proc. Twenty-Ninth AAAI Conference on Artificial Intelligence (AAAI-15)*. 10, 29

Buchman, D. and Poole, D. (2016). Negation without negation in probabilistic logic programming. In *Proc. 15th International Conference on Principles of Knowledge Representation and Reasoning.* 31

Bui, H., Huynh, T., and Riedel, S. (2013a). Automorphism groups of graphical models and lifted variational inference. In *Proceedings of the 29th Annual Conference on Uncertainty in Artificial Intelligence (UAI)*. 172, 176

Bui, H., Huynh, T., and Sontag, D. (2014). Lifted tree-reweighted variational inference. In *Proceedings of the 30th Conference on Uncertainty in Artificial Intelligence (UAI)*. 304

Bui, H. B., Huynh, T. N., and de Salvo Braz, R. (2012). Exact lifted inference with distinct soft evidence on every object. In *AAAI*. 164, 170, 174, 184, 185, 193, 194, 239, 315

Bui, H. H., Huynh, T. N., and Riedel, S. (2013b). Automorphism groups of graphical models and lifted variational inference. In *UAI*, 132. 89, 164

Buntine, W. L. (1994). Operations for learning with graphical models. *J. Artif. Intell. Res. (JAIR)*, 2:159–225. 8, 53, 115

Burgers, G., van Leeuwen, P. J., Evensen, G., Burgers, G., and Burgers, G. (1998). On the analysis scheme in the ensemble kalman filter. *Monthly Weather Review*, 126:1719–1724. 327

Cai, J.-Y., Fürer, M., and Immerman, N. (1992). An optimal lower bound on the number of variables for graph identification. *Combinatorica*, 12:389–410. 352

Carbonetto, P., Kisyński, J., Chiang, M., and Poole, D. (2009). Learning a contingently acyclic, probabilistic relational model of a social network. TR-2009-08, Univ of British Columbia, Dept of Comp Sci. 126

Carbonetto, P., Kisynski, J., de Freitas, N., and Poole, D. (2005). Nonparametric Bayesian logic. In *Proc. 21st Conference on Uncertainty in AI (UAI)*. 37

Cardon, A. and Crochemore, M. (1982). Partitioning a graph in $O(|A| \log_2 |V|)$. *Theoretical Computer Science*, 19(1):85 – 98. 350, 353

Celler, F., Leedham-Green, C. R., Murray, S. H., Niemeyer, A. C., and O'brien, E. (1995). Generating random elements of a finite group. *Communications in Algebra*, 23(13):4931–4948. 185, 188, 189

Ceylan, I. I., Darwiche, A., and Van den Broeck, G. (2016). Open-world probabilistic databases. In *Description Logics*. 101, 106

Chang, C. and Keisler, J. (1990). *Model Theory*. Elsevier, Amsterdam, Holland. 381

Chang, C. L. and Lee, R. C. T. (1973). *Symbolic Logic and Mechanical Theorem Proving*. Academic Press. 13

Chavira, M. and Darwiche, A. (2008). On probabilistic inference by weighted model counting. *Artificial Intelligence*, 172(6):772–779. 11, 95

Chavira, M., Darwiche, A., and Jaeger, M. (2006). Compiling relational Bayesian networks for exact inference. *International Journal of Approximate Reasoning*, 42(1-2):4–20. 95

Chen, F., Lovász, L., and Pak, I. (1999). Lifting Markov chains to speed up mixing. In *Proceedings of the thirty-first annual ACM symposium on Theory of computing*, STOC '99, 275–281. 198

Chen, Y., Ruozzi, N., and Natarajan, S. (2019). Lifted message passing for hybrid probabilistic inference. In *Proceedings of the Twenty-Eighth International Joint Conference on Artificial Intelligence, IJCAI-19*, 5701–5707. International Joint Conferences on Artificial Intelligence Organization. 343

Chen, Y., Yang, Y., Natarajan, S., and Ruozzi, N. (2020). Hybrid lifted variational inference. In *Proceedings of the Twenty-Eighth International Joint Conference on Artificial Intelligence, IJCAI-20*. 344

Cheng, Q., Liu, Q., Chen, F., and Ihler, A. T. (2013). Variational planning for graph-based MDPs. In *Proc. Advances in Neural Information Processing Systems*, 2976–2984. 376, 377

Chieu, H. and Lee, W. (2009). Relaxed survey propagation for the weighted maximum satisfiability problem. *Journal of Artificial Intelligence Research (JAIR)*, 36:229–266. 226

Choi, A., Chavira, M., and Darwiche, A. (2007). Node splitting: A scheme for generating upper bounds in Bayesian networks. In *Proceedings of the 23rd Conference on Uncertainty in Artificial Intelligence (UAI)*, 57–66. 273

Choi, A. and Darwiche, A. (2006). An edge deletion semantics for belief propagation and its practical impact on approximation quality. In *Proceedings of the 21st AAAI Conference on Artificial Intelligence,*, 1107–1114. 261, 262, 273

Choi, A. and Darwiche, A. (2008). Approximating the partition function by deleting and then correcting for model edges. In *Proceedings of the 24th Conference in Uncertainty in Artificial Intelligence (UAI)*, 79–87. 273

Choi, A. and Darwiche, A. (2011). Relax, compensate and then recover. In Onada, T., Bekki, D., and McCready, E. (eds.), *New Frontiers in Artificial Intelligence*, volume 6797 of *Lecture Notes in Computer Science*, 167–180. Springer Berlin / Heidelberg. 259, 263, 273

Choi, J. and Amir, E. (2012). Lifted relational variational inference. In *Proceedings of the 28th Conference on Uncertainty in Artificial Intelligence (UAI)*, 196–206. 256, 257, 318

Choi, J., Amir, E., Xu, T., and Valocchi, A. J. (2015). Learning relational Kalman filtering. In *Proceedings of the Twenty-Fifth AAAI Conference on Artificial Intelligence*, 2539–2546. 318

Choi, J., de Salvo Braz, R., and Bui, H. H. (2011a). Efficient methods for lifted inference with aggregate factors. In *AAAI*. 89

Choi, J., Guzman-Rivera, A., and Amir, E. (2011b). Lifted relational kalman filtering. In *Proc. of the 22nd Int. Joint Conf. on Artificial Intelligence(IJCAI)*, 2092–2099. 318

Choi, J., Hill, D., and Amir, E. (2010). Lifted inference for relational continuous models. In *Proceedings of the 26th Conference on Uncertainty in Artificial Intelligence (UAI)*, 126–134. 57, 318, 327, 331, 336

Clark, K. L. (1978). Negation as failure. In Gallaire, H. and Minker, J. (eds.), *Logic and Databases*, 293–322. Springer. 6, 30, 142

Clautiaux, F., Moukrim, A., Nègre, S., and Carlier, J. (2004). Heuristic and metaheuristic methods for computing graph treewidth. *RAIRO-Operations Research*, 38(1):13–26. 99

Cozman, F. G. (2004). Axiomatizing noisy-or. In *Proceedings of European Conference on Artificial Intelligence (ECAI)*, 979–980. 143

Craven, M. and Slattery, S. (2001). Relational learning with statistical predicate invention: Better models for hypertext. *Machine Learning Journal*, 43(1/2):97–119. 196

Crawford, J. (1992). A theoretical analysis of reasoning by symmetry in first-order logic. In *Proceedings of the Workshop on Tractable Reasoning*. 226

Crawford, J. M., Ginsberg, M. L., Luks, E. M., and Roy, A. (1996). Symmetry-breaking predicates for search problems. In *Proceedings of the 5th Conference on Principles of Knowledge Representation and Reasoning*, 148–159. 221

Cui, H., Keller, T., and Khardon, R. (2019). Stochastic planning with lifted symbolic trajectory optimization. In *ICAPS*. 394

Cui, H., Marinescu, R., and Khardon, R. (2018). From stochastic planning to marginal MAP. In *NIPS*, 3085–3095. 376, 394

Dalvi, N. and Suciu, D. (2012). The dichotomy of probabilistic inference for unions of conjunctive queries. *Journal of the ACM (JACM)*, 59(6):30. 106

Dalvi, N. N. and Suciu, D. (2004). Efficient query evaluation on probabilistic databases. In Nascimento, M. A., Özsu, M. T., Kossmann, D., Miller, R. J., Blakeley, J. A., and Schiefer, K. B. (eds.), *Proceedings of the 30th International Conference on Very Large Databases (VLDB-04)*, 864–875. Morgan Kaufmann Publishers. 21

Damien, P. and Walker, S. G. (2001). Sampling truncated normal, beta, and gamma densities. *Journal of Computational and Graphical Statistics*, 10(2):206–215. 133

Dantsin, E. (1991). Probabilistic logic programs and their semantics. In Voronkov, A. (ed.), *Proceedings of the First Russian Conference on Logic Programming*, volume 592 of *Lecture Notes in Computer Science*, 152–164. Springer. 21

Darwiche, A. (2001). Recursive conditioning. *Artificial Intelligence*, 126(1-2):5–41. 12, 89

Darwiche, A. (2003). A differential approach to inference in Bayesian networks. *Journal of the ACM (JACM)*, 50(3):280–305. 147, 149

Darwiche, A. (2009). *Modeling and Reasoning with Bayesian Networks*. Cambridge University Press. 99, 161, 165

Darwiche, A. and Marquis, P. (2002). A knowledge compilation map. *J. Artif. Intell. Res. (JAIR)*, 17:229–264. 147, 148, 149

Davis, J. and Domingos, P. (2009). Deep transfer via second-order Markov logic. In *Proc. International Conference on Machine Learning (ICML)*. ACM Press, Montréal, Canada. 48

Davis, J. and Domingos, P. (2011). Deep transfer: A Markov logic approach. *AI Magazine*, 32(1):51–52. 48

Davis, M., Logemann, G., and Loveland, D. (1962). A machine program for theorem proving. *Communications of the ACM*, 5(7):394–397. 12

De Raedt, L., Frasconi, P., Kersting, K., and Muggleton, S. (eds.) (2008). *Probabilistic Inductive Logic Programming — Theory and Applications*, volume 4911 of *Lecture Notes in Artificial Intelligence*. Springer. 57, 174

De Raedt, L., Kersting, K., Natarajan, S., and Poole, D. (2016). *Statistical Relational Artificial Intelligence: Logic, Probability, and Computation*. Morgan & Claypool. 3, 16, 28, 39

De Raedt, L. and Kimmig, A. (2015). Probabilistic (logic) programming concepts. *Machine Learning*, 100(1):5–47. 16, 17, 21, 24, 29

De Raedt, L., Kimmig, A., and Toivonen, H. (2007). ProbLog: A probabilistic Prolog and its application in link discovery. In Veloso, M. M. (ed.), *Proceedings of the 20th International Joint Conference on Artificial Intelligence (IJCAI-07)*, 2462–2467. Morgan Kaufmann Publishers. 3, 16, 17, 21, 24, 40, 53, 101, 137, 141

de Salvo Braz, R., Amir, E., and Roth, D. (2005). Lifted first-order probabilistic inference. In *Proceedings of the 19th International Joint Conference on Artificial Intelligence (IJCAI)*, 1319–1325. 57, 89, 105, 107, 174, 270, 275, 276, 318, 319, 320, 327, 335

de Salvo Braz, R., Amir, E., and Roth, D. (2006). MPE and partial inversion in lifted probabilistic variable elimination. In *Proceedings of the 21st AAAI Conference on Artificial Intelligence (AAAI)*, 1123–1130. 106, 319

de Salvo Braz, R., Amir, E., and Roth, D. (2007). Lifted first-order probabilistic inference. In Getoor, L. and Taskar, B. (eds.), *An Introduction to Statistical Relational Learning*, 433–451. MIT Press. 66, 71, 72, 79, 85, 113, 118, 120, 125

de Salvo Braz, R., Natarajan, S., Bui, H., Shavlik, J., and Russell, S. (2009). Anytime lifted belief propagation. *Proc. SRL-09*. 275

de Salvo Braz, R., O'Reilly, C., Gogate, V., and Dechter, R. (2016). Probabilistic inference modulo theories. *arXiv preprint arXiv:1605.08367*. 101

Dean, T. and Kanazawa, K. (1989). A Model for Reasoning about Persistence and Causation. *Computational Intelligence*, 5:142–150. 392

Dearden, R. and Boutilier, C. (1997). Abstraction and approximate decision theoretic planning. *Artificial Intelligence*, 89(1-2):219–283. 392

Dechter, R. (1996). Bucket elimination: A unifying framework for probabilistic inference. In Horvitz, E. and Jensen, F. (eds.), *Proc. Twelfth Conference on Uncertainty in Artificial Intelligence (UAI-96)*, 211–219. 12

Dechter, R. (2003). *Constraint Processing*. Morgan Kaufmann. 12, 67, 99

Dechter, R., Kask, K., and Mateescu, R. (2002). Iterative join-graph propagation. In *Proceedings of the 18th Conference in Uncertainty in Artificial Intelligence (UAI)*, 128–136. 273

Dechter, R. and Mateescu, R. (2007). And/or search spaces for graphical models. *Artif. Intell.*, 171(2-3):73–106. 149

Dechter, R. and Rish, I. (2003). Mini-buckets: A general scheme for bounded inference. *Journal of the ACM (JACM)*, 50(2):107–153. 273

Dempster, A. P., Laird, N. M., and Rubin, D. B. (1977). Maximum likelihood from incomplete data via the em algorithm. *Journal of the Royal Statistical Society B*, 39:1–38. 238

Derisavi, S., Hermanns, H., and Sanders, W. H. (2003). Optimal state-space lumping in Markov chains. *Inf. Process. Lett.*, 87(6):309–315. 187, 188

Diaconis, P. and Freedman, D. (1980). De Finetti's generalizations of exchangeability. In *Studies in Inductive Logic and Probability*, volume II. 162, 165

Diez, F. J. (1993). Parameter adjustment in Bayes networks. the generalized noisy or-gate. In *Proc. ninth UAI*, 99–105. 117

Díez, F. J. and Galán, S. F. (2003). Efficient computation for the noisy max. *International Journal of Intelligent Systems*, 18(2):165–177. 32, 121, 125, 129, 130, 132, 143

Domingos, P. (2015). *The Master Algorithm: How the Quest for the Ultimate Learning Machine Will Remake Our World*. Basic Books, New York:. 11

Domingos, P., Kok, S., Lowd, D., Poon, H., Richardson, M., and Singla, P. (2008). Markov logic. In Raedt, L. D., Frasconi, P., Kersting, K., and Muggleton, S. (eds.), *Probabilistic Inductive Logic Programming*, 92–117. Springer. 93

Domingos, P. and Lowd, D. (2009). *Markov Logic: An Interface Layer for Artificial Intelligence*. Synthesis Lectures on Artificial Intelligence and Machine Learning. Morgan & Claypool. 3, 10, 260

Domshlak, C. and Hoffmann, J. (2006). Fast probabilistic planning through weighted model counting. In *Proceedings of the International Conference on Automated Planning and Scheduling*. 376

Dos Martires, P. Z., Dries, A., and De Raedt, L. (2019). Exact and approximate weighted model integration with probability density functions using knowledge compilation. In *Proceedings of the AAAI Conference on Artificial Intelligence*, volume 33, 7825–7833. 101

Doshi, F. and Roy, N. (2008). The permutable POMDP: Fast solutions to POMDPs for preference elicitation. In *Proceedings of the Seventh International Conference on Autonomous Agents and Multiagent Systems (AAMAS 2008)*. Estoril, Portugal. 394

Doucet, A., de Freitas, N., and Gordon, N. (eds.) (2001). *Sequential Monte Carlo in Practice*. Springer-Verlag. 13

Driessens, K. and Dzeroski, S. (2002). Integrating experimentation and guidance in relational reinforcement learning. In *International Conference on Machine Learning (ICML)*, 115–122. 394

Duchi, J., Tarlow, D., Elidan, G., and Koller, D. (2007). Using combinatorial optimization within max-product belief propagation. In *Advances in Neural Information Processing Systems 19 (NIPS)*, 369–376. 227

Duris, D. (2012). Some characterizations of γ and β-acyclicity of hypergraphs. *Information Processing Letters*, 112(16):617–620. 156

Dyer, M. and Greenhill, C. (2000). On Markov chains for independent sets. *Journal of Algorithms*, 35(1):17–49. 196

Dzeroski, S., DeRaedt, L., and Driessens, K. (2001). Relational reinforcement learning. *Machine Learning Journal (MLJ)*, 43:7–52. 394

Eaton, F. and Ghahramani, Z. (2009). Choosing a variable to clamp. In *International Conference on Artificial Intelligence and Statistics*, 145–152. 237

Elidan, G. and Globerson, A. (2010). Summary of the 2010 UAI approximate inference challenge. http://www.cs.huji.ac.il/project/UAI10/. 259, 273

Erdős, P. and Rényi, A. (1963). Asymmetric graphs. *Acta Mathematica Hungarica*, 14(3):295–315. 194

Esseen, C.-G. (1942). On the liapunoff limit of error in the theory of probability. *Arkiv foer Matematik, Astronomi, och Fysik*, A28(9):1–19. 135

Evensen, G. (1994). Sequential data assimilation with a nonlinear quasi-geostrophic model using monte carlo methods to forecast error statistics. *Journal of Geophysical Research*, 99:10143–10162. 327

Fages, F. (1994). Consistency of Clark's completion and existence of stable models. *Journal of Methods of Logic in Computer Science*, 1:51–60. 142

Fagin, R. (1983). Degrees of acyclicity for hypergraphs and relational database schemes. *Journal of the ACM (JACM)*, 30(3):514–550. 158

Feynman, R. P. (1987). Negative probability. *Quantum implications: essays in honour of David Bohm*, 235–248. 143

Fierens, D., Van den Broeck, G., Renkens, J., Shterionov, D., Gutmann, B., Thon, I., Janssens, G., and De Raedt, L. (2015). Inference and learning in probabilistic logic programs using weighted Boolean formulas. *Theory and Practice of Logic Programming*, 15(03):358–401. 21, 24, 101, 141

Fierens, D., Van den Broeck, G., Thon, I., Gutmann, B., and De Raedt, L. (2011). Inference in probabilistic logic programs using weighted CNFs. In Cozman, F. G. and Pfeffer, A. (eds.), *Proceedings of the 27th Conference on Uncertainty in Artificial Intelligence (UAI-11)*, 211–220. AUAI Press. 95

Fikes, R. E. and Nilsson, N. J. (1971). STRIPS: A new approach to the application of theorem proving to problem solving. *AI Journal*, 2:189–208. 374

Freedman, R., de Salvo Braz, R., Bui, H., and Natarajan, S. (2012). Initial empirical evaluation of anytime lifted belief propagation. *Proceedings of Statistical Relational AI (StaRAI) workshop*. 275

Friedman, N., Getoor, L., Koller, D., and Pfeffer, A. (1999). Learning probabilistic relational models. In *Proc. of the Sixteenth International Joint Conference on Artificial Intelligence*, 1300–1307. Sweden: Morgan Kaufmann. 39, 319, 327

Friedman, T. and Van den Broeck, G. (2018). Approximate knowledge compilation by online collapsed importance sampling. In *Advances in Neural Information Processing Systems 31 (NeurIPS)*. 95

Friedman, T. and Van den Broeck, G. (2019). On constrained open-world probabilistic databases. In *Proceedings of the 28th International Joint Conference on Artificial Intelligence (IJCAI)*. 101, 106

Friedman, T. and Van den Broeck, G. (2020). Symbolic querying of vector spaces: Probabilistic databases meets relational embeddings. In *Ninth International Workshop on Statistical Relational AI (StarAI)*. 101

Fuhr, N. (2000). Probabilistic Datalog: Implementing logical information retrieval for advanced applications. *Journal of the American Society for Information Science (JASIS)*, 51(2):95–110. 21

Furmston, T. and Barber, D. (2010). Variational methods for reinforcement learning. In *Proceedings of the International Conference on Artificial Intelligence and Statistics, AISTATS*, 241–248. 376, 377

GAP (2008). *GAP – Groups, Algorithms, and Programming, Version 4.4.12*. The GAP Group. 181, 189

Gartner, T., Driessens, K., and Ramon, J. (2006). Graph kernels and Gaussian processes for relational reinforcement learning. *Machine Learning Journal (MLJ)*, 64:91–119. 394

Gartner, T., Flach, P., and Wrobel, S. (2003). On graph kernels: Hardness results and efficient alternatives. In Scholkopf, I. M. W. B. (ed.), *Proceedings of the 16th Annual Conference on Computational Learning Theory and the 7th Kernel Workshop*, 129–143. 368

Gehrke, M., Braun, T., and Möller, R. (2018). Lifted dynamic junction tree algorithm. In *International Conference on Conceptual Structures*, 55–69. Springer. 147

Getoor, L., Friedman, N., Koller, D., and Pfeffer, A. (2001). Learning probabilistic relational models. *Relational Data Mining, S. Dzeroski and N. Lavrac, Eds.* 39, 40, 41

Getoor, L. and Taskar, B. (eds.) (2007). *An Introduction to Statistical Relational Learning*. MIT Press. 14, 39, 57, 174, 319

Geweke, J. (1991). Efficient simulation from the multivariate normal and student-t distributions subject to linear constraints and the evaluation of constraint probabilities. In *Computer Sciences and Statistics Proceedings the 23rd Symposium on the Interface between*, 571–578. 133

Gkirtzou, K., Honorio, J., Samaras, D., Goldstein, R. Z., and Blaschko, M. B. (2013). fMRI analysis with sparse weisfeiler-lehman graph statistics. In *MLMI*, volume 8184 of *Lecture Notes in Computer Science*, 90–97. Springer. 370

Globerson, A. and Jaakkola, T. (2007). Fixing max-product: Convergent message passing algorithms for map LP-relaxations. In *Proc. of the 21st Annual Conf. on Neural Inf. Processing Systems (NIPS)*. 238, 305

Gogate, V. and Domingos, P. (2010). Exploiting logical structure in lifted probabilistic inference. In *Proceedings of the first international workshop on statistical relational AI (StarAI)*, 1–8. 89, 105, 140, 148

Gogate, V. and Domingos, P. (2011). Probabilistic theorem proving. In *Proceedings of the 27th Conference on Uncertainty in Artificial Intelligence (UAI)*, 256–265. 101

Gogate, V., Jha, A. K., and Venugopal, D. (2012). Advances in lifted importance sampling. In *Proceedings of the 26th AAAI Conference on Artificial Intelligence (AAAI)*, 8–14. 174, 185

Goldberg, L. A. (2001). Computation in permutation groups: counting and randomly sampling orbits. In *Surveys in Combinatorics*, 109–143. Cambridge University Press. 188

Gomes, C., Hoffmann, J., Sabharwal, A., and Selman, B. (2007). From sampling to model counting. In *Proceedings of the 20th International Joint Conference on Artificial Intelligence (IJCAI)*, 2293–2299. 224, 234

Gomes, T. and Costa, V. S. (2012). Evaluating inference algorithms for the prolog factor language. In *International Conference on Inductive Logic Programming*, 74–85. Springer. 101

Goodman, N. D., Mansinghka, V. K., Roy, D. M., Bonawitz, K., and Tenenbaum, J. B. (2008). Church: A language for generative models. In McAllester, D. A. and Myllymäki, P. (eds.), *Proceedings of the 24th Conference on Uncertainty in Artificial Intelligence (UAI-08)*, 220–229. AUAI Press. 36, 53

Gotze, F. (1991). On the rate of convergence in the multivariate clt. *The Annals of Probability*, 19(2):724–739. 135

Gretton, C. and Thiebaux, S. (2004). Exploiting first-order regression in inductive policy selection. In *Uncertainty in Artificial Intelligence (UAI-04)*, 217–225. Banff, Canada. 394

Gribkoff, E., Suciu, D., and Van den Broeck, G. (2014a). Lifted probabilistic inference: A guide for the database researcher. *Bulletin of the Technical Committee on Data Engineering*, 37(3):6–17. 101

Gribkoff, E., Van den Broeck, G., and Suciu, D. (2014b). Understanding the complexity of lifted inference and asymmetric weighted model counting. In *Conference on Uncertainty in Artificial Intelligence (UAI)*. 101

Grohe, M. (1998). Fixed-point logics on planar graphs. In *Proceedings of the 13th IEEE Symposium on Logic in Computer Science*, 6–15. 358

Grohe, M. (2017). *Descriptive Complexity, Canonisation, and Definable Graph Structure Theory*, volume 47 of *Lecture Notes in Logic*. Cambridge University Press. 352

Grohe, M., Kersting, K., Mladenov, M., and Selman, E. (2014). Dimension reduction via colour refinement. In Schulz, A. and Wagner, D. (eds.), *Proceedings of the 22nd Annual European Symposium on Algorithms*, volume 8737, 505–516. 365, 366, 367, 368

Grohe, M. and Lindner, P. (2019). Probabilistic databases with an infinite open-world assumption. In *Proceedings of the 38th ACM SIGMOD-SIGACT-SIGAI Symposium on Principles of Database Systems*, 17–31. 101

Grohe, M. and Otto, M. (2015). Pebble games and linear equations. *Journal of Symbolic Logic*, 80(3):797–844. 364

Gupta, R., Diwan, A. A., and Sarawagi, S. (2007). Efficient inference with cardinality-based clique potentials. In *Proc. 24th ICML*, 329–336. 118, 174

Gutmann, B. and Kersting, K. (2006). Tildecrf: Conditional random fields for logical sequences. In *ECML*. 40, 45

Hadiji, F., Ahmadi, B., and Kersting, K. (2011). Efficient sequential clamping for lifted message passing. In *Proceedings of the 34th Annual German Conference on AI (KI)*. 275

Häggström, O. (2002). *Finite Markov chains and algorithmic applications*. London Mathematical Society student texts. Cambridge University Press. 190, 191

Halpern, J. Y. (2003). *Reasoning about Uncertainty*. MIT Press. 90

Hamilton, W. L., Ying, R., and Leskovec, J. (2017). Inductive representation learning on large graphs. In *Proceeding of the 30th Annual Conference on Neural Information Processing Systems*, 1025–1035. 370

Hammersley, J. M. and Clifford, P. (1971). Markov fields on finite graphs and lattices. *Unpublished*. 7

Hazan, T. and Shashua, A. (2008). Convergent message-passing algorithms for inference over general graphs with convex free energies. In *Proceedings of the Twenty-Forth Conference in Uncertainty in Artificial Intelligence, (UAI)*, 264–273. 303

Heckerman, D., Chickering, D., Meek, C., Rounthwaite, R., and Kadie, C. (2000). Dependency networks for inference, collaborative filtering, and data visualization. *J. Mach. Learn. Res.*, 1:49–75. 12

Heckerman, D., Meek, C., and Koller, D. (2004). Probabilistic models for relational data. Technical Report MSR-TR-2004-30, Microsoft Research. 40

Hescott, B. and Khardon, R. (2015). The complexity of reasoning with FODD and GFODD. *Artificial Intelligence*. 393

Heskes, T. (2006). Convexity arguments for efficient minimization of the Bethe and Kikuchi free energies. *Journal of Artificial Intelligence Research (JAIR)*, 26:153–190. 281

Hill, D. J., Minsker, B. S., and Amir, E. (2009). Real-time Bayesian anomaly detection in streaming environmental data. *Water Resources Research*, 45:W00D28. 318

Hoey, J., St-Aubin, R., Hu, A., and Boutilier, C. (1999). SPUDD: Stochastic planning using decision diagrams. In *Uncertainty in Artificial Intelligence (UAI-99)*, 279–288. Stockholm. 393

Hofmann, T., Schölkopf, B., and Smola, A. J. (2008). Kernel methods in machine learning. *Ann. Statist.*, 36(3):1171–1220. 368

Hölldobler, S., Karabaev, E., and Skvortsova, O. (2006). FluCaP: A heuristic search planner for first-order MDPs. *Journal of Artificial Intelligence Research (JAIR)*, 27:419–439. 393

Holtzen, S., Millstein, T., and Van den Broeck, G. (2019a). Generating and sampling orbits for lifted probabilistic inference. In *Proceedings of the 35th Conference on Uncertainty in Artificial Intelligence (UAI)*. 198

Holtzen, S., Millstein, T., and Van den Broeck, G. (2019b). Symbolic exact inference for discrete probabilistic programs. *arXiv.cs.PL*, 1904.02079. 95

Holtzen, S., Van den Broeck, G., and Millstein, T. (2018). Sound abstraction and decomposition of probabilistic programs. In *Proceedings of the 35th International Conference on Machine Learning (ICML)*. 95

Hopcroft, J. (1971). An n log n algorithm for minimizing states in a finite automaton. In Kohavi, Z. and Paz, A. (eds.), *Theory of Machines and Computations*, 189–196. Academic Press. 353

Howard, R. A. (1988). Decision analysis: Practice and promise. *Management Science*, 34(6):679–695. 6

Huang, J. and Darwiche, A. (2007). The language of search. *Journal of Artificial Intelligence Research (JAIR)*, 29:191–219. 148

Huth, M. and Ryan, M. (2004). *Logic in Computer Science: Modelling and Reasoning About Systems*. Cambridge University Press. 30

Huynh, T. N. and Mooney, R. J. (2008). Discriminative structure and parameter learning for Markov logic networks. In *Proc. of the international conference on machine learning*. 44

Ihler, A., Fisher III, J., and Willsky, A. (2005). Loopy belief propagation: Convergence and effects of message errors. *Journal of Machine Learning Research (JMLR)*, 6:905–936. 208, 242, 243

Immerman, N. and Lander, E. (1990). Describing graphs: A first-order approach to graph canonization. In Selman, A. (ed.), *Complexity theory retrospective*, 59–81. Springer-Verlag. 352, 355, 358, 360

Issakkimuthu, M., Fern, A., Khardon, R., Tadepalli, P., and Xue, S. (2015). Hindsight optimization for probabilistic planning with factored actions. In *ICAPS*. 376

Jaeger, M. (1997). Relational Bayesian networks. In Geiger, D. and Shenoy, P. P. (eds.), *Proceedings of the 13th Conference on Uncertainty in Artificial Intelligence (UAI-97)*, 266–273. Morgan Kaufmann Publishers. 33, 39

Jaeger, M. (2002). Relational Bayesian networks: a survey. *Electronic Articles in Computer and Information Science*, 6. 117

Jaeger, M. (2015). Lower complexity bounds for lifted inference. *Theory and Practice of Logic Programming*, 15(2):246–263. 158, 159

Jaeger, M. and Van den Broeck, G. (2012). Liftability of probabilistic inference: Upper and lower bounds. In *Proceedings of the 2nd International Workshop on Statistical Relational AI (StaRAI)*, 55–62. 143, 158, 159, 164, 173, 174

Jaimovich, A., Meshi, O., and Friedman, N. (2007). Template-based Inference in Symmetric Relational Markov Random Fields. In *Proceedings of the Conference on Uncertainty in Artificial Intelligence (UAI)*, 191–199. 274, 304

Jain, A., Friedman, T., Kuzelka, O., Van den Broeck, G., and De Raedt, L. (2019). Scalable rule learning in probabilistic knowledge bases. In *The 1st Conference On Automated Knowledge Base Construction (AKBC)*. 101

Janhunen, T. (2004). Representing normal programs with clauses. In *Proceedings of European Conference on Artificial Intelligence (ECAI)*, volume 16, 358. 142

Jensen, F. V., Lauritzen, S. L., and Olesen, K. G. (1990). Bayesian updating in causal probabilistic networks by local computations. *Computational statistics quarterly 4*, 269–282. 147

Jernite, Y., Rush, A. M., and Sontag, D. (2015). A fast variational approach for learning Markov random field language models. In *Proc. International Conference on Machine Learning (ICML)*. 89

Jha, A., Gogate, V., Meliou, A., and Suciu, D. (2010). Lifted inference seen from the other side : The tractable features. In *Proceedings of the 23rd Annual Conference on Neural Information Processing Systems (NIPS)*, 973–981. 89, 105, 163, 174

Jha, A. and Suciu, D. (2012). Probabilistic databases with MarkoViews. *Proceedings of the VLDB Endowment*, 5(11):1160–1171. 143

Jordan, M. I. (2010). Bayesian nonparametric learning: Expressive priors for intelligent systems. In Dechter, R., Geffner, H., and Halpern, J. Y. (eds.), *Heuristics, Probability and Causality: A Tribute to Judea Pearl*, 167–186. College Publications. 53

Jordan, M. I., Ghahramani, Z., Jaakkola, T. S., and Saul, L. K. (1997). An introduction to variational methods for graphical models. Technical report, MIT Computational Cognitive Science. 13

Joshi, S., Kersting, K., and Khardon, R. (2010). Self-taught decision theoretic planning with first order decision diagrams. In *Proc. of ICAPS*, 89–96. 393

Joshi, S., Kersting, K., and Khardon, R. (2011). Decision theoretic planning with generalized first order decision diagrams. *AIJ*, 175:2198–2222. 381, 383, 385, 387, 390, 393

Joshi, S. and Khardon, R. (2008). Stochastic planning with first order decision diagrams. In *Proc. of ICAPS*, 156–163. 393

Joshi, S., Khardon, R., Raghavan, A., Tadepalli, P., and Fern, A. (2013). Solving relational MDPs with exogenous events and additive rewards. In *ECML*. 381, 383, 386, 390, 393

Joshi, S., Schermerhorn, P. W., Khardon, R., and Scheutz, M. (2012). Abstract planning for reactive robots. In *ICRA*, 4379–4384. 394

Kalman, R. E. (1960). A new approach to linear filtering and prediction problems. *Transactions of the ASME–Journal of Basic Engineering*, 82(Series D):35–45. 327

Kang, B. K. and Kim, K. (2012). Exploiting symmetries for single- and multi-agent partially observable stochastic domains. *Artif. Intell.*, 182-183:32–57. 394

Karabaev, E. and Skvortsova, O. (2005). A heuristic search algorithm for solving first-order MDPs. In *Uncertainty in Artificial Intelligence (UAI-05)*, 292–299. Edinburgh, Scotland. 393

Karp, R. (1972). Reducibilities among combinatorial problems. In Miller, R. and Thatcher, J. (eds.), *Complexity of Computer Computations*, 85–103. Plenum Press, New York. 355

Kask, K. and Dechter, R. (2001). A general scheme for automatic generation of search heuristics from specification dependencies. *Artificial Intelligence*, 129(1-2):91–131. 273

Kazemi, S. M., Buchman, D., Kersting, K., Natarajan, S., and Poole, D. (2014). Relational logistic regression. In *Proc. 14th International Conference on Principles of Knowledge Representation and Reasoning (KR-2014)*. 10, 31, 32, 33

Kazemi, S. M., Kimmig, A., Van den Broeck, G., and Poole, D. (2016). New liftable classes for first-order probabilistic inference. In *Advances in Neural Information Processing Systems*, 3117–3125. xvii, 89, 109, 147, 149, 152, 153, 155, 156, 157, 158, 164, 392

Kazemi, S. M., Kimmig, A., Van den Broeck, G., and Poole, D. (2017). Domain recursion for lifted inference with existential quantifiers. In *Workshop on Statistical Relational Artificial Intelligence (StaRAI)*. 110, 155, 157, 158, 159, 392

Kazemi, S. M. and Poole, D. (2014). Elimination ordering in first-order probabilistic inference. In *AAAI*. 89, 110

Kersting, K., Ahmadi, B., and Natarajan, S. (2009). Counting belief propagation. In *Proceedings of the 25th Conference on Uncertainty in Artificial Intelligence (UAI)*, 277–284. 89, 185, 194, 206, 211, 274, 277, 304, 338, 339, 343

Kersting, K. and De Raedt, L. (2001). Bayesian logic programs. *CoRR*, cs.AI/0111058. 40, 41, 51

Kersting, K. and De Raedt, L. (2008). Basic principles of learning Bayesian logic programs. In *Probabilistic Inductive Logic Programming - Theory and Applications*, 189–221. Springer-Verlag. 45

Kersting, K., De Raedt, L., and Raiko, T. (2006). Logical hidden Markov models. *Journal of Artificial Intelligence Research (JAIR)*, 25:425–456. 31, 33, 36, 331

Kersting, K. and Driessens, K. (2008). Non-parametric policy gradients: A unified treatment of propositional and relational domains. In *Proc. International Conference on Machine Learning (ICML)*. 394

Kersting, K., El Massaoudi, Y., Hadiji, F., and Ahmadi, B. (2010). Informed lifting for message-passing. In *AAAI*. 248, 275

Kersting, K., Mladenov, M., Garnett, R., and Grohe, M. (2014). Power iterated color refinement. In *Proceedings of the Twenty-Eighth AAAI Conference on Artificial Intelligence*, 1904–1910. 193, 194

Kersting, K., Mladenov, M., and Tokmakov, P. (2015). Relational linear programs. *Artificial Intelligence Journal*. 256

Kersting, K., van Otterlo, M., and de Raedt, L. (2004). Bellman goes relational. In *International Conference on Machine Learning (ICML-04)*, 465–472. ACM Press. 393

Khardon, R. (1999a). Learning action strategies for planning domains. *Artificial Intelligence*, 113(1-2):125–148. 394

Khardon, R. (1999b). Learning to take actions. *Machine Learning*, 35:57–90. 394

Khosravi, H., Schulte, O., Hu, J., and Gao, T. (2012). Learning compact Markov logic networks with decision trees. *Machine Learning*, 89(3):257–277. 47

Khot, T., Natarajan, S., Kersting, K., and Shavlik, J. (2011). Learning Markov logic networks via functional gradient boosting. In *ICDM*. 47

Khot, T., Natarajan, S., Kersting, K., and Shavlik, J. (2015). Gradient-based boosting for statistical relational learning: the Markov logic network and missing data cases. *Machine Learning*, 100(1):75–100. 47

Kiefer, S., Ponomarenko, I., and Schweitzer, P. (2017). The Weisfeiler-Leman dimension of planar graphs is at most 3. In *Proceedings of the 32nd ACM-IEEE Symposium on Logic in Computer Science*. 358

Kimmig, A., Bach, S., Broecheler, M., Huang, B., and Getoor, L. (2012). A short introduction to probabilistic soft logic. In *Proceedings of the NIPS Workshop on Probabilistic Programming: Foundations and Applications*, 1–4. 28

Kimmig, A., Mihalkova, L., and Getoor, L. (2015). Lifted graphical models: A survey. *Machine Learning*, 99(1):1–45. 16, 17

Kimmig, A., Van den Broeck, G., and De Raedt, L. (2011). An algebraic Prolog for reasoning about possible worlds. In Burgard, W. and Roth, D. (eds.), *Proceedings of the 25th AAAI Conference on Artificial Intelligence (AAAI-11)*, 209–214. AAAI Press. 21

Kipf, T. N. and Welling, M. (2017). Semi-supervised classification with graph convolutional networks. In *Proceedings of the 5th International Conference on Learning Representations*. 370

Kisa, D., Van den Broeck, G., Choi, A., and Darwiche, A. (2014). Probabilistic sentential decision diagrams. In *Proceedings of the 14th International Conference on Principles of Knowledge Representation and Reasoning (KR)*, 1–10. 95, 147

Kisynski, J. and Poole, D. (2009). Constraint processing in lifted probabilistic inference. In *Proceedings of the 25th Conference on Uncertainty in Artificial Intelligence (UAI)*, 293–302. 58, 68, 89, 101, 108

Kjaerulff, U. (1990). Triangulation of graphs–algorithms giving small total state space. Technical Report Research Report R-90-09, Aalborg University. 99

Kok, S. and Domingos, P. (2005). Learning structure of Markov logic networks. In *Proc. International Conference on Machine Learning (ICML)*. 44, 45

Kok, S. and Domingos, P. (2007). Statistical predicate invention. In *Proc. International Conference on Machine Learning (ICML)*, 433–440. Corvallis, OR. 48

Kok, S. and Domingos, P. (2010). Learning Markov logic networks using structural motifs. In *Proc. International Conference on Machine Learning (ICML)*. 45

Koller, D. and Friedman, N. (2009). *Probabilistic Graphical Models - Principles and Techniques*. MIT Press. 7, 99, 161

Koller, D. and Pfeffer, A. (1997). Object-oriented Bayesian networks. In *Proceedings of the 13th Annual Conference on Uncertainty in AI (UAI)*, 302–313. 319

Kondor, R. and Lafferty, J. D. (2002). Diffusion kernels on graphs and other discrete input spaces. In *Proc. International Conference on Machine Learning (ICML)*, 315–322. Morgan Kaufmann. 368

Kopp, T., Singla, P., and Kautz, H. (2015). Lifted symmetry detection and breaking for map inference. In *Advances in Neural Information Processing Systems*, 1315–1323. 89

Koren, Y., Bell, R., and Volinsky, C. (2009). Matrix factorization techniques for recommender systems. *IEEE Computer*, 42(8):30–37. 14

Kowalski, R. A. (2014). *Logic for Problem Solving, Revisited*. Books on Demand. 13

Kroc, L., Sabharwal, A., and Selman, B. (2007). Survey propagation revisited. In *Proceedings of the 23rd Conference on Uncertainty in Artificial Intelligence (UAI)*, 217–226. 213

Kroc, L., Sabharwal, A., and Selman, B. (2009). Messagepassing and local heuristics as decimation strategies for satisfiability. In *Proceedings of the ACM Symposium on Applied Computing (SAC)*, 1408–1414. 213

Kroc, L., Sabharwal, A., and Selman, B. (2011). Leveraging belief propagation, backtrack search, and statistics for model counting. *Annals of Operations Research*, 184(1):209–231. 216

Kschischang, F. R., Frey, B. J., and Loeliger, H.-A. (2001). Factor graphs and the sum-product algorithm. *IEEE Transactions on Information Theory*, 47(2):498–519. 59, 66

Kumar, A. and Zilberstein, S. (2010). Map estimation for graphical models by likelihood maximization. In *Advances in Neural Information Processing Systems 23 (NIPS)*, 1180–1188. 212, 238, 239, 240, 242, 249, 252

Kumaraswamy, R., Odom, P., Kersting, K., Leake, D., and Natarajan, S. (2015). Transfer learning across relational and uncertain domains: A language-bias approach. In *ICDM*. 48, 51, 52

Kuzelka, O. and Kungurtsev, V. (2019). Lifted weight learning of markov logic networks revisited. *arXiv preprint arXiv:1903.03099*. 159

Kuzelka, O. and Wang, Y. (2020). Domain-liftability of relational marginal polytopes. *arXiv preprint arXiv:2001.05198*. 159

Lang, M. and Toussaint, M. (2009). Approximate inference for planning in stochastic relational worlds. In *Proc. International Conference on Machine Learning*. 376, 377

Laplace, P. S. (1814). *Essai philosophique sur les probabilités*. Courcier. Reprinted (1812) in English, F.W. Truscott amd F. L. Emory (Trans.) by Wiley, New York. 15

Larranaga, P., Kuijpers, C. M., Poza, M., and Murga, R. H. (1997). Decomposing Bayesian networks: triangulation of the moral graph with genetic algorithms. *Statistics and Computing*, 7(1):19–34. 99

Lasserre, J. (2002). An explicit equivalent positive semidefinite program for nonlinear 0-1 programs. *SIAM Journal on Optimization*, 12(3):756–769. 364

Lauritzen, S. and Spiegelhalter, D. (1988). Local computations with probabilities on graphical structures and their application to expert systems. *Journal of the Royal Statistical Society. Series B (Methodological)*, 157–224. 147, 165

Lauritzen, S. L. (1996). *Graphical Models*. Oxford University Press. 161

Lauritzen, S. L., Barndorff-Nielsen, O. E., Dawid, A. P., Diaconis, P., and Johansen, S. (1984). Extreme point models in statistics. *Scandinavian Journal of Statistics*, 11(2). 165

Lee, J., Marinescau, R., and Dechter, R. (2014). Applying marginal map search to probabilistic conformant planning. In *Fourth International Workshop on Statistical Relational AI (StarAI)*. 376

Lee, J., Marinescau, R., and Dechter, R. (2016). Applying search based probabilistic inference algorithms to probabilistic conformant planning: Preliminary results. In *Proceedings of the International Symposium on Artificial Intelligence and Mathematics (ISAIM)*. 376

Leeuwen, J. (1990). *Handbook of theoretical computer science: Algorithms and complexity*, volume 1. Elsevier. 159

Lesner, B. and Zanuttini, B. (2011). Efficient policy construction for MDPs represented in probabilistic PDDL. In Bacchus, F., Domshlak, C., Edelkamp, S., and Helmert, M. (eds.), *ICAPS*. AAAI. 393

Levine, S. (2018). Reinforcement learning and control as probabilistic inference: Tutorial and review. *arXiv*, 1805.00909. 377

Li, W., Saidi, H., Sanchez, H., Ṣchäf, M., and Schweitzer, P. (2016). Detecting similar programs via the weisfeiler-leman graph kernel. In *ICSR*, volume 9679 of *Lecture Notes in Computer Science*, 315–330. Springer. 370

Limketkai, B., Liao, L., and Fox, D. (2005). Relational object maps for mobile robots. In *Proceedings of the Nineteenth International Joint Conference on Artificial Intelligence, IJCAI 2005*, 1471–1476. 318, 327

Liu, Q. and Ihler, A. T. (2012). Belief propagation for structured decision making. In *Proceedings of the Conference on Uncertainty in Artificial Intelligence (UAI)*, 523–532. 376, 377

Lloyd, J. (1987). *Foundations of Logic Programming*. Springer Verlag. Second Edition. 381

Lovász, L. and Schrijver, L. (1991). Cones of matrices and set-functions and 0–1 optimization. *Cones of Matrices and Set-Functions and 0–1 Optimization SIAM Journal on Optimization*, 1(2):166–190. 364

Loveland, D. W. (1978). *Automated Theorem Proving: A Logical Basis*. Fundamental Studies in Computer Science. North-Holland. 13

Lowd, D. and Domingos, P. (2007). Efficient weight learning for Markov logic networks. In *Proceedings of the Eleventh European Conference on Principles and Practice of Knowledge Discovery in Databases*, 200–211. Springer, Warsaw, Poland. 44

Luby, M. and Vigoda, E. (1999). Fast convergence of the glauber dynamics for sampling independent sets. *Random Struct. Algorithms*, 15(3-4):229–241. 196

Madan, G., Anand, A., Singla, P., et al. (2018). Block-value symmetries in probabilistic graphical models. *arXiv preprint arXiv:1807.00643*. 198

Malkin, P. (2014). Sherali–adams relaxations of graph isomorphism polytopes. *Discrete Optimization*, 12:73–97. 364

Maneva, E., Mossel, E., and Wainwright, M. (2007). A new look at survey propagation and its generalizations. *Journal of the ACM (JACM)*, 54:2–41. 213

Marshall, A., Olkin, I., and Arnold, B. (2011). *Inequalities: Theory of majorization and its applications*. Springer series in statistics. Springer, New York. 302

McCarthy, J. (1958). Programs with common sense. In *Proceedings of the Symposium on the Mechanization of Thought Processes*, volume 1, 77–84. National Physical Laboratory. Reprinted in R. Brachman and H. Levesque (Eds.), Readings in Knowledge Representation, 1985, Morgan Kaufmann, Los Altos, CA. 375

McKay, B. D. (1981). Practical graph isomorphism. *Congressus Numerantium*, 30:45–87. 353, 356

Meert, W., Van den Broeck, G., and Darwiche, A. (2014). Lifted inference for probabilistic logic programs. In *Workshop on Probabilistic Logic Programming (PLP)*. 101

Meert, W., Vlasselaer, J., and Van den Broeck, G. (2016). A relaxed tseitin transformation for weighted model counting. In *Proceedings of the Sixth International Workshop on Statistical Relational AI (StarAI)*, 1–7. 101, 143

Meshi, O., Jaimovich, A., Globerson, A., and Friedman, N. (2009). Convexifying the Bethe free energy. In *Proceedings of the Twenty-Fifth Conference on Uncertainty in Artificial Intelligence (UAI)*, 402–410. 281, 303, 304

Mézard, M. and Montanari, A. (2009). *Information, Physics, and Computation*. Oxford University Press. 227, 229

Mézard, M., Parisi, G., and Zecchina, R. (2002). Analytic and algorithmic solution of random satisfiability problems. *Science*, 297:812–815. 213, 220

Mihalkova, L., Huynh, T., and Mooney, R. J. (2007). Mapping and revising Markov logic networks for transfer learning. In *Proc. AAAI National Conference on Artificial intelligence*, 608–614. Proc. AAAI Conference on Artificial Intelligence, Vancouver, Canada. 45, 48

Milch, B., Marthi, B., Russell, S. J., Sontag, D., Ong, D. L., and Kolobov, A. (2005). BLOG: Probabilistic models with unknown objects. In Kaelbling, L. P. and Saffiotti, A. (eds.), *Proceedings of the 19th International Joint Conference on Artificial Intelligence (IJCAI-05)*, 1352–1359. Professional Book Center. 37, 40, 319, 327

Milch, B. and Russell, S. J. (2006). First-order probabilistic languages: Into the unknown. In *International Conference on Inductive Logic Programming (ILP)*. 318, 319, 320, 327

Milch, B., Zettlemoyer, L. S., Kersting, K., Haimes, M., and Kaelbling, L. P. (2008). Lifted probabilistic inference with counting formulas. In *Proceedings of the 23rd AAAI Conference on Artificial Intelligence (AAAI)*, 1062–1608. 57, 58, 66, 68, 71, 73, 75, 76, 79, 81, 82, 85, 89, 101, 105, 107, 113, 118, 120, 125, 127, 174, 276, 319

Mladenov, M., Ahmadi, B., and Kersting, K. (2012). Lifted linear programming. In *Proceedings of the 15th International Conference on Artificial Intelligence and Statistics (AISTATS)*, 788–797. 164, 208

Mladenov, M., Globerson, A., and Kersting, K. (2014). Lifted message passing as reparametrization of graphical models. In *Proceedings of the 30th Conference on Uncertainty in Artificial Intelligence (UAI)*, 603–612. 256

Mladenov, M. and Kersting, K. (2013). Lifted inference via k-locality. In *Proceedings of the 3rd International Workshop on Statistical Relational AI*. 185

Mohan, K. and Pearl, J. (2014). Graphical models for recovering probabilistic and causal queries from missing data. In Welling, M., Ghahramani, Z., Cortes, C., and Lawrence, N. (eds.), *Advances of Neural Information Processing 27 (NIPS Proceedings)*, 1520–1528. 14

Montanari, A., Ricci-Tersenghi, F., and Semerjian, G. (2007). Solving constraint satisfaction problems through belief propagation-guided decimation. In *Proceedings of the 45th Allerton Conference on Communications, Control and Computing*, 352–359. 213, 216, 217, 227

Mooij, J. M. (2008). *Understanding and Improving Belief Propagation*. Ph.D. thesis, Radboud University Nijmegen. 252

Mooij, J. M. (2010). libDAI: A free and open source C++ library for discrete approximate inference in graphical models. *Journal of Machine Learning Research*, 11:2169–2173. 224, 253

Morettin, P., Passerini, A., and Sebastiani, R. (2019). Advanced SMT techniques for weighted model integration. *Artificial Intelligence*, 275:1–27. 101

Morik, K. and Kietz, J. (1989). A bootstrapping approach to concept clustering. In *Proceedings of the 6th International Workshop on Machine Learning (ML)*, 503–504. 239

Morris, C., Kriege, N. M., Kersting, K., and Mutzel, P. (2016). Faster kernels for graphs with continuous attributes via hashing. *CoRR*, abs/1610.00064. 370

Morris, C., Ritzert, M., Fey, M., Hamilton, W., Lenssen, J., rattan, G., and Grohe, M. (2019). Weisfeiler and leman go neural: Higher-order graph neural networks. In *Proceedings of the 33rd AAAI Conference on Artificial Intelligence*. 370

Muggleton, S. (1996). Stochastic logic programs. In *Advances in Inductive Logic Programming*, 254–264. 25, 35

Muggleton, S. and de Raedt, L. (1994). Special issue: Ten years of logic programming inductive logic programming: Theory and methods. *The Journal of Logic Programming*, 19:629 – 679. 40

Murphy, K., Weiss, Y., and Jordan, M. (1999). Loopy Belief Propagation for Approximate Inference: An Empirical Study. In *Proc. of the Conf. on Uncertainty in Artificial Intelligence (UAI-99)*, 467–475. 206

Natarajan, S., Khot, T., Kersting, K., Gutmann, B., and Shavlik, J. (2012). Gradient-based boosting for statistical relational learning: The Relational Dependency Network case. *Machine Learning Journal*. 45, 46, 47

Natarajan, S., Khot, T., Kersting, K., and Shavlik, J. (2015). *Boosted Statistical Relational Learners: From Benchmarks to Data-Driven Medicine*. SpringerBriefs in Computer Science. 46

Natarajan, S., Tadepalli, P., Altendorf, E., Dietterich, T. G., Fern, A., and Restificar, A. C. (2005). Learning first-order probabilistic models with combining rules. In De Raedt, L. and Wrobel, S. (eds.), *Proceedings of the 22nd International Conference on Machine Learning (ICML-05)*, volume 119 of *ACM International Conference Proceeding Series*, 609–616. ACM. 10, 33, 41, 51

Natarajan, S., Tadepalli, P., Dietterich, T. G., and Fern, A. (2009). Learning first-order probabilistic models with combining rules. *Special Issue on Probabilistic Relational Learning, AMAI*. 36

Nath, A. and Domingos, P. (2010). Efficient lifting for online probabilistic inference. In *Proceedings of the 24th AAAI Conference on Artificial Intelligence (AAAI)*, 1193–1198. 228, 275, 341

Neville, J. and Jensen, D. (2007). Relational dependency networks. *Journal of Machine Learning Research*, 8:653–692. 12, 31, 40, 45

Ng, R. and Subrahmanian, V. S. (1992). Probabilistic logic programming. *Information and Computation*, 101(2):150–201. 319

Ngo, L. and Haddawy, P. (1995). Probabilistic logic programming and Bayesian networks. In *Proceedings ACSC95*. 39

Nickel, M., Murphy, K., Tresp, V., and Gabrilovich, E. (2016). A review of relational machine learning for knowledge graphs. *Proceedings of the IEEE*, 104(1):11–33. 3

Niemira, M. P. and Saaty, T. L. (2004). An analytic network process model for financial-crisis forecasting. *International Journal of Forecasting*, 20(4):573–587. 318

Niepert, M. (2012a). Lifted probabilistic inference: An MCMC perspective. In *Proceedings of the 2nd International Workshop on Statistical Relational AI (StaRAI)*. 172, 198

Niepert, M. (2012b). Markov chains on orbits of permutation groups. In *Proceedings of the 28th Conference on Uncertainty in Artificial Intelligence (UAI)*, 624–633. 89, 164, 170, 174, 176, 184, 185, 186, 188, 189, 193, 195, 196, 198

Niepert, M. (2013). Symmetry-aware marginal density estimation. In *Proceedings of the 27th Conference on Artificial Intelligence (AAAI)*. 164, 174, 184, 185

Niepert, M. and Domingos, P. (2014). Exchangeable variable models. In *Proceedings of the International Conference on Machine Learning (ICML)*. 174

Niepert, M. and Van den Broeck, G. (2014). Tractability through exchangeability: A new perspective on efficient probabilistic inference. In *AAAI*, 2467–2475. 175, 185

Nitti, D., De Laet, T., and De Raedt, L. (2013). A particle filter for hybrid relational domains. In Amato, N. (ed.), *Proceedings of the IEEE/RSJ International Conference on Intelligent Robots and Systems (IROS-13)*, 2764–2771. IEEE. 36

Noessner, J., Niepert, M., and Stuckenschmidt, H. (2013). RockIt: Exploiting Parallelism and Symmetry for MAP Inference in Statistical Relational Models. In *Proceedings of the 27th Conference on Artificial Intelligence (AAAI)*. 164, 185

Paige, R. and Tarjan, R. (1987). Three partition refinement algorithms. *SIAM Journal on Computing*, 16(6):973–989. 350, 353

Pak, I. (2000). The product replacement algorithm is polynomial. In *Proceedings of the 41st Annual Symposium on Foundations of Computer Science*, 476–485. 189

Pasula, H., Marthi, B., Milch, B., Russell, S. J., and Shpitser, I. (2003). Identity uncertainty and citation matching. In *Advances in Neural Information Processing Systems*, 1425–1432. 37, 38

Pearl, J. (1986). Fusion, propagation and structuring in belief networks. *Artificial Intelligence*, 29(3):241–288. 117

Pearl, J. (1988). *Probabilistic reasoning in intelligent systems: Networks of plausible inference.* Morgan Kaufmann Publishers. 3, 7, 10, 22, 26, 161, 174, 205, 263, 273

Pearl, J. (2009). *Causality: Models, Reasoning and Inference.* Cambridge University Press, 2nd edition. 8, 14, 15

Pearson, M. and Michell, L. (2000). Smoke Rings: social network analysis of friendship groups, smoking and drug-taking. *Drugs: education, prevention and policy*, 7:21–37. 126

Pednault, E. P. D. (1989). ADL: Exploring the middle ground between STRIPS and the situation calculus. In *International Conference on Principles of Knowledge Representation and Reasoning (KR)*, 324–332. 374

Perlich, C. and Provost, F. J. (2003). Aggregation-based feature invention and relational concept classes. In *ACM SIGKDD international conference on Knowledge discovery and data mining (KDD)*. 10, 33

Pfeffer, A. (2007). The design and implementation of IBAL: A general-purpose probabilistic language. In Getoor, L. and Taskar, B. (eds.), *Statistical Relational Learning*. MIT Press. 53

Pfeffer, A., Koller, D., Milch, B., and Takusagawa, K. T. (1999). Spook: A system for probabilistic object-oriented knowledge representation. In *Proceedings of the Fifteenth Conference on Uncertainty in Artificial Intelligence, UAI 1999*, 541–550. 39, 319

Piṅgala (200 BC). *Chandah-sûtra.* 123

Poole, D. (1991). Representing Bayesian networks within probabilistic Horn abduction. In *Proc. Seventh Conference on Uncertainty in Artificial Intelligence (UAI-91)*, 271–278. 15, 21

Poole, D. (1993). Probabilistic Horn abduction and Bayesian networks. *Artificial Intelligence*, 64:81–129. 3, 4, 15, 21, 25, 27, 35, 39

Poole, D. (1997). The independent choice logic for modelling multiple agents under uncertainty. *Artificial Intelligence*, 94:7–56. Special issue on economic principles of multi-agent systems. 21, 25

Poole, D. (2000). Abducing through negation as failure: Stable models within the independent choice logic. *Journal of Logic Programming*, 44(1–3):5–35. 21

Poole, D. (2003). First-order probabilistic inference. In Gottlob, G. and Walsh, T. (eds.), *Proceedings of the Eighteenth International Joint Conference on Artificial Intelligence (IJCAI)*, 985–991. Morgan Kaufmann Publishers. 17, 18, 57, 71, 79, 81, 89, 94, 101, 105, 113, 126, 267, 270, 275, 318, 319, 320, 327, 335

Poole, D. (2007). Logical generative models for probabilistic reasoning about existence, roles and identity. In *22nd AAAI Conference on AI (AAAI-07)*. 37

Poole, D. (2008). The independent choice logic and beyond. In *De Raedt et al (Eds) Probabilistic Inductive Logic Programming*, 222–243. Springer-Verlag. 37

Poole, D., Bacchus, F., and Kisynski, J. (2011). Towards completely lifted search-based probabilistic inference. *CoRR*, abs/1107.4035. 89, 105, 155, 156, 270

Poole, D., Buchman, D., Kazemi, S. M., Kersting, K., and Natarajan, S. (2014). Population size extrapolation in relational probabilistic modelling. In *Proc. of the Eighth International Conference on Scalable Uncertainty Management*, volume LNAI 8720, 292–305. 31, 32, 33

Poole, D. L. and Mackworth, A. K. (2017). *Artificial Intelligence: foundations of computational agents*. Cambridge University Press, 2nd edition. 3

Poon, H. and Domingos, P. (2006). Sound and efficient inference with probabilistic and deterministic dependencies. In Gil, Y. and Mooney, R. J. (eds.), *Proceedings of the 21st National Conference on Artificial Intelligence (AAAI-06)*, 458–463. AAAI Press. 190, 195, 212

Poon, H. and Domingos, P. (2011). Sum-product networks: A new deep architecture. In *Computer Vision Workshops (ICCV Workshops), 2011 IEEE International Conference on*, 689–690. IEEE. 149

Poon, H., Domingos, P., and Summer, M. (2008). A general method for reducing the complexity of relational inference and its application to MCMC. In *Proceedings of the 23rd AAAI Conference on Artificial Intelligence (AAAI)*, 1075–1080. 212

Puterman, M. L. (1994). *Markov Decision Processes: Discrete Stochastic Dynamic Programming*. Wiley, New York. 384

Raedt, L. D., Kimmig, A., and Toivonen, H. (2007). Problog: A probabilistic prolog and its application in link discovery. In *Proc. International Joint Conference on Artificial Intelligence (IJCAI)*, 2462–2467. 382

Raghavan, A., Joshi, S., Fern, A., Tadepalli, P., and Khardon, R. (2012). Planning in factored action spaces with symbolic dynamic programming. In *Proc. AAAI Conference on Artificial Intelligence*. 393

Raghavan, A., Khardon, R., Fern, A., and Tadepalli, P. (2013). Symbolic opportunistic policy iteration for factored-action MDPs. In *Proc. Advances in Neural Information Processing Systems*, 2499–2507. 393

Ramakrishnan, R. and Gehrke, J. (2003). *Database management systems (3. ed.)*. McGraw-Hill. 66

Ramanan, N., Kunapuli, G., Khot, T., Fatemi, B., Kazemi, S. M., Poole, D., Kersting, K., and Natarajan, S. (2018). Structure learning for relational logistic regression: An ensemble approach. In *Sixteenth International Conference on Principles of Knowledge Representation and Reasoning*. 40

Rao, C. R., Rao, C. R., Statistiker, M., Rao, C. R., and Rao, C. R. (1973). *Linear statistical inference and its applications*, volume 2. Wiley New York. 13

Ravkic, I., Ramon, J., and Davis, J. (2015). Learning relational dependency networks in hybrid domains. *Machine Learning*, 100(2-3):217–254. 47

Read, R. and Corneil, D. (1977). The graph isomorphism disease. *Journal of Graph Theory*, 1(4):339–363. 350

Reiter, R. (2001). *Knowledge in Action: Logical Foundations for Specifying and Implementing Dynamical Systems*. MIT Press. 385

Rice, J. A. (2006). *Mathematical Statistics and Data Analysis*. Duxbury Press. 129

Richards, B. and Mooney, R. (1995). Automated refinement of first-order horn-clause domain theories. *Machine Learning*, 19(2):95–131. 51

Richardson, M. and Domingos, P. (2006). Markov logic networks. *Machine Learning*, 62(1–2):107–136. 17, 19, 20, 40, 41, 42, 44, 51, 52, 101, 137, 139, 163, 206, 250, 315, 318, 319, 327, 382

Riguzzi, F., Bellodi, E., Zese, R., Cota, G., and Lamma, E. (2017). A survey of lifted inference approaches for probabilistic logic programming under the distribution semantics. *International Journal of Approximate Reasoning*, 80:313–333. 101

Ristoski, P. and Paulheim, H. (2016). RDF2Vec: RDF graph embeddings for data mining. In *International Semantic Web Conference (1)*, volume 9981 of *Lecture Notes in Computer Science*, 498–514. 370

Robertson, N. and Seymour, P. (1986). Graph minors. II. Algorithmic aspects of tree-width. *Journal of algorithms*, 7(3):309–322. 161

Rose, D. J., Tarjan, R. E., and Lueker, G. S. (1976). Algorithmic aspects of vertex elimination on graphs. *SIAM Journal on computing*, 5(2):266–283. 99

Russell, S. and Norvig, P. (2010). *Artificial Intelligence: A Modern Approach*. Prentice Hall, 3rd edition. 3, 381, 384

Sahs, J. and Khan, L. (2012). A machine learning approach to android malware detection. In *EISIC*, 141–147. IEEE Computer Society. 370

Sanner, S. (2008). *First-order Decision-theoretic Planning in Structured Relational Environments*. Ph.D. thesis, University of Toronto, Toronto, ON, Canada. 386, 390, 393

Sanner, S. and Boutilier, C. (2005). Approximate linear programming for first-order MDPs. In *Uncertainty in Artificial Intelligence (UAI-05)*, 509–517. Edinburgh, Scotland. 394

Sanner, S. and Boutilier, C. (2006). Practical linear evaluation techniques for first-order MDPs. In *Uncertainty in Artificial Intelligence (UAI-06)*. Boston, Mass. 394

Sanner, S. and Boutilier, C. (2007). Approximate solution techniques for factored first-order MDPs. In *Proceedings of the Seventeenth International Conference on Automated Planning and Scheduling, ICAPS 2007*, 288–295. 386, 390

Sanner, S. and Boutilier, C. (2009). Practical solution techniques for first-order MDPs. *Artif. Intell.*, 173:748–488. 381, 384, 393, 394

Sanner, S. and Kersting, K. (2010). Symbolic dynamic programming for first-order POMDPs. In *Proceedings of the Twenty-Fourth AAAI Conference on Artificial Intelligence, AAAI 2010, Atlanta, Georgia, USA, July 11-15, 2010*. 394

Sarkhel, S., Venugopal, D., Singla, P., and Gogate, V. G. (2014). An integer polynomial programming based framework for lifted map inference. In *Advances in Neural Information Processing Systems*, 3302–3310. 106

Sato, N. and Tinney, W. F. (1963). Techniques for exploiting the sparsity of the network admittance matrix. *Power Apparatus and Systems, IEEE Transactions on*, 82(69):944–950. 99

Sato, T. (1995). A statistical learning method for logic programs with distribution semantics. In *Proceedings of the 12th International Conference on Logic Programming (ICLP95)*, 715–729. 15, 21, 25, 141

Sato, T. and Kameya, Y. (1997). PRISM: A symbolic-statistical modeling language. In *Proceedings of the 15th International Joint Conference on Artificial Intelligence (IJCAI-97)*, 1330–1335. 3

Sato, T. and Kameya, Y. (2001). Parameter learning of logic programs for symbolic-statistical modeling. *Journal of Artificial Intelligence Research (JAIR)*, 15:391–454. 21, 25, 35, 36, 40

Sato, T., Kameya, Y., and Zhou, N.-F. (2005). Generative modeling with failure in PRISM. In Kaelbling, L. P. and Saffiotti, A. (eds.), *Proceedings of the 19th International Joint Conference on Artificial Intelligence (IJCAI-05)*, 847–852. Professional Book Center. 21

Saul, L. K., Jaakkola, T., and Jordan, M. I. (1996). Mean field theory for sigmoid belief networks. *arXiv preprint cs/9603102*. 28

Savicky, P. and Vomlel, J. (2007). Exploiting tensor rank-one decomposition in probabilistic inference. *Kybernetika*, 43(5):747–764. 121

Selman, B., Kautz, H., and Cohen, B. (1995). Local search strategies for satisfiability testing. In *DIMACS Series in Discrete Mathematics and Theoretical Computer Science*. 235

Sen, P., Deshpande, A., and Getoor, L. (2009a). Bisimulation-based approximate lifted inference. In *Proceedings of the twenty-fifth Conference on Uncertainty in Artificial Intelligence*, 496–505. AUAI Press. 57, 274

Sen, P., Deshpande, A., and Getoor, L. (2009b). PrDB: managing and exploiting rich correlations in probabilistic databases. *VLDB Journal*, 18(5):1065–1090. 185

Shafer, G. R. and Shenoy, P. P. (1990). Probability propagation. *Annals of Mathematics and Artificial Intelligence*, 2(1-4):327–351. 147

Shariff, R., György, A., and Szepesvári, C. (2015). Exploiting symmetries to construct efficient mcmc algorithms with an application to slam. In *Artificial Intelligence and Statistics*, 866–874. 198

Sheldon, D. and Dietterich, T. (2011). Collective graphical models. In *Advances in Neural Information Processing Systems (NIPS)*, 1161–1169. 174

Shental, O., Bickson, D., Siegel, P. H., Wolf, J. K., and Dolev, D. (2008). Gaussian belief propagation solver for systems of linear equations. In *IEEE Int. Symp. on Inform. Theory (ISIT)*. Toronto, Canada. 337, 338

Sherali, H. D. and Adams, W. P. (1990). A hierarchy of relaxations between the continuous and convex hull representations for zero-one programming problems. *SIAM Journal on Discrete Mathematics*, 3(3):411–430. 364

Shervashidze, N., Schweitzer, P., van Leeuwen, E. J., Mehlhorn, K., and Borgwardt, K. M. (2011). Weisfeiler-lehman graph kernels. *Journal of Machine Learning Research*, 12:2539–2561. 369, 370

Singla, P. and Domingos, P. (2006a). Entity resolution with Markov logic. In *Proceedings of the 6th IEEE International Conference on Data Mining (ICDM-06)*, 572–582. 37

Singla, P. and Domingos, P. (2006b). Memory-efficient inference in relational domains. In *Proceedings of the 21st National Conference on Artificial Intelligence (AAAI)*, 488–493. 212

Singla, P. and Domingos, P. (2008). Lifted first-order belief propagation. In *Proceedings of the 23rd AAAI Conference on Artificial Intelligence (AAAI)*, 1094–1099. 89, 128, 185, 195, 206, 210, 211, 220, 260, 274, 276, 277, 304, 319, 327

Singla, P., Nath, A., and Domingos, P. (2010). Approximate Lifted Belief Propagation. In *Proceedings of the 1st International Workshop on Statistical Relation AI (StaRAI)*, 92–97. 248, 275

Singla, P., Nath, A., and Domingos, P. (2014). Approximate lifting techniques for belief propagation. In *Proceedings of AAAI*. 194

Smith, D. B. and Gogate, V. G. (2015). Bounding the cost of search-based lifted inference. In *Advances in Neural Information Processing Systems*, 946–954. 110

Sontag, D., Globerson, A., and Jaakkola, T. (2008a). Clusters and coarse partitions in LP relaxations. In *Proceedings of the 22nd Annual Conference on Neural Information Processing Systems (NIPS)*, 1537–1544. 305

Sontag, D., Meltzer, T., Globerson, A., Jaakkola, T., and Weiss, Y. (2008b). Tightening LP relaxations for map using message passing. In *Proceedings of the 24th Conference in Uncertainty in Artificial Intelligence (UAI)*, 503–510. 305

St-Aubin, R., Hoey, J., and Boutilier, C. (2000). APRICODD: Approximate policy construction using decision diagrams. In *Advances in Neural Information Processing 13 (NIPS-00)*, 1089–1095. Denver. 393

Suciu, D., Olteanu, D., Ré, C., and Koch, C. (2011). *Probabilistic Databases*, volume 16 of *Synthesis Lectures on Data Management*. Morgan & Claypool Publishers. 101, 106, 143

Taghipour, N. and Davis, J. (2012). Generalized counting for lifted variable elimination. In *Proceedings of the 2nd International Workshop on Statistical Relational AI (StaRAI)*, 1–8. 57, 174

Taghipour, N., Davis, J., and Blockeel, H. (2013a). Generalized counting for lifted variable elimination. In *International Conference on Inductive Logic Programming*, 107–122. Springer. 57

Taghipour, N., Fierens, D., Davis, J., and Blockeel, H. (2013b). Lifted variable elimination: Decoupling the operators from the constraint language. *Journal of Artificial Intelligence Research*, 47:393–439. 57

Taghipour, N., Fierens, D., Van den Broeck, G., Davis, J., and Blockeel, H. (2013c). Completeness results for lifted variable elimination. In *Proceedings of the 16th International Conference on Artificial Intelligence and Statistics (AISTATS)*. (Under review). 57, 89, 105, 109, 164

Tarjan, R. E. and Mihalis, Y. (1984). Simple linear-time algorithms to test chordality of graphs, test acyclicity of hypergraphs, and selectively reduce acyclic hypergraphs. *SIAM Journal on computing*, 13(3):566–579. 99

Taskar, B., Abbeel, P., and Koller, D. (2002). Discriminative Probabilistic Models for Relational Data. In Darwiche, A. and Friedman, N. (eds.), *Proceedings of the Eighteenth Conference on Uncertainty in Artificial Intelligence (UAI-02)*, 485–492. 40

Thon, I., Landwehr, N., and De Raedt, L. (2011). Stochastic relational processes: Efficient inference and applications. *Machine Learning*, 82(2):239–272. 36

Tierney, L. (1994). Markov chains for exploring posterior distributions. *The Annals of Statistics*, 22(4):1701–1728. 190

Tinhofer, G. (1991). A note on compact graphs. *Discrete Applied Mathematics*, 30:253–264. 362

Toussaint, M., Charlin, L., and Poupart, P. (2008). Hierarchical POMDP controller optimization by likelihood maximization. In *Proceedings of the 24th Annual Conference on Uncertainty in Artificial Intelligence (UAI)*, 562–570. 238, 241

Toussaint, M. and Storsky, A. (2006). Probabilistic inference for solving discrete and continuous state Markov decision processes. In *Proc. International Conference on Machine Learning*. 376, 377

Tseitin, G. (1968). On the complexity of derivation in propositional calculus. *Studies in Constrained Mathematics and Mathematical Logic*. 96, 101

Valiant, L. G. (1979). The complexity of enumeration and reliability problems. *SIAM Journal on Computing*, 8(3):410–421. 159

van Bremen, T. and Kuzelka, O. (2020). Approximate weighted first-order model counting: Exploiting fast approximate model counters and symmetry. *arXiv preprint arXiv:2001.05263*. 101

van de Meent, J., Paige, B., Tolpin, D., and Wood, F. (2016). Black-box policy search with probabilistic programs. In *Proceedings of the International Conference on Artificial Intelligence and Statistics, AISTATS*, 1195–1204. 376, 377

Van den Broeck, G. (2011). On the completeness of first-order knowledge compilation for lifted probabilistic inference. In *Proceedings of the 24th Annual Conference on Advances in Neural Information Processing Systems(NIPS)*, 1386–1394. 109, 155, 156, 164, 264, 276, 392

Van den Broeck, G. (2013). *Lifted Inference and Learning in Statistical Relational Models*. Ph.D. thesis, KU Leuven. 100, 103, 113, 147, 272, 392

Van den Broeck, G. (2015). Towards high-level probabilistic reasoning with lifted inference. In *Proceedings of the AAAI Spring Symposium on KRR*. xvi, 100

Van den Broeck, G. (2016). First-order model counting in a nutshell. In *Proceedings of the 25th International Joint Conference on Artificial Intelligence (IJCAI), Early Career Spotlight Track*. 11

Van den Broeck, G., Choi, A., and Darwiche, A. (2012). Lifted relax, compensate and then recover: From approximate to exact lifted probabilistic inference. In *Conference on Uncertainty in Artificial Intelligence (UAI)*. 239

Van den Broeck, G. and Darwiche, A. (2013). On the complexity and approximation of binary evidence in lifted inference. In *Advances in Neural Information Processing Systems*, 2868–2876. 89, 172, 192, 194, 197, 254

Van den Broeck, G. and Davis, J. (2012). Conditioning in first-order knowledge compilation and lifted probabilistic inference. In *Proceedings of the 26th AAAI Conference on Artificial Intelligence (AAAI)*, 1–7. 149, 173

Van den Broeck, G., Meert, W., and Darwiche, A. (2014). Skolemization for weighted first-order model counting. In Baral, C., Giacomo, G. D., and Eiter, T. (eds.), *Proceedings of the 14th International Conference on Principles of Knowledge Representation and Reasoning (KR-14)*, 111–120. AAAI Press. 89, 103, 104, 139, 155, 156, 164

Van den Broeck, G., Meert, W., and Davis, J. (2013). Lifted generative parameter learning. In *Statistical Relational AI (StaRAI) workshop*. 159

Van den Broeck, G. and Niepert, M. (2015). Lifted probabilistic inference for asymmetric graphical models. In *AAAI*, 3599–3605. 198

Van den Broeck, G. and Suciu, D. (2017). Query processing on probabilistic data: A survey. *Foundations and Trends® in Databases*, 7(3-4):197–341. 95, 101

Van den Broeck, G., Taghipour, N., Meert, W., Davis, J., and De Raedt, L. (2011). Lifted probabilistic inference by first-order knowledge compilation. In Walsh, T. (ed.), *Proc. International Joint Conference on AI (IJCAI)*, 2178–2185. AAAI Press. 89, 90, 101, 105, 140, 147, 149, 174, 275, 276, 382

Van Gelder, A., Ross, K. A., and Schlipf, J. S. (1991). The well-founded semantics for general logic programs. *Journal of the ACM (JACM)*, 38(3):619–649. 141

Van Haaren, J., Kolobov, A., and Davis, J. (2015). Todtler: Two-order-deep transfer learning. In *Proceedings of the Twenty-Ninth AAAI Conference on Artificial Intelligence*, 3007–3015. 48, 49

Van Haaren, J., Van den Broeck, G., Meert, W., and Davis, J. (2016). Lifted generative learning of Markov logic networks. *Machine Learning*, 103(1):27–55. 155

Vennekens, J., Denecker, M., and Bruynooghe, M. (2009). CP-logic: A language of causal probabilistic events and its relation to logic programming. *Theory and Practice of Logic Programming (TPLP)*, to appear. 25, 26

Vennekens, J., Verbaeten, S., and Bruynooghe, M. (2004). Logic programs with annotated disjunctions. In Demoen, B. and Lifschitz, V. (eds.), *Proceedings of the 20th International Conference on Logic Programming (ICLP-04)*, volume 3132 of *Lecture Notes in Computer Science*, 431–445. Springer. 25, 26

Venugopal, D. and Gogate, V. (2012). On lifting the gibbs sampling algorithm. In *Proceedings of the 26th Annual Conference on Advances in Neural Information Processing Systems (NIPS)*, 1–6. 185

Venugopal, D. and Gogate, V. (2014a). Evidence-based clustering for scalable inference in Markov logic. In *ECML PKDD*, 258–273. 89, 194

Venugopal, D. and Gogate, V. G. (2014b). Scaling-up importance sampling for Markov logic networks. In *Advances in Neural Information Processing Systems*, 2978–2986. 89

Vishwanathan, S. V. N., Schraudolph, N. N., Kondor, R., and Borgwardt, K. M. (2010). Graph kernels. *Journal of Machine Learning Research*, 11:1201–1242. 368

Vlasselaer, J., Kimmig, A., Dries, A., Meert, W., and De Raedt, L. (2016a). Knowledge compilation and weighted model counting for inference in probabilistic logic programs. In *AAAI Workshop: Beyond NP*. 101

Vlasselaer, J., Meert, W., Van den Broeck, G., and De Raedt, L. (2016b). Exploiting local and repeated structure in dynamic bayesian networks. *Artificial Intelligence*, 232:43 – 53. 95

Wainwright, M., Jaakkola, T., and Willsky, A. (2005). Map estimation via agreement on trees: message-passing and linear programming. *IEEE Transactions on Information Theory*, 51(11):3697–3717. 305, 306

Wainwright, M. and Jordan, M. (2008). Graphical models, exponential families, and variational inference. *Found. Trends Mach. Learn.*, 1(1-2):1–305. 281, 285, 286

Wang, C., Joshi, S., and Khardon, R. (2008). First order decision diagrams for relational MDPs. *Journal of Artificial Intelligence Research (JAIR)*, 31:431–472. 385, 387, 393

Wang, C. and Khardon, R. (2007). Policy iteration for relational MDPs. In *Uncertainty in Artificial Intelligence (UAI-07)*. Vancouver, Canada. 394

Wang, C. and Khardon, R. (2010). Relational partially observable MDPs. In *Proceedings of the Twenty-Fourth AAAI Conference on Artificial Intelligence, AAAI 2010, Atlanta, Georgia, USA, July 11-15, 2010*. 394

Wang, J. and Domingos, P. (2008). Hybrid Markov logic networks. In *Proceedings of the Twenty-Third AAAI Conference on Artificial Intelligence, AAAI 2008*. 327

Weiss, Y. and Freeman, W. (2001). Correctness of belief propagation in gaussian graphical models of arbitrary topology. *Neural Computation*, 13(10):2173–330. 337

Wellman, M. P., Breese, J. S., and Goldman, R. P. (1992). From knowledge bases to decision models. *The Knowledge Engineering Review*, 7(1):35–53. 19

Yedidia, J. S., Freeman, W. T., and Weiss, Y. (2003). Understanding belief propagation and its generalizations. In Lakemeyer, G. and Nebel, B. (eds.), *Exploring Artificial Intelligence in the New Millennium*, chapter 8, 239–269. Morgan Kaufmann. 263, 273

Yoon, S., Fern, A., and Givan, R. (2002). Inductive policy selection for first-order Markov decision processes. In *Uncertainty in Artificial Intelligence (UAI-02)*, 569–576. Edmonton. 394

Yoon, S., Fern, A., and Givan, R. (2006). Approximate policy iteration with a policy language bias: Learning to solve relational Markov decision processes. *Journal of Artificial Intelligence Research (JAIR)*, 25:85–118. 394

Younes, H. L. S., Littman, M. L., Weissman, D., and Asmuth, J. (2005). The first probabilistic track of the international planning competition. *Journal of Artificial Intelligence Research (JAIR)*, 24:851–887. 393

Zeng, Z. and Van den Broeck, G. (2019). Efficient search-based weighted model integration. In *Proceedings of the 35th Conference on Uncertainty in Artificial Intelligence (UAI)*. 101

Zhang, H. and Stickel, M. E. (1996). An efficient algorithm for unit propagation. In *Proceedings of the 4th International Symposium on Artificial Intelligence and Mathematics (AI-MATH)*, 166–169. 217

Zhang, N. L. and Poole, D. (1994). A simple approach to Bayesian network computations. In *Proceedings of the 10th Canadian Conference on AI*, 171–178. 12, 115

Zhang, N. L. and Poole, D. (1996). Exploiting causal independence in Bayesian network inference. *Journal of Artificial Intelligence Research (JAIR)*, 5:301–328. 117

Index